Asymptotics beyond All Orders

NATO ASI Series

Advanced Science Institutes Series

A series presenting the results of activities sponsored by the NATO Science Committee, which aims at the dissemination of advanced scientific and technological knowledge, with a view to strengthening links between scientific communities.

The series is published by an international board of publishers in conjunction with the NATO Scientific Affairs Division

A	**Life Sciences**	Plenum Publishing Corporation
B	**Physics**	New York and London
C	**Mathematical and Physical Sciences**	Kluwer Academic Publishers
D	**Behavioral and Social Sciences**	Dordrecht, Boston, and London
E	**Applied Sciences**	
F	**Computer and Systems Sciences**	Springer-Verlag
G	**Ecological Sciences**	Berlin, Heidelberg, New York, London,
H	**Cell Biology**	Paris, Tokyo, Hong Kong, and Barcelona
I	**Global Environmental Change**	

Recent Volumes in this Series

Volume 278—Painlevé Transcendents
edited by Decio Levi and Pavel Winternitz

Volume 279—Fundamental Aspects of Inert Gases in Solids
edited by S. E. Donnelly and J. H. Evans

Volume 280—Chaos, Order, and Patterns
edited by Roberto Artuso, Predrag Cvitanović, and Giulio Casati

Volume 281—Low-Dimensional Structures in Semiconductors: From Basic Physics to Applications
edited by A. R. Peaker and H. G. Grimmeiss

Volume 282—Quantum Measurements in Optics
edited by Paolo Tombesi and Daniel F. Walls

Volume 283—Cluster Models for Surface and Bulk Phenomena
edited by Gianfranco Pacchioni, Paul S. Bagus, and Fulvio Parmigiani

Volume 284—Asymptotics beyond All Orders
edited by Harvey Segur, Saleh Tanveer, and Herbert Levine

Volume 285—Highlights in Condensed Matter Physics and Future Prospects
edited by Leo Esaki

Series B: Physics

Asymptotics beyond All Orders

Edited by

Harvey Segur

University of Colorado
Boulder, Colorado

Saleh Tanveer

Ohio State University
Columbus, Ohio

and

Herbert Levine

University of California, San Diego
La Jolla, California

Plenum Press
New York and London
Published in cooperation with NATO Scientific Affairs Division

Proceedings of a NATO Advanced Research Workshop
on Asymptotics beyond All Orders,
held January 7-11, 1991,
in La Jolla, California

Library of Congress Cataloging-in-Publication Data

Asymptotics beyond all orders / edited by Harvey Segur, Saleh Tanveer, and Herbert Levine.
p. cm. -- (NATO ASI series. Series B, Physics ; v. 284)
Proceedings of a NATO Advanced Research Workshop on Asymptotics beyond All Orders, held January 7-11, 1991, in La Jolla, California.
"Published in cooperation with: NATO Scientific Affairs Division."
Includes bibliographical references and index.

1. Mathematical physics--Asymptotic theory--Congresses. I. Segur, Harvey. II. Tanveer, Saleh. III. Levine, Herbert, 1955- . IV. North Atlantic Treaty Organization. Scientific Affairs Division. V. NATO Advanced Research Workshop on Asymptotics beyond All Orders (1991 : La Jolla, San Diego, Calif.) VI. Series.
QC20.7.A85A86 1991
530.1'5--dc20 91-43647
CIP

ISBN 978-1-4757-0437-2 ISBN 978-1-4757-0435-8 (eBook)
DOI 10.1007/978-1-4757-0435-8

Softcover reprint of the hardcover 1st edition 1991
A Division of Plenum Publishing Corporation
233 Spring Street, New York, N.Y. 10013

SPECIAL PROGRAM ON CHAOS, ORDER, AND PATTERNS

This book contains the proceedings of a NATO Advanced Research Workshop held within the program of activities of the NATO Special Program on Chaos, Order, and Patterns.

Volume 208—MEASURES OF COMPLEXITY AND CHAOS
edited by Neal B. Abraham, Alfonso M. Albano, Anthony Passamante, and Paul E. Rapp

Volume 225—NONLINEAR EVOLUTION OF SPATIO-TEMPORAL STRUCTURES IN DISSIPATIVE CONTINUOUS SYSTEMS
edited by F. H. Busse and L. Kramer

Volume 235—DISORDER AND FRACTURE
edited by J. C. Charmet, S. Roux, and E. Guyon

Volume 236—MICROSCOPIC SIMULATIONS OF COMPLEX FLOWS
edited by Michel Mareschal

Volume 240—GLOBAL CLIMATE AND ECOSYSTEM CHANGE
edited by Gordon J. MacDonald and Luigi Sertorio

Volume 243—DAVYDOV'S SOLITON REVISITED: Self-Trapping of Vibrational Energy in Protein
edited by Peter L. Christiansen and Alwyn C. Scott

Volume 244—NONLINEAR WAVE PROCESSES IN EXCITABLE MEDIA
edited by Arun V. Holden, Mario Markus, and Hans G. Othmer

Volume 245—DIFFERENTIAL GEOMETRIC METHODS IN THEORETICAL PHYSICS: Physics and Geometry
edited by Ling-Lie Chau and Werner Nahm

Volume 256—INFORMATION DYNAMICS
edited by Harald Atmanspacher and Herbert Scheingraber

Volume 260—SELF-ORGANIZATION, EMERGING PROPERTIES, AND LEARNING
edited by Agnessa Babloyantz

Volume 263—BIOLOGICALLY INSPIRED PHYSICS
edited by L. Peliti

Volume 264—MICROSCOPIC ASPECTS OF NONLINEARITY IN CONDENSED MATTER
edited by A. R. Bishop, V. L. Pokrovsky, and V. Tognetti

Volume 268—THE GLOBAL GEOMETRY OF TURBULENCE: Impact of Nonlinear Dynamics
edited by Javier Jiménez

Volume 270—COMPLEXITY, CHAOS, AND BIOLOGICAL EVOLUTION
edited by Erik Mosekilde and Lis Mosekilde

Volume 272—PREDICTABILITY, STABILITY, AND CHAOS IN N-BODY DYNAMICAL SYSTEMS
edited by Archie E. Roy

Volume 276—GROWTH AND FORM: Nonlinear Aspects
edited by M. Ben Amar, P. Pelcé, and P. Tabeling

SPECIAL PROGRAM ON CHAOS, ORDER, AND PATTERNS

Volume 278—PAINLEVÉ TRANSCENDENTS: Their Asymptotics and Physical Applications
edited by Decio Levi and Pavel Winternitz

Volume 280—CHAOS, ORDER, AND PATTERNS
edited by Roberto Artuso, Predrag Cvitanović, and Giulio Casati

Volume 284—ASYMPTOTICS BEYOND ALL ORDERS
edited by Harvey Segur, Saleh Tanveer, and Herbert Levine

PREFACE

An *asymptotic expansion* is a series that provides a sequence of increasingly accurate approximations to a function in a particular limit. The formal definition, given by Poincare (1886, Acta Math. 8:295), is as follows. Given a function, $\phi(\varepsilon)$, the series $\sum_0^\infty \phi_n \varepsilon^n$ is said to be *asymptotic to* $\phi(\varepsilon)$ *as* $\varepsilon \to 0$ for every nonnegative integer N,

$$\lim_{\varepsilon \to 0} \left[\frac{\phi(\varepsilon) - \sum_{n=0}^{N} \phi_n \varepsilon^n}{\varepsilon^N} \right] = 0. \tag{1}$$

Note:

(i) ϕ might also depend on another parameter, in the form $\phi(x, \varepsilon)$. Then ϕ_n should be replaced by $\phi_n(x)$, and one tests the asymptoticity of the series at each fixed x.

(ii) Asymptotic series can be more complicated than simple power series in ε, but they are sufficient to illustrate our main points.

At the simplest level, $N = 0$, (1) implies that $\phi(\varepsilon) \to \phi_0$ as $\varepsilon \to 0$. A more accurate approximation is obtained at $N = 1$:

$$\frac{\phi(\varepsilon) - \phi_0}{\varepsilon} \to \phi_1 \quad \text{as} \quad \varepsilon \to 0,$$

and so on. If the series is asymptotic to $\phi(\varepsilon)$, we write

$$\phi(\varepsilon) \sim \sum_0^\infty \phi_n \varepsilon^n. \tag{2}$$

The important limit in deciding whether the series is asymptotic is $\{\varepsilon \to 0, N \text{ fixed}\}$. This limit differs from that used for testing the convergence of the series, $\{N \to \infty, \varepsilon \text{ fixed}\}$, and an asymptotic series need not converge for $\varepsilon \neq 0$. In fact, one advantage of asymptotic analysis is that one can accurately approximate a nonanalytic function, which has no convergent series representation, using a few terms of its asymptotic series. As an example, Copson (1965, Asymptotic Expansions, Cambridge University Press) mentions a series due to Euler:

$$\sum_1^M \frac{1}{n} \sim ln\ M + \gamma + \frac{1}{2M} + \sum_1^\infty \frac{B_{2k}}{(2k)M^{2k}}, \tag{3}$$

where γ is Euler's constant $(0.5772\dots)$, the coefficients $\{B_{2k}\}$ are Bernoulli numbers, defined by a generating function:

$$\frac{z}{e^z - 1} = \sum_0^\infty B_k \frac{z^k}{k!},$$

and the small parameter (ε) is $1/M$. The infinite series in (3) diverges for every finite M, but Euler used it with $M = 10$ to calculate γ correctly to 15 decimal places. This example illustrates an important aspect of asymptotic series: asymptoticity is decided in the limit $(\varepsilon) \to 0$, but the series provide useful information at finite ε. [The article by M. Berry, in these Proceedings, discusses how to truncate an asymptotic series to obtain optimal accuracy at finite ε.]

An important feature of an asymptotic series like $\sum \phi_n \varepsilon^n$ is that every term in the series is algebraic in ε. Transcendentally small terms like $\exp\{-1/\varepsilon^2\}$ are smaller than every term in the series as $\varepsilon \to 0$, and are not captured by it. Thus if (2) is valid, then

$$\phi(\varepsilon) + \exp\{-1/\varepsilon^2\} \sim \sum_{n=0}^{\infty} \phi_n \varepsilon^n \tag{4}$$

is valid as well. Such transcendentally small terms are said to lie *beyond all orders* of the asymptotic expansion.

In most applications, these tiny corrections are insignificant and they can safely be neglected. However, exceptional problems in which these very small terms have great practical interest are known in many branches of science, including dendritic crystal growth, viscous fluid flow, quantum tunneling, KAM theory, and others. For these exceptional problems, conventional asymptotic analysis is simply inadequate. These problems require improved methods, designed to obtain meaningful corrections that lie beyond all orders of a conventional asymptotic expansion. The phrase *asymptotics beyond all orders* refers to the collection of such methods.

The reader might wonder how a transcendentally small term, hiding behind all orders of a (divergent) asymptotic series, could have any practical effect. We mention here two ways in which this apparently contradictory situation can occur.

(i) Symmetry

Suppose that $\phi(x, \varepsilon)$ satisfies a differential equation in x, with ε as a small parameter in the equation. Further, suppose that conventional asymptotic analysis leads to a divergent asymptotic series for $\phi(x, \varepsilon)$ in the form $\sum \phi_n(x)\varepsilon^n$. Finally, suppose that each term of the asymptotic series exhibits a symmetry in x about $x = 0$, so that $\phi_n(0) = 0$ for all n. Then at $x = 0$, and only there, the asymptotic series converges (trivially, to zero), and one can meaningfully ask for transcendentally small corrections to the convergent series at $x = 0$. Often the question takes the form: At $x = 0, \phi(0, \varepsilon)$ vanishes to all orders; is $\phi(0, \varepsilon) = 0$? (Or: $\phi(x, \varepsilon)$ is symmetric to all orders about $x = 0$; is $\phi(x, \varepsilon)$ symmetric?) Symmetry is a qualitative property, so any deviation, no matter how small, is enough to destroy it. In this way, transcendentally small effects can be important because they change a qualitative feature of $\phi(x, \varepsilon)$.

Several of the problems analyzed in this book involve a symmetry in just this way, although the notion of "symmetry" is sometimes more general than that described here. Moreover, many of the problems exhibit a *selection mechanism* (i.e., the problem has no solution unless certain external parameters satisfy some constraint), and the selection criterion usually arises as a requirement that the solution be symmetric in a particular sense.

(ii) Finite ε

As mentioned above, the practical use of an asymptotic series occurs at finite (but small) values of ε. For example, one might compare an asymptotic theory with an experiment, where the experiment is performed at a fixed (small) value of ε. Then $\exp\{-1/\varepsilon^2\}$ is smaller than any power of ε as $\varepsilon \to 0$, but for $\varepsilon = 1/2$, $\exp\{-1/\varepsilon^2\} > \varepsilon^6$, and $60\exp\{-1/\varepsilon^2\} > 1$. In this way, transcendentally small terms can be numerically important in practical applications.

A small revolution in asymptotic analysis has occurred within the last decade: outstanding problems requiring asymptotics beyond all orders have been solved in several scientific fields. A number of apparently unrelated methods, with varying levels of rigor, have been devised to solve these problems. In order to identify and to clarify the relations that might exist between these new methods of analysis, a Workshop on *Asymptotics Beyond All Orders* was held in La Jolla, CA, during Jan 7-11, 1991. Sponsored by NATO and by NSF, the Workshop was attended by about 50 mathematicians, physicists, and engineers from 12 countries. This book is the Proceedings of that Workshop.

The Workshop was organized according to subject matter, as are these Proceedings. The subject areas, and the papers presented within each area, are as follows.

a) *Asymptotology and Borel Summation*

Two papers by M. Berry and by V. Hakim show how to "sum" divergent series using Borel summation, and how to recover information that lies beyond all orders of the asymptotic expansion in this way.

b) *The Geometric Model of Crystal Growth*

This model has been the testing ground for many of the methods developed in this subject. In two papers, H. Segur describes the model, and J. M. Hammersley discusses some numerical methods used in its study.

c) *Dendritic Crystal Growth*

The geometric model was devised to mimic the growth of dendritic crystals. H. Levine surveys the prevailing theory, which is based on the assumptions of a contimuum model, and that the steady growth of the tip region can be studied using steady far-field boundary conditions. Then J.P. Gollub assesses how well that theory compares with currently available experimental results. Finally, E.A. Coutsias and H. Segur present an alternative formulation of the problem, which formulation does not require steady far-field boundary conditions.

d) *Directional Solidification of Solids*

In this set-up, the crystal grows in a particular direction because the liquid is moved through a spatial temperature gradient as it cools. It provides a convenient laboratory in which to study various kinds of pattern formation. K. Kassner, C. Misbah, H. Muller-Krumbhaar, Y. Saito, and D.E. Temkin study aspects of this problem analytically and numerically, while A. Libchaber, A. J. Simon, and J.M. Flesselles study other aspects experimentally.

e) *Flow in a Hele-Shaw Cell (also known as Viscous Fingering)*

If surface tension is small, then asymptotics beyond all orders plays an essential role in the selection of steady flows in a Hele-Shaw cell, and in their stability. S. Tanveer surveys the theory of these flows, both as steady-states and as initial-value problems. Two papers, by M. BenAmar and R. Combescot and by Y. Tu, discuss novel flows in a wedge-shaped geometry.

f) *The Rapidly Forced Pendulum*

The motion of a frictionless pendulum, subject to weak, high- frequency, external forcing, is often studied as a prototype of problems that arise in KAM theory. The question is whether the homoclinic orbit that exists without forcing persists when forcing is applied. The first important result on this problem was due to J. Scheurle, J.E. Marsden, and P. J. Holmes, who review their earlier work here, and discuss some other aspects of the problem. Their result is improved in a second paper, by M. Kummer, J.A. Ellison, and A.W. Saenz, using a similar method. Finally, Y.-H. Chang shows that this problem can be stated in terms of a symmetry (as described above), and proves that the formal expansion is indeed asymptotic.

g) *Ordinary Differential Equations*
Several papers study particular, singularly perturbed, ordinary differential equations, and whether these equations admit homoclinic solutions in singular limits of the equations. Typically these ODEs arise as reductions of PDEs obtained, for example, by seeking travelling waves. The three papers by J.G.B. Byatt-Smith and A.M. Davie, by Y. Pomeau, and by J.B. McLeod all fall into this category. In addition, J. Hu and M.D. Kruskal generalize the classic result of Pokrovskii and Khalatnikov on the reflection coefficient in the Schrodinger equation to include singular potentials. H.S. Dumas uses the results of Nekhoroshev to obtain estimates on the (exponentially long) channelling time of particle in a crystal.

h) *Solitary Water Waves in the Presence of Small Surface Tension*
The existence of a continuous family of solitary water waves has been proven without surface tension, or with strong surface tension. Oddly, no proof is available for the usual situation, in which surface tension is present but weaker than gravity, and in this case true solitary waves might not even exist! J.M. VandenBroeck presents numerical evidence that truly solitary waves might exist only for discrete values of the parameters; i.e, that surface tension provides a selection mechanism, as discussed above. Supporting this idea, J.T. Beale proves the existence of almost-solitary waves that are coupled to tiny, oscillatory wavetrains fore and aft of the main wave. Similarly, M.C. Shen and S.M. Sun prove a similar result for internal waves in a density-stratified fluid.

i) *Problems in Optics*
Two papers, by A.K. Hobbs, W.L. Kath, and G.A. Kreigsmann and by A.D. Wood, study problems related to the transcendentally small energy loss of an electromagnetic wave as it propagates down a slightly curved optical fiber. Separately, O. Martin and S.V. Branis study the phenomenon of self-induced transparency in a dielectric medium. They show that without the usual slowly-varying-envelope approximation, the continuous family of solitary waves is replaced by a discrete set of parameter values (another selection mechanism).

j) *Potpourri*
The following four papers do not fit into any of the tidy categories listed above. R.E. Meyer, who did some of the earliest work on exponentially small asymptotic effects, shows how such effects arise in a linear, nonseparable, partial differential equation. V. Eleonsky summarizes the work done by his group over several years on the search for "breather" solutions in nonlinear Klein-Gordon equations. R.S. Mackay proves a conjecture of J.M. Greene on analytic, area- preserving, twist maps. N. Goldenfeld, O. Martin and Y. Oono study the relation between asymptotic analysis and use of the renormalization group, using a variation of heat equation as an example.

We take this opportunity to thank NATO (Scientific Affairs Division) and NSF (DMS-9010990) for their financial support for this Workshop. We also thank Martin Kruskal and Yves Pomeau, the other members of the Organizing Committee besides the two of us (H.S. and H.L.), for their efforts in putting the Workshop together, and David Campbell, who originally conceived of the Workshop. Equally important were the staff members who did most of the dirty work: Andreah Hennessy and Jan Horn (at U. Colorado), Penny Lang and Terry England (at O.S.U.) and Alice McCutchen and Terry Peters (at UCSD).

Finally, we mention two of our collegues, whose premature deaths tragically prevented their attending the Workshop. Charles Amick (1952-1991) and Stephanos Pnevmatikos (1957-1990) had made important contributions to mathematics and to physics in their respective short careers. We regret the deaths of both.

Harvey Segur
Saleh Tanveer
Herbert Levine

CONTENTS

FLOW IN A HELE-SHAW CELL (Also known as VISCOUS FINGERING)

THE RAPIDLY FORCED PENDULUM

ORDINARY DIFFERENTIAL EQUATIONS

SOLITARY WATER WAVES IN THE PRESENCE OF SMALL SURFACE TENSION

PROBLEMS IN OPTICS

POTPOURRI

ASYMPTOTICS, SUPERASYMPTOTICS, HYPERASYMPTOTICS...

MICHAEL BERRY

H.H.Wills Physics Laboratory
Tyndall Avenue
Bristol BS8 1TL
United Kingdom

'Divergent series are the invention of the devil, and it is shameful to base on them any demonstration whatever' (Abel, 1828)

'2+2=5, for sufficiently large values of 2' (Princeton asymptotic graffito)

1. INTRODUCTION

My purpose is to describe several recent developments in our understanding of divergent series and the accurate calculation of the functions they represent. All the work has been[1,2] or is being published[3], so this will be an informal account, emphasising the new concepts and illustrating them with pictures.

It is useful to introduce some terminology. Typically, an asymptotic series for a function depending on a large parameter k and several variables $X=(X_1,X_2,...)$ has the form

$$y(k;X) = M(k;X)\exp\{k\phi(X)\} \sum_{r=0}^{\infty} Y_r(k,X), \quad \text{where } Y_0 = 1 \quad \text{and } Y_r \propto k^{-r} \tag{1}$$

(Often k - which will not always be written explicitly - serves simply as a book-keeping parameter, to order the terms in the series.) In the cases we are interested in, the series diverges and so is meaningless when interpreted conventionally. The usual 'asymptotics' is the study of the series truncated at fixed order $r=N$: according to Poincaré's definition[4], the series is asymptotic if the error is of order $k^{-(N+1)}$. However, as was known to Stokes [5] nearly half a century before Poincaré, much more accurate approximations can be obtained by truncating not at fixed order but at the least term, which typically increases with k. It is common to achieve errors of order $\exp(-k)$ with such optimal truncation, which therefore constitutes 'asymptotics beyond all orders' or, as I will call it, 'superasymptotics'. (After introducing this term I felt at first shamed by Barbara Levi's gentle satire[6] on physicists' predilection for terminological 'super'iority, but was later made unashamed by reading Lord Kelvin's memorial appreciation[7], in which he described Stokes' early work on divergent series as 'mathematical supersubtlety'.) We shall also require a term for systematic improvements to the exponentially small remainder of a optimally truncated series . I will call these 'hyperasymptotics'. Thus hypersaymptotics goes 'beyond asymptotics beyond all orders.'

Underlying the recent work are two ideas. First, that an asymptotic series such as (1) is a compact encoding of a function, and its divergence should be regarded not as a deficiency but as a source of information about the function. In particular, divergence usually indicates the presence of exponentially small terms which the bare asymptotic series, uninterpreted, cannot capture. This is why superasymptotics can yield exponential accuracy. A consequence is that the late terms of the asymptotic series associated with one exponential are frequently related by 'resurgence' to the early terms of the series associated with another exponential. Second, that the divergences of the series obtained by a variety of methods, and representing a variety of functions, follow a common pattern: factorial divided by a power. Recognition of this universality and its cause leads to powerful resummation techniques enabling the asymptotics to be decoded to yield precise (hyperasymptotic) numerical information. These principles were systematically explored and exploited by Dingle in the 1950s, and summarised in his 1973 book[8], but are only now becoming widely known.

In the new results I describe here, Dingle's work is extended in two ways. The first concerns Stokes' phenomenon[9], namely rapid jumps, as the variables X are changed, in the multipliers M of a small (subdominant) exponential whilst hidden behind a big (dominant) one. In a sense this is the very heart of asymptotics, because such changes in form necessarily accompany the divergence of the asymptotic series associated with each exponential, reflecting its inability to describe the other exponentials. In my opinion, the persistent failure to understand Stokes' phenomenon (still evident in the literature) is in large measure responsible for what Littlewood[10] called the 'aroma of paradox and audacity' that has hung about the whole subject of divergent series, connection formulae in WKB theory, etc. By appropriate magnification and resummation, however, a precise description of the change in the subdominant multiplier can be obtained, in terms of a universal scaling function. This refinement and demystification of Stokes' phenomenon can be regarded as the consequence of just the first step into hyperasymptotics.

The second result goes much further. By systematically exploiting resurgence, the remainder in an optimally truncated expansion can itself be expressed as an asymptotic series, which has its own remainder,..... Iteration of this hyperasymptotic process leads to an intricate sequence of hyperseries in which the original asymptotic coefficients (the Y_r in (1)) are renormalised by certain universal functions. At the end, after hyperasymptotics has come to a natural halt, the error is reduced, not to zero but to less than the square of the superasymptotic error. For integrals of exponentials with several saddles, there is a remarkable resurgence identity connecting the expansions about the different saddles; this can be employed to refine the method of steepest descent into an exact technique, whose hyperasymptotics can be accomplished without resumming divergent series.

I will illustrate these general ideas with the Airy function:

$$\mathrm{Ai}(z) \equiv \frac{1}{2\pi}\int_C du \exp\left\{i\left(\frac{u^3}{3}+zu\right)\right\} \qquad (2)$$

Here the infinite contour C runs from $\infty\exp(5\pi i/6)$ to $\infty\exp(\pi i/6)$, so that the integral converges for all complex z. For real z, C can be deformed to the real axis, and Ai(z) is real. For any z, Ai depends on two real quantities: the modulus and phase of z. In terms of the general theory for the series (1), $|z|$ (or rather $|z|^{3/2}$) will be the large parameter k and arg z will be the variable X. In § 2 we show the ordinary asymptotics and superasymptotics of Ai, with and without the Stokes jump, which is represented as a discontinuity. In §3 the universal smoothing is described. §4 contains an account of resurgence and hyperasymptotics, again with Ai as an example.

Having these new techniques, I would like to hear from anybody who needs the Airy function to twenty decimals, but am not expecting an early call. Probably no application requires such accuracy. This being so, it is important to reveal the motivation for this renewed interest in the oldest and simplest problems of asymptotics. This I leave to the concluding §5.

2. DOMINANT AND SUBDOMINANT SERIES; STOKES' JUMP

We shall display several approximations to Ai(z) for large $|z|$, the aim being to understand the asymptotics in the upper half-plane (fig. 1), that is as θ=arg z varies from 0 to π. For large $|z|$ it is appropriate to approximate (2) by the saddle-point method[11]. There are two saddles, at $u=\pm i z^{1/2}$, at which the integrand is

$$\exp\left\{\pm\tfrac{1}{2}F\right\} \equiv \exp\left\{\mp\tfrac{2}{3}z^{3/2}\right\} \tag{3}$$

(note the signs- see fig. 1). Following Dingle[8], we have introduced the 'singulant' F, namely the difference between the two exponents. The full significance of this quantity will emerge later. In our first numerical illustrations we shall take the 'large parameter' as $|F|=3$, i.e. $|z|=1.7171$.

Study of the topology of the phase in (2) shows that when $\theta<120°$ the contour C can be deformed into a steepest path passing through only one of the saddles, yielding the lowest approximation

$$\mathrm{Ai}(z) \approx \mathrm{Ai1}(z) = \frac{1}{2z^{1/4}\sqrt{\pi}}\exp\left\{\frac{F}{2}\right\} \tag{4}$$

Fig. 2 is an Argand plot comparing Ai1 with the exact Ai (computed to high precision, e.g. by the convergent series) in the upper half-plane. Agreement is reasonable for small θ, but rapidly deteriorates, becoming worst on the negative real axis, where Ai1 is complex whereas Ai is real.

The natural next step is to include higher-order corrections to Ai1, giving the approximation[8]

$$\mathrm{Ai}(z) \approx \mathrm{Ai2}(z,N) = \frac{1}{2z^{1/4}\sqrt{\pi}}\exp\left\{\frac{F}{2}\right\}\sum_{r=1}^{N} Y_r \quad \text{where } Y_r = \frac{\Gamma\left(r+\frac{1}{6}\right)\Gamma\left(r+\frac{5}{6}\right)}{2\pi F^r\,\Gamma(r+1)} \tag{5}$$

For large orders ('the asymptotics of the asymptotics'), the coefficients are

$$Y_r \xrightarrow[r\to\infty]{} \frac{(r-1)!}{2\pi F^r} \tag{6}$$

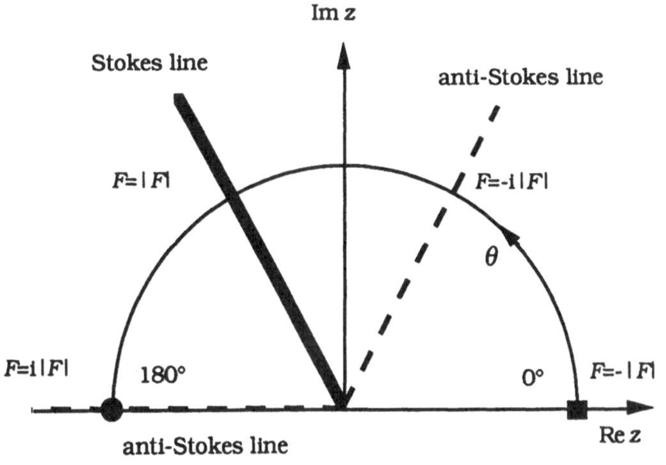

Fig. 1 Upper half-plane of argument z of the Airy function Ai(z).

Thus the smallest term, corresponding to optimal truncation, i.e. superasymptotics, is near $N=r^*=\mathrm{Int}|F|$. The superasymptotic Ai2 is shown in fig.2. The agreement is much better for small θ, but no better for $\theta=180°$.

The reason for the poor agreement near the negative real axis is the neglect of *Stokes' phenomenon*: for $\theta>120°$ the steepest-descent deformation of C passes through both saddles, so that the contribution of the second exponential in (3) should also be included. $\theta=120°$ is the *Stokes line* for Ai, defined as the locus of greatest disparity between the two exponentials, where F is positive real and the terms Y_r in the dominant series (5) all have the same sign, so that the divergence of the series is most severe. Stokes[9,12] argued that the extra exponential should be regarded as being born on this line, where it is smallest. Incorporating it into the lowest-order approximation gives

$$\begin{aligned}\mathrm{Ai}(z)\approx \mathrm{Ai3}(z)&=\frac{1}{2z^{1/4}\sqrt{\pi}}\left(\exp\left\{+\frac{F}{2}\right\}+i\exp\left\{-\frac{F}{2}\right\}\mathrm{H}(\theta-120°)\right)\\ &=\frac{1}{(-z)^{1/4}\sqrt{\pi}}\sin\left\{\tfrac{2}{3}(-z)^{3/2}+\tfrac{1}{4}\pi\right\}\quad \text{if}\quad \theta>120°\end{aligned} \tag{7}$$

where H denotes the unit step. Note the factor i in the new exponential. This birth 'in quadrature' not only makes the jump most unobtrusive but also ensures that Ai3 is real on the negative axis, which is an anti-Stokes line for Ai, that is the locus of equal magnitude of the two exponentials. Fig. 3 shows the considerable improvement that this produces: the overall agreement is much better, and the discontinuity is indeed unobtrusive.

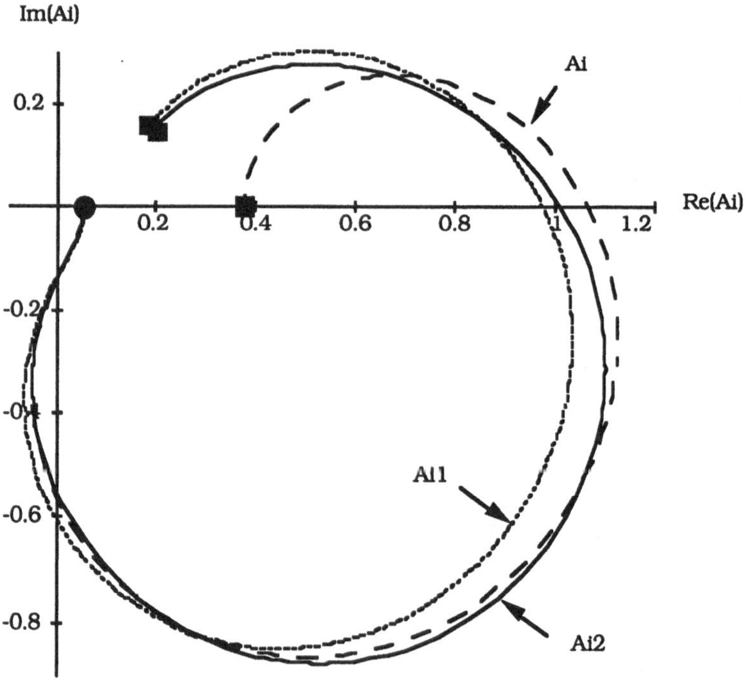

Fig. 2. Argand plot of exact Airy function Ai (dashed line) in the upper half-plane, along the semicircular path shown in fig.1, for $|F|=3$ (i.e. $|z|=1.7171$), compared with lowest-order (dominant exponential) asymptotics Ai1 (dotted line) and superasymptotics Ai2 with $N=3$ (dominant exponential x optimally truncated series) (full line). l marks $\theta=0$, n marks $\theta=180°$.

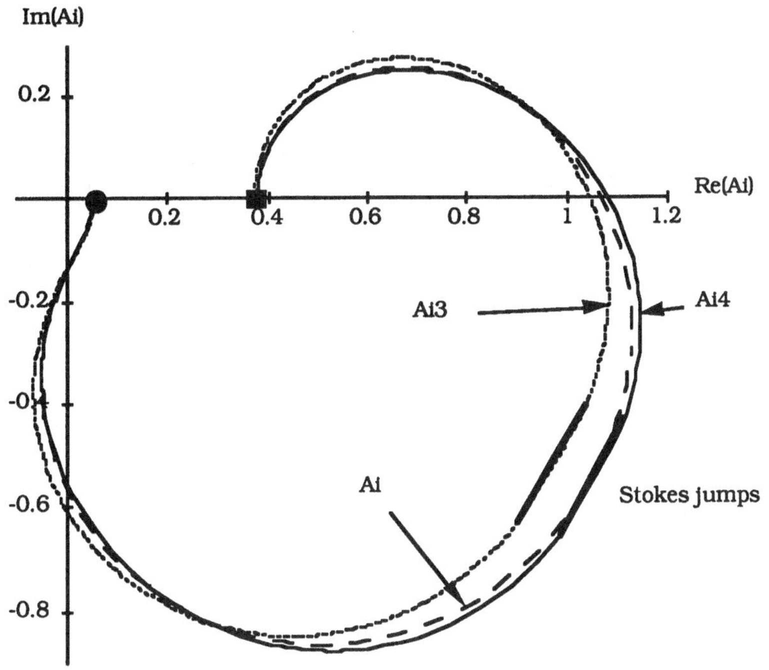

Fig. 3. As fig. 2, but comparing the exact Ai (dashed line) with Ai3 (dotted line) (Stokes jump included to lowest order) and Ai4 with N=3 (full line)(Stokes jump included superasymptotically). The jumps are shown as bold lines.

Stokes' analysis indicated that the best approximation is obtained by including the jump at the superasymptotic level, i.e. in the optimally truncated series. This gives

$$\mathrm{Ai}(z) \approx \mathrm{Ai4}(z,N) = \frac{1}{2z^{1/4}\sqrt{\pi}}\left[\exp\left\{+\frac{F}{2}\right\}\sum_{r=0}^{N} Y_r + \mathrm{i}\exp\left\{-\frac{F}{2}\right\}\sum_{r=1}^{N}(-1)^r Y_r\ \mathrm{H}(\theta-120°)\right] \tag{8}$$

As fig. 3 shows, Ai4 indeed gives a dramatic improvement, even here where $|F|$=3 and the new exponential appears with relative magnitude exp(-3)=0.0498, which is hardly small.

Figs. 2 and 3 extend a numerical experiment of Stokes[9] demonstrating the reality of his phenomenon. Since then, many people(e.g.[13, 14]) have rediscovered the increased accuracy achieved by correctly including small exponentials. For this to be a consistent procedure, it is essential to go to the superasymptotic level, where the first neglected term is exponentially small. Ordinary Poincaré asymptotics is inadequate because with this, as has often been remarked, the small exponentials $\exp(-|F|)$ are lost in the truncation errors $|F|^{-N}$.

3. SMOOTHING STOKES' DISCONTINUITY

The exact Airy function changes smoothly, so that any discontinuity at the Stokes line, where the subdominant exponential appears suddenly, must be an artefact of poor resolution. To get an approximation without discontinuity, it is necessary to go beyond superasymptotics. The first step into hyperasymptotics is sufficient to resolve the structure near the Stokes line. This can be accomplished by taking seriously what superasymptotics neglects, namely the tail of the series beyond the optimal truncation limit $N=r^*$. An important observation is that the late terms formula (6) (factorial divided by a power) is not restricted to Ai but has a very wide range of validity when modified as follows:

$$Y_r \xrightarrow[r\to\infty]{} \frac{M_-}{M}\frac{(r-1)!}{2\pi F^r} \tag{9}$$

Here F denotes the difference between the exponent $k\phi$ in (1) and the leading subdominant exponent, which we call $k\phi_-$, and M_- the multiplier appropriate to the subdominant series (for Ai, $M_-=M$). The simplicity of (9) is remarkable. Anybody who has computed asymptotic corrections in realistic (i.e. not textbook) applications knows that the algebra gets very heavy and generates enormous formulae (see e.g. pp 119-121 of Dingle's book[8]).

The underlying reason for the simplification, well explained by Dingle[8], is that all asymptotic methods (saddle-point and end-point integration, WKB solution of differential equations, etc.) are generated by *local expansions*. Thus, successive terms in the expansions involve successive derivatives, and late terms correspond to high derivatives. But by Darboux's theorem[8] the high derivatives of a function $f(t)$ at, say, $t=0$ are dominated by the nearest singularity, at $t=t^*$, say. Typically this will be a pole or branch point, and then the high derivatives will indeed have the form factorial/power. A common case is the simple pole

$$f(t) \xrightarrow[t\to t^*]{} \frac{A}{(t-t^*)} \tag{10}$$

Successive differentiation swells the range of validity of this formula from the neighbourhood of t^* to $t=0$, so that

$$\frac{d^r}{dt^r} f(0) \xrightarrow[r\to\infty]{} -A\frac{r!}{(t^*)^{r+1}} \tag{11}$$

For example[8], if

$$f(t) = \frac{1}{1+\log(1+t)} \tag{12}$$

Darboux's principle gives, from the pole at $t^*=e^{-1}-1$,

$$r!\frac{d^r}{dt^r} f(0) \xrightarrow[r\to\infty]{} \frac{(-1)^r}{e\left(1-e^{-1}\right)^{r+1}} \tag{13}$$

If $r=0$, this formula gives 0.58, in poor agreement with the exact value 1, but for the 'late' term $r=8$ it gives 22.8300, close to the exact value 38371/1680=22.8399.

Formally, we can write the expansion (1) as the optimally truncated series plus the divergent tail with its terms approximated by (9), in the form

$$y \approx M\exp\{k\phi\}\sum_{r=0}^{r^*} Y_r + iM_- S\exp\{k\phi_-\}, \quad \text{where}$$
$$S(F) \equiv \frac{-i}{2\pi}\exp(F)\sum_{r^*+1}^{\infty}\frac{(r-1)!}{F^r} \tag{14}$$

In coded form, $S(F)$ is the Stokes multiplier, describing the appearance of the subdominant exponential across the Stokes line F positive real. To decode it, we employ *Borel summation*, that is[4,8] writing the factorial in the familiar integral representation and then evaluating the sum. This replaces (14) by a convergent integral, which must be approximated for large $|F|$ (this is 'the asymptotics of the asymptotics of the asymptotics'). I have done this elsewhere[1], and do not repeat the details here. The important point is that the evaluation of the Borel integral is greatly simplified by optimal truncation, because then (and only then) a pole and saddle in its integrandcoincide.

In the case (e.g. Ai) where there is no subdominant exponential 'before' the Stokes line, i.e. for Im F<<0, the multiplier takes the very simple form

$$S(F) \approx \frac{1}{2}\left[1+\mathrm{Erf}\left\{\frac{\mathrm{Im}F}{\sqrt{2\,\mathrm{Re}\,F}}\right\}\right] \tag{15}$$

where Erf denotes the familiar error function[15]. As the argument of Erf increases from $-\infty$ to ∞ (i.e. between the two anti-Stokes lines adjacent to the Stokes line Im F=0), $S(F)$ increases from -1 to 1. Recalling that F is proportional to the large parameter k, we see that the 'width' of the Stokes line, that is the range in the space of variables X over which the subdominant exponential enters, scales as $k^{-1/2}$.

I wish to make four remarks about the error-function smoothing formula. The first concerns its generality. Although it provides a refined description of the Stokes phenomenon in Airy, Bessel, hypergeometric, and Mathieu functions (and even the error function itself), it is not restricted to these special functions, nor to the solutions of certain differential equations, nor to integrals with a large parameter. Its range of applicability is all functions whose asymptotic series diverge as (factorial/power).

The second remark concerns the extent to which the derivation of the error-function smoothing requires the resummation of divergent series. Until now, resummation provides the most direct and context-free route to the formula. It does not however seem to be popular amongst mathematicians - certainly not those who have taken up the important question of providing a rigorous justification for the smoothing, with error bounds. In particular cases where this has been possible[16,18,19], it was achieved by using special methods, appropriate to particular classes of integrals, where the remainder can be expressed in closed form rather than as a divergent series (see §4 for a wide generalisation of such cases). Other problems for which the smoothing can be obtained without resummation, albeit sill non-rigorously, are certain second-order differential equations[20], or equivalent first-order systems[21].

The third remark concerns the importance of optimal truncation. Without this, the Stokes multiplier is still defined as in (14), but with a different summation limit. This change seems innocuous but actually makes a big difference[20-22]. For non-optimal truncation (which means that N lies outside the range $|F|-\sqrt{|F|}$ to $|F|+\sqrt{|F|}$), $S(F)$ still increases from 0 to 1, but with exponentially large oscillations and over an X-range bigger than $k^{-1/2}$.

The fourth remark is that the smoothing has applications in physics. In wave theory, small exponentials represent complex evanescent rays, so that Stokes' phenomenon describes the gentle birth of rays[12]- in contrast to caustics, which represent the violent transformation of real into complex rays. Mathematically, rays correspond to saddles of diffraction integrals, Stokes' phenomenon to two saddles having the same (imaginary part of) height, and caustics to two saddles colliding. For integrals more complicated than that describing Ai, Stokes' phenomenon can occur on surfaces in the space of real parameters X, and can take interesting forms[23]. One direct application of the smoothing formula is to the generation of exponentially weak reflections[20], for example above a potential barrier in quantum mechanics. Another is to the history of a quantal transition between two states, induced by a slowly changing field; in this case, optimal truncation corresponds to a particular choice of basis states, and suggests new experiments[21,22] to detect the Stokes phenomenon.

Applied to Ai, the smoothing (15) spreads the Stokes jump over the region between the anti-Stokes lines at θ=60° and θ=180°, and gives the approximation

$$\mathrm{Ai}(z) \approx \mathrm{Ai5}(z,N)$$

$$= \frac{1}{2z^{1/4}\sqrt{\pi}} \left(\exp\left\{+\frac{F}{2}\right\} \sum_{r=0}^{N} Y_r + \mathrm{i}\exp\left\{-\frac{F}{2}\right\} \frac{\left[1+\mathrm{Erf}\left\{-\sin\left(\frac{3}{2}\theta\right)\sqrt{\frac{-|F|}{2\cos\left(\frac{3}{2}\theta\right)}}\right\}\right]}{2} \mathrm{H}(\theta-60°) \right) \tag{16}$$

(despite the step on the antiStokes line, there is no discontinuity). In fig. 4 this is compared with the exact Ai. Evidently the agreement is again much improved: the curves can hardly be distinguished over the whole range of θ. (Actually the approximation Ai5 is defective in that it is not real when θ=180°, but the cure, which is to include in the subdominant contribution the first N terms of its asymptotic series (cf. (8)) - a procedure for which there is theoretical justification[2] - leads to a curve which cannot be distinguished from Ai5 in fig. 4.)

A more discriminating test is shown in fig. 5. Here the error-function smoothing (15) is compared with the exact multiplier defined by (14) and (5), namely

$$S(F) = -2\mathrm{i}z^{1/4}\sqrt{\pi}\exp\left(\frac{F}{2}\right)[\mathrm{Ai}(z) - \mathrm{Ai2}(z,r^{*})] \tag{17}$$

This multiplier, predicted to be of order unity, is the difference between two quantities which near the Stokes line are both exponentially large: the exact Ai and its superasymptotic approximation. Even under this magnification, the agreement is excellent, and, as expected, is better for the larger singulant $|F|=10$ (refinement of the general theory[1] shows that the error in the error-function smoothing is of order $F^{-1/2}$ for Im S, and of order F^{-1} for Re S).

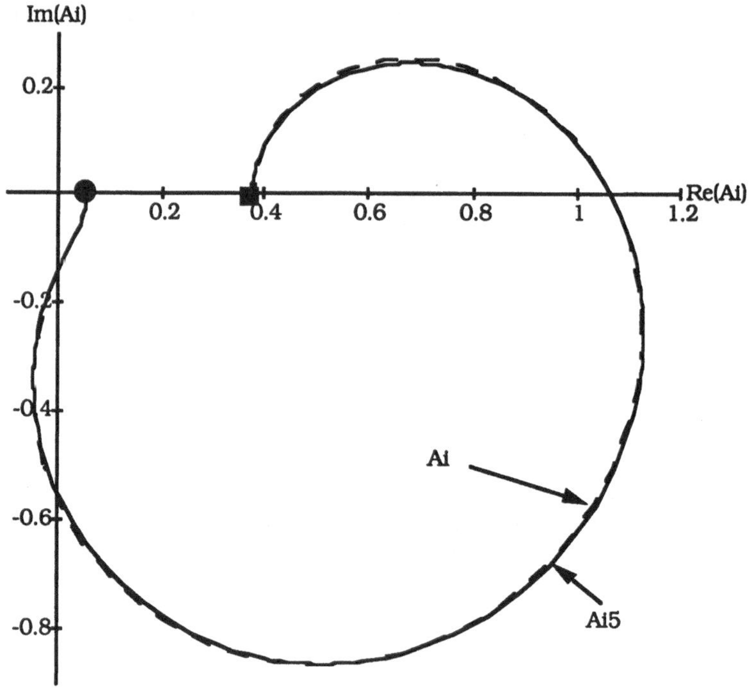

Fig. 4. As fig. 2, but comparing the exact Ai (dashed line)with Ai5 with N=3 (full line) (superasymptotics + Stokes smoothing of subdominant exponential).

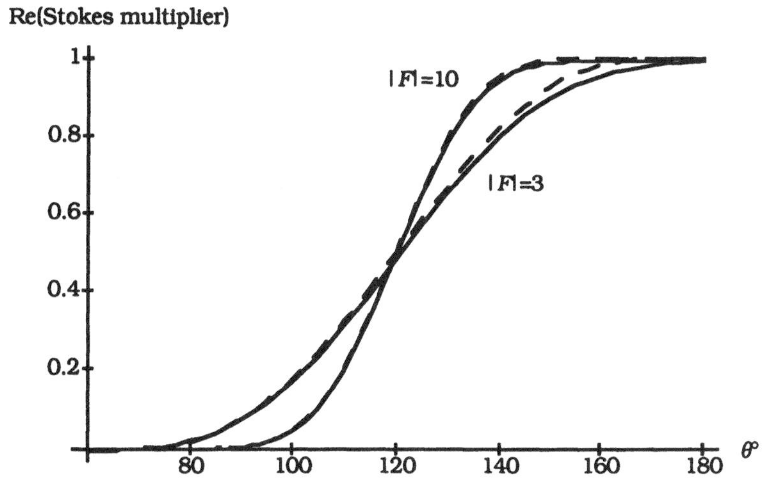

Fig.5. Real part of Stokes multiplier for the Airy function, across the Stokes line θ=120°. The full lines are the exact $S(F)$ (equation 17); the dashed lines are the theoretical smoothing (15).

4. RESURGENCE

Deeper penetration of the asymptotic series (1) requires more accurate asymptotics of the asymptotics' than the leading term (9). Resurgence is a principle that greatly assists the determination of such higher-order approximations to the late terms. The idea is that if (1) is regarded as a complete asymptotic expansion which can represent the function $y(k;X)$ exactly, resummation of the late terms must yield not only the leading-order subdominant exponential but also the corrections terms in *its* asymptotic series. This must hold for all component asymptotic series, so each must contain, encoded in its late terms, all the terms of all the other series. Systematic exploration of this requirement of mutual consistency is still not completed. Dingle[8] gave several examples of resurgence; Écalle[24] described it at length (as well as inventing the term); Voros[25] applied it to differential equations (calling it 'analytic bootstrap'); and Flagolet and Odlyzko[26] examined applications to generating functions with exotic singularities.

Before illustrating resurgence, I should point out that for some simple functions it occurs only in rudimentary form. One such class (which includes the integrals Ei and Erf) is where the form (factorial/power) holds for all the terms Y_r, not just as $r\to\infty$; then a single resummation terminates the series exactly. Another class (which includes log $\Gamma(z)$[32]) is where the Y_r are given by an infinite convergent series of (factorial/power) terms, each of which can be exactly terminated by a single resummation. (A curiosity is that the superasymptotics of $n!$ requires summing to the least term $r\approx\pi n$, which involves $[\mathrm{Int}(2\pi n)]!$ - a case of runaway self-reference, if not resurgence.)

Usually, though, we can expect resurgence to arise in all its glory, which will be illustrated now with a brief description of a new result obtained with Howls[3]. Consider the integral

$$I_j(k) = \int_{C_j(k)} dz G(z) \exp\{-k\phi(z)\} \tag{18}$$

where $G(z)$ and $\phi(z)$ are nonsingular and $\phi(z)$ has saddles at a number of points z_j. $C_j(k)$ is one of the two infinite oriented steepest-descent contours through z_j. It is convenient, although not necessary, to think of k as complex, with $|k|$ as the large parameter and arg k as the variable X. Standard steepest-descents[8] yields the following series of the form (1), in which for convenience the prefactor and the coefficients Y_r have been amalgamated:

$$I_j(k) = \exp\{-k\phi_j\} \sum_{r=0}^{\infty} T_{jr}(k) \tag{19}$$

The terms are

$$T_{jr}(k) = \frac{\left(r-\frac{1}{2}\right)!}{2\pi i} \oint_j dz \frac{G(z)}{\left[F_j(z)\right]^{r+1/2}} \text{ , where}$$

$$F_j(z) \equiv k\left[\phi(z) - \phi_j\right] \tag{20}$$

Here the subscript j means 'evaluated at z_j', and the contour is a positive circuit of z_j. The integrals can all be evaluated explicitly in terms of derivatives of ϕ and G at z_j. For example,

$$T_{j0}(k) = \left(\frac{2\pi}{k\phi_j''}\right)^{1/2} G_j \tag{21}$$

(primes denote z derivatives).

The series (19) is a local expansion about the saddle j, and diverges because of the other saddles l. An explicit and exact expression, whose derivation[3], (involving P2C2E[27] here) has been obtained for this resurgence, showing how the integrals through certain other saddles give the remainder of the truncated series for a given saddle:

$$I_j(k) = \exp\{-k\phi_j\} \sum_{r=0}^{N-1} T_{jr}(k) +$$

$$+ \frac{\exp\{-k\phi_j\}}{2\pi i} \sum_l \frac{(-1)^{\gamma_{jl}}}{\left(F_{jl}\right)^{N-1/2}} \int_0^{\infty} dv \frac{v^{N-1/2}\exp(-v)}{F_{jl} - v} \left\{\exp\left(\frac{kv}{F_{jl}}\right) I_l\left(\frac{kv}{F_{jl}}\right)\right\} \tag{22}$$

Here

$$F_{jl} \equiv k\left(\phi_l - \phi_j\right) \tag{23}$$

denotes the singulant giving the exponent difference beteween the two saddles. The contributing other saddles l are determined by a topological rule: They are the saddles reached by lines of constant phase of $F_j(z)$ issuing from z_j. The sign of each contribution is determined by γ_{jl}, which is zero if the expanded loop contour through z_j has the same sense at z_l as $C_l(kv/F_{jl})$, and unity otherwise. Being exact, (22) contains the Stokes phenomenon (contribution from the pole at $v=F_{jl}$), and the higher terms of the original expansion (from the expression of the v integral as a factorial for large N).

Substitution of the series (19) converts (22) into a formally exact resurgence relation between the coefficients for the saddle j and the contributing other saddles l:

$$T_{jr} = \frac{1}{2\pi i} \sum_l (-1)^{\gamma_{jl}} \sum_{s=0}^{\infty} \frac{(r-s-1)!}{\left(F_{jl}\right)^{r-s}} T_{ls} \tag{24}$$

For large r, the leading contribution is the term $s=0$ from the saddle l for which $|F_{jl}|$ is smallest. Denoting this singulant by F, we obtain

$$T_{jr} \xrightarrow[r\to\infty]{} \frac{(-1)^{\gamma_{jl}}}{2\pi}\left(\frac{2\pi}{-k\phi_l''}\right)^{1/2} G_l \frac{(r-1)!}{F^r} \tag{25}$$

With appropriate identifications, this is the same as the previous late-terms approximation (9). Now, however, we have the complete expression, giving all the corrections, from all the contributing saddles. Of course it is only formal, because in the terms $s>r-1$ the factorials diverge; but it can be made to converge by resummation, which reproduces (22).

Hyperasymptotics consists of iterating (22), with optimal truncations, and substituting at each stage the truncated asymptotic series for the I_l, a procedure that can be regarded as multiple scattering among the saddles. Each iteration produces an exponential improvement. The result is, in the general case[3], an intricate sequence of truncated asymptotic series, eventually involving all the saddles (i.e. not just those which contribute to (22) at a given stage). These series involve the original asymptotic coefficients T_{jr}, which of course depend on the particular function being approximated, and certain 'generalised terminants', in the form of multiple integrals, which are universal functions of the singulants F_{jl}. The advantage of the integral (22) over the 'pure asymptotic' resurgence (24) is that at every stage of hyperasymptotics there is an explicit expression for the remainder, which will, we anticipate, be indispensable in constructing rigorous error bounds.

Hyperasymptotics comes to a natural halt, because each hyperseries is shorter than its predecessor, and eventually contains only one term. The decreasing length is a consequence of the 'live now, pay later' philosophy, natural in asymptotics, that the terms must continue to decrease, not only within each hyperseries but from each hyperseries to the next. A typical result[2] is that hyperasymptotics reduces the ultimate error from $\exp\{-|F|\}$ (superasymptotics) to $\exp\{-(1+2\log 2)|F|\}\approx\exp\{-2.386|F|\}$; thus the error is reduced to less than its square.

To illustrate hyperasymptotics, we again employ Ai, defined by (2) This example is special because there are only two saddles (and therefore only one singulant $F_{jl}= -F_{lj} \equiv F$), and hyperasymptotic multiple scattering is simply back and forth between them. Moreover, the pre-exponential factor is $G=1$. The terms in the dominant and subdominant series, including the prefactors M in (1), are the same (apart from signs) and the general resurgence (24) reduces in the notation of (1) and (5) to

$$Y_r = \frac{1}{2\pi F^r}\sum_{s=0}^{\infty}(r-s-1)!(-F)^s Y_s \tag{26}$$

This relation holds not only for integrals with two saddles but also - as was discovered by Dingle[8] - for the coefficients in the asymptotic expansion of solutions of second-order linear ordinary differential equations with a single transition point[2,17]. Its compactified form - a special case of (22) - is equivalent to the self-Stieltjes transform relation for Airy functions, as noticed and exploited by Boyd[18]. In this case, each hyperseries is half the length of its predecessor, so hyperasymptotics stops after $\mathrm{Int}\log_2|F|$ stages, with a total of $2\mathrm{Int}|F|$ terms (as opposed to $\mathrm{Int}|F|$ terms in the zeroth stage of hyperasymptotics, namely superasymptotics).

We express the numerical results in terms of $Y(z)$, defined (cf.5) by

$$\mathrm{Ai}(z) \equiv \frac{1}{2z^{1/4}\sqrt{\pi}}\exp\left\{\frac{F}{2}\right\}Y(F) \tag{27}$$

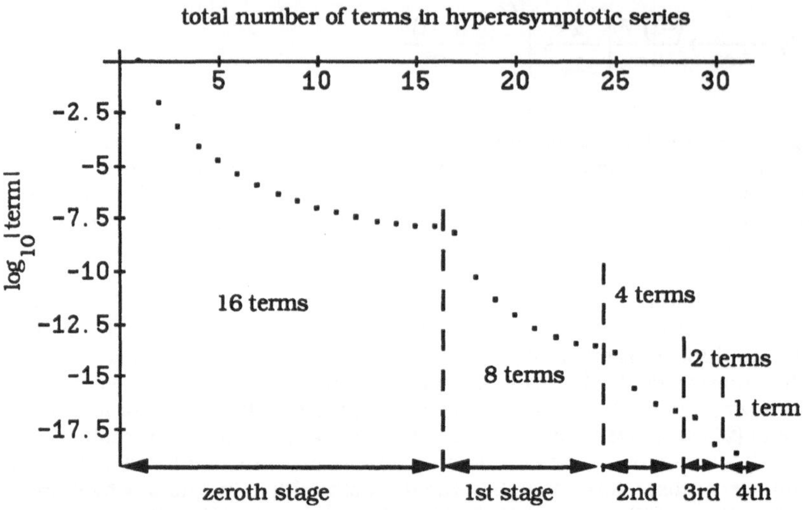

Fig. 6. Decrease of the terms in the five hyperseries constituting hyperasymptotics for $Y(-16)$ for the Airy function.

For $F=-3$ (i.e. $z=1.7171$), the exact value is $Y(-3)=0.96419...$. Thus the lowest approximation $Y\sim 1$ (i.e. Ai1 - cf. (4)) is in error by 0.036. Superasymptotics, i.e.Ai2 (equation 5) with $N=4$, gives $Y\approx 0.95895...$, an error of -0.00524, about ten times better. With a single stage of hyperasymptotics we do much better: $Y\approx 0.96410...$, an error of -0.00009, about 60 times better than superasymptotics

The improvements are even more dramatic for larger $|F|$. Fig. 6 (taken from [2]) shows the magnitudes of the terms in the five hyperseries for Ai when $F=-16$, that is $z=+5.2414827884177932413...$, corresponding to $\theta=0$ (fig. 1). The exact value of $Y(-16)$ is 0.99183679918826259891....Thus the lowest approximation $Y\sim 1$ is in error by 8.163×10^{-3}. Superasymptotics, i.e. Ai2 with 16 terms, gives $Y\approx 0.99183679351132345911...$, an error of -5.677×10^{-9}. With hyperasymptotics, taken to its natural halt, we obtain $Y\approx 0.99183679918826260060....$, an error of 1.151×10^{-18} (close to that predicted theoretically). Similar accuracy is obtained on the Stokes line $\theta=120°$ (fig.1) and on the anti-Stokes line $\theta=180°$.(where, for example, hyperasymptotics can be employed to solve the 'eigenvalue problem' of determing the zeros of Ai).

5. OUTLOOK

There are several scientific reasons, as well as purely mathematical ones, for seeking such detailed understanding of the relatively simple asymptotic problems I have been discussing. One is that asymptotics is often deeply involved in the conections between physical theories[28]. It is common for a more general theory to 'reduce' to a less general theory when some parameter vanishes. For example, special relativity reduces to Newtonian mechanics as the particle speed vanishes; wave optics reduces to ray optics as the wavelength vanishes; quantum mechanics reduces to classical mechanics as Planck's constant vanishes; Navier-Stokes fluid motion reduces to Eulerian flow as the viscosity vanishes; statistical mechanics reduces to thermodynamics as the reciprocal number of particles vanishes, etc., etc. Only in the first of these examples is the limit regular and the expansion in the small parameter convergent. In all the other cases, the limits are singular and lead to divergent series. Associated with the singularities are important phenomena such as ray caustics, turbulence and critical behaviour. This connection between asymptotics and theory reduction, not appreciated by philosophers (at least in my experience), is sufficient reason to try to understand divergent series as deeply as possible.

A more concrete reason is that in practice we often encounter asymptotics which is intrinsically more complicated than what I have been describing here, and there is no hope of understanding the complicated problems unless we penetrate the simpler ones first. One important extension would be to the case of *many exponentials*. In Stokes' phenomenon and its smoothing (§§3 and 4) only two exponentials are essentially involved: the dominant and leading subdominant ones. With the resurgence relation (22) we make the first step towards the consistent treatment of many exponentials . However, the integral (18) only involves a single variable, and the topology involved in the derivation[3] of (22) suggests that the results might not be the same when many exponentials arise in the multiple integrals of diffraction theory, or the infinite-dimensional functional integrals of quantum mechanics, statistical mechanics and field theory.

Another complication is that when many exponentials appear they need not always be ordered in a dominance hierarchy: the singulants can all be imaginary. An important class of such problems occurs in quantum chaology[28], that is the semiclassical asymptotics of quantum systems whose classical counterparts have chaotic trajectories. There, the exponentials appearing in the asymptotic expansion of (for example) the density of energy levels are associated with classical periodic orbits. Even in lowest order, where each exponential is included bare - that is, without its correction terms - the proliferation of periodic orbits makes the sum diverge, in ways that are still mysterious and probably related to the Riemann hypothesis of arithmetic. There are strong hints[28,29] that at this higher level resurgence may again prove to be an important guiding principle, this time relating the exponentials associated with long and short orbits rather than the late and early terms of the series associated with individual orbits.

There are many directions for further research. One is to break through the $\exp\{-2.386|F|\}$ barrier (ultraasymptotics?). Another is to extend hyperasymptotics to multiple integrals, and to the Schrödinger equation wiith many transition points (the work of Balian and Bloch[30], Knoll and Schaeffer[31], and Voros[25] could be helpful here). Yet another is to provide rigorous mathematical underpinning for the results already obtained, hopefully going beyond proofs for particular functions[16,18,19] to genericity theorems which would delineate the universality class for which the results (for example the smoothing (15)) are valid. Going further, it would be helpful to know something about the late terms of the asymptotic series corresponding to each of the classical closed orbits in quantum chaology (for example whether the divergences of these individual series are related to the divergence of the sum over lowest-order exponentials for all the orbits).

REFERENCES

1. M.V. Berry, Uniform asymptotic smoothing of Stokes's discontinuities, *Proc. Roy. Soc. Lond,* A422: 7-21(1989).
2. M.V. Berry and C.J.Howls, Hyperasymptotics,*Proc. Roy. Soc. Lond,* A430: 653-668 (1990).
3. M.V. Berry and C.J. Howls, Hyperasymptotics for integrals with saddles, submitted to *Proc. Roy. Soc. Lond.* (1991).
4. G.H. Hardy, "Divergent Series", Clarendon Press, Oxford (1949).
5. G.G.Stokes, On the numerical calculation of a class of definite integrals and infinite series, *Trans. Camb. Phil. Soc.* 9: 379-407 (1847).
6. B.G. Levi, A super time to renormalize, *Physics Today*, April 1988: 25.
7. Lord Kelvin, The scientific work of Sir George Stokes, *Nature* 67: 337-338 (1903).
8. R.B.Dingle, "Asymptotic Expansions: their Derivation and Interpretation", Academic Press, New York and London (1973).
9. G.G.Stokes, On the discontinuity of arbitrary constants which appear in divergent developments, *Trans. Camb. Phil. Soc.* 10: 106-128 (1864).
10. J.E. Littlewood , preface to ref.4.
11. N.G. de Bruijn , "Asymptotic methods in analysis", North-Holland, Amsterdam (1958).
12. M.V. Berry, Stokes' phenomenon; smoothing a Victorian discontinuity, *Publ. Math.of the Institut des Hautes Études scientifique*, 68: 211-221 (1989).
13. R.Balian, G.Parisi, and A.Voros, Discrepancies from asymptotic series and their relation to complex classical trajectories, *Phys. Rev, Lett.* 41: 141-1144(1978).
14. F.W.J. Olver, "Asymptotics and special functions" Academic Press, New York and London (1974).

15. M.Abramowitz and I.A.Stegun, 1964, "Handbook of mathematical functions", National Bureau of Standards, Washington (1964).
16. F.W.J. Olver, On Stokes' phenomenon and converging factors, in "Proceedings of International Symposium on Asymptotic and Computational Analysis (Manitoba, Winnipeg 1989)", R.Wong, ed., Marcel Dekker, New York (1990), pp329-355.
17. M.J. Rakovic, and E.A.Solov'ev, Higher orders of semiclassical expansion for the one-dimensional Schrödinger equation, *Phys. Rev.* A40: 6692-6694 (1989).
18. W.G.C. Boyd , Stieltjes transforms and the Stokes phenomenon, *Proc. Roy. Soc. Lond.*, A429: 227-246 (1990).
19. D.S.Jones, Uniform asymptotic remainders, in "Proceedings of International Symposium on Asymptotic and Computational Analysis (Manitoba, Winnipeg 1989)" R. Wong, ed., Marcel Dekker, New York (1990), pp241-264.
20. M.V. Berry, Waves near Stokes lines, *Proc. Roy. Soc. Lond*, A427: 265-280 (1990).
21. M.V. Berry, Histories of adiabatic quantum transitions, *Proc. Roy. Soc. Lond*, A429: 61-72 (1990).
22. R. Lim and M.V. Berry, Submitted to *J.Phys.A.* (1991).
23. M.V. Berry and C.J. Howls, Stokes surfaces of diffraction catastrophes with codimension three, *Nonlinearity*, 3:281-291 (1990).
24. J.Écalle, "Les fonctions résurgentes " (3 volumes) Publ. Math. Université de Paris-Sud (1981), and "Cinq applications des fonctions résurgentes ", Preprint 84T62, Orsay (1984).
25. A.Voros, The return of the quartic oscillator. The complex WKB method, *Ann. Inst H. Poincaré*, 39:211-338 (1983).
26. P. Flagolet and A.M. Odlyzko, Singularity analysis of generating functions, *SIAM J. Discrete Math.*, 3 (2): 216-240 (1990).
27. P2C2E=processes too complicated to explain, see S. Rushdie, "Haroun and the sea of stories", Granta, London (1990).
28. M.V. Berry , Some quantum-to-classical asymptotics, in "Chaos and quantum physics",Les Houches Lecture Series 52, M J Giannoni and A Voros, eds. ,North-Holland, Amsterdam (1991).
29. M.V. Berry and J.P. Keating, A rule for quantizing chaos?, *J. Phys. A.*, 23: 4839-4849 (1990).
30. R. Balian and C. Bloch, Solution of the Schrödinger equation in terms of classical paths, *Ann. Phys. (N.Y.)* 85: 514-545 (1974).
31. J. Knoll and R. Schaeffer, Semiclassical scattering theory with complex trajectories. 1. Elastic waves, *Ann. Phys. (N.Y.)*, 97: 307 (1976).
32. M.V. Berry, Infinitely many Stokes smoothings in the Gamma Function, submitted to *Proc. Roy. Soc. Lond*, (1990).

Computation of Transcendental Effects in Growth Problems: Linear Solvability Conditions and Nonlinear Methods-The Example of the Geometric Model

Vincent Hakim
L.P.S., C.N.R.S. and Universities Paris VI and VII
Ecole Normale Supérieure 24, rue Lhomond 75231 Paris

I. Introduction

It is now well understood that in many moving interface problems, surface tension plays a singular role(1). The basic reason is that surface tension multiplies a term that contains the highest number of derivatives. This realization(2)has made possible the analysis of several problems, including the velocity selection for a parabolic needle crystal moving in an undercooled melt (3) and the shape selection of viscous fingers moving in a linear Hele-Shaw cell (4). The first general analytic method for treating this kind of singular perturbation problems was found by J. Langer(5) in a simple case. He proposed to linearize the equations around a zero-surface tension solution while keeping in the linear equation terms with the highest number of derivatives although they were formally of higher order in the expansion parameter. Exponentially small terms could then be obtained from a linear solvability condition. It was however clear since its very proposal that Langer's method could not be completely correct since the magnitude of the exponentially small term was about half the result obtained by solving the equation numerically (5). A more satisfactory way of handling the problem was described by M. Kruskal and H. Segur(6) who generalized to nonlinear ODE a well-known idea(7) in the linear case. It consists in extending the equation into the complex plane and analyzing it carefully in the neighborhhood of a singularity of the terms of the regular perturbation expansion. Contrary to Langer's treatment it is not entirely analytic but reduces to solving a parameter free nonlinear problem numerically. Therefore the method of ref.(5) has continued to be utilized under one form or another by some authors, in spite of its somewhat unclear basis. Our aim in this paper is to show exactly what is wrong in Langer's method and how it can be corrected. We take as an example the geometric model of growth because of its simplicity

Asymptotics beyond All Orders, Edited by H. Segur *et al.*
Plenum Press, New York, 1991

and because it originally served to develop the competing methods. The linearization is generally thought as the reason for the incorrectness of Langer's method. We show that, although incorrect, this step is not the one responsible for the main numerical discrepancy and find that the usual treatment of the linear problem itself needs correction. We then show how Langer's technique can be systematically improved to take the nonlinearities into account. This gives a series of analytical approximations that converges to the correct result. This suggests that in the method of ref(6) the asymptotic behavior of the solution of the nonlinear inner problem, which is usually obtained numerically, depends in fact only on tbe coefficients of the regular asymptotic expansion. We show directly that this is the case by performing a Borel summation of the asymptotic expansion.

It is probably apparent from this short summary that the content of this paper is rather technical. Nevertheless, we hope that it brings some clarification and we tried to write it as pedagogically as possible so that it can be useful as a simple introduction to the field.

II. The Geometric Growth Model

The geometric model [8] is a very simple model of interface motion. It mimicks more realistic models in that outward-pointing parts tend to move faster than flat ones, but it is much simpler because non-local interactions between different parts of the interface are completely neglected. Specifically, one postulates that the normal velocity v_n is proportional to the curvature K of the interface. One adds a term that stabilizes the interface at short distances and plays the role of surface tension in a more realistic description. In an adimensioned form

$$v_n = K + \gamma \frac{d^2K}{ds^2} \tag{1}$$

The simple question that we are going to analyze is the non-existence of steady state shapes moving at velocity **v** that have the form of a symmetric needle around the velocity axis. Let us consider a steady state shape moving at velocity **v**. If θ denotes the angle between the velocity and the local normal to the interface then

$$v_n = v \cos(\theta)\,, \qquad K = \frac{d\theta}{ds}$$

Therefore, the equation satisfied by the interface profile is

$$\delta^2 \frac{d^3\theta}{dz^3} + \frac{d\theta}{dz} = \cos(\theta) \tag{2}$$

where δ^2 denotes γv^2 and we have rescaled the curvilinear abscissa $s = z/v$. A simple shooting argument can be given that makes it plausible that such shapes do not exist [2c,d]. When equation (2) is linearized in the large z region two divergent modes are found on each side. This is however an equation of the third order and there are only three arbitrary constants. When two of them have been fixed by the annulation of the divergent modes at $z=-\infty$, one constant is missing to kill the two diverging modes at $z=+\infty$. In an equivalent way, one can use the last constant to choose $\theta(0)=0$. There is then no reason for $\theta''(0)$ to vanish. We are going to use this last formulation and compute $\theta''(0)$ in the small δ limit for the solution of eq.2 such that $\theta(-\infty) = -\pi/2$ and $\theta(0) = 0$ (here and in the following a prime denotes differentiation with respect to z).

The difficulty of this problem comes from the fact that the usual asymptotic perturbation expansion in δ^2 does not give any obvious hint that $\theta''(0)$ is non zero. Let us first recall how this works. θ is searched for under the form

$$\theta = \theta_0 + \delta^2 \theta_1 + \ldots + \delta^{2(n-1)} \theta_{n-1} + \delta^{2n} \theta_n + \ldots \tag{3}$$

When this is substituted in eq. 2, one obtains

$$\theta_0 = \cos(\theta_0) \tag{4.a}$$

$$\theta_1 + \theta_1 \sin(\theta_0) = -\theta_0''' \tag{4.b}$$

$$\ldots$$

$$\theta_n + \theta_n \sin(\theta_0) = -\theta_{n-1}''' + NL_n(\theta_0, \ldots, \theta_{n-1}) \tag{4.c}$$

The first few terms are easily computed

$$\theta_0 = -\frac{\pi}{2} + 2\,\mathrm{Arctg}(e^z), \quad \theta_1 = -\frac{z}{\mathrm{ch} z} + 2\,\frac{\mathrm{sh}(z)}{\mathrm{ch}^2(z)} \tag{5}$$

One notices that θ_0 and θ_1 are odd and satisfy the correct boundary conditions $\theta_0(\pm\infty) = \pm\pi/2$, $\theta_1(\pm\infty) = 0$. This is easily proved by induction for all θ_n. The nonlinear term of the r.h.s. can be written

$$NL_n(\theta_0,\ldots,\theta_{n-1}) = \sin\theta_0 \, P(\theta_1,\ldots,\theta_{n-1}) + \cos\theta_0 \, Q(\theta_1,\ldots,\theta_{n-1}) \tag{6}$$

with

$$\sin\theta_0 = \mathrm{th}(z), \quad \cos\theta_0 = 1/\mathrm{ch}(z).$$

P is a polynomial with all its monomial of total odd degree and Q a polynomial with all its monomial of total even degree. Therefore, if $\theta_1, \ldots, \theta_{n-1}$ are odd and tend to zero at $\pm\infty$, the inhomogenous term of the nth equation (eq. (4c)) is even and tend to zero at $\pm\infty$. Hence, if $\theta_n(z)$ is a solution of this equation that tends to zero at $-\infty$, it also tends to zero at $+\infty$ since the operator L on the l.h.s.($L = \partial_z + th(z)$) has no diverging mode at infinity. Moreover, $\theta_n(z)$ is odd since L commutes with parity and the inhomogenous term is even. Therefore one can find a satisfactory solution to all orders in the asymptotc expansion and $\theta''(0)$ is smaller than any power of δ in the small δ limit.

III. Langer's Linearization Method

One clear cause of the behavior of the regular perturbation expansion is that by performing the expansion we have replaced a linear operator with two divergent modes at infinity by one that has none. Langer's proposal is to perform the linearization, taking care of preserving the correct behavior at infinity, even if that means keeping terms that are formally subdominant. Therefore instead of equation (4.b) we write[5]:

$$L_\delta(\tilde{\theta}_1) = \delta^2 \tilde{\theta}_1''' + \tilde{\theta}_1' + \tilde{\theta}_1 \sin(\theta_0) = -\theta_0''', \quad \tilde{\theta}_1(0) = \tilde{\theta}_1(-\infty) = 0 \tag{7}$$

In this approach, the solution θ of eq.(2) is approximated by $\theta_0+\delta^2\tilde{\theta}_1$. Now, let us suppose we have determined an even function ϕ in the kernel of L_δ^+ the adjoint of L_δ (a shooting argument analogous to the one following eq.(2) shows that this is possible). That is:

$$L_\delta^+(\phi) = -\delta^2 \phi''' - \phi' + \phi \sin(\theta_0) = 0, \quad \phi'(0) = \phi(-\infty) = 0 \tag{8}$$

Then, $\theta''(0)$ can be estimated by computing in two different ways[10] the following integral I_{sc} (its annulation gives the so-called 'solvability condition'):

$$I_{sc} = \int_{-\infty}^{0} L_\delta(\tilde{\theta}_1)\, \phi\, dz$$

We can first compute I_{sc} directly by using eq.7 and obtain:

$$I_{sc} = -\int_{-\infty}^{0} \theta_0'''\, \phi\, dz$$

We can also integrate the derivative terms by parts in L_δ, keeping carefully the boundary contributions (this is where the parity condition on ϕ is useful). We get in this way :

$$I_{sc} = \left[\delta^2(\phi\tilde{\theta}_1'' - \phi'\tilde{\theta}_1' + \phi''\tilde{\theta}_1) + \phi\tilde{\theta}_1\right]_{-\infty}^{0} + \int_{-\infty}^{0} L_\delta^+(\phi)\,\tilde{\theta}_1\,dz = \delta^2\phi(0)\tilde{\theta}_1''(0)$$

In order to get the last equalities, eq.8 and the different boundary conditions on ϕ and θ_1 have been used. By comparing the two results one obtains the expression that we are searching for:

$$\delta^2\tilde{\theta}_1''(0) = -\frac{1}{2}\int_{-\infty}^{\infty} \phi\,\tilde{\theta}_0'''\,dz \tag{9}$$

The symmetry properties of ϕ and θ_0 have been used to extend the range of integration over the whole real axis and ϕ has been normalized in such a way that $\phi(0)=1$. Up to this point, no approximations have been made other than the somewhat heuristic linearization. In order to evaluate eq.(9), we have to find an expression of ϕ, an even function in the kernel of $L_\delta{}^+$. It was proposed in ref.(5) to look for it in the W.K.B form:

$$\phi(z) = \frac{1}{2}\left\{g(z)\exp\left(\frac{iz}{\delta}\right) + g^*(z)\exp\left(-\frac{iz}{\delta}\right)\right\} \tag{10}$$

(g^* is the complex conjugate of g). When this is substituted in eq.(8), g is found to found to satisfy:

$$\delta^2 g'''(z) + 3\,i\,\delta\,g''(z) - 2\,g'(z) - g(z)\,\mathrm{th}(z) = 0, \quad g(0) = 1 \tag{11}$$

Clearly g(z) can be expanded in powers of δ:

$$g(z) = g_0(z) + \delta\,g_1(z) + \ldots + \delta^n\,g_n(z) + \ldots \tag{12}$$

It may seem plausible that to get the dominant contribution in δ^2 to $\theta''(0)$ (eq.9) it is sufficient to estimate g to dominant order in δ^2 (i.e. to approximate g by g_0). One obtains from eq.(10):

$$g_0(z) = \frac{1}{\sqrt{\mathrm{ch}(z)}}, \quad \theta_0'''(z) = \frac{1}{\mathrm{ch}(z)} - \frac{2}{\mathrm{ch}(z)^3}$$

and this gives:

$$\delta^2 \tilde{\theta}_1''(0) = -\frac{1}{2}\int_{-\infty}^{\infty} \frac{\exp(i\frac{z}{\delta})}{\sqrt{ch(z)}}\left[\frac{1}{ch(z)} - \frac{2}{ch(z)^3}\right] dz$$

Since the integrand consists of an analytic function multiplied by a rapidly oscillating factor, the result is exponentially small. Its leading behavior depends only on the behavior of the integrand in the neighborhood of the singularity closest to the real axis, in $z=i\pi/2$. By deforming the integration contour, one easily obtains easily :

$$I_\alpha(\delta) = \int_{-\infty}^{\infty} \frac{\exp(iz/\delta)\ dz}{ch(z)^\alpha} \sim \frac{2\pi}{\Gamma(\alpha)} \frac{\exp(-\frac{\pi}{2\delta})}{\delta^{(\alpha-1)}}, \tag{13}$$

Finally, this gives Langer's result ($\Gamma(7/2)= 15\sqrt{\pi}/8$):

$$\theta''(0) = \delta^2 \tilde{\theta}_1''(0) \sim \frac{16\sqrt{\pi}}{15} \frac{\exp(-\frac{\pi}{2\delta})}{\delta^{5/2}}, \tag{14}$$

It is important to notice that this result comes entirely from the term in the integrand that is the most divergent near $i\pi/2$. The dependence of $\theta''(0)$ on δ in eq.(14) appears to be the right one but the prefactor is definitely not correct and is about half too small. For the problem considered, this is not so important since the result shows that there is no steady state shape as long as the prefactor is non zero. However, in more complicated and more realistic cases, the equation can depend on an explicit or on an hidden parameter, like surface tension anisotropy (3)or the width of viscous fingers(4). Then, the prefactor is not a constant but a full function and its zeroes determine the possible steady state shapes. It is therefore of some interest to compute it correctly. In the forthcoming two sections we are therefore going to show where we have erred in the previous derivation.

IV. A Refined Solution of the Linear Problem

One obvious approximation of the previous treament is that we have analyzed the linear equation (7) instead of the real non-linear equation (2). An easy way to improve on this will be given in the next section. In the present section, our aim is to show that even if we restrict ourselves to the analysis of the linear eq.(7), the previous treatment is not entirely correct, a fact that is usually not so well appreciated. The error comes from the estimation of the function ϕ (= $1/2(g \exp(iz/\delta)+ c.c.)$) in the kernel of the adjoint operator $L_\delta{}^+$. It was argued that in order to evaluate the solvability integral (eq.(9)) to lowest order in δ it was sufficient to compute g to lowest order in δ. Although this may appear quite convincing at first thought, the trouble is that the higher corrections $g_1, \ldots, g_n$ can be, and in fact are, more divergent than g_0 around $i\pi/2$. Therefore, when the solvability condition is evaluated, their contributions get divided by higher powers of δ (see eq.(13)) and they contribute as much as g_0 to the final result.

Let us detail this reasoning. We have seen previously that the behavior of $g_0(z)$ near $i\pi/2$ is

$$g_0 \sim \gamma_0\,(iz + \pi/2)^{-1/2}\ , \qquad \text{with } \gamma_0 = 1 \tag{15}$$

Similarly, the behavior of $g_n(z)$ near $i\pi/2$ is

$$g_n \sim \gamma_n\,(iz + \pi/2)^{-(n+1/2)} \tag{16}$$

Therefore, the dominant contribution C_n of $\delta^n\, g_n(z)$ to $\delta^2\, \theta_1''(0)$ is from eq.(13)

$$C_n \sim \gamma_n\, \delta^n\ I_{n+7/2}\,(\delta) \sim \frac{2\pi\,\gamma_n}{\Gamma(\frac{7}{2}+n)}\ \frac{\exp(-\frac{\pi}{2\delta})}{\delta^{5/2}} \tag{17}$$

As explained above all C_n are of the same order in δ. The previous result (eq.14) is given by C_0 only. Thus if one wants to obtain the complete result for the linear equation a possible strategy is now apparent: instead of considering only g_0, one should keep track of the most diverging term of each g_n near $i\pi/2$. In order to do it, we find it convenient to introduce a variable w that measures the distance of z to $i\pi/2$:

$$z = i\pi/2 + w\,\delta$$

Then, we denote by g_{md} the sum of the most divergent terms near $i\pi/2$:

$$g_{md}(z) \equiv \sqrt{\delta} \sum_{n=0}^{\infty} \delta^n\ g_n(w)$$

It is readily seen from eq.(11) that g_{md} satisfies (since $th(z) \sim 1/w\delta$):

$$g_{md}'''(w) + 3\,i\ g_{md}''(w) - 2\,g_{md}'(w) - g_{md}(w)/w = 0 \tag{18}$$

Using a Laplace transform, one gets the integral representation:

$$g_{md}(w) = \frac{2}{\sqrt{\pi i w}} \int_0^\infty du\, e^{-u} \frac{\sqrt{u\,(1 - \frac{u}{2i\,w})}}{1 - \frac{u}{iw}} \qquad (19)$$

where $g_{md}(w)$ has been normalized in such a way that the value of the first term of its Laurent series in w agrees with eq.(15). This integral representation is convenient for performing the summation of the C_n without explicitly computing them. Let us write:

$$\frac{\sqrt{1 - \frac{u}{2i\,w}}}{1 - \frac{u}{iw}} = \sum_{n=0}^{\infty} \beta_n \left(\frac{u}{iw}\right)^n \qquad (20)$$

Then, eq.(19) gives

$$g_{md}(w) = \frac{2}{\sqrt{\pi}} \sum_{n=0}^{\infty} \frac{\beta_n\, \Gamma(n+3/2)}{(iw)^{n+1/2}} \qquad (21)$$

By comparing eq.(21) and eq.(16) we obtain

$$\gamma_n = \frac{2}{\sqrt{\pi}}\, \beta_n\, \Gamma(n+3/2)$$

The sum of the contributions C_n's to $\theta''(0)$ can now be performed (see eq.(17)):

$$\sum_{n=0}^{\infty} C_n = \frac{4\sqrt{\pi}}{\delta^{5/2}} \exp\left(-\frac{\pi}{2\delta}\right) \sum_{n=0}^{\infty} \beta_n \frac{\Gamma(n+3/2)}{\Gamma(n+7/2)}$$

$$= \frac{4\sqrt{\pi}}{\delta^{5/2}} \exp\left(-\frac{\pi}{2\delta}\right) \sum_{n=0}^{\infty} \frac{\beta_n}{(n+3/2)(n+5/2)} \qquad (22)$$

This sum can be computed from the integral representation of g_{md} (eq.(19)) and the definition of the β_n's (eq.(20)):

$$\sum_{n=0}^{\infty} \frac{\beta_n}{(n+3/2)(n+5/2)} = \int_0^1 dt \int_0^t ds\, \frac{\sqrt{s(1 - s/2)}}{1 - s} = \frac{\pi}{4\sqrt{2}}$$

Finally, we obtain the complete and correct result [11] that replaces eq.(14):

$$\delta^2 \tilde{\theta}_1''(0) \sim \frac{\pi^{3/2}}{\sqrt{2}} \frac{\exp(-\frac{\pi}{2\delta})}{\delta^{5/2}} , \tag{23}$$

We have obtained a prefactor equal to $\pi^{3/2}/\sqrt{2}$ (= 3.9374...) which is almost two times bigger than the one obtained previously ($16\sqrt{\pi}/15$ = 1.89..). It differs by less than ten percent from the result for the full non-linear equation (2) (see below). It is in fact historically amusing to notice that this result is in good agreement with Langer's first numerical estimate[5] of this last quantity.

V. A Systematic Inclusion of the Nonlinearities

In the previous section, the treatment of the linear equation (7) has been analyzed. Of course, the goal is to analyze the full non-linear equation (2). The results differ for the same reason as the one that forbids use of the lowest order W.K.B. approximation: although higher order terms are formally of higher order in δ^2 they diverge more strongly near $i\pi/2$ and therefore cannot be neglected. There is a simple way to take them into account within the approach of ref.(5) that we have been following. Langer's proposal was to modify the first equation of the usual asymptotic expansion (4). In some sense it is like closing the hierarchy of equations (4) at the first level. We can improve on this by closing it at the nth level instead. Specifically, we propose to analyze the system:

$$\theta_1' + \theta_1 \sin(\theta_0) = -\theta_0''' \tag{24.a}$$

.....

$$\theta_{n-1}' + \theta_{n-1} \sin(\theta_0) = -\theta_{n-2}''' + NL_{n-1}(\theta_0,\ldots,\theta_{n-2}) \tag{24.b}$$

$$L_\delta(\tilde{\theta}_n) = \delta^{2n} \tilde{\theta}_n''' + \tilde{\theta}_n' + \tilde{\theta}_n \sin(\theta_0) = -\theta_{n-1}''' + NL_n(\theta_0,\ldots,\theta_{n-1}) \tag{24.c}$$

where the nth equation has been modified instead of the first. The second derivative $\theta''(0)$ is now approximated by $\delta^{2n}\,\tilde{\theta}_n''(0)$ since $\theta_1''(0),\ldots, \theta_{n-1}''(0)$ vanish. It can be computed exactly as in the previous section by using a function in the kernel of L_δ^+. Since the inhomogenous term on the r.h.s. of eq.(24.c) is a function of the regular perturbation expansion, the analog of eq.(9) can be written as

$$\delta^{2n} \tilde{\theta}_n''(0) = \frac{\delta^{2(n-1)}}{2} \int_{-\infty}^{\infty} \phi \left[\theta_n' + \theta_n \sin\theta_0\right] dz$$

The leading behavior of the integral on the r.h.s. comes from the most diverging term of the integrand near $z=i\pi/2$. As explained below, the behavior of θ_n near $i\pi/2$ is ($z=i\pi/2+v$):

$$\theta_n \sim i\, a_n\, v^{-2n} \tag{25}$$

where a_n is a real constant. We are interested in the limit $n \to \infty$ where the nonlinearities are fully taken into account. In this case it is sufficient to approximate ϕ by the lowest order W.K.B. approximation $g_0 \cos(s/\delta)$ (eq.12). This is not because higher approximations give higher powers of δ but because they are suppressed by powers of n (see below). Therefore we get using the definition of I_α (eq.13):

$$\theta''(0) \sim \underset{n\to\infty}{\mathrm{Lim}}\left[\delta^{2n}\, \tilde{\theta}_n''(0)\right] = \underset{n\to\infty}{\mathrm{Lim}}\left[(-1)^n \frac{\delta^{2(n-1)}}{2}(2n-1)\, a_n\, I_{2n+3/2}(\delta)\right]$$

When the asymptotic expansion of $I_{2n+3/2}$ (eq.13) is substituted in this equation one gets the simple formula:

$$\theta''(0) \sim \underset{n\to\infty}{\mathrm{Lim}}\left[(-1)^n \frac{2\pi n\, a_n}{\Gamma(2n+3/2)}\right] \frac{\exp\left(\frac{-\pi}{2\delta}\right)}{\delta^{5/2}} \tag{26}$$

The important point is that the prefactor is expressed explicitly in function of the regular perturbation expansion (eq.4) and depends only on the limiting behavior of the most diverging term of θ_n near $i\pi/2$. The use of g_0 is sufficient because the contributions of the higher order contributions have exactly the same form but with n in the argument of the Γ function replaced by a larger integer. Therefore as mentionned above these contributions are suppressed by powers of n compared to the one that has been retained and do not contribute to the result in the limit $n\to\infty$. Before obtaining the final expression for the nonlinear case let us show that eq.(26) gives in a simple way the correct result for the linear equation (eq.23). If we omit the nonlinear terms in equations (24), the system is strictly equivalent to the linear eq.7 with

$$\delta^2 \tilde{\theta}_1 = \delta^2 \theta_1 + \ldots + \delta^{2(n-1)} \theta_{n-1} + \delta^{2n} \tilde{\theta}_n$$

Moreover, the generic equation reduces to

$$\theta'_j + \theta_j \sin\theta_0 = -\theta'''_{j-1}$$

This shows that in the linear case the behavior of θ_n near $i\pi/2$ is (since $\sin\theta_0 \sim 1/v$):

$$\theta_n \sim i\, b_n\, v^{-2n}, \qquad b_n = -4\, n\, (n-1)\, b_{n-1}, \qquad b_1 = -2 \tag{27}$$

Therefore ,

$$b_n = \frac{1}{2n}(-4)^n\, n!^2 \tag{28}$$

When one replaces a_n by b_n in eq.(26) and uses Stirling Formula ($\Gamma(x) \sim (2\pi)^{1/2} x^{x-1/2} e^{-x}$) one obtains the previous result for the linear case (eq.23) :

$$\theta''(0) \sim \frac{\pi^{3/2}}{\sqrt{2}} \frac{\exp(-\frac{\pi}{2\delta})}{\delta^{5/2}} .$$

In the nonlinear case, one can write linear recursion relations for the coefficients a_n and one can convince oneself that they are dominated by the linear behavior (eq.27) as n gets large(9). That is

$$\lim_{n \to \infty} \left(\frac{a_n}{b_n}\right) = \kappa \tag{29}$$

This gives the main and final result of this section that relates the leading behavior of $\theta''(0)$ to the behavior near $i\pi/2$ of the successive terms of the asymptotic expansion:

$$\theta''(0) \sim \kappa\, \frac{\pi^{3/2}}{\sqrt{2}} \frac{\exp(-\frac{\pi}{2\delta})}{\delta^{5/2}} . \tag{30}$$

The first few approximations to κ are :

$$\kappa_n \equiv \frac{a_n}{al_n}\,;\ \kappa_1 = 1,\ \kappa_2 = \frac{25}{24} = 1.041..,\ \kappa_3 = \frac{763}{720} = 1.059..,$$
$$\kappa_4 = 1.067..,\ \kappa_6 = 1.07279..,\ \kappa_8 = 1.07455..$$

V. 'Asymptotics Beyond All Orders' From the Regular Perturbation Expansion: The Borel Summation Method.

The result summarized by eq.(30) raises a question concerning the approach developed in ref.(6) for the computation of exponentially small terms. There, a nonlinear inner problem around $i\pi/2$ was defined. It reads

$$\frac{d^3}{dz^3}\Phi + \frac{d}{dz}\Phi = \exp(\Phi) \,, \qquad \Phi \sim -\ln(-z) \quad \text{when } z \to \infty \tag{31}$$

The function Φ can be thought as a generating function for the coefficients a_n (eq.(25)) since, as explained in ref.(6), the asymptotic behavior of Φ for large negative z is simply

$$\Phi(z) + \ln(-z) = \sum_{n=1}^{\infty} a_n z^{-2n} \tag{32}$$

It is argued in ref.(6) that there are terms beyond all orders of this asymptotic expansion and that their functional form can be determined by linearizing the equation around its asymptotic series. By following this procedure one finds on the negative imaginary axis(6):

$$\mathrm{Im}\big(\Phi(z) + \ln(-z)\big) \sim \Gamma |z|^{1/2} \exp(-|z|) \tag{33}$$

since only terms beyond all orders contribute to the result. By matching eq.(33) to an outer expansion, it can then be shown(6) that the prefactor in eq.(30) is equal to 2Γ. In ref.(6), Γ was determined by solving numerically the nonlinear equation (31). Therefore, it could have been thought that the number Γ was a new number not related in any simple way to the a_n's. We have now suggested that this not the case since our previous result (eq.(30)) gives

$$2\Gamma = \kappa \frac{\pi^{3/2}}{\sqrt{2}} \tag{34}$$

where κ is defined by eq.(29). Can this be seen directly ? An approach to answer this question was proposed in ref.(9) in the context of viscous fingering. Let us show how it works in the present simple case. In the previous section, we have seen that the coefficients a_n of the asymptotic series of $\Phi(z)$ have a factorial type of growth (eqs.(28), (29)). The asymptotic series (eq.(32)) is therefore strongly divergent. In order to proceed we follow Borel's strategy(12) and define a new series:

$$B(x) = \sum_{n=1}^{\infty} \frac{a_n}{(2n)!} x^n \tag{35}$$

The coefficients of this new series are no longer growing too quickly and their asymptotic behavior follows from eq.(29):

$$\frac{a_n}{(2n)!} \sim \frac{\kappa(-1)^n}{2}\sqrt{\frac{\pi}{n}} \tag{36}$$

B(x) is therefore a convergent series for $|x| < 1$ and it defines an analytic function there. If we analytically continue B(x), eq.(36) shows that the closest singularity to zero is located at $x = -1$. The behavior of the singular part $B_S(x)$ of B(x) near $x = -1$ is given by:

$$B_s(x) \sim \kappa \frac{\pi}{2}(1+x)^{-1/2} \tag{37}$$

In order to go backward from B(x) to $\Phi(z)+\ln(-z)$, we define

$$\Psi(z) = \int_0^\infty dt\, e^{-t} B\left(\frac{t^2}{z^2}\right) \tag{38}$$

We then identify $\Psi(z)$ and $\Phi(z)+\ln(-z)$ which satisfy the same differential equation and have the same asymptotic series on the real negative axis. From this explicit representation for Ψ, we can find the asymptotic behavior of its imaginary part on the negative imaginary z axis. We analytically continue $\Psi(z)$ from the negative real axis to the negative imaginary axis (staying along the way in the lower half complex z plane). In this way, we obtain on the negative imaginary z axis:

$$\Psi(z) = \int_0^{|z|} dt\, e^{-t} B\left(\frac{t^2}{z^2}\right) + \int_{|z|}^\infty dt\, e^{-t} B\left(\frac{t^2}{z^2}\right) \tag{39}$$

The integral has been split into two parts. In the first integral the integration is along the real t axis from 0 to $|z|$ and the argument of B goes from zero to -1. Therefore B is purely real there and the first integral does not contribute to the imaginary part of Ψ. In the second integral the integration contour goes from $|z|$ to $+\infty$ along a contour that is determined by the singularities of B(x) outside the unit circle. Moreover the asymptotic behavior of the second integral is dominated by the neighborhood of $t = |z|$ (13) so that it is possible to replace B(x) by its singular behavior $B_S(x)$ near x=-1. One obtains therefore

$$\mathrm{Im}\{\Psi(z)\} \sim \frac{k\pi}{2} \int_{|z|}^\infty dt\, e^{-t}\left(\frac{t^2}{|z|^2} - 1\right)^{-1/2} \sim \frac{\kappa\pi^{3/2}}{2\sqrt{2}}\sqrt{|z|}\, e^{-|z|} \tag{40}$$

The functional form of $\mathrm{Im}\{\Psi(z)\}$ is indeed the one predicted by a linearization around the asymptotic series (eq.33). Moreover, the prefactor is determined by the asymptotic behavior of the coefficient a_n of the asymptotic series in agreement with eq.(36).

Acknowledgments: This study is a byproduct of a joint study(4c, 9, 10) of the viscous fingering problem. I am very grateful to R. Combescot, Y. Pomeau and A. Pumir for generously sharing their insights into the problem with me. I owe a special debt to T. Dombre for innumerable discussions during the academic year 85-86 that shaped my whole understanding of the subject.

REFERENCES

(1) Reviews can be found in D.A. Kessler, J. Koplik and H. Levine 1988, Adv. Phys.**37**, 255; Pelcé P. *Dynamics of curved fronts* (Academic, Orlando, Fl, 1988)

(2) a) McLean J.W., P.G.Saffman 1981 , J.Fluid Mech. **102**, 445
b)Vanden-Broeck J.M. 1983, Phys.Fluids **26**, 2033
c)Ben-Jacob E., N.D. Goldenfeld, B.G. Kotliar and J.S. Langer 1983 Phys.Rev.Lett. **53**, 2110
d)Kessler D. A., J. Koplik, H. Levine (1985) Phys. Rev. **A31**, 1712
e)Pelcé P. and Y. Pomeau 1986, Stud. App. Math. **74**, 245

(3) Benamar M., Y. Pomeau 1986, Europhys. Lett. **2**, 307
Barbieri A., D. Hong and J.S. Langer 1986 , Phys.Rev. **A35**, 1802

(4) a) Shraiman B.I. 1986 , Phys.Rev.Lett. **56**, 2028;
b)Hong D.C., J.S. Langer 1986 Phys.Rev.Lett. **56**,2032
c)Combescot R., T. Dombre, V. Hakim, Y. Pomeau, A. Pumir 1986, Phys.Rev.Lett. **56**, 2036

(5) Langer J.S. 1986 , Phys.Rev. **A33**, 435

(6) M.Kruskal and H. Segur 1985, Aero. Res. Ass. Tech. Memo. 85-25

(7) Pokrovsky V.L. and I.M. Khalatnikov 1961, Sov. Phys. J.E.T.P. **13**, 1207
Meyer R.E. 1980 SIAM Review **22**, 213

(8) Brower R., D.A. Kessler, J. Koplik, H. Levine 1984, Phys.Rev. **A29**, 1335

(9) Combescot R., V. Hakim, T. Dombre, Y. Pomeau, A. Pumir 1988, Phys.Rev. **A37**, 1270

(10) Our exposition of Langer's method differs slightly from the original one and follows closely Dombre T., V. Hakim, Y. Pomeau 1986, C. R. Acad. Sci. (Paris) **302**, 803

(11) This result was first obtained by a different method in Daschen R., D.A. Kessler, H. Levine, R. Savit 1986, Physica **21D** 371

(12) Borel E., 1928 *Leçons sur les séries divergentes* (Gauthier-Villars, Paris)

(13) This is in fact true only if B(x) has no other singularity inside the curve of equation (in polar coordinates) $\rho^{1/2}\sin(\theta/2)=-1$ in the lower half plane. This is more restrictive that the absence of singularity inside the unit circle ($\rho=1$) that we have obtained and one would need a finer analysis of the asymptotic series to prove it. Nevertheless we will suppose that it is the case

THE GEOMETRIC MODEL OF CRYSTAL GROWTH - AN OVERVIEW

Harvey Segur

Program in Applied Mathematics
University of Colorado, Boulder 80309-0526

ABSTRACT

The *geometric model* is a phenomenological model of growing dendritic crystals. It is too simple to be accurate physically, but it has played an important role in identifying some of the delicate mathematical issues in the problem. The purpose of this overview is to explain:
(a) what the model is; (b) how to interpret its results physically; and
(c) its significance in a workshop on Asymptotics Beyond All Orders.

A. WHAT IS THE GEOMETRIC MODEL?

Since its invention by Brower, Kessler, Koplik & Levine (1983), the name "geometric model" has come to refer to one of two third-order, nonlinear, ordinary differential equations; either

$$\varepsilon^2 \frac{d^3\theta}{ds^3} + \frac{d\theta}{ds} = \cos\theta, \tag{1}$$

in which $\theta(s)$ also depends on a (small, positive) parameter, ε, or its generalization

$$\varepsilon^2 \frac{d^3\theta}{ds^3} + \frac{d\theta}{ds} = (1 + \alpha\cos\{4\theta\})^{-1}\cos\theta. \tag{2}$$

in which $\theta(s)$ depends on both ε ($0 < \varepsilon \ll 1$) and α ($0 \le \alpha < 1$). The original model was more complicated, but the essential issue exposed by the geometric model is found in these two versions of it. In either case, one wants a solution of the differential equation that satisfies boundary conditions: that $\theta \to -\pi/2$ as $s \to -\infty$, and $\theta \to \pi/2$ as $s \to \infty$. The question is: *Can one choose either ε, for (1), or (ε, α), for (2), so that the differential equation has a solution satisfying these boundary conditions?*

Asymptotics beyond All Orders, Edited by H. Segur *et al.*
Plenum Press, New York, 1991

The question was answered, using a variety of methods, by Dashen, Kessler, Levine, & Savit (1986), Hammersley & Mazzarino (1989), Amick & McLeod (1990), Troy (1990), and Kruskal & Segur (1985, 1991). In addition, the interested reader should consult papers by Langer (1986) and Hakim (1991, in these Proceedings).

Some of these papers are summarized by their authors in these Proceedings, so there in little need for additional discussion of mathematical methods. However, a brief classification of the papers might be appropriate, and several classifications are possible.

(i) The papers by Dashen *et al.* (1986), Kruskal & Segur (1985, 1991), Langer (1986) and Hakim (1991) all apply for small ε, the region of primary physical interest in the model. Troy's (1990) analysis is valid for ε near 1, and the results of Hammersley & Mazzarino (1989), and Amick & McLeod (1990) hold for all $\varepsilon > 0$.

(ii) All of these papers show that for small, positive ε, (1) has no solution satisfying the required boundary condtions. In addition, the papers estimate a certain function, $\theta''(0; \varepsilon)$, whose nonzero value demonstrates that no solution exists. The method of calculation outlined by Amick & McLeod (1990) is similar to that used by Kruskal & Segur (1985, 1991). Langer's (1986) method, which can be considered a simplified version of this, obtains the correct scaling, but is numerically inaccurate. Hammersley and Mazzarino (1989) obtain a positive lower bound.

(iii) Hakim's (1991) analysis suggests that Borel summation might relate both the work of Dashen *et al.* (1986), and that of Hammersley & Mazzarino (1989) to the other analyses.

(iv) Dashen *et al.* (1986) and Kruskal & Segur (1985, 1991) also show that (2) does admit solutions satisfying the boundary conditions, for particular choices of (ε, α).

Regardless of how one analyzes it, the *geometric model*, at least in these Proceedings, refers to one of two nonlinear ordinary differential equations, with either one (ε) or two (ε, α) parameters that act like eigenvalues. The model was designed to mimic the growth of dendritic crystals. Questions about the model fall into two categories.

i) *mathematics*

Does the differential equation have a solution satisfying the boundary conditions? Can one prove existence or nonexistence?

(Answering these questions may require that one accurately evaluate exponentially small quantities, and this may require asymptotics beyond all orders.)

ii) *physics*

Can one use measurements of concrete physical quantities to discriminate between zero, and something that is exponentially small but nonzero? Clearly, if a mathematical theory depends on a distinction that cannot be measured physically, so that the theory cannot be tested in a physical experiment, then it has limited physical significance.

For the geometric model, it turns out that if $\theta''(0; \varepsilon)$ is (exponentially small but) nonzero, then it induces spatial oscillations that are easily observable. Thus in the geometric model, one can effectively distinguish "zero" from something that is exponentially small but nonzero, by direct measurement. The same question arises for other models discussed in this Workshop, and it must be answered on a case-by-case basis.

B. PHYSICAL MEANING OF THE GEOMETRIC MODEL

Some background information about dendritic crystals is necessary to explain the physical meaning of the geometric model. In the paper by Gollub (1991) in these Proceedings, Figure 1 is a composite of a sequence of photographs, taken at regular intervals, showing the motion of the solid-liquid interface of a growing dendritic crystal. [Actually, the figure shows a cross-section; the real crystal is three-dimensional.] The Figure shows how complicated the growth process of dendritic crystals is: as the crystal grows, sidebranches form, these grow at different rates, some grow and then remelt, etc. However, the Figure also shows that the growth pattern in the neighborhood of the *tip of the crystal* (at the top of each interface in the Figure) seems to be much simpler. The shape and speed of the tip apparently remain fairly constant as the crystal grows.

The reader can observe how remarkably constant are the shape and speed of the tip by the following procedure: xerox onto a transparency the page on which Figure 1 appears, and then lay the xeroxed copy over the original Figure. Naturally, the patterns on the two copies all line up. Now displace the copy upwards or downwards, so that the Nth interface on the original lies beneath the Mth interface on the copy. The reader should now observe that every *tip region* on the copy lines up precisely with the *tip region* beneath it, even though the two curves now represent the shape of the interface at different times. This shows that the shape of the tip of a growing dendritic crystal is constant in time (to within the accuracy of these photographs), and because the photographs were taken at regular intervals, it also shows that the tip speed is approximately constant.

To summarize, dendritic crystal growth is inherently unsteady, and complicated, but in the immediate neighborhood of the tip of the crystal, the shape and velocity of the moving interface are approximately constant. Thus, the simplest question one might hope to answer about the formation of growing dendritic crystals is this:
Which physical effects select the steady velocity and steady shape of the tip of a growing dendritic crystal?

Important work on the problem was done by Ivantsov (1947), who constructed and solved a 2-dimensional model of a steadily growing, self-similar crystal, without side-branching (called a "needle crystal"). In Ivantsov's model, latent heat is released as the material freezes at the interface, and the rate at which the crystal grows is controlled by the rate at which this excess heat diffuses away from the interface. The shape of the interface always turns out to be parabolic in this model, and the *Peclet number* (a dimensionless product of the tip radius and the tip speed) is uniquely determined by the “undercooling” (the dimensionless temperature of the liquid far from the interface). The 2-dimensional model was generalized to certain 3-dimensional shapes by Ivantsov (1947) and by Horvay & Cahn (1961).

Experiments by Huang & Glicksman (1981) confirm the relation between Peclet number and undercooling predicted by Ivantsov. However, the experiments also indicate that for a given material, the undercooling determines *both* the tip radius and the tip speed, not just their product. Thus Ivantsov's model is apparently too simple, in the sense that some physical effect omitted from Ivantsov's model selects a particular tip speed. The question is:
Which physical effects should be added to Ivantsov's model, to create a more complicated model in which both the tip speed and the tip radius are determined by the undercooling?
Candidates include surface tension along the interface, with or without crystalline anisotropy, interfacial kinetics, and others.

The problem is difficult, and efforts were made to gain intuition by constructing simplified mathematical models that could be analyzed in detail. Two models that were very helpful were the *boundary layer model* of Ben-Jacob, Goldenfeld, Langer, & Schön (1983, 1984), and the *geometric model* of Brower, Kessler, Koplik & Levine (1983, 1984). Neither of these models was derived systematically from the full equations of crystal growth. Instead, they are intentionally simplified, *ad hoc* models, invented to provide hypotheses about the growth of real crystals. Both models have successfully provided fruitful ideas about the growth of real crystals.

The logic of the (steady-state version of the) geometric model is as follows. One seeks a steadily growing needle crystal, whose tip moves with an unknown speed, V. The objective of the model is to predict V; *i.e.*, to determine which speed has been selected. Figure 1 of this paper shows the geometry of the model. On the side of the crystal, at a distance (s) from the tip, the local normal vector makes an angle (θ) with the direction of growth of the tip. If the shape is to remain steady, then the normal velocity there must be [$V \cos \theta$]. The essential assumption of the geometric model is that the normal velocity is a function of the local curvature, $d\theta/ds$, and its (even) spatial derivatives:

$$V \cos \theta = f(\frac{d\theta}{ds}, \frac{d^3\theta}{ds^3}, \ldots).$$

In a simple version, f is linear in the first two derivatives:

$$V \cos \theta = A \frac{d\theta}{ds} + B \frac{d^3\theta}{ds^3}, \tag{3}$$

where A and B are fixed parameters. To the extent that (3) has physical meaning, the parameter (B) represents "surface tension", in the sense that it is the coefficient of the highest derivative in the equation.

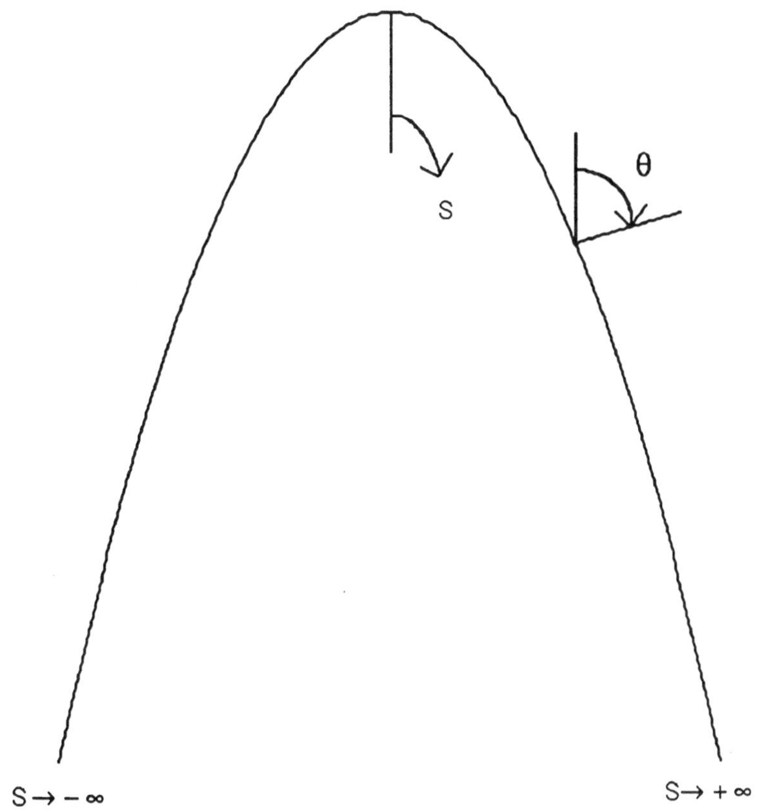

Coordinate system for the geometric model, (1) and/or (2).

To put (3) into the form of (1), divide by V, and rescale (s) to eliminate (A/V). Then the coefficient of last term becomes (BV^2/A^3), which is now renamed "ε^2". Thus, the parameter (ε^2) represents surface tension, but it also contains the unknown velocity, V. In this version of the model, "velocity selection" would mean that certain values of ε^2 are selected by the requirement that the differential equation have a needle-crystal solution, with $\theta \to -\pi/2$ as $s \to -\infty$, and $\theta \to \pi/2$ as $s \to +\infty$. [If one also includes crystalline anisotropy in the surface tension, then a similar line of reasoning leads to (2). More details can be found in the paper by Brower *et al.* (1984).]

Notice that without surface tension ($\varepsilon = 0$), (1) admits a needle crystal:

$$\theta(s; 0) = -\pi/2 + 2 \tan^{-1}(e^s). \qquad (4)$$

Moreover, since V has been scaled into (s), this solution is valid for any speed (V), just as in Ivantsov's model. The important question is: *What happens for* $\varepsilon > 0$*?*

Three choices are logically possible:

i) Eq'n (1) admits a needle-crystal solution for every value of ε in some interval: $0 \leq \varepsilon \leq \varepsilon_0$. Then (1) successfully generalizes (the analogue of) Ivantsov's model to include surface tension, but surface tension by itself provides no selection mechanism.

ii) Eq'n (1) admits no needle-crystal solution for any finite value of ε in some interval, $0 < \varepsilon \leq \varepsilon_0$. Then (1) is not a self-consistent model, and it certainly provides no selection mechanism.

iii) Eq'n (1) admits needle-crystal solutions only for selected values of ε in some interval, $0 < \varepsilon \leq \varepsilon_0$. Then according to this model, surface tension does select a set of possible growth velocities for a crystal, with each selected value of ε providing one such velocity.

Note that the key issue here is simply the existence of a needle-crystal solution. Knowledge of the detailed shape of the solution is less important than knowing that it exists.

C. DOES A SOLUTION EXIST FOR $\varepsilon > 0$?

At this point we need a careful formulation of the problem. To this end, we fix $\varepsilon > 0$, and impose some of the boundary conditions, so that (1) admits a unique solution, $\theta(s; \varepsilon)$, satisfying these conditions. Because $\theta(s; \varepsilon)$ is now uniquely defined, it must be the needle crystal, if one exists. Then we simply check whether $\theta(s; \varepsilon)$ satisfies the remaining conditions.

Note first that both (1) and the boundary conditions are invariant under a translation: $s \rightarrow s + s_0$. Hence we may identify $s = 0$ by requiring that $\theta(0; \varepsilon) = 0$. Next, we require that $\theta(s; \varepsilon)$ satisfy the boundary condition that $\theta \rightarrow -\pi/2$ as $s \rightarrow -\infty$. Kruskal & Segur (1991) show that these two conditions uniquely define $\theta(s; \varepsilon)$.

It remains to decide whether this uniquely defined function represents a needle crystal. This can be done in three equivalent ways.

i) Does $\theta(s; \varepsilon) \rightarrow \pi/2$ as $s \rightarrow +\infty$?

In this version of the problem, nothing is beyond all orders. If there is no needle crystal for the fixed value of ε, then $\theta(s; \varepsilon)$ oscillates extremtely rapidly for large s. In fact, when Kruskal and I first learned of this problem from Goldenfeld, Kotliar & Langer in Santa Barbara in 1984, their numerical experiments already showed this wild behaviour for every value of ε that they had tried.

ii) Is $\theta(s; \varepsilon)$ antisymmetric in s?

Necessarily $\theta(s; \varepsilon)$ satisfies the boundary condition as $s \rightarrow -\infty$. If $\theta(s; \varepsilon)$ happens to be antisymmetric, then it must also satisfy the boundary condition as $s \rightarrow +\infty$, so it must be a needle crystal. A more careful argument shows that this is the only possibility: if a needle crystal exists, then it must be antisymmetric (cf. Kruskal & Segur, 1991).

iii) Is $\theta''(0; \varepsilon) = 0$?

If $\theta(s; \varepsilon)$ is antisymmetric in s, then all of its even derivatives vanish at $s = 0$, including $\theta''(0; \varepsilon)$. Conversely, if $\theta(0; \varepsilon)$ and $\theta''(0; \varepsilon)$ both vanish, then one can show by differentiating (1) repeatedly that all of the even derivatives of $\theta(s; \varepsilon)$ vanish at $s = 0$. Thus, (1) admits a needle crystal if and only if $\theta''(0; \varepsilon) = 0$.

This last formulation is where asymptotics beyond all orders finally appears. Careful analysis of (1) shows that as $\varepsilon \to 0$,

$$\theta''(0; \varepsilon) \sim 2\,\Gamma\, \varepsilon^{-5/2}\, e^{-\pi/2\varepsilon}, \tag{5}$$

where Γ is a number ($\Gamma \approx 2.11$). Thus for small ε, $\theta''(0; \varepsilon)$ is exponentially small, but it is nonzero, so (1) admits no needle crystal in this limit.

Can such a tiny effect be measured using instruments with finite precision? In this problem, the answer is affirmative. Because $\theta''(0; \varepsilon)$ is nonzero, no matter how small it might be, $\theta(s; \varepsilon)$ is not antisymmetric in s, and it oscillates wildly as $s \to +\infty$. Thus an exponentially small effect at $s = 0$ induces large oscillations as $s \to +\infty$, where they can be detected easily.

ACKNOWLEDGEMENTS

The author is grateful to the Scientific Affairs Division of NATO, and to NSF for supporting this Workshop. Some of the work reported herein was supported by NSF, grant #DMS-9096156.

REFERENCES

Amick, C.J., and McLeod, B.J., 1990, Arch. Rat. Mech. Anal., 109, 139-171,
Ben-Jacob, E., Goldenfeld, N., Langer, J.S., and Schön, G., 1983, Phys. Rev. Lett., 51, 1930-1932
__, 1984, Phys. Rev., 29A, pp. 330-340
R. Brower, D. Kessler, J. Koplik, & H. Levine, *Phys. Rev. Lett.*, **51**, pp. 1111-1114, 1983
___, *Phys. Rev.*, **29A**, pp. 1335-1342, 1984
R.F. Dashen, D.A. Kessler, H. Levine & R. Savit, *Physica*, **21D**, pp. 371-380, 1986
J.P. Gollub, "Experimental evidence for (and against) microscopic solvability", in these Proceedings, 1991; see also: A. Dougherty, P.D. Kaplan, & J.P. Gollub, *Phys Rev. Lett.*, **58**, pp. 1652-1655, 1987
J.M. Hammersley & G. Mazzarino, *IMA J. of App. Math.*, **42**, pp.43-75, 1989
V. Hakim, "Asymptotics beyond all orders from the asymptotic expansion: the Borel summation method", these Proceedings, 1991
G. Horvay & J.W. Cahn, *Acta Metall.*, **9**, pp. 965-705, 1961
S.C. Huang & M.E. Glicksman, *Acta Metall.*, **29**, pp. 701-715, 1981
G.P. Ivantsov, *Doklady Akad. Nauk* (Russian), **58**, pp. 567-569, 1947
M.D. Kruskal & H. Segur, *Stud. App. Math.*, to appear, 1991
___, ARAP Tech. Memo (unpublished), 1985
J.S. Langer, *Phys. Rev A*, **33**, pp. 435-441, 1986
W.C. Troy, *Q. Appl. Math.*, **48**, pp. 209-216, 1990

NUMERICAL ANALYSIS OF THE GEOMETRIC MODEL FOR DENDRITIC GROWTH OF CRYSTALS

J.M. Hammersley[1] and G. Mazzarino[2]

[1] Trinity College, Oxford OX1 3BH, UK.
[2] Institute of Economics and Statistics, University of Oxford
Oxford OX1 3UL, UK

1. INTRODUCTION AND SUMMARY

This paper investigates the numerical properties of solutions $\theta = \theta(t)$ of the third–order equation

$$\epsilon\theta_3 + \theta_1 = \frac{\cos\theta}{1+\alpha\cos 4\theta}, \; \theta(\pm\infty) = \pm\pi/2 . \tag{1.1}$$

Suffices to θ in (1.1) denote differentiation with respect to t; that is to say $\theta_n = d^n\theta/dt^n$; the boundary conditions $\theta(\pm\infty) = \pm\pi/2$ are abbreviations for $\theta(t) \to \pm\pi/2$ as $t \to \pm\infty$ respectively; and ϵ and α are prescribed parameters satisfying $\epsilon > 0$ and $0 \leq \alpha < 1$. It is convenient to write $\epsilon = 2^k$, and to tabulate results as functions of α and $k = \log_2\epsilon$. Kruskal and Segur [1] give references to the appearance of (1.1) as a model for the dendritic growth of crystals in a supercooled liquid. [Warning: there is some variation of notation in the literature; and, in particular, Kruskal and Segur write ϵ^2 for the coefficient of θ_3, thus entailing $k = 2\log_2\epsilon$ for their use of ϵ.] A strictly monotonic solution of (1.1) is called a <u>needle crystal solution</u>; and interest centres upon the question of the existence or non–existence of needle solutions. Our earlier paper [2] proved that needle solutions could not exist for $\alpha = 0$. Kruskal and Segur [1] concluded that, for sufficiently small ϵ, needle solutions would exist for certain discrete values of $\alpha = \alpha(k) > 0$. Thus we have an eigenvalue problem with a discrete spectrum. Our analysis of this problem is incomplete, and several interesting questions remain unresolved.

Provided that ϵ is not too small (say $\epsilon > 0.003$), equation (1.1) can be handled satisfactorily by a stock library routine for solving ordinary differential equations. On the other hand for smaller values of ϵ, stock routines are not accurate enough to countenance the exponentially small effects that lie beyond all orders of magnitude in an asymptotic expansion; and special methods are then required. In [3] we discussed such a special method for a class of autonomous differential equations, including the geometric model of dendritic growth; and in the present paper we apply it to (1.1). While this method worked well enough for the examples in [3], it is only partially successful for (1.1) because it is limited to the range $0 \leq \alpha < 1/7$. Consequently we have not been able to determine any eigenvalues $\alpha(k) \geq 1/7$ when ϵ is very small. Needle crystal solutions of (1.1) do not exist for $k > -2.78$, that is to say for $\epsilon > 0.146$.

Equation (1.1) is invariant under translations $t \mapsto t + t_0$ for any constant t_0; and (1.1) cannot have any solution unless $\theta(t_0) = 0$ for at least one value of t_0. Thus we look for solutions of

Asymptotics beyond All Orders, Edited by H. Segur *et al.*
Plenum Press, New York, 1991

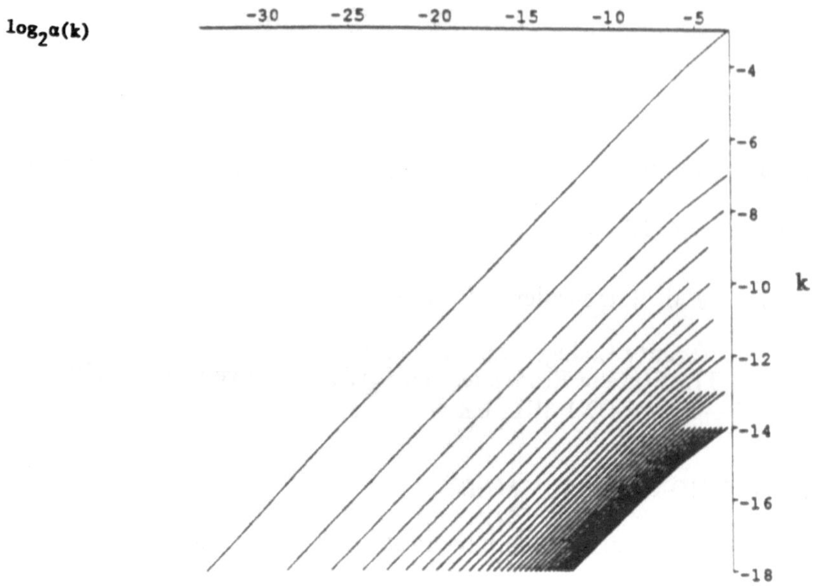

Figure 1. Eigenvalues from STIFF

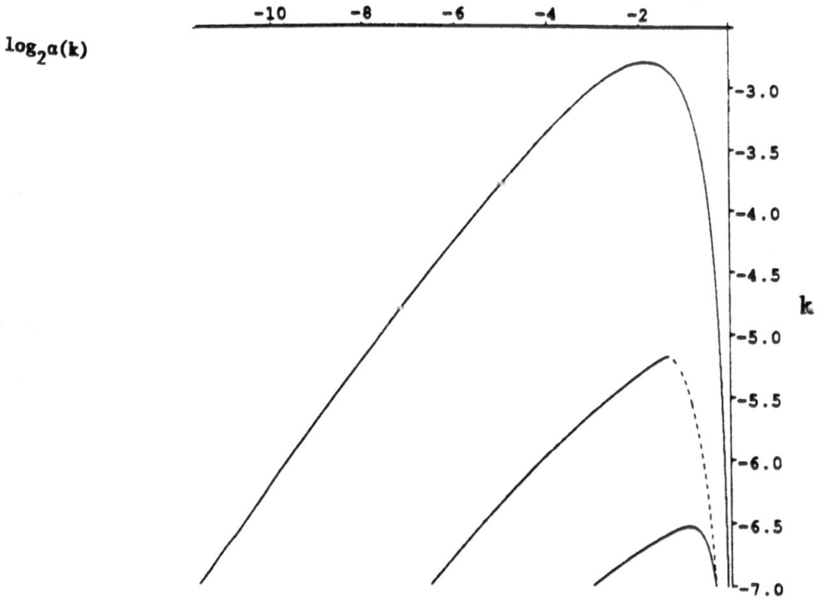

Figure 2. Eigenvalues from DEND

$$\epsilon\theta_3 + \theta_1 = \frac{\cos\theta}{1+\alpha\cos4\theta}\,,\ \theta(0) = 0\ ,\ \theta(+\infty) = +\pi/2\ ,\ (0 \le t < \infty)\,. \quad (1.2)$$

Let $\theta(t)$ be any particular solution of (1.2), and consider the function

$$\theta^*(t) = \begin{Bmatrix} \theta(t) & , (t \geq 0) \\ -\theta(-t) & , (t \leq 0) \end{Bmatrix}\,. \quad (1.3)$$

Since (1.1) is invariant under the involution $(\theta,t) \mapsto (-\theta\,, -t)$, we see that θ^* will be a solution of (1.1) for all $t \neq 0$. At $t = 0$, θ^* is continuous and continuously differentiable, but it will have a discontinuity $2\theta_2(0)$ in its second derivative if $\theta_2(0) \neq 0$. On the other hand, as we shall prove in section 2, $\theta_2(0) = 0$ is a sufficient condition for θ^* to be a solution of (1.1) throughout $-\infty < t < \infty$. If further $\theta(t)$ is a strictly monotone solution of (1.2), then θ^* will be a needle solution of the problem. As we shall discover later, it may well happen (depending upon the values of k and α) that (1.2) has no monotone solution.

Linearizing (1.2) in the immediate neighbourhood of $\theta = \pi/2$ shows that there is a non–zero constant K such that

$$\frac{\pi}{2} - \theta(t) \sim Ke^{-\lambda t} \text{ as } t \to \infty \quad (1.4)$$

where λ is the unique real root of

$$\epsilon\lambda^3 + \lambda = 1/(1+\alpha)\,. \quad (1.5)$$

The case $K < 0$ cannot lead to a monotonic solution of (1.2). Accordingly we define the <u>primary solution</u> of (1.2) to be one such that $K > 0$ and $\theta(t) > 0$ for all $t > 0$. In [2] we proved that the primary solution exists when $\alpha = 0$, and is unique and strictly monotonic for all $\epsilon > 0$. The argument used in [2] also proves, subject to relatively minor amendments, that the primary solution (if it exists) is unique for any $0 \leq \alpha < 1$ and any $\epsilon > 0$; but, as already remarked, the primary solution need not be monotonic. We do not have a proof of the existence of the primary solution for all $0 \leq \alpha < 1$; but it is likely that it does exist for all $0 \leq \alpha < 1$ and all $\epsilon > 0$. If the primary solution does not exist, then there can be no solution of (1.1); so it is reasonable to assume its existence in a study of (1.1). Accordingly we define

$$\Delta = \Delta(k,\alpha) = \theta_2(0) \quad (1.6)$$

where θ is the primary solution.

We have calculated $\Delta(k,\alpha)$ for selected values of k and α, using two programmes (respectively here called STIFF and DEND) on the mainframe VAX cluster 8800 at Oxford University. These two programmes are quite different in character and are intended to cover different ranges of the parameters k and α, though there is an overlap of these ranges which permits a useful check on the accuracy of the resulting $\Delta(k,\alpha)$. STIFF is one of a battery of 57 library routines prepared by the (British) National Algorithm Group for the solution of ordinary differential equations. STIFF accepts (1.2) as a system of three simultaneous first order differential equations for the three components of the vector $(\theta,\theta_1,\theta_2)$; and starting from a sufficiently large value of t with the asymptotic linearization (1.4), it proceeds stepwise to reduce t until θ first vanishes. STIFF uses a Runge–Kutta–Merson backward differentiation algorithm, fed (via a subroutine) with the <u>analytical</u> formulation for the Jacobian matrix. At each step it estimates the <u>local</u> error committed in taking that step, and it accepts this step if the local error is less than a preset tolerance. We chose to set this tolerance at 10^{-13}. If the local error exceeds the tolerance, STIFF rejects the current step and successively replaces it by shorter steps until eventually a short enough step passes the local tolerance test. Of course local errors cumulate as STIFF runs along the curve to the

Table 1.1 Overall picture of STIFF results

```
                                 alpha
        0.0        0.2       0.4       0.6        0.8       0.98
   k     |          |         |         |          |         |
  0.0    --------------------------------------------------
- 0.2    --------------------------------------------------
- 0.4    --------------------------------------------------
- 0.6    --------------------------------------------------
- 0.8    --------------------------------------------------
- 1.0    --------------------------------------------------
- 1.2    --------------------------------------------------
- 1.4    --------------------------------------------------
- 1.6    --------------------------------------------------
- 1.8    --------------------------------------------------
- 2.0    --------------------------------------------------
- 2.2    --------------------------------------------------
- 2.4    --------------------------------------------------
- 2.6    --------------------------------------------------
- 2.8    ------------++++----------------------------------
- 3.0    -------+++++++++++++++++--------------------------
- 3.2    -----+++++++++++++++++++++++----------------------
- 3.4    ---++++++++++++++++++++++++++++-------------------
- 3.6    ---++++++++++++++++++++++++++++++-----------------
- 3.8    --+++++++++++++++++++++++++++++++++---------------
- 4.0    --+++++++++++++++++++++++++++++++++++-------------
- 4.2    -+++++++++++++++++++++++++++++++++++++------------
- 4.4    -+++++++++++++++++++++++++++++++++++++++----------
- 4.6    -++++++++++++++++++++++++++++++++++++++++---------
- 4.8    -+++++++++++++++++++++++++++++++++++++++++--------
- 5.0    -++++++++++++++++++++++++++++++++++++++++++-------
- 5.2    -+++++++++++++++++-==++++++++++++++++++++++-------
- 5.4    -++++++++++---------======++++++++++++++++++------
- 5.6    -++++++-------------=========++++++++++++++++-----
- 5.8    -++++----------------==========++++++++++++++-----
- 6.0    -++------------------============+++++++++++++----
- 6.2    -+--------------------=============+++++++++++----
- 6.4    -+---------------------=============+++++++++++---
- 6.6    -------------------+++++*********=====+++++++++---
- 6.8    ----------++++++++++++++*************==++++++++---
- 7.0    ------+++++++++++++++++++***************++++++++--
- 7.2    ----++++++++++++++++++++++***************+++++++--
- 7.4    ---++++++++++++++++++++++++***************++++++--
- 7.6    --++++++++++++++----------====************++++++--
- 7.8    --+++++++-----------------======***********++++++-
- 8.0    -++++---------------------========**********+++++-
- 8.2    -+++----------------------+******===********+++++-
- 8.4    -++---------++++++++++++++++*****************++++-
- 8.6    -+-----+++++++++++++++++++++*****************++++-
- 8.8    -+--+++++++++++++++++++-----==****************+++-
- 9.0    -+-+++++++-------------------=====************+++-
- 9.2    +-++++-----------------------=======***********++-
- 9.4    -+++---------+++++++++++++++++*****************+++
- 9.6    +++----++++++++++++++++++++++++****************+++
- 9.8    -++--++++++++------------------====*************++
-10.0    -++++++------------------------======***********++
```

Table 1.2 Overall picture of STIFF results

```
                                alpha
       0.0       0.2       0.4       0.6       0.8      0.98
  k     |         |         |         |         |         |
 10.0  hhhhhhhhhhhhhhhhhhhggggggggggggggggggggggfffffffeee
  9.0  hhhhhhhhhhhhhhhhhgggggggggggggggggggggggffffffffeee
  8.0  hhhhhhhhhhhhhhhhggggggggggggggggggggggggffffffffeee
  7.0  hhhhhhhhhhhhhhgggggggggggggggggggggggggfffffffffeed
  6.0  hhhhhhhhhhhhhgggggggggggggggggggggggggfffffffffeeed
  5.0  hhhhhhhhhhhhggggggggggggggggggggggggggfffffffffeeed
  4.0  hhhhhhhhhhgggggggggggggggggggggggggggffffffffffeeed
  3.0  hhhhhhhhggggggggggggggggggggggggggggffffffffffeeeed
  2.0  hhhhhhhgggggggggggggggggggggggggggggffffffffffeeeed
  1.0  hhhhhgggggggggggggggggggggggggggggfffffffffffeeeed
  0.0  hhhgggggggggggggggggggggggggggggggfffffffffffeeeeed
- 1.0  ggggggggggggggggggggggggggggggffffffffffffeeeeeeddc
- 2.0  hhhhhhhhhhhhhhhhhgggggggggggggffffffffeeeeeeeedddcc

- 2.4  hhhhhhhhhhhhhhhhhhhhhhhhggggggggffffffeeeeeeddddcc
- 2.6  hhhhhiiiiiiiiiiiiiiiihhhhhgggggffffffeeeeeedddccb
- 2.8  hhhiiijjjklmQNMNrmlkjjiiihhhggggffffeeeeeddddccb
- 3.0  hhiijkmLJJIIIIIIIIIIIIJJLljiihhggggfffeeeeeddddccb
- 3.2  hiijmKJIIHHHHHHHHHHHHHHHIIJKkihhgggfffeeeedddddccb
- 3.4  hijNJIIHHHGGGGGGGGGGGGGGGHHHIKjihggfffeeeedddcccb
- 3.6  ijmJIHHHGGGGGGGGFFFFFFFGGGGGGHHIkhggfffeeedddcccb
- 3.8  ijKIHHGGGGGFFFFFFFFFFFFFFFFFFGGGHIjhgffeeedddccbb
- 4.0  ilJHHGGGGFFFFFFFFFFFFFEEEFFFFFFFFFGGHJigffeedddccbb
- 4.2  jLIHHGGGFFFFFFFEEEEEEEEEEEEEEEEEEEFFFFGHjgffeeddccbb
- 4.4  jKIHGGGFFFFFFEEEEEEEEEEEEEEEEEEEEEEEEFFFGKgfeeddccba
- 4.6  jJIHGGGFFFFFEEEEEEEEEEEDDDDDDDDEEEEEEEEEFGIgfeddccba
- 4.8  kJIHGGGGFFFFEEEEEEEEDDDDDDDDDDDDDDDDDDEEEFFIgeddccba
- 5.0  lJIHHGGGGFFFFFEEEEEEDDDDDDDDDDDDDDDDDDDEEEFJfedccba
- 5.2  lJIHHHHGGGGGGGGGGHgedDDDDDDDDDDDDDDDDDDDDEEFhedccba
- 5.4  mJIIIHHIIIKihgffeeeddddddCCCCCCCCCCCCDDDDEFfdccba
- 5.6  nKJJJJKkihggffeeeedddddddddddCCCCCCCCCCCDDDEHedcba
- 5.8  nKKKMkihhggfffeeeeeddddddddddCCCCCCCCCCCDDEfdcba
- 6.0  oLLnkihhggffffeeeeeeeeddddddddddBBBBBCCCCCDDFdcba
- 6.2  pMrkjihhgggfffeeeeeeeeeeeeeeeeeddddBBBBBBCCCDEecba
- 6.4  rOmkjiihhggggfffffffffffffffffeeeedBBBBBBBCCDGcba
- 6.6  svmkjjiihhhhhhhhiikJIHHGGGGGGGGHLhgfeeBBBBBBBCDdba
- 6.8  tpmlkjjjjkNJIHHGGFFFFEEEEEEEEEEEEEEEFGhfAABBBBBCeba
- 7.0  vpnmlmQKJIHHGGFFFEEEEDDDDDDDDDDDDDDDDEFIAAAABBCFba
- 7.2  wpooPLKJIIHGGGFFEEEEDDDDDCCCCCCCCCCCCDDDEAAAABBDba
- 7.4  yqqPMLKJIIHHGGGFFFEEEDDDCCCCCCCBBBBCCCCCDDAAAABCca
- 7.6  zsSOMLKKJJJIIIJLjhgffeddcccccBBBBBBBBBBBCCAAAAABda
- 7.8  zuRONMMMOlkjihggffeeeddddccccccBBBBBBBBBBCAAAABFa
- 8.0  zWRQPunmkjjihhggfffeeeeeddddddddddAAAAAABBBBAAAACa
- 8.2  zWTTqomllkjjiihhhhhhgghhhijJIHGGGHlgfAAAAAAABAAAABa
- 8.4  zXWsqonnmmmtLKJIHGGFFEEEDDDDDCCCCCDDAAAAAAABAAAAa
- 8.6  zZwtrqrPNMLKJIHHGGFFEEEDDDCCCCBBBBBBBBCAAAAAAAAAAa
- 8.8  zZxvYRPONMLKKJJIIIIIIILigfedcbBBBBBABBBBAAAAAAAAAAa
- 9.0  zZzXTSQPPPrnlkjiihggffeedddccccccbAAAAAAAAAAAAAAAb
- 9.2  ZzZXVUurpomllkjjihhgggfffffeeeeeeeedAAAAAAAAAAAAAh
- 9.4  zZZZzvsrqpoopOMLJIIHGGFFEEDDDCCCCCCCCCAAAAAAAAAAAB
- 9.6  ZZZzywvWSQPNMLKKJIIHHGFFFEEDCCBBBBAAAABAAAAAAAAAAAA
- 9.8  zZZzzZVUSRQQQromlkjihggfeeddcccbbbbAAAAAAAAAAAAAAA
-10.0  zZZZZZYzvsrponmlkkjjihhhgggggffgggffeAAAAAAAAAAAAA
```

Table 1.3

```
+A  a-
320e+2
+B  b-
160e+2
+C  c-
800e+1
+D  d-
400e+1
+E  e-
200e+1
+F  f-
100e+1
+G  g-
500e+0
+H  h-
250e+0
+I  i-
125e+0
+J  j-
625e-1
+K  k-
313e-1
+L  l-
156e-1
+M  m-
781e-2
+N  n-
391e-2
+O  o-
195e-2
+P  p-
977e-3
+Q  q-
488e-3
+R  r-
244e-3
+S  s-
122e-3
+T  t-
610e-4
+U  u-
305e-4
+V  v-
153e-4
+W  w-
763e-5
+X  x-
381e-5
+Y  y-
191e-5
+Z  z-
```

final point t = 0 where $\theta(0) = 0$; so the output value of $\theta_2(0) = \Delta(k,\alpha)$ is subject to errors much greater than 10^{-13}. Comparison of the results of STIFF and DEND in the region where they overlap indicates that typically STIFF commits errors of the order 10^{-8} in $\Delta(k,\alpha)$. STIFF also monitors the sign of θ_1 at each step, and sets a flag if the solution ceases to be monotone.

For the three simultaneous first–order equations used by STIFF, the Jacobian of the system has a pair of conjugate complex roots whose imaginary parts can become large on certain stretches of the curve if ϵ is small and/or if α is near 1. Under these circumstances, the wanted solution may be endangered by unwanted high frequency transients. From the available battery of NAG routines, we selected the routine DO2EJF for use as STIFF, because this particular routine is designed to constrain the corrupting influence of high–frequency transients. In this respect STIFF performed creditably well on the whole. Nevertheless a few of the values of $\Delta(k,\alpha)$ for α near 1 or for the larger negative values of k should be treated with considerable caution.

The overall picture presented by STIFF is summarized in Tables 1.1, 1.2, and 1.3. In Table 1.1, α runs horizontally left to right from $\alpha = 0.00$ to $\alpha = 0.98$ in steps of 0.02, and k runs vertically downwards from k = 0.0 to k = − 10.0 in steps of 0.2. The tabular entries for the resulting array indicate whether $\Delta(k,\alpha)$ is positive or negative and whether or not a monotonic solution exists, according to the following symbols:–

+ means Δ is positive and the solution is monotonic,
* means Δ is positive but no monotonic solution exists,
− means Δ is negative and the solution is monotonic,
= means Δ is negative but no monotonic solution exists.

In Table 1.2, α runs horizontally left to right from 0.00 to 0.98 in steps of 0.02 and k runs vertically downwards from k = 10.0 to k = − 2.0 in steps of 1.0 and from k = − 2.4 to k = − 10.0 in steps of 0.2. Lower case letters denote negative values of Δ, and upper case letters denote positive Δ. The letters a,A signal $|\Delta| \geq 2^5$; the letters b,B signal $2^4 \leq |\Delta| < 2^5$; and so on through the alphabet by powers of 2 until y,Y signal $2^{-19} \leq |\Delta| < 2^{-18}$ and z,Z signal $|\Delta| < 2^{-19}$. Table 1.3 is a key for Table 1.2 translating letters into decimal ranges. All the letters on the left–hand margin of Table 1.2 should be lower case, because $\Delta(k,0) < 0$ for all k, as proved in [2]. So the false entry Z for $k = -9.2$, $\alpha = 0$ should have been z; in fact the actual value $\Delta(-9.2,0) = -3.209 \times 10^{-13}$ (obtained by DEND) has been swamped by STIFF's output error of magnitude 6×10^{-8}. Not only do the true values of Δ become much smaller as ϵ decreases to zero (for example $\Delta(-25,0) = -2.389 \times 10^{-3942}$), but STIFF also finds it harder to control the increasing frequencies of spurious solutions (for example STIFF's output errors have risen to magnitude 10^{-6} when k = − 15); so there is little point in running STIFF for k < − 10.

Our other programme DEND is designed to calculate $\Delta(k,\alpha)$ correct to 4 or 5 significant decimal digits, even when Δ is very small. It is an extension of the method used in [2] to calculate $\Delta(k,0)$. Unfortunately it is always restricted to the range $0 \leq \alpha < 1/7$, and moreover for some values of k it fails to work for $\alpha \geq 0.1225$. We shall describe it in detail later; but in this introductory summary we only outline the underlying idea of the method, namely to express θ_1^2 as a suitable multiple of the expansion

$$B(\cos^2\theta) = \sum_{n=0}^{\infty} b_n \cos^{2n}\theta . \tag{1.7}$$

It is found that

$$b_n = B_\infty n^{-3/2} + 0(n^{-5/2}) \text{ as } n \to \infty \tag{1.8}$$

Table 2. Specimen comparisons of STIFF and DENO

alpha=	0.00	0.01	0.02	0.03	0.04	0.05	0.06	0.07
k=-6	17350	16900	17050	17000	17000	17000	17050	17050
	-2047e-002	1195e-001	2049e-001	2332e-001	2016e-001	1076e-001	-5165e-002	-2791e-001
	-2047e-002	1196e-001	2049e-001	2332e-001	2016e-001	1076e-001	-5165e-002	-2791e-001
-7	17300	17000	17350	17400	17400	17400	17350	17350
	-2903e-004	3829e-004	-1429e-002	-4020e-002	-7302e-002	-1082e-001	-1409e-001	-1661e-001
	-2888e-004	3836e-004	-1429e-002	-4021e-002	-7303e-002	-1082e-001	-1409e-001	-1661e-001
-8	17350	17300	17050	17050	17050	16900	17000	17050
	-1067e-006	-8150e-005	9095e-005	1051e-003	2997e-003	5748e-003	8779e-003	1117e-002
	-4613e-007	-7882e-005	9208e-005	1053e-003	2995e-003	5750e-003	8780e-003	1117e-002
-9	27050	26550	28050	27150	27050	27050	26500	27200
	-9830e-007	1239e-006	6670e-007	-7972e-006	-1453e-005	-9392e-006	5449e-005	2341e-004
	-3438e-011	1427e-007	-9830e-007	-1664e-005	-1632e-005	-8776e-006	5487e-005	2361e-004
-10	147000	36500	36500	37000	37350	37100	36450	36600
	-1063e-007	3811e-006	3110e-006	-2453e-008	4062e-006	-2405e-006	4494e-006	-2369e-007
	-3395e-017	4012e-012	2250e-010	-3099e-009	-2078e-008	-3527e-008	8817e-008	7125e-007

alpha=	0.08	0.09	0.10	0.11	0.12	0.13	0.14
k=-6	17050	17000	16950	17000	17000	17000	16900
	-5777e-001	-9503e-001	-1400e+000	-1929e+000	-2540e+000	-3235e+000	-4016e+000
	-5777e-001	-9503e-001	-1400e+000	-1929e+000	-2540e+000	-3235e+000	-4016e+000
-7	17400	17400	17350	17350	17050	17050	17100
	-1787e-001	-1732e-001	-1442e-001	-8614e-002	6841e-003	1405e-001	3206e-001
	-1787e-001	-1732e-001	-1442e-001	-8614e-002	6841e-003	1405e-001	3206e-001
-8	16900	17050	17400	17400	17050	16950	17400
	1163e-002	8458e-003	-4229e-004	-1751e-002	-4572e-002	-8836e-002	-1492e-001
	1163e-002	8462e-003	-4202e-004	-1751e-002	-4572e-002	-8836e-002	8332e+284
-9	27200	26700	26500	27150	26650	17350	17400
	6147e-004	1280e-003	2322e-003	3794e-003	5727e-003	8064e-003	1064e-002
	6146e-004	1280e-003	2319e-003	3796e-003	5731e-003	1360e+124	1363e+447
-10	36450	36450	42100	42100	41800	33850	34000
	2827e-006	7940e-006	1001e-005	2163e-005	3231e-005	2661e-005	-3972e-006
	2487e-006	6236e-006	1253e-005	2076e-005	2752e-005	-1904e+391	4533e+996

Table 3.1 STIFF output for delta

	alpha									
k	0.00	0.02	0.04	0.06	0.08	0.10	0.12	0.14	0.16	0.18
0.0	-492e+0	-494e+0	-497e+0	-501e+0	-505e+0	-509e+0	-514e+0	-520e+0	-526e+0	-533e+0
- 0.2	-506e+0	-509e+0	-511e+0	-515e+0	-519e+0	-523e+0	-528e+0	-534e+0	-540e+0	-547e+0
- 0.4	-519e+0	-521e+0	-523e+0	-526e+0	-530e+0	-534e+0	-539e+0	-545e+0	-552e+0	-559e+0
- 0.6	-529e+0	-530e+0	-532e+0	-534e+0	-538e+0	-542e+0	-547e+0	-553e+0	-559e+0	-567e+0
- 0.8	-536e+0	-536e+0	-537e+0	-539e+0	-542e+0	-546e+0	-550e+0	-556e+0	-562e+0	-569e+0
- 1.0	-540e+0	-539e+0	-539e+0	-540e+0	-541e+0	-544e+0	-548e+0	-553e+0	-559e+0	-566e+0
- 1.2	-540e+0	-537e+0	-536e+0	-535e+0	-536e+0	-537e+0	-540e+0	-544e+0	-550e+0	-556e+0
- 1.4	-537e+0	-531e+0	-527e+0	-525e+0	-523e+0	-524e+0	-525e+0	-528e+0	-532e+0	-538e+0
- 1.6	-529e+0	-520e+0	-513e+0	-508e+0	-504e+0	-502e+0	-502e+0	-503e+0	-506e+0	-510e+0
- 1.8	-516e+0	-503e+0	-493e+0	-484e+0	-477e+0	-472e+0	-469e+0	-467e+0	-468e+0	-471e+0
- 2.0	-499e+0	-481e+0	-465e+0	-452e+0	-441e+0	-432e+0	-425e+0	-421e+0	-418e+0	-418e+0
- 2.2	-477e+0	-452e+0	-431e+0	-412e+0	-395e+0	-381e+0	-370e+0	-361e+0	-355e+0	-351e+0
- 2.4	-450e+0	-418e+0	-389e+0	-363e+0	-339e+0	-319e+0	-302e+0	-287e+0	-276e+0	-267e+0
- 2.6	-419e+0	-378e+0	-339e+0	-305e+0	-273e+0	-245e+0	-220e+0	-198e+0	-180e+0	-165e+0
- 2.8	-384e+0	-332e+0	-283e+0	-238e+0	-196e+0	-158e+0	-123e+0	-929e-1	-662e-1	-434e-1
- 3.0	-347e+0	-282e+0	-220e+0	-163e+0	-109e+0	-589e-1	-131e-1	286e-1	659e-1	989e-1
- 3.2	-307e+0	-228e+0	-153e+0	-811e-1	-131e-1	509e-1	111e+0	166e+0	216e+0	262e+0
- 3.4	-266e+0	-173e+0	-824e-1	505e-2	890e-1	169e+0	245e+0	317e+0	383e+0	445e+0
- 3.6	-226e+0	-118e+0	-116e-1	925e-1	194e+0	292e+0	387e+0	477e+0	563e+0	644e+0
- 3.8	-187e+0	-651e-1	563e-1	177e+0	297e+0	414e+0	530e+0	641e+0	749e+0	853e+0
- 4.0	-150e+0	-172e-1	118e+0	254e+0	390e+0	527e+0	664e+0	799e+0	931e+0	106e+1
- 4.2	-117e+0	236e-1	168e+0	316e+0	467e+0	621e+0	777e+0	934e+0	109e+1	125e+1
- 4.4	-887e-1	552e-1	204e+0	358e+0	517e+0	682e+0	853e+0	103e+1	121e+1	139e+1
- 4.6	-647e-1	764e-1	222e+0	374e+0	533e+0	699e+0	873e+0	106e+1	125e+1	145e+1
- 4.8	-454e-1	868e-1	222e+0	363e+0	508e+0	661e+0	823e+0	994e+0	118e+1	138e+1
- 5.0	-306e-1	871e-1	205e+0	323e+0	443e+0	566e+0	693e+0	826e+0	966e+0	112e+1
- 5.2	-197e-1	793e-1	173e+0	262e+0	345e+0	423e+0	496e+0	563e+0	623e+0	677e+0
- 5.4	-121e-1	660e-1	133e+0	187e+0	228e+0	253e+0	261e+0	248e+0	210e+0	141e+0
- 5.6	-708e-2	500e-1	900e-1	111e+0	111e+0	862e-1	337e-1	-516e-1	-175e+0	-344e+0
- 5.8	-392e-2	341e-1	510e-1	444e-1	114e-1	-510e-1	-146e+0	-279e+0	-452e+0	-670e+0
- 6.0	-205e-2	205e-1	202e-1	-517e-2	-578e-1	-140e+0	-254e+0	-402e+0	-584e+0	-801e+0
- 6.2	-101e-2	102e-1	-426e-3	-342e-1	-921e-1	-175e+0	-284e+0	-417e+0	-575e+0	-754e+0
- 6.4	-462e-3	354e-2	-110e-1	-440e-1	-950e-1	-163e+0	-247e+0	-345e+0	-454e+0	-572e+0
- 6.6	-198e-3	-182e-4	-137e-1	-395e-1	-755e-1	-119e+0	-168e+0	-219e+0	-268e+0	-311e+0
- 6.8	-789e-4	-136e-2	-114e-1	-274e-1	-460e-1	-636e-1	-762e-1	-797e-1	-699e-1	-428e-1
- 7.0	-290e-4	-143e-2	-730e-2	-141e-1	-179e-1	-144e-1	684e-3	321e-1	844e-1	162e+0
- 7.2	-984e-5	-992e-3	-344e-2	-388e-2	135e-2	161e-1	442e-1	897e-1	157e+0	250e+0
- 7.4	-304e-5	-517e-3	-904e-3	156e-2	947e-2	253e-1	513e-1	897e-1	142e+0	311e+0
- 7.6	-887e-6	-195e-3	242e-3	294e-2	909e-2	194e-1	340e-1	524e-1	735e-1	956e-1
- 7.8	-313e-6	-395e-4	462e-3	213e-2	498e-2	831e-2	106e-1	966e-2	211e-2	-163e-1
- 8.0	-107e-6	909e-5	300e-3	878e-3	116e-2	-423e-4	-457e-2	-149e-1	-343e-1	-664e-1
- 8.2	-293e-7	123e-4	109e-3	822e-4	-641e-3	-294e-2	-793e-2	-169e-1	-311e-1	-517e-1
- 8.4	-438e-7	603e-5	107e-4	-154e-3	-770e-3	-211e-2	-428e-2	-711e-2	-990e-2	-113e-1
- 8.6	-100e-6	184e-5	-122e-4	-108e-3	-312e-3	-496e-3	-254e-3	122e-2	528e-2	139e-1
- 8.8	-661e-7	175e-6	-693e-5	-267e-4	379e-5	265e-3	109e-2	299e-2	661e-2	127e-1
- 9.0	-983e-7	667e-7	-145e-5	545e-5	615e-4	232e-3	573e-3	106e-2	150e-2	135e-2
- 9.2	630e-7	-136e-6	153e-6	544e-5	216e-4	331e-4	-347e-4	-369e-3	-133e-2	-354e-2
- 9.4	-869e-7	139e-6	199e-6	100e-5	-191e-5	-299e-4	-123e-3	-345e-3	-756e-3	-137e-2
- 9.6	188e-7	391e-7	252e-6	-331e-6	-315e-5	-112e-4	-179e-4	979e-5	167e-3	667e-3
- 9.8	-388e-6	275e-6	236e-6	-340e-6	-367e-6	190e-5	164e-4	603e-4	163e-3	352e-3
-10.0	-106e-7	311e-6	406e-6	449e-6	283e-6	138e-5	291e-5	-397e-6	-298e-4	-140e-3

where B_∞ is a constant (depending on k and α); and this implies that θ_1^2 is a known multiple of $\sin\theta$ together with a differentiable function of $\cos^2\theta$. Differentiation of θ_1^2 with respect to θ at $\theta = 0$ then yields the value of $\theta_2(0) = \Delta(k,\alpha)$. The coefficients b_n are generated sequentially from a rather complicated recurrence relation, and B_∞ is estimated from $\lim_{n\to\infty} n^{3/2}b_n$. In [2] we gave a rigorous proof of (1.8) for $\alpha = 0$. In the present extension to $0 \leq \alpha < 1/7$ a rigorous proof of (1.8) has eluded us, and we have had to rely on the observed numerical behaviour of the coefficients generated by the recurrence relation as empirical evidence supporting (1.8). Convergence to the asymptotic relation (1.8) is rather slow, sometimes necessitating as many as a quarter of a million coefficients b_n to get a reliable estimate of B_∞; and consequently DEND takes a few minutes, or in the worst cases several hours, to find each value of $\Delta(k,\alpha)$, whereas STIFF delivered each value of Δ in a couple of seconds. The following outstanding problems connected with DEND deserve further research: (i) to find a means of accelerating the convergence in (1.8), (ii) to produce a rigorous theoretical proof of (1.8); and (iii) to extend the method to the range $1/7 \leq \alpha < 1$. Nevertheless STIFF and DEND can both be used in their overlapping region $-10 \leq k \leq 0 \leq \alpha < 1/7$, and there is comforting agreement between the two programmes. Table 2 contains specimen numerical values of Δ. There are three entries in each cell of Table 2: the middle entry gives the STIFF determination of Δ, the bottom entry is the DEND determination of Δ, and the top entry is the number of coefficients b_n used by DEND to obtain B_∞. In Table 2, the typical notation –2047e–002 is an abbreviation for $-$ 0.2047e–002 = $-$ 0.2047 $\times 10^{-2}$. For compactness, similar notation is used elsewhere in other tables. Some specimen values in Table 2, for example 8332e+284 = 0.8332 $\times 10^{284}$ when $k = -8$, $\alpha = 0.14$ indicate failure of DEND, as discussed further in Section 3. Similar failures are evident in Tables 4.1 and 4.2 for the larger values of α. Tables 3.1 to 3.5 give more extensive calculations of Δ by STIFF, and Tables 4.1 and 4.2 calculations of Δ by DEND. In Tables 3.1 to 3.5 the letter d replaces e whenever the solution is not monotonic.

The exceptionally small values of Δ, associated with small $\epsilon > 0$, are beyond all orders of magnitude of an asymptotic expansion in powers of ϵ, and indeed exemplify the subject matter of this conference. The associated eigenvalue problem of calculating the discrete spectrum $\alpha = \alpha(k)$ such that $\Delta = 0$ is of course interesting in its own right as a piece of pure mathematics; but it should not divert us from asking to what extent it has physical meaningfulness. Often enough in applied mathematics, differential equations serve as models of physical phenomena because they are adequate approximations to more realistic difference equations. In the present case the term $\epsilon\theta_3$ in (1.1) arises from capillarity, which is a discrete stochastic process on the molecular scale. So, if the discontinuity 2Δ in the second derivative of $\theta^*(t)$ at $t = 0$ is, at most, of comparable magnitude to intermolecular distances in the liquid crystal, then it could be reasonable to accept the near solution (1.3) as an answer to the problem. Moreover, it is debatable to what extent equation (1.1) itself is an adequate model of the physics of dendritic growth. This is an issue which physicists must judge for themselves, in the light of the numerical evidence provided by Tables 3 and 4.

However, in the present paper, we shall merely concentrate upon the pure mathematical question of the existence or non–existence of solutions of (1.1), and the consequent numerical calculation of Δ and the occasions when $\Delta = 0$. One might expect the zeros of Δ to lie on the boundaries that separate lower case letters from the upper case letters in the overall picture Table 1.2. Certainly the zeros of Δ lie on the separating boundary; but the converse is false, because $\Delta(k,\alpha)$ is not a continuous function of k and α, and hence not every point on the separating boundary yields an eigenvalue $\alpha(k)$.

Table 5 illustrates a typical case of a discontinuity of the separating boundary.

Table 3.2 STIFF output for delta

	alpha									
k	0.20	0.22	0.24	0.26	0.28	0.30	0.32	0.34	0.36	0.38
0.0	-540e+0	-548e+0	-557e+0	-566e+0	-576e+0	-587e+0	-599e+0	-612e+0	-625e+0	-640e+0
- 0.2	-555e+0	-564e+0	-573e+0	-583e+0	-594e+0	-605e+0	-618e+0	-632e+0	-646e+0	-662e+0
- 0.4	-567e+0	-576e+0	-586e+0	-596e+0	-608e+0	-620e+0	-634e+0	-648e+0	-664e+0	-681e+0
- 0.6	-575e+0	-584e+0	-594e+0	-605e+0	-618e+0	-631e+0	-645e+0	-661e+0	-678e+0	-696e+0
- 0.8	-578e+0	-587e+0	-598e+0	-609e+0	-622e+0	-636e+0	-651e+0	-668e+0	-686e+0	-705e+0
- 1.0	-575e+0	-584e+0	-595e+0	-607e+0	-620e+0	-635e+0	-651e+0	-668e+0	-687e+0	-708e+0
- 1.2	-564e+0	-574e+0	-584e+0	-596e+0	-610e+0	-625e+0	-642e+0	-660e+0	-680e+0	-702e+0
- 1.4	-545e+0	-554e+0	-565e+0	-576e+0	-590e+0	-605e+0	-623e+0	-642e+0	-663e+0	-686e+0
- 1.6	-516e+0	-524e+0	-534e+0	-545e+0	-559e+0	-574e+0	-591e+0	-611e+0	-633e+0	-657e+0
- 1.8	-475e+0	-482e+0	-490e+0	-501e+0	-514e+0	-529e+0	-546e+0	-566e+0	-588e+0	-613e+0
- 2.0	-420e+0	-425e+0	-431e+0	-441e+0	-452e+0	-467e+0	-483e+0	-503e+0	-526e+0	-551e+0
- 2.2	-350e+0	-351e+0	-356e+0	-363e+0	-372e+0	-385e+0	-401e+0	-420e+0	-442e+0	-468e+0
- 2.4	-262e+0	-259e+0	-260e+0	-264e+0	-271e+0	-281e+0	-295e+0	-313e+0	-334e+0	-359e+0
- 2.6	-154e+0	-146e+0	-142e+0	-142e+0	-145e+0	-152e+0	-163e+0	-178e+0	-198e+0	-222e+0
- 2.8	-246e-1	-993e-2	633e-3	697e-2	898e-2	656e-2	-414e-3	-121e-1	-286e-1	-501e-1
- 3.0	127e+0	151e+0	170e+0	184e+0	193e+0	198e+0	196e+0	190e+0	178e+0	160e+0
- 3.2	303e+0	338e+0	368e+0	392e+0	411e+0	424e+0	431e+0	432e+0	426e+0	414e+0
- 3.4	500e+0	551e+0	595e+0	633e+0	664e+0	689e+0	707e+0	718e+0	721e+0	717e+0
- 3.6	718e+0	787e+0	849e+0	905e+0	953e+0	993e+0	103e+1	105e+1	107e+1	107e+1
- 3.8	951e+0	104e+1	113e+1	121e+1	127e+1	134e+1	139e+1	143e+1	146e+1	149e+1
- 4.0	119e+1	131e+1	142e+1	153e+1	162e+1	171e+1	179e+1	186e+1	191e+1	196e+1
- 4.2	141e+1	156e+1	171e+1	185e+1	198e+1	211e+1	222e+1	232e+1	241e+1	249e+1
- 4.4	158e+1	177e+1	196e+1	214e+1	232e+1	249e+1	265e+1	280e+1	294e+1	306e+1
- 4.6	166e+1	188e+1	211e+1	234e+1	258e+1	281e+1	303e+1	325e+1	345e+1	363e+1
- 4.8	159e+1	182e+1	207e+1	234e+1	263e+1	293e+1	323e+1	354e+1	384e+1	412e+1
- 5.0	128e+1	147e+1	167e+1	191e+1	219e+1	252e+1	290e+1	333e+1	379e+1	425e+1
- 5.2	722e+0	758e+0	782e+0	790e+0	777e+0	730e+0	623e+0	363e+0	-599e+0	-285d+1
- 5.4	319e-1	-129e+0	-357e+0	-677e+0	-111e+1	-169e+1	-239e+1	-314e+1	-386e+1	-450e+1
- 5.6	-567e+0	-851e+0	-120e+1	-163e+1	-211e+1	-264e+1	-318e+1	-370e+1	-418e+1	-461e+1
- 5.8	-935e+0	-125e+1	-160e+1	-199e+1	-240e+1	-282e+1	-322e+1	-360e+1	-395e+1	-425e+1
- 6.0	-105e+1	-133e+1	-163e+1	-194e+1	-225e+1	-256e+1	-285e+1	-312e+1	-336e+1	-357e+1
- 6.2	-951e+0	-116e+1	-138e+1	-159e+1	-180e+1	-200e+1	-218e+1	-234e+1	-248e+1	-260e+1
- 6.4	-693e+0	-814e+0	-931e+0	-104e+1	-113e+1	-122e+1	-128e+1	-132e+1	-135e+1	-137e+1
- 6.6	-346e+0	-369e+0	-376e+0	-367e+0	-338e+0	-290e+0	-222e+0	-137e+0	-363e-1	771e-1
- 6.8	533e-2	778e-1	177e+0	305e+0	462e+0	648e+0	862e+0	110e+1	136e+1	163e+1
- 7.0	270e+0	412e+0	593e+0	814e+0	108e+1	139e+1	175e+1	215e+1	259e+1	306e+1
- 7.2	373e+0	532e+0	731e+0	976e+0	127e+1	163e+1	206e+1	256e+1	313e+1	380e+1
- 7.4	298e+0	405e+0	535e+0	690e+0	876e+0	110e+1	136e+1	168e+1	207e+1	257e+1
- 7.6	116e+0	131e+0	136e+0	125e+0	905e-1	210e-1	-985e-1	-290e+0	-586e+0	-104e+1
- 7.8	-511e-1	-109e+0	-199e+0	-330e+0	-515e+0	-768e+0	-111e+1	-155e+1	-210e+1	-279e+1
- 8.0	-116e+0	-188e+0	-288e+0	-423e+0	-598e+0	-818e+0	-109e+1	-141e+1	-178e+1	-219e+1
- 8.2	-799e-1	-116e+0	-161e+0	-213e+0	-271e+0	-332e+0	-393e+0	-448e+0	-493e+0	-519e+0
- 8.4	-914e-2	-110e-3	203e-1	578e-1	120e+0	215e+0	352e+0	543e+0	798e+0	113e+1
- 8.6	299e-1	569e-1	996e-1	164e+0	257e+0	389e+0	568e+0	810e+0	113e+1	155e+1
- 8.8	221e-1	356e-1	537e-1	766e-1	104e+0	135e+0	168e+0	198e+0	222e+0	231e+0
- 9.0	-427e-3	-563e-2	-171e-1	-390e-1	-772e-1	-140e+0	-239e+0	-387e+0	-604e+0	-912e+0
- 9.2	-791e-2	-158e-1	-288e-1	-493e-1	-800e-1	-124e+0	-185e+0	-266e+0	-371e+0	-503e+0
- 9.4	-205e-2	-239e-2	-152e-2	219e-2	115e-1	305e-1	655e-1	125e+0	222e+0	370e+0
- 9.6	190e-2	453e-2	960e-2	186e-1	338e-1	581e-1	953e-1	150e+0	230e+0	343e+0
- 9.8	622e-3	872e-3	798e-3	-289e-3	-371e-2	-118e-1	-284e-1	-596e-1	-115e+0	-206e+0
-10.0	-448e-3	-117e-2	-269e-2	-558e-2	-107e-1	-192e-1	-327e-1	-531e-1	-827e-1	-124e+0

Table 3.3 STIFF output for delta

	alpha									
k	0.40	0.42	0.44	0.46	0.48	0.50	0.52	0.54	0.56	0.58
0.0	-655e+0	-672e+0	-690e+0	-709e+0	-729e+0	-751e+0	-775e+0	-801e+0	-828e+0	-858e+0
- 0.2	-679e+0	-697e+0	-716e+0	-737e+0	-759e+0	-783e+0	-809e+0	-837e+0	-867e+0	-900e+0
- 0.4	-699e+0	-719e+0	-740e+0	-762e+0	-786e+0	-813e+0	-841e+0	-871e+0	-904e+0	-940e+0
- 0.6	-715e+0	-736e+0	-759e+0	-784e+0	-810e+0	-838e+0	-869e+0	-902e+0	-938e+0	-977e+0
- 0.8	-726e+0	-749e+0	-773e+0	-800e+0	-828e+0	-859e+0	-893e+0	-929e+0	-968e+0	-101e+1
- 1.0	-730e+0	-755e+0	-781e+0	-809e+0	-840e+0	-874e+0	-910e+0	-949e+0	-991e+0	-104e+1
- 1.2	-726e+0	-752e+0	-780e+0	-811e+0	-844e+0	-880e+0	-919e+0	-962e+0	-101e+1	-106e+1
- 1.4	-711e+0	-739e+0	-769e+0	-802e+0	-837e+0	-876e+0	-918e+0	-964e+0	-101e+1	-107e+1
- 1.6	-684e+0	-713e+0	-745e+0	-780e+0	-818e+0	-860e+0	-905e+0	-955e+0	-101e+1	-107e+1
- 1.8	-641e+0	-671e+0	-705e+0	-742e+0	-783e+0	-828e+0	-877e+0	-930e+0	-988e+0	-105e+1
- 2.0	-580e+0	-611e+0	-647e+0	-686e+0	-729e+0	-777e+0	-829e+0	-886e+0	-949e+0	-102e+1
- 2.2	-497e+0	-530e+0	-566e+0	-607e+0	-653e+0	-703e+0	-759e+0	-820e+0	-888e+0	-962e+0
- 2.4	-388e+0	-422e+0	-460e+0	-502e+0	-550e+0	-603e+0	-662e+0	-727e+0	-799e+0	-879e+0
- 2.6	-250e+0	-283e+0	-322e+0	-365e+0	-415e+0	-470e+0	-532e+0	-601e+0	-678e+0	-763e+0
- 2.8	-769e-1	-109e+0	-147e+0	-191e+0	-242e+0	-299e+0	-364e+0	-437e+0	-518e+0	-609e+0
- 3.0	136e+0	106e+0	693e-1	258e-1	-251e-1	-838e-1	-151e+0	-227e+0	-312e+0	-409e+0
- 3.2	395e+0	368e+0	334e+0	293e+0	243e+0	184e+0	115e+0	368e-1	-525e-1	-154e+0
- 3.4	704e+0	684e+0	655e+0	617e+0	569e+0	511e+0	443e+0	363e+0	271e+0	165e+0
- 3.6	107e+1	106e+1	104e+1	101e+1	963e+0	908e+0	841e+0	761e+0	668e+0	559e+0
- 3.8	150e+1	150e+1	149e+1	147e+1	143e+1	138e+1	132e+1	124e+1	115e+1	104e+1
- 4.0	199e+1	201e+1	202e+1	201e+1	199e+1	195e+1	189e+1	182e+1	173e+1	162e+1
- 4.2	255e+1	259e+1	262e+1	264e+1	263e+1	261e+1	257e+1	250e+1	242e+1	232e+1
- 4.4	316e+1	324e+1	331e+1	335e+1	337e+1	337e+1	335e+1	331e+1	324e+1	314e+1
- 4.6	379e+1	393e+1	405e+1	414e+1	421e+1	425e+1	426e+1	424e+1	419e+1	411e+1
- 4.8	438e+1	461e+1	481e+1	498e+1	511e+1	521e+1	527e+1	530e+1	529e+1	524e+1
- 5.0	470e+1	510e+1	546e+1	577e+1	603e+1	623e+1	638e+1	648e+1	653e+1	653e+1
- 5.2	-421d+1	427e+1	536e+1	613e+1	670e+1	714e+1	748e+1	772e+1	788e+1	797e+1
- 5.4	-505d+1	-551d+1	-591d+1	-625d+1	-655d+1	-682d+1	806e+1	877e+1	921e+1	949e+1
- 5.6	-499d+1	-532d+1	-560d+1	-586d+1	-609d+1	-632d+1	-653d+1	-676d+1	-699d+1	107e+2
- 5.8	-453e+1	-477d+1	-498d+1	-517d+1	-535d+1	-553d+1	-571d+1	-590d+1	-611d+1	-635d+1
- 6.0	-376e+1	-392d+1	-406d+1	-420d+1	-433d+1	-446d+1	-460d+1	-475d+1	-493d+1	-515d+1
- 6.2	-270e+1	-278e+1	-286d+1	-292d+1	-299d+1	-307d+1	-316d+1	-328d+1	-342d+1	-360d+1
- 6.4	-137e+1	-136e+1	-135e+1	-134d+1	-133d+1	-134d+1	-137d+1	-143d+1	-153d+1	-167d+1
- 6.6	199e+0	325e+0	450e+0	567e+0	671d+0	753d+0	807d+0	825d+0	797d+0	714d+0
- 6.8	191e+1	220e+1	248e+1	275e+1	299d+1	321d+1	339d+1	352d+1	359d+1	360d+1
- 7.0	355e+1	406e+1	457e+1	506e+1	553e+1	595d+1	632d+1	663d+1	686d+1	701d+1
- 7.2	454e+1	534e+1	619e+1	706e+1	790e+1	869e+1	941d+1	100d+2	105d+2	109d+2
- 7.4	321e+1	407e+1	521e+1	668e+1	839e+1	101e+2	117e+2	131d+2	142d+2	151d+2
- 7.6	-172e+1	-275e+1	-424e+1	-614e+1	-816e+1	-100d+2	-115d+2	-128d+2	-139d+2	181d+2
- 7.8	-360e+1	-451e+1	-549e+1	-648e+1	-742e+1	-828e+1	-905d+1	-973d+1	-103d+2	-109d+2
- 8.0	-263e+1	-308e+1	-353e+1	-395e+1	-435e+1	-469e+1	-500d+1	-527d+1	-551d+1	-576d+1
- 8.2	-523e+0	-498e+0	-441e+0	-351e+0	-230e+0	-838e-1	782e-1	243d+0	393d+0	507d+0
- 8.4	154e+1	204e+1	262e+1	329e+1	402e+1	479e+1	559e+1	636e+1	709d+1	774d+1
- 8.6	209e+1	278e+1	366e+1	477e+1	614e+1	776e+1	958e+1	115e+2	134d+2	151d+2
- 8.8	214e+0	152e+0	158e-1	-249e+0	-746e+0	-171e+1	-374e+1	-788e+1	-132d+2	-175d+2
- 9.0	-134e+1	-192e+1	-268e+1	-366e+1	-485e+1	-623e+1	-773e+1	-927e+1	-107e+2	-121d+2
- 9.2	-663e+0	-850e+0	-106e+1	-129e+1	-151e+1	-173e+1	-193e+1	-209e+1	-221e+1	-230d+1
- 9.4	588e+0	900e+0	133e+1	190e+1	264e+1	356e+1	466e+1	592e+1	728e+1	867e+1
- 9.6	499e+0	714e+0	101e+1	141e+1	196e+1	275e+1	389e+1	564e+1	838e+1	125e+2
- 9.8	-355e+0	-586e+0	-937e+0	-146e+1	-222e+1	-330e+1	-478e+1	-675e+1	-919e+1	-119e+2
-10.0	-179e+0	-250e+0	-339e+0	-444e+0	-562e+0	-687e+0	-807e+0	-912e+0	-985e+0	-102e+1

This gives the STIFF values for $\Delta(k,0.61)$ for closely spaced values of k in three places near to the separating boundary. The uniformity of the first differences of $\Delta(k,0.61)$ with respect to k show that Δ is continuous and has a zero near $k = -3.41206$ and also near $k = -6.56531$; but, on the other hand, Δ suffers a discontinuity and has no zero when crossing the separating boundary between $k = -5.80554$ and $k = -5.80555$. The explanation of this lies in the fact that the primary solution $\theta(t)$ is monotone for $k = -5.80554$ and $\alpha = 0.61$, but not monotone for $k = -5.80555$ and $\alpha = 0.61$. Table 6 compares the values of $(\theta,\theta_1,\theta_2)$ against an independent variable x = constant − t, where the constant translation is chosen to exhibit the close similarity of the two trajectories: the trajectory for $k = -5.80555$ just fails to reach $\theta = 0$ near $x = 9.2$ where θ_2 is positive and has to wait until $x = 9.72405$ before attaining $\theta = 0$ where θ_2 is negative; whereas the curve for $k = -5.80554$ attains $\theta = 0$ at $x = 9.21686$ where θ_2 is positive.

Thus we have to distinguish between discontinuities and genuine zeros of Δ on the boundary. This can be done by estimating $\partial\Delta/\partial k$ in terms of first differences of θ_2 as k varies for fixed α; or alternatively, $\partial\Delta/\partial\alpha$ in terms of first differences of θ_2 as α varies for fixed k. The full curves in Figure 1 give true zeros of Δ, while the broken curves represent discontinuities of Δ at which $\Delta \neq 0$. Table 7 lists eigenvalues $\alpha(k)$ for which $\Delta = 0$; this list is incomplete, partly because DEND cannot cope with $\alpha \geq 1/7$, and partly because the existence of zeros of Δ are sometimes difficult to locate and some may have been overlooked. In Table 7, the eigenvalues for $k = -3$ to $k = -7$ were found from the STIFF calculations, while those for $k = -8$ to $k = -18$ originate from DEND calculations, and consequently are restricted to the range $0 \leq \alpha < 1/7$. Further discussion of these eigenvalues appears in Section 6. Table 8.1 quotes the values of $-\log_2\alpha$ for each eigenvalue α; and Tables 8.2 to 8.3 give the horizontal and vertical first differences of the entries in Table 8.1. From this it appears that the smallest eigenvalue is approximately $5\epsilon^2$ as $\epsilon \to 0$, while the next two eigenvalues are approximately $135\epsilon^2$ and $831\epsilon^2$ as $\epsilon \to 0$.

2. SOLUTION OF (1.1) VIA SOLUTIONS OF (1.2)

Equation (1.1) has the general form

$$\epsilon\theta_3 + \theta_1 = F(\theta) , (-\infty < t < \infty) , \tag{2.1}$$

where F is an infinitely differentiable even function of θ. We write F_n for the nth derivative of F with respect to θ at $\theta = 0$; so

$$F_1 = F_3 = F_5 = \ldots = 0 . \tag{2.2}$$

Let $\theta(t) = P(t)$ and $\theta(t) = Q(t)$ be two (not necessarily distinct) solutions of

$$\epsilon\theta_3 + \theta_1 = F(\theta) , (0 \leq t < \infty) , \theta(0) = 0; \tag{2.3}$$

and consider

$$\theta^*(t) = \begin{cases} P(t) , & (t \geq 0) \\ -Q(-t) , & (t \leq 0) \end{cases} \tag{2.4}$$

as a candidate solution of (2.1), inasmuch as (2.1) is invariant under the involution $(\theta,t) \mapsto (-\theta,-t)$. By hypothesis, P(t) is thrice–differentiable for $t \geq 0$; so $F(\theta)$ is a differentiable function of t for $t \geq 0$. Hence the left–hand side of (2.3) is differentiable and consequently $\theta(t)$ is differentiable four times. Continuing to differentiate (2.3) and arguing by induction on n, we see that P_n, the nth right–hand derivative of P with respect to t at $t = 0$, exists for all $n > 0$; and similarly Q_n, the nth right–hand derivative of Q with respect to t at $t = 0$, exists for all $n > 0$. If L_n and R_n denote the left–hand and right–hand nth derivatives of θ^* with respect to t at $t = 0$, (2.4) will be a solution of (2.1) if and only if

$$L_j = (-1)^{j-1}Q_j = P_j = R_j \qquad (j > 0) \tag{2.5}$$

Table 3.4 STIFF output for delta

	alpha									
k	0.60	0.62	0.64	0.66	0.68	0.70	0.72	0.74	0.76	0.78
0.0	-890e+0	-925e+0	-963e+0	-101e+1	-105e+1	-110e+1	-116e+1	-122e+1	-129e+1	-137e+1
- 0.2	-935e+0	-973e+0	-102e+1	-106e+1	-111e+1	-117e+1	-123e+1	-130e+1	-137e+1	-146e+1
- 0.4	-978e+0	-102e+1	-107e+1	-112e+1	-117e+1	-123e+1	-130e+1	-138e+1	-146e+1	-156e+1
- 0.6	-102e+1	-106e+1	-112e+1	-117e+1	-123e+1	-130e+1	-137e+1	-145e+1	-155e+1	-165e+1
- 0.8	-106e+1	-111e+1	-116e+1	-122e+1	-129e+1	-136e+1	-144e+1	-153e+1	-163e+1	-175e+1
- 1.0	-109e+1	-114e+1	-120e+1	-127e+1	-134e+1	-142e+1	-151e+1	-161e+1	-172e+1	-185e+1
- 1.2	-111e+1	-117e+1	-124e+1	-131e+1	-139e+1	-148e+1	-157e+1	-168e+1	-181e+1	-195e+1
- 1.4	-113e+1	-119e+1	-126e+1	-134e+1	-143e+1	-153e+1	-163e+1	-175e+1	-189e+1	-204e+1
- 1.6	-113e+1	-120e+1	-128e+1	-137e+1	-146e+1	-157e+1	-168e+1	-181e+1	-196e+1	-213e+1
- 1.8	-112e+1	-120e+1	-128e+1	-138e+1	-148e+1	-159e+1	-172e+1	-187e+1	-203e+1	-221e+1
- 2.0	-109e+1	-118e+1	-127e+1	-137e+1	-148e+1	-161e+1	-175e+1	-190e+1	-208e+1	-228e+1
- 2.2	-104e+1	-113e+1	-123e+1	-134e+1	-147e+1	-160e+1	-175e+1	-193e+1	-212e+1	-234e+1
- 2.4	-967e+0	-106e+1	-117e+1	-129e+1	-142e+1	-157e+1	-174e+1	-193e+1	-214e+1	-238e+1
- 2.6	-858e+0	-963e+0	-108e+1	-121e+1	-135e+1	-151e+1	-170e+1	-190e+1	-213e+1	-240e+1
- 2.8	-710e+0	-823e+0	-948e+0	-109e+1	-124e+1	-142e+1	-162e+1	-184e+1	-209e+1	-238e+1
- 3.0	-516e+0	-637e+0	-772e+0	-923e+0	-109e+1	-128e+1	-150e+1	-174e+1	-201e+1	-233e+1
- 3.2	-268e+0	-396e+0	-541e+0	-703e+0	-886e+0	-109e+1	-132e+1	-159e+1	-189e+1	-223e+1
- 3.4	450e-1	-911e-1	-245e+0	-419e+0	-615e+0	-837e+0	-109e+1	-138e+1	-170e+1	-208e+1
- 3.6	433e+0	290e+0	127e+0	-585e-1	-269e+0	-508e+0	-780e+0	-109e+1	-145e+1	-186e+1
- 3.8	910e+0	760e+0	589e+0	392e+0	167e+0	-895e-1	-383e+0	-719e+0	-110e+1	-155e+1
- 4.0	149e+1	133e+1	116e+1	948e+0	709e+0	435e+0	120e+0	-243e+0	-661e+0	-115e+1
- 4.2	218e+1	203e+1	184e+1	163e+1	138e+1	108e+1	746e+0	356e+0	-959e-1	-622e+0
- 4.4	302e+1	286e+1	267e+1	245e+1	219e+1	188e+1	152e+1	110e+1	613e+0	434e-1
- 4.6	400e+1	385e+1	366e+1	344e+1	316e+1	284e+1	246e+1	202e+1	149e+1	877e+0
- 4.8	515e+1	502e+1	484e+1	461e+1	434e+1	400e+1	360e+1	313e+1	257e+1	191e+1
- 5.0	648e+1	638e+1	622e+1	601e+1	573e+1	539e+1	498e+1	448e+1	388e+1	317e+1
- 5.2	799e+1	794e+1	782e+1	764e+1	738e+1	704e+1	662e+1	610e+1	547e+1	471e+1
- 5.4	965e+1	970e+1	966e+1	953e+1	930e+1	898e+1	856e+1	803e+1	737e+1	657e+1
- 5.6	113e+2	116e+2	117e+2	117e+2	115e+2	112e+2	108e+2	103e+2	964e+1	880e+1
- 5.8	-662d+1	132e+2	138e+2	140e+2	140e+2	139e+2	135e+2	130e+2	123e+2	115e+2
- 6.0	-540d+1	-571d+1	-607d+1	162e+2	167e+2	168e+2	166e+2	161e+2	155e+2	146e+2
- 6.2	-384d+1	-413d+1	-448d+1	-492d+1	-546d+1	198e+2	200e+2	197e+2	192e+2	183e+2
- 6.4	-187d+1	-213d+1	-248d+1	-291d+1	-345d+1	-412d+1	233e+2	237e+2	234e+2	226e+2
- 6.6	565d+0	339d+0	228d-1	-398d+0	-939d+0	-162d+1	-246d+1	-350d+1	280e+2	276e+2
- 6.8	352d+1	336d+1	309d+1	270d+1	218d+1	149d+1	617d+0	-467d+0	-180d+1	330e+2
- 7.0	706d+1	700d+1	681d+1	648d+1	599d+1	531d+1	443d+1	330d+1	189d+1	143d+0
- 7.2	112d+2	113d+2	112d+2	110d+2	106d+2	997d+1	910d+1	794d+1	645d+1	459d+1
- 7.4	158d+2	162d+2	164d+2	164d+2	161d+2	156d+2	148d+2	136d+2	121d+2	101d+2
- 7.6	201d+2	214d+2	222d+2	226d+2	226d+2	222d+2	215d+2	204d+2	189d+2	168d+2
- 7.8	-114d+2	-120d+2	274d+2	292d+2	299d+2	300d+2	295d+2	286d+2	271d+2	250d+2
- 8.0	-603d+1	-636d+1	-679d+1	-736d+1	368d+2	385d+2	388d+2	382d+2	369d+2	348d+2
- 8.2	558d+0	513d+0	333d+0	-288d-1	-628d+0	-153d+1	485d+2	492d+2	484d+2	465d+2
- 8.4	827d+1	866d+1	886d+1	882d+1	849d+1	780d+1	668d+1	596d+2	615d+2	603d+2
- 8.6	166d+2	178d+2	187d+2	192d+2	193d+2	190d+2	182d+2	167d+2	145d+2	759d+2
- 8.8	207d+2	257d+2	287d+2	306d+2	318d+2	322d+2	318d+2	307d+2	287d+2	256d+2
- 9.0	-132d+2	-143d+2	-152d+2	-162d+2	440d+2	467d+2	477d+2	473d+2	458d+2	430d+2
- 9.2	-236d+1	-243d+1	-256d+1	-282d+1	-329d+1	-408d+1	638d+2	662d+2	659d+2	636d+2
- 9.4	100d+2	113d+2	123d+2	131d+2	135d+2	134d+2	127d+2	113d+2	882d+2	878d+2
- 9.6	177e+2	227d+2	269d+2	301d+2	324d+2	338d+2	342d+2	336d+2	317d+2	114d+3
- 9.8	-148e+2	-174d+2	-197d+2	-217d+2	-235d+2	545d+2	583d+2	596d+2	589d+2	563d+2
-10.0	-100e+1	-958d+0	-907d+0	-903d+0	-103d+1	-143d+1	-228d+1	864d+2	902d+2	895d+2

Table 3.5 STIFF output for delta

	alpha									
k	0.80	0.82	0.84	0.86	0.88	0.90	0.92	0.94	0.96	0.98
0.0	-146e+1	-157e+1	-169e+1	-185e+1	-203e+1	-227e+1	-259e+1	-305e+1	-382e+1	-548e+1
- 0.2	-156e+1	-168e+1	-182e+1	-198e+1	-219e+1	-245e+1	-280e+1	-331e+1	-415e+1	-598e+1
- 0.4	-167e+1	-180e+1	-195e+1	-213e+1	-236e+1	-264e+1	-303e+1	-359e+1	-451e+1	-653e+1
- 0.6	-178e+1	-192e+1	-208e+1	-228e+1	-253e+1	-285e+1	-327e+1	-389e+1	-490e+1	-712e+1
- 0.8	-188e+1	-204e+1	-222e+1	-244e+1	-272e+1	-306e+1	-353e+1	-421e+1	-532e+1	-776e+1
- 1.0	-200e+1	-217e+1	-237e+1	-261e+1	-291e+1	-329e+1	-381e+1	-455e+1	-577e+1	-845e+1
- 1.2	-211e+1	-229e+1	-252e+1	-278e+1	-311e+1	-353e+1	-410e+1	-492e+1	-626e+1	-920e+1
- 1.4	-222e+1	-242e+1	-267e+1	-296e+1	-332e+1	-378e+1	-440e+1	-530e+1	-678e+1	-100e+2
- 1.6	-232e+1	-255e+1	-282e+1	-314e+1	-354e+1	-404e+1	-473e+1	-572e+1	-734e+1	-109e+2
- 1.8	-242e+1	-267e+1	-296e+1	-332e+1	-375e+1	-431e+1	-506e+1	-615e+1	-794e+1	-119e+2
- 2.0	-252e+1	-279e+1	-311e+1	-350e+1	-398e+1	-459e+1	-542e+1	-661e+1	-858e+1	-129e+2
- 2.2	-260e+1	-289e+1	-325e+1	-367e+1	-420e+1	-488e+1	-578e+1	-710e+1	-926e+1	-140e+2
- 2.4	-266e+1	-299e+1	-337e+1	-384e+1	-442e+1	-516e+1	-616e+1	-760e+1	-998e+1	-152e+2
- 2.6	-270e+1	-306e+1	-349e+1	-400e+1	-463e+1	-545e+1	-655e+1	-813e+1	-107e+2	-165e+2
- 2.8	-272e+1	-311e+1	-358e+1	-414e+1	-484e+1	-573e+1	-694e+1	-868e+1	-116e+2	-178e+2
- 3.0	-270e+1	-313e+1	-364e+1	-426e+1	-502e+1	-601e+1	-733e+1	-925e+1	-124e+2	-193e+2
- 3.2	-263e+1	-310e+1	-366e+1	-434e+1	-518e+1	-626e+1	-772e+1	-983e+1	-133e+2	-209e+2
- 3.4	-252e+1	-303e+1	-364e+1	-439e+1	-531e+1	-650e+1	-809e+1	-104e+2	-142e+2	-226e+2
- 3.6	-234e+1	-290e+1	-357e+1	-438e+1	-539e+1	-670e+1	-845e+1	-110e+2	-152e+2	-244e+2
- 3.8	-207e+1	-269e+1	-342e+1	-431e+1	-542e+1	-685e+1	-878e+1	-116e+2	-162e+2	-263e+2
- 4.0	-171e+1	-238e+1	-319e+1	-416e+1	-538e+1	-695e+1	-907e+1	-121e+2	-172e+2	-283e+2
- 4.2	-124e+1	-197e+1	-285e+1	-392e+1	-525e+1	-697e+1	-930e+1	-127e+2	-183e+2	-305e+2
- 4.4	-627e+0	-142e+1	-238e+1	-355e+1	-501e+1	-690e+1	-945e+1	-132e+2	-193e+2	-327e+2
- 4.6	150e+0	-715e+0	-176e+1	-304e+1	-464e+1	-671e+1	-951e+1	-136e+2	-203e+2	-351e+2
- 4.8	112e+1	183e+0	-954e+0	-235e+1	-410e+1	-637e+1	-945e+1	-139e+2	-213e+2	-376e+2
- 5.0	233e+1	131e+1	708e-1	-145e+1	-337e+1	-585e+1	-923e+1	-142e+2	-223e+2	-401e+2
- 5.2	380e+1	270e+1	136e+1	-303e+0	-240e+1	-512e+1	-882e+1	-142e+2	-232e+2	-428e+2
- 5.4	559e+1	441e+1	295e+1	114e+1	-114e+1	-412e+1	-818e+1	-141e+2	-240e+2	-455e+2
- 5.6	776e+1	648e+1	491e+1	294e+1	451e+0	-281e+1	-726e+1	-138e+2	-246e+2	-483e+2
- 5.8	104e+2	899e+1	729e+1	516e+1	245e+1	-111e+1	-599e+1	-131e+2	-250e+2	-511e+2
- 6.0	135e+2	120e+2	102e+2	788e+1	492e+1	103e+1	-431e+1	-122e+2	-252e+2	-539e+2
- 6.2	171e+2	156e+2	137e+2	112e+2	796e+1	372e+1	-213e+1	-107e+2	-251e+2	-566e+2
- 6.4	215e+2	199e+2	178e+2	151e+2	117e+2	704e+1	647e+0	-880e+1	-246e+2	-592e+2
- 6.6	265e+2	249e+2	228e+2	199e+2	161e+2	111e+2	413e+1	-622e+1	-235e+2	-616e+2
- 6.8	323e+2	308e+2	286e+2	256e+2	215e+2	161e+2	846e+1	-288e+1	-219e+2	-638e+2
- 7.0	388e+2	377e+2	355e+2	323e+2	280e+2	221e+2	138e+2	136e+1	-195e+2	-655e+2
- 7.2	228d+1	454e+2	435e+2	403e+2	357e+2	293e+2	203e+2	668e+1	-162e+2	-668e+2
- 7.4	760d+1	450d+1	527e+2	496e+2	448e+2	380e+2	281e+2	133e+2	-118e+2	-675e+2
- 7.6	142d+2	108d+2	623e+2	605e+2	556e+2	483e+2	376e+2	214e+2	-612e+1	-673e+2
- 7.8	222d+2	186d+2	140d+2	727e+2	682e+2	605e+2	490e+2	313e+2	115e+1	-660e+2
- 8.0	319d+2	281d+2	231d+2	166d+2	828e+2	750e+2	626e+2	434e+2	103e+2	-635e+2
- 8.2	436d+2	396d+2	342d+2	271d+2	990e+2	920e+2	788e+2	579e+2	217e+2	-594e+2
- 8.4	576d+2	535d+2	477d+2	401d+2	299d+2	112e+3	980e+2	753e+2	357e+2	-534e+2
- 8.6	740d+2	700d+2	640d+2	558d+2	447d+2	134e+3	121e+3	961e+2	528e+2	-450e+2
- 8.8	927d+2	895d+2	835d+2	748d+2	627d+2	462d+2	147e+3	121e+3	737e+2	-338e+2
- 9.0	386d+2	112d+3	107d+3	976d+2	845d+2	663d+2	178e+3	150e+3	988e+2	-191e+2
- 9.2	595d+2	533d+2	134d+3	125d+3	111d+3	909d+2	627d+2	185e+3	129e+3	-306e+0
- 9.4	845d+2	784d+2	692d+2	157d+3	142d+3	121d+3	894d+2	226e+3	165e+3	234e+2
- 9.6	114d+3	108d+3	992d+2	194d+3	180d+3	157d+3	122d+3	273e+3	208e+3	529e+2
- 9.8	145d+3	144d+3	135d+3	121d+3	224d+3	200d+3	162d+3	104d+3	260e+3	894e+2
-10.0	855d+2	182d+3	177d+3	163d+3	274d+3	251d+3	210d+3	146d+3	321e+3	134e+3

We shall show that (2.5) follows from the simpler relations

$$P_1 = Q_1 \text{ and } P_2 = -Q_2 \tag{2.6}$$

Now $\epsilon P_3 + P_1 = F(0) = \epsilon Q_3 + Q_1$; so $P_3 = Q_3$. Proceeding by induction on n, and differentiating (2.4) on the left and right n times with respect to t at t = 0, we have

$$\epsilon L_{n+3} + L_{n+1} = (-1)^n \sum_{r,\delta} C(r,\delta) F_r \prod_{j=1}^{n} \left[-Q_j\right]^{\delta(j)} \tag{2.7}$$

where r is a positive integer, $\delta(j)$ is a non-negative integer, and $C(r,\delta)$ is an integer depending only on $r, \delta(1),\ldots,\delta(n)$ and independent of the functions F and Q. Also

$$\sum_{j=1}^{n} \delta(j) = r \text{ and } \sum_{j=1}^{n} j\delta(j) = n\,. \tag{2.8}$$

Hence, by the inductive hypothesis ($0 \leq j < n + 3$) on (2.5), and by (2.2) and (2.8),

$$\begin{aligned} \epsilon L_{n+3} + L_{n+1} &= (-1)^n \sum_{r,\delta} C(r,\delta) F_r \prod_{j=1}^{n} \left\{(-1)^j P_j\right\}^{\delta(j)} \\ &= (-1)^n \sum_{r,\delta} C(r,\delta)(-1)^n F_r \prod_{j=1}^{n} P_j^{\delta(j)} \\ &= \sum_{r,\delta} C(r,\delta) F_r \prod_{j=1}^{n} P_j^{\delta(j)} = \epsilon R_{n+3} + R_{n+1} \end{aligned} \tag{2.9}$$

Hence, $L_{n+3} = R_{n+3}$, closing the inductive loop.

In particular, when P = Q, (2.6) reduces to the single statement $\Delta = 0$.

In principle there is a countable infinity of ways of fitting together solutions P and Q in (2.4) to produce candidate solutions θ_*. How then may these P and Q be classified? To answer this question consider solutions to the equation

$$\epsilon\theta_3 + \theta_1 = \frac{\cos\theta}{1+\alpha\cos 4\theta}\,, \; (-\infty < t < \infty)\,, \; \theta(+\infty) = +\,\pi/2\,, \tag{2.10}$$

which is simply (1.1) omitting the boundary condition $\theta(-\infty) = -\,\pi/2$. Equation (2.10) has the trivial solution $\theta(t) \equiv \pi/2$. All non-trivial solutions of (2.10) are of two kinds: <u>upper solutions</u> in which $\theta(t) \downarrow \pi/2$ as $t \to \infty$, and <u>lower solutions</u> in which $\theta(t) \uparrow \pi/2$ as $t \to \infty$. An upper solution is the reflection of a lower solution in the trivial solution, and vice versa, since (2.10) is invariant under the involution $\theta \longleftrightarrow \pi - \theta$. Also (2.10) is an autonomous equation; and therefore, for any arbitrary t_0, the translate $\theta(t+t_0)$ of a solution $\theta(t)$ is also a solution. Given any particular upper solution of (2.10), all possible upper solutions of (2.10) are translates of this particular upper solution; and all possible lower solutions (2.10) are translates of the reflection in the trivial solution of this particular upper solution: this fact was proved in [2] for the special case $\alpha = 0$; but the proof in [2] remains valid (with minor amendments) for any fixed α in $0 \leq \alpha < 1$. We conjecture that any non-trivial solution $\theta(t)$ of (2.10) possesses a countable set of discrete roots $t = \tau$ for which $\theta(\tau) = 0$. On the assumption that this conjecture is correct, we may define a solution $\theta(t) = P(t)$ to be an <u>mth order solution</u> of (2.3) if it has precisely m non-negative roots τ. Figure 3 (kindly provided by Dr. C.R. Prior) illustrates the very irregular behaviour of solutions of (2.10) for values of t below the largest zero of θ. The curves (arbitrarily translated for the sake of clarity) are all for $k = -\,5.2$ and deal respectively with $\alpha = 0.0(0.1)0.9$ reading from right to left in the positions of their largest zeros.

Table 4.1 DEND output for delta

alpha	k= -7	-8	-9	-10	-11	-12
0.0000	-2888e-004	-4613e-007	-3438e-011	-3395e-017	-7471e-026	-2938e-038
0.0025	1534e-003	-1006e-006	-4109e-009	8298e-014	1891e-020	5882e-030
0.0050	2213e-003	-2364e-005	1529e-008	-1118e-012	-1803e-019	-6658e-028
0.0075	1807e-003	-5395e-005	7464e-008	-3274e-012	-4857e-018	1045e-026
0.0100	3836e-004	-7882e-005	1427e-007	4012e-012	4704e-018	7298e-026
0.0125	-2000e-003	-8646e-005	1553e-007	3697e-011	1083e-016	-9414e-025
0.0150	-5285e-003	-6640e-005	2970e-008	1037e-010	3409e-016	-5148e-024
0.0175	-9400e-003	-9479e-006	-3226e-007	1865e-010	3577e-016	3432e-024
0.0200	-1429e-002	9208e-005	-9830e-007	2250e-010	-9887e-016	1010e-022
0.0225	-1988e-002	2447e-004	-2014e-006	1053e-010	-5761e-015	3912e-022
0.0250	-2610e-002	4535e-004	-3449e-006	-3418e-010	-1619e-014	6525e-022
0.0275	-3290e-002	7222e-004	-5280e-006	-1331e-009	-3250e-014	-6007e-022
0.0300	-4021e-002	1053e-003	-1664e-005	-3099e-009	-4896e-014	-7437e-021
0.0325	-4794e-002	1447e-003	-9857e-006	-5863e-009	-4811e-014	-2665e-020
0.0350	-5603e-002	1903e-003	-1231e-005	-9768e-009	5742e-015	-6358e-020
0.0375	-6442e-002	2420e-003	-1458e-005	-1482e-008	1728e-013	-1075e-019
0.0400	-7303e-002	2995e-003	-1632e-005	-2078e-008	5413e-013	-1005e-019
0.0425	-8178e-002	3622e-003	-1715e-005	-2709e-008	1223e-012	1146e-019
0.0450	-9060e-002	4296e-003	-1656e-005	-3276e-008	2340e-012	8585e-019
0.0475	-9942e-002	5008e-003	-1399e-005	-3623e-008	3996e-012	2672e-018
0.0500	-1082e-001	5750e-003	-8776e-006	-3527e-008	6222e-012	6300e-018
0.0525	-1168e-001	6511e-003	-1635e-007	-2683e-008	8889e-012	1250e-017
0.0550	-1251e-001	7279e-003	1268e-005	-6951e-009	1162e-011	2151e-017
0.0575	-1332e-001	8040e-003	3069e-005	2934e-008	1360e-011	3196e-017
0.0600	-1409e-001	8780e-003	5487e-005	8817e-008	1344e-011	3892e-017
0.0625	-1481e-001	9482e-003	8630e-005	1768e-007	8867e-012	3084e-017
0.0650	-1547e-001	1013e-002	1262e-004	3037e-007	-3433e-012	-1487e-017
0.0675	-1608e-001	1070e-002	1756e-004	4786e-007	-2820e-011	-1377e-016
0.0700	-1661e-001	1117e-002	2361e-004	7125e-007	-7186e-011	-4009e-016
0.0725	-1706e-001	1153e-002	3088e-004	1017e-006	-1427e-010	-8985e-016
0.0750	-1743e-001	1174e-002	3951e-004	1405e-006	-2515e-010	-1760e-015
0.0775	-1770e-001	1178e-002	4966e-004	1891e-006	-4110e-010	-3154e-015
0.0800	-1787e-001	1163e-002	6146e-004	2487e-006	-6367e-010	-5270e-015
0.0825	-1792e-001	1125e-002	7504e-004	3207e-006	-9452e-010	-8308e-015
0.0850	-1785e-001	1062e-002	9057e-004	4065e-006	-1356e-009	-1237e-014
0.0875	-1766e-001	9702e-003	1082e-003	5071e-006	-1889e-009	-1745e-014
0.0900	-1732e-001	8462e-003	1280e-003	6236e-006	-2563e-009	-2317e-014
0.0925	-1684e-001	6868e-003	1502e-003	7564e-006	-3390e-009	-2861e-014
0.0950	-1620e-001	4882e-003	1748e-003	9059e-006	-4384e-009	-3187e-014
0.0975	-1540e-001	2466e-003	2020e-003	1072e-005	-5541e-009	-2960e-014
0.1000	-1442e-001	-4202e-004	2319e-003	1253e-005	-6850e-009	-1633e-014
0.1025	-1327e-001	-3817e-003	2645e-003	1447e-005	-8262e-009	1665e-014
0.1050	-1192e-001	-7768e-003	3000e-003	1652e-005	-9719e-009	8214e-014
0.1075	-1037e-001	-1232e-002	3384e-003	1864e-005	-1111e-008	1988e-013
0.1100	-8614e-002	-1751e-002	3796e-003	2076e-005	-1227e-008	3936e-013
0.1125	-6639e-002	-2340e-002	4238e-003	2281e-005	-1295e-008	7001e-013
0.1150	-4437e-002	-3003e-002	4708e-003	2470e-005	-1285e-008	1168e-012
0.1175	-1999e-002	-3745e-002	5206e-003	2632e-005	-1154e-008	1859e-012
0.1200	6841e-003	-4572e-002	5731e-003	2752e-005	-8465e-009	2859e-012
0.1225	3621e-002	-5488e-002	6282e-003	2812e-005	3097e+002	3290e+005
0.1250	6821e-002	-6501e-002	6856e-003	-1311e+035	2190e+061	-2128e+058
0.1275	1029e-001	-7615e-002	-1867e+038	7354e+116	6897e+140	-1089e+125
0.1300	1405e-001	-8836e-002	1360e+124	-1904e+391	1182e+385	9955e+353
0.1325	1809e-001	1818e+032	-4674e+199	-2869e+276	-2443e+301	3893e+283
0.1350	2244e-001	2442e+118	-1912e+287	1198e+370	1168e+383	-2289e+360
0.1375	2709e-001	5420e+192	1000e+355	8314e+421	6879e+443	-1048e+416
0.1400	3206e-001	8332e+284	1363e+447	4533e+996	-9408e+913	1199e+866

Table 4.2 DEND output for delta

alpha	k=-13	-14	-15	-16	-17	-18
0.0000	-5768e-056	-3586e-081	-5980e-117	-1006e-167	-1082e-239	-1193e-341
0.0025	-9077e-044	-4995e-063	-2164e-091	-1324e-130	5253e-185	-7522e-264
0.0050	3920e-041	-5281e-059	6882e-085	6489e-124	-1449e-173	1748e-247
0.0075	-7285e-040	-1839e-056	4242e-081	-1633e-118	-1589e-166	5544e-237
0.0100	1486e-037	3451e-054	3824e-078	-6885e-112	-3991e-160	-6368e-228
0.0125	-4093e-036	-1701e-052	-1293e-075	-1258e-109	-3347e-155	-3340e-221
0.0150	6339e-038	1696e-051	1582e-073	1277e-105	8952e-152	5513e-215
0.0175	4212e-034	1055e-049	-1356e-071	2926e-104	-2118e-147	-5767e-211
0.0200	9929e-034	-2113e-048	8952e-070	-2969e-100	-2256e-143	-1424e-204
0.0225	-1321e-032	-1252e-047	-1597e-068	-3692e-099	-5967e-141	-3004e-200
0.0250	-8853e-032	2758e-046	-3073e-067	-5200e-096	2213e-137	-3284e-196
0.0275	-1269e-031	1830e-045	1603e-065	-1979e-096	-1046e-134	-5160e-193
0.0300	9915e-031	-1054e-044	-3973e-065	4905e-094	2830e-132	1049e-188
0.0325	6852e-030	-1428e-043	-4301e-063	4208e-092	-5622e-130	5861e-186
0.0350	2090e-029	-3510e-043	5446e-063	4591e-089	1091e-127	-2513e-182
0.0375	2408e-029	2668e-042	6396e-061	9051e-090	-1741e-125	1921e-179
0.0400	-9447e-029	2493e-041	2590e-060	-3766e-086	1986e-123	-7861e-177
0.0425	-6606e-028	8140e-041	-4608e-059	-4389e-086	-1064e-121	-2031e-175
0.0450	-2275e-027	-3347e-041	-4284e-058	1483e-083	-1061e-120	5885e-172
0.0475	-5336e-027	-1640e-039	7201e-059	2562e-083	5605e-118	-1380e-169
0.0500	-7952e-027	-8747e-039	2802e-056	-4553e-081	-2922e-116	-3476e-167
0.0525	-3175e-028	-2495e-038	1499e-055	-3583e-081	-1868e-115	2475e-164
0.0550	4641e-026	-2341e-038	1524e-055	9508e-079	5209e-113	-7940e-162
0.0575	1965e-025	1714e-037	-3935e-054	-4526e-079	-6761e-112	1420e-159
0.0600	5630e-025	1166e-036	-4059e-053	-1070e-076	-4808e-110	-1974e-157
0.0625	1289e-024	4287e-036	-1105e-052	2655e-076	7980e-109	9179e-157
0.0650	2433e-024	1093e-035	4830e-053	1400e-075	2895e-107	1267e-153
0.0675	3635e-024	1655e-035	2789e-051	-3758e-074	-3845e-106	-1248e-151
0.0700	3359e-024	-8970e-036	2196e-050	7682e-073	-1537e-104	1958e-150
0.0725	-2558e-024	-1731e-034	5920e-050	7457e-074	6092e-104	5071e-148
0.0750	-2332e-023	-7615e-034	7545e-050	-4655e-071	5196e-102	-2489e-146
0.0775	-7676e-023	-2341e-033	-4817e-049	1247e-070	2625e-101	-7825e-145
0.0800	-1932e-022	-5731e-033	-5284e-048	-1023e-069	-1069e-099	1002e-142
0.0825	-4187e-022	-1097e-032	-1803e-047	-2950e-069	-1552e-098	-4161e-140
0.0850	-8131e-022	-1411e-032	-5298e-047	1077e-067	3230e-098	-9061e-137
0.0875	-1433e-021	2651e-033	-8731e-047	-2995e-067	2912e-096	2567e-133
0.0900	-2287e-021	9168e-032	1627e-046	1158e-066	2759e-095	1712e-129
0.0925	-3237e-021	3787e-031	1686e-045	4146e-066	-9890e-095	-6347e-127
0.0950	-3805e-021	1133e-030	8118e-045	-1061e-064	-4109e-093	2420e-122
0.0975	-2835e-021	2809e-030	2760e-044	2879e-064	-1275e-091	-5678e-120
0.1000	2093e-021	6440e-030	9545e-044	-1758e-063	2051e-088	-1952e-115
0.1025	1542e-020	1156e-029	1290e-043	-3486e-064	4374e-086	2746e-111
0.1050	4505e-020	1865e-029	5476e-044	4314e-062	5778e-083	3221e-107
0.1075	1039e-019	2244e-029	-8232e-043	-1854e-061	-6662e-081	-1460e-103
0.1100	2147e-019	7947e-030	-6287e-042	1320e-060	8981e-077	-2884e-099
0.1125	3995e-019	-6615e-029	-1720e-041	-3214e-059	9223e-074	-1968e-095
0.1150	7075e-019	-2894e-028	-5070e-041	7621e-057	-3612e-070	-4287e-091
0.1175	1190e-018	-8407e-028	-1250e-040	2736e-054	-4241e-067	-2131e-087
0.1200	1934e-018	-2191e-027	-3411e-040	-6999e-052	-9451e-063	-4793e-080
0.1225	1143e-015	-4552e-027	-3749e-040	-6294e-048	3322e-059	3406e-074
0.1250	-1214e+041	-7108e+006	2364e-037	1713e-045	-5160e-054	2292e-068
0.1275	3374e+092	1900e+060	4366e+023	-1953e-038	2868e-048	3242e-061
0.1300	4971e+320	-3184e+231	-1226e+103	-6484e+051	8706e-002	-7095e-047
0.1325	2699e+241	2610e+185	-2805e+137	-4119e+095	1403e+056	-2565e+025
0.1350	3577e+305	6097e+259	1005e+205	-4805e+157	1160e+146	-2642e+075
0.1375	-5815e+374	2461e+317	1716e+268	6086e+221	2134e+176	-1845e+179
0.1400	-6680e+873	1245e+768	-7329e+449	-1439e+400	1920e+320	5237e+266

Consider next the question of fitting together two solutions of (2.10), P(t) and Q(t), to form a candidate function

$$\theta^*(t) = \begin{cases} P(t), & (t > 0) \\ -Q(-t), & (t \leq 0) \end{cases}. \tag{2.11}$$

At first sight, it seems that we have two conditions, namely (2.6), to meet, whereas for fixed k we have only the single variable α at our disposal when searching for an eigenvalue $\alpha = \alpha(k)$ that will make the candidate function θ^* a solution of (1.1). On the other hand the pairs of conditions (2.6) reduce to the single condition $\Delta = 0$ in the special case P = Q. In any case, any solution of (1.1) must have an odd number of roots τ at which $\theta(\tau) = 0$; and we can find a translation that will ensure that $\tau = 0$ holds for the central zero of these zeros of θ. Thus we can classify the possible solutions of (1.1) by the odd number $2m - 1$ of their zeros. We may take P(t) to be an mth order solution of (1.2) choosing $\alpha = \alpha(k,m)$ to satisfy $P_2(0) = 0$. Then θ^* will be a solution of (1.1) with Q = P in (2.11). However, having found this solution θ^* with its central zero at t = 0, we may translate it so that any other of its zeros occurs at t = 0. We conclude that an m_1th order lower solution may be married successfully with an m_2th order lower solution provided that $m_1 + m_2 = 2m$ is even. Similarly a pair of upper solutions can be married provided that the sum of their orders is even. A pair of solutions, of opposite parity, can never be fitted together; and it is unlikely that an upper solution can be fitted to a lower solution inasmuch as this would demand a pair of conditions in (2.6).

To sum up the foregoing, the possible solutions of (1.1) comprise two families, according to whether $\theta(t) \downarrow \pi/2$ or $\theta(t) \uparrow \pi/2$ as $t \to \infty$. In each family, there is a countable infinity of possible solutions, classified by the number $2m - 1$ (m=1,2,...) of zeros possessed by the solution; and for each value of m and each value of k in each of the two families, there is a discrete spectrum of eigenvalues $\alpha = \alpha(k,m)$. The spectrum may be empty if $k \geq k_0(m)$.

A needle crystal solution requires that $\theta(t)$ is monotonic, and hence belongs to the family $\theta(t) \uparrow \pi/2$ as $t \to \infty$ and class number m = 1. In this paper, we shall deal only with this particular case, hereafter called the primary solution.

3. THE BASIS FOR DEND

Although DEND is intended to work for $0 \leq \alpha < 1/7$, it fails for α too near 1/7: for example when k = −8, it succeeds for $0 \leq \alpha \leq 0.1300$ but fails for $0.1325 < \alpha < 0.143$. When DEND does succeed, it provides an automatic check that the primary solution (1.2) is strictly monotone, and therefore any eigenvalue $\alpha(k)$ produced by DEND must yield a needle solution (1.3). One might guess that failure of DEND for α too near 1/7 results from non–monotonicity of the primary solution; but this guess is wrong, because STIFF confirms that the primary solution when k = −8 and $0.1325 < \alpha < 0.143$ is strictly monotone. We have not managed to pinpoint the reasons for failures of DEND when α nears 1/7 from below.

The linearization (1.4) shows that the primary solution $\theta(t)$ is monotone increasing for sufficiently large $t > t_0$ for some t_0, or in other words for some $\theta_0 < \theta(t) < \frac{1}{2}\pi$. For the time being we confine attention to this range in which $\theta_1 > 0$, and in which θ_1 is consequently a well–defined function of $\cos\theta$; but it will eventually appear that $t_0 = \theta_0 = 0$ whenever DEND succeeds. The linearization of (1.4) implies that

$$\theta_1 \sim \lambda \cos\theta \quad \text{as} \quad \theta \to \tfrac{1}{2}\pi \tag{3.1}$$

where λ is the real root of (1.5). Since $\theta_1 > 0$, we can write

$$\theta_3 = \theta_1 d^2(\tfrac{1}{2}\theta_1^2)/d\theta^2 . \tag{3.2}$$

We use the abbreviation

Table 5 Specimen indications of discontinuities

k	Δ(k,0.61)	First difference of Δ
-3.41201	-8217e-4	1748e-4
-3.41202	-6468e-4	1749e-4
-3.41203	-4719e-4	1749e-4
-3.41204	-2970e-4	1749e-4
-3.41205	-1222e-4	1749e-4
-3.41206	5274e-5	1749e-4
-3.41207	2276e-4	1749e-4
-3.41208	4025e-4	1749e-4
-3.41209	5774e-4	1749e-4
-3.41210	7523e-4	1749e-4
-5.80550	1190e+2	-3899e-2
-5.80551	1189e+2	-4222e-2
-5.80552	1189e+2	-4637e-2
-5.80553	1188e+2	-5198e-2
-5.80554	1188e+2	-6018e-2
-5.80555	-6742e+1	-1862e+2
-5.80556	-6742e+1	5384e-4
-5.80557	-6742e+1	5384e-4
-5.80558	-6742e+1	5384e-4
-5.80559	-6742e+1	5384e-4
-6.56527	-5480e-3	1310e-3
-6.56528	-4170e-3	1310e-3
-6.56529	-2860e-3	1310e-3
-6.56530	-1550e-3	1310e-3
-6.56531	-2404e-4	1310e-3
-6.56532	1070e-3	1310e-3
-6.56533	2379e-3	1310e-3
-6.56534	3689e-3	1310e-3
-6.56535	4999e-3	1310e-3
-6.56536	6309e-3	1310e-3

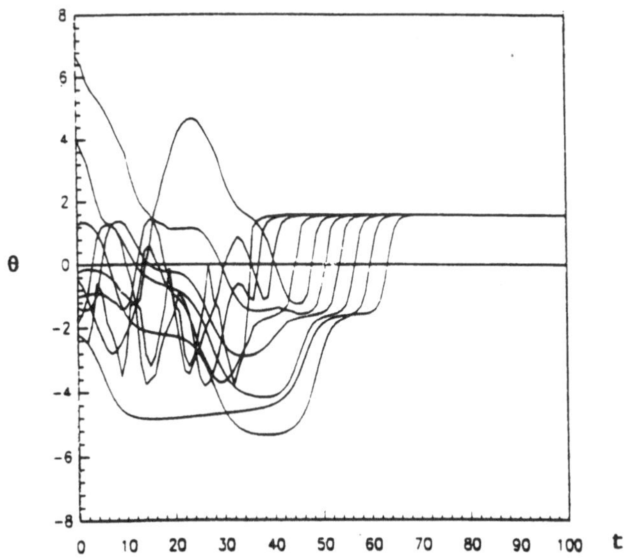

Figure 3. Specimen behavior of θ as a function of t

Table 6 Where a pair of values is given in Table 6, the upper entry of the pair relates to the case k=-5.80555 and the lower entry to the case k=-5.80554. Entries for x=9.22 to x=9.70 all relate to k=-5.80555.

x	θ	θ_1	θ_2
0.00	1568703417e+1	1291157636e-2	-7965409915e-3
	1568703417e+1	1291157944e-2	-7965411450e-3
2.00	1563607952e+1	4435270953e-2	-2737416389e-2
	1563607950e+1	4435271759e-2	-2737416765e-2
4.00	1546090264e+1	1526632407e-1	-9463687647e-2
	1546090260e+1	1526632597e-1	-9463688375e-2
6.00	1485192235e+1	5384414816e-1	-3517729183e-1
	1485192226e+1	5384415135e-1	-3517729166e-1
8.00	1237666707e+1	2724632505e+0	-3419256490e+0
	1237666716e+1	2724631689e+0	-3419253340e+0
9.00	2768936938e+0	2249898309e+1	4256399869e+1
	2768951088e+0	2249903556e+1	4256278776e+1
9.20	1914480931e-2	2163746453e+0	1235653917e+2
	1912597389e-2	2163981166e+0	1235654292e+2
9.22	2354268620e-4	-2531665626e-1	1176751577e+2
9.24	2830841797e-2	-2525728523e+0	1091571803e+2
9.26	9996292002e-2	-4603187426e+0	9820099803e+1
9.28	2108245306e-1	-6439115688e+0	8504842276e+1
9.30	3556418131e-1	-7992388727e+0	6998679432e+1
9.40	1359514428e+0	-1054482672e+1	-2160102400e+1
9.50	2155173073e+0	-3982918262e+0	-1039343566e+2
9.60	1957772382e+0	8405164186e+0	-1320330936e+2
9.70	5028790007e-1	1993755353e+1	-8717243421e+1
9.72405	0000000000e+0	2180170355e+1	-6741964407e+1
9.21686	0000000000e+0	1186528451e-1	1187791426e+2

$$c = \cos\theta\,, \tag{3.3}$$

and define the function

$$A(c^2) = 1 - \sum_{n=1}^{\infty} a_n c^{2n} = \lambda c/\theta_1 \tag{3.4}$$

in accordance with (3.1). Likewise we define

$$B(c^2) = \sum_{n=0}^{\infty} b_n c^{2n} = \theta_1^2/\lambda^2 c^2 \tag{3.5}$$

The series in (3.4) and (3.5) are formal series, whose convergence will be considered presently. From (3.4) and (3.5) we get

$$A^2 B = 1\,,\ b_0 = 1\,. \tag{3.6}$$

Hence comparing coefficients in a differentiated version of (3.6), namely $2A'B + AB' = 0$, we find

$$a_1 = \tfrac{1}{2}b_1\,;\ a_n = \tfrac{1}{2}(b_n - W_n)\,,\ (n = 2,3,\ldots)\,, \tag{3.7}$$

where
$$W_n = \sum_{r=1}^{n-1} (1+r/n) a_r b_{n-r}\,. \tag{3.8}$$

We define the parameters η, μ and β by

$$\eta = \epsilon\lambda^2\,,\ \mu = 1 + \eta^{-1}\,,\ \beta = 8\alpha/(1+\alpha)\,. \tag{3.9}$$

From (3.9) and (1.5), we find that η is the unique real root of

$$\eta(1+\eta)^2 = \epsilon/(1+\alpha)^2\,. \tag{3.10}$$

Also (3.3) and (3.9) give the identity

$$1 + \alpha\cos 4\theta = (1+\alpha)(1-\beta c^2+\beta c^4)\,. \tag{3.11}$$

By substituting (3.2), (3.3), (3.4), (3.5), (3.9) and (3.11) into (1.2) we get

$$\mu - 1 + \frac{d^2}{d\theta^2}[\tfrac{1}{2}c^2 B(c^2)] = \frac{\mu A(c^2)}{1-\beta c^2+\beta c^4}\,. \tag{3.12}$$

Since $d^2(\tfrac{1}{2}c^{2n+2})/d\theta^2 = (n+1)(2n+1)c^{2n} - 2(n+1)^2 c^{2n+2}$, (3.13)

we can compare coefficients in

$$\mu A(c^2) = \mu\left\{1 - \sum_{n=1}^{\infty} a_n c^{2n}\right\} = \mu\left\{1 - \tfrac{1}{2}b_1 c^2 + \tfrac{1}{2}\sum_{n=2}^{\infty} (W_n - b_n) c^{2n}\right\}$$

$$= (1-\beta c^2+\beta c^4)\left\{\mu - \sum_{n=1}^{\infty} [2n^2 b_{n-1} - (n+1)(2n+1) b_n] c^{2n}\right\} \tag{3.14}$$

This gives

$$b_1 = \frac{4+2\beta\mu}{12+\mu}\,, \tag{3.15}$$

Table 7 First 30 eigenvalues

k=-8	k=-9	k=-10	k=-11	k=-12	k=-13	k=-14	k=-15	k=-16	k=-17	k=-18	
7662e-4	1909e-4	4764e-5	1190e-5	2974e-6	7432e-7	1858e-7	4644e-8	1161e-8	2903e-9	7257e-10	1
2310e-2	5453e-3	1326e-3	3271e-4	8123e-5	2024e-5	5052e-6	1262e-6	3153e-7	7882e-8	1970e-8	2
1779e-1	3705e-2	8546e-3	2057e-3	5047e-4	1250e-4	3112e-5	7762e-6	1938e-6	4843e-7	1210e-7	3
9966e-1	1531e-1	3214e-2	7436e-3	1792e-3	4401e-4	1091e-4	2715e-5	6772e-6	1691e-6	4226e-7	4
	5254e-1	9283e-2	2024e-2	4756e-3	1154e-3	2844e-4	7059e-5	1758e-5	4388e-6	1096e-6	5
		2335e-1	4659e-2	1058e-2	2529e-3	6183e-4	1530e-4	3804e-5	9484e-6	2368e-6	6
		5559e-1	9646e-2	2093e-2	4907e-3	1190e-3	2930e-4	7270e-5	1811e-5	4519e-6	7
			1868e-1	3814e-2	8733e-3	2095e-3	5133e-4	1271e-4	3161e-5	7884e-6	8
			3485e-1	6552e-2	1457e-2	3452e-3	8411e-4	2076e-4	5158e-5	1286e-5	9
			6447e-1	1078e-1	2315e-2	5405e-3	1308e-3	3218e-4	7983e-5	1988e-5	10
				1721e-1	3537e-2	8120e-3	1950e-3	4782e-4	1184e-4	2946e-5	11
				2690e-1	5241e-2	1180e-2	2810e-3	6862e-4	1696e-4	4215e-5	12
				4163e-1	7578e-2	1668e-2	3935e-3	9567e-4	2359e-4	5858e-5	13
				6445e-1	1074e-1	2305e-2	5379e-3	1302e-3	3202e-4	7943e-5	14
				1015e+0	1500e-1	3123e-2	7204e-3	1734e-3	4255e-4	1054e-4	15
					2070e-1	4162e-2	9478e-3	2268e-3	5552e-4	1373e-4	16
					2833e-1	5470e-2	1228e-2	2920e-3	7127e-4	1761e-4	17
					3859e-1	7104e-2	1570e-2	3708e-3	9021e-4	2225e-4	18
					5254e-1	9132e-2	1982e-2	4650e-3	1128e-3	2777e-4	19
					7182e-1	1164e-1	2464e-2	5767e-3	1393e-3	3426e-4	20
					9926e-1	1473e-1	3069e-2	7082e-3	1705e-3	4184e-4	21
						1852e-1	3769e-2	8620e-3	2067e-3	5063e-4	22
						2319e-1	4596e-2	1041e-2	2485e-3	6075e-4	23
						2893e-1	5563e-2	1247e-2	2964e-3	7233e-4	24
						3602e-1	6698e-2	1485e-2	3512e-3	8551e-4	25
						4480e-1	8021e-2	1757e-2	4135e-3	1004e-3	26
						5577e-1	9560e-2	2067e-2	4840e-3	1173e-3	27
						6961e-1	1135e-1	2420e-2	5635e-3	1362e-3	28
						8732e-1	1342e-1	2820e-2	6528e-3	1574e-3	29
						1105e+0	1582e-1	3271e-2	7526e-3	1809e-3	30

	k=-3	k=-4	k=-5	k=-6	k=-7
1	1261e+0	2257e-1	5185e-2	1253e-2	3087e-3
2	4705e+0	7269e+0	8410e+0	5718e-1	1048e-1
3				9044e+0	1194e+0
4					7815e+0
5					9417e+0

$$b_2 = \frac{16b_1+\mu W_2+2\beta\left[6b_1-\mu-2\right]}{30+\mu}, \tag{3.16}$$

and, for $n \geq 3$,

$$[(n+1)(2n+1)+\tfrac{1}{2}\mu]b_n$$
$$= 2n^2b_{n-1} + \tfrac{1}{2}\mu W_n + \beta\{n(2n-1)b_{n-1} - (n-1)(4n-5)b_{n-2} + 2(n-2)^2b_{n-3}\} . \tag{3.17}$$

The recurrence relations (3.7) and (3.17) allow us to calculate b_1, a_1, b_2, $a_2,\dots,b_n$, $a_n,\dots$ in that order; and hence to obtain the quantities

$$A_n = n^3/^2a_n,\ B_n = n^3/^2b_n,\ (n=1,2,\dots) . \tag{3.18}$$

We say that DEND <u>succeeds</u> if and only if the sequences $\{A_n\}$, $\{B_n\}$ are bounded as $n \to \infty$; and otherwise DEND <u>fails</u>. In [2] we gave a rigorous proof that DEND must succeed when $\alpha = 0$. We have not managed to find a rigorous argument for $\alpha > 0$.

However we prove presently that, if

$$A_n = 0(1),\ B_n = 0(1) \text{ as } n \to \infty \tag{3.19}$$

then there exists a constant B_∞ (depending on k and α) such that

$$B_n = B_\infty + 0(n^{-1}) \text{ as } n \to \infty . \tag{3.20}$$

In practice it is easy to tell from the computer output whether (3.20) holds or whether the sequence $\{B_n\}$ diverges to $\pm\infty$. Before proceeding to the proof that (3.19) implies (3.20), we examine the consequences of (3.19) and (3.20).

Firstly (3.19) implies that

$$a_n = 0(n^{-3}/^2),\ b_n = 0(n^{-3}/^2) \text{ as } n \to \infty , \tag{3.21}$$

and therefore the functions A(z) and B(z) are analytic for $|z| < 1$ and continuous for $|z| \leq 1$. Moreover we cannot have $A(z) = 0$ anywhere in $|z| \leq 1$ for that would contradict (3.6). Since $A(0) = 1$, it follows that $A(c^2) > 0$ for $0 \leq c \leq 1$; and hence $\theta_1 > 0$ for whenever $0 \leq 0 < \tfrac{1}{2}\pi$. Thus the quantities t_0 and θ_0 introduced in the second paragraph of §3 are both zero whenever DEND succeeds; and the resulting primary solution $\theta(t)$ is strictly monotonic and (1.3) will give a needle solution whenever $\Delta = 0$. Secondly because A(z) and B(z) are analytic for $|z| < 1$, the formal manipulations leading to (3.7) and (3.17) are validated. Thirdly, $1 - \beta z + \beta z^2$ cannot vanish for $|z| \leq 1$ for otherwise B(z) would not be continuous for $|z| \leq 1$. The zeros of $1 - \beta z + \beta z^2$ are

$$\zeta_1,\zeta_2 = \tfrac{1}{2} \pm i\surd(\beta^{-1} - \tfrac{1}{4}) ; \tag{3.22}$$

and therefore $|\zeta|^2 = \beta^{-1} > 1$, which implies $0 \leq \alpha < 1/7$. Hence DEND cannot succeed for $\alpha \geq 1/7$. Fourthly, (3.20) implies that

$$b_n = B_\infty n^{-3}/^2 + 0(n^{-5}/^2) \text{ as } n \to \infty . \tag{3.23}$$

Now

$$\sin\theta = \surd(1-c^2) = 1 - \sum_{n=1}^{\infty} \frac{\gamma_n c^{2n}}{2n-1}, \tag{3.24}$$

where
$$\gamma_n = 4^{-n}\binom{2n}{n} = (\pi n)^{-\frac{1}{2}}[1+0(n^{-1})] \text{ as } n \to \infty. \tag{3.25}$$

Table 8.1 Logarithmic Spectra

k=-8	k=-9	k=-10	k=-11	k=-12	k=-13	k=-14	k=-15	k=-16	k=-17	k=-18	
13.6719	15.6769	17.6794	19.6807	21.6813	23.6816	25.6818	27.6818	29.6819	31.6819	33.6819	1
8.7578	10.8406	12.8805	14.8999	16.9096	18.9144	20.9168	22.9180	24.9186	26.9189	28.9190	2
5.8125	8.0763	10.1925	12.2474	14.2741	16.2873	18.2939	20.2971	22.2987	24.2995	26.2999	3
3.3268	6.0295	8.2815	10.3931	12.4460	14.4717	16.4844	18.4908	20.4939	22.4955	24.4963	4
	4.2505	6.7512	8.9483	11.0378	13.0808	15.1017	17.1120	19.1173	21.1198	23.1211	5
		5.4204	7.7457	9.8846	11.9493	13.9813	15.9964	18.0041	20.0080	22.0099	6
		4.1690	6.6958	8.9004	10.9930	13.0373	15.0590	17.0696	19.0749	21.0776	7
			5.7424	8.0345	10.1612	12.2208	14.2497	16.2640	18.2710	20.2746	8
			4.8428	7.2538	9.4231	11.5001	13.5373	15.5557	17.5647	19.5692	9
			3.9551	6.5352	8.7548	10.8534	12.9006	14.9234	16.9347	18.9403	10
				5.8610	8.1431	10.2662	12.3240	14.3521	16.3660	18.3729	11
				5.2161	7.5758	9.7269	11.7971	13.8310	15.8476	17.8559	12
				4.5864	7.0440	9.2275	11.3113	13.3516	15.3713	17.3811	13
				3.9556	6.5405	8.7612	10.8603	12.9075	14.9306	16.9420	14
				3.3009	6.0591	8.3229	10.4389	12.4936	14.5205	16.5336	15
					5.5946	7.9084	10.0431	12.1063	14.1368	16.1522	16
					5.1417	7.5141	9.6695	11.7415	13.7764	15.7934	17
					4.6956	7.1371	9.3154	11.3971	13.4363	15.4556	18
					4.2505	6.7748	8.9785	11.0705	13.1145	15.1361	19
					3.7994	6.4253	8.6648	10.7599	12.8090	14.8330	20
					3.3326	6.0855	8.3480	10.4635	12.5181	14.5447	21
						5.7546	8.0516	10.1801	12.2404	14.2697	22
						5.4306	7.7655	9.9083	11.9747	14.0069	23
						5.1113	7.4899	9.6471	11.7200	13.7551	24
						4.7952	7.2221	9.3955	11.4752	13.5135	25
						4.4803	6.9620	9.1528	11.2397	13.2813	26
						4.1643	6.7087	8.9181	11.0126	13.0577	27
						3.8446	6.4616	8.6908	10.7933	12.8420	28
						3.5175	6.2197	8.4703	10.5811	12.6337	29
						3.1782	5.9824	8.2561	10.3759	12.4323	30

	k=-3	k=-4	k=-5	k=-6	k=-7
1	2.9879	5.4697	7.5914	9.6404	11.6615
2				4.1283	6.5768
3					3.0663
4					

Consequently from (3.5), (3.9), (3.23) and (3.24)

$$\tfrac{1}{2}\theta_1^2 = \tfrac{1}{2}\lambda^2 c^2 B(c^2) = -\frac{\eta\pi^{\frac{1}{2}}B_\infty}{\epsilon}\sin\theta + D(c^2) \tag{3.26}$$

where $D(z)$ is continuously differentiable for $|z| \leq 1$. Hence

$$\Delta = \theta_2(0) = \lim_{\theta\to 0}\frac{d}{d\theta}(\tfrac{1}{2}\theta_1^2) = -\eta\pi^{\frac{1}{2}}B_\infty/\epsilon . \tag{3.27}$$

4. PROOF THAT (3.19) IMPLIES (3.20)

We fix k and α, thereby fixing μ and β. Let

$$m_r = 1 + \frac{3}{2r} + \frac{\mu+2}{4r^2}; \quad M_n = \prod_{r=1}^{n} m_r . \tag{4.1}$$

Then there is a strictly positive constant $L = L(\mu)$ such that

$$M_n = Ln^{3/2}[1+0(n^{-1})] \text{ as } n \to \infty . \tag{4.2}$$

We define $$g_n = M_n b_n . \tag{4.3}$$

Now (3.19) implies that $a_n = 0(n^{-3/2})$ as $n \to \infty$, and hence $W_n = 0(n^{-3/2})$ by virtue of (3.7). So

$$g_n = 0(1) , \ \mu W_n M_{n-1} = 0(1) \ \text{ as } n \to \infty . \tag{4.4}$$

$$\begin{aligned} g_n &= g_{n-1} + \frac{\mu W_n M_{n-1}}{4n^2} + \beta\left\{\left[1 - \frac{1}{2n}\right]g_{n-1} - \left[1 - \frac{1}{n}\right]\left[2 - \frac{5}{2n}\right]m_{n-1}g_{n-2} + \right. \\ &\quad \left. + \left[1 - \frac{2}{n}\right]^2 m_{n-1}m_{n-2}g_{n-3}\right\} \\ &= g_{n-1} + \beta\left\{\left[1 - \frac{1}{2n}\right]g_{n-1} - \left[2 - \frac{3}{2n}\right]g_{n-2} + \left[1 - \frac{1}{n}\right]g_{n-3}\right\} + 0(n^{-2}) . \end{aligned} \tag{4.5}$$

Hence

$$\begin{aligned} g_n - \beta\left\{\left[1 - \frac{1}{2n}\right]g_{n-1} - \left[1 - \frac{1}{n}\right]g_{n-2}\right\} &= g_{n-1} - \beta\left\{\left[1 - \frac{1}{2n-2}\right]g_{n-2} - \left[1 - \frac{1}{n-1}\right]g_{n-3}\right\} + 0(n^{-2}) \\ &= \sum_{r=1}^{n} 0(r^{-2}) = q + 0(n^{-1}) , \end{aligned} \tag{4.6}$$

for some constant q. Then (4.4) and (4.6) give

$$h_n = g_n - \beta g_{n-1} + \beta g_{n-2} = q + 0(n^{-1}) \text{ as } n \to \infty . \tag{4.7}$$

Next define

$$G(z) = \sum_{n=0}^{\infty} g_n z^n , \ H(z) = \sum_{n=0}^{\infty} h_n z^n \tag{4.8}$$

These functions are analytic for $|z| < 1$ because of (4.4) and (4.7). Also by (4.7), provided we define $g_{-1} = g_{-2} = 0$, we have

$$H(z) = (1-\beta+\beta z^2)G(z) \tag{4.9}$$

Then we can compare coefficients in

$$G(z) = (1-\beta z+\beta z^2)^{-1}H(z) . \tag{4.10}$$

According to (3.22), the roots of $1 - \beta z + \beta z^2$ take the form $\beta^{-\frac{1}{2}}e^{\pm i\phi}$ where ϕ is a constant (depending on the fixed β) such that $0 < \phi < \frac{1}{2}\pi$. Hence

$$\begin{aligned}(1-\beta z+\beta z^2)^{-1} &= \frac{1}{e^{i\phi}-e^{-i\phi}}\left[\frac{e^{i\phi}}{1-\beta^{\frac{1}{2}}e^{i\phi}z} - \frac{e^{-i\phi}}{1-\beta^{\frac{1}{2}}e^{-i\phi}z}\right] \\ &= \frac{1}{\sin\phi}\sum_{n=0}^{\infty}(\beta^{\frac{1}{2}}z)^n\sin(n+1)\phi \end{aligned} \tag{4.11}$$

This converges for $z = 1$ since $\beta < 1$, and therefore

$$\frac{1}{\sin\phi}\sum_{n=0}^{\infty}\beta^{n/2}\sin(n+1)\phi = 1 , \tag{4.12}$$

and

$$\frac{1}{\sin\phi}\sum_{r=0}^{n}\beta^{r/2}\sin(r+1)\phi = 1 + 0(\beta^{n/2}) . \tag{4.13}$$

From (4.7), (4.10), (4.11) and (4.13) we have

$$\begin{aligned} g_n &= \sum_{r=0}^{n} h_{n-r}\beta^{r/2}\sin(r+1)\phi\,\mathrm{cosec}\,\phi \\ &= q[1+0(\beta^{n/2})] + 0\left[\sum_{r=0}^{n}\beta^{r/2}/(n-r+1)\right] \\ &= q + 0(n^{-1}) \quad \text{as } n \to \infty . \end{aligned} \tag{4.14}$$

Then (4.2), (4.3) and (4.14) yield (3.20) with $B_\infty = q/L$, as required.

5. RATES OF CONVERGENCE OF DEND

When DEND succeeds, Δ is given by (3.27) where

$$B_\infty = \lim_{n\to\infty} B_n . \tag{5.1}$$

In practice the computer only provides a finite number of terms $B_1, B_2, \ldots, B_N$ from which B_∞ has to be obtained by some sort of extrapolation. For small ϵ the sequence $\{B_n\}$ converges slowly and methods of improving the rate of convergence are desirable. One method is a first–order Richardson extrapolate, replacing B_n by

$$B_n^* = nB_n - (n-1)B_{n-1} . \tag{5.2}$$

If

$$B_n = B_\infty + \sigma n^{-1} + \tau n^{-2} \tag{5.3}$$

for constants σ and τ, then

$$B_N^* = B_\infty - \tau/N(N-1) \tag{5.4}$$

Table 8.2 Horizontal Differences

k=-9\8	-10\9	-11\10	-12\11	-13\12	-14\13	-15\14	-16\15	-17\16	-18\17	
2.0051	2.0025	2.0012	2.0006	2.0003	2.0002	2.0001	2.0000	2.0000	2.0000	1
2.0829	2.0398	2.0195	2.0097	2.0048	2.0024	2.0012	2.0006	2.0003	2.0001	2
2.2638	2.1162	2.0549	2.0267	2.0132	2.0066	2.0032	2.0016	2.0008	2.0004	3
2.7027	2.2519	2.1116	2.0529	2.0258	2.0127	2.0063	2.0031	2.0016	2.0008	4
	2.5008	2.1970	2.0896	2.0430	2.0208	2.0104	2.0053	2.0025	2.0013	5
		2.3252	2.1390	2.0646	2.0320	2.0151	2.0077	2.0038	2.0019	6
		2.5268	2.2046	2.0926	2.0442	2.0217	2.0106	2.0053	2.0027	7
			2.2921	2.1267	2.0596	2.0290	2.0142	2.0071	2.0035	8
			2.4111	2.1693	2.0770	2.0372	2.0184	2.0090	2.0045	9
			2.5801	2.2196	2.0986	2.0472	2.0228	2.0114	2.0056	10
				2.2821	2.1231	2.0578	2.0281	2.0139	2.0069	11
				2.3597	2.1511	2.0701	2.0339	2.0167	2.0083	12
				2.4577	2.1835	2.0838	2.0402	2.0197	2.0098	13
				2.5849	2.2207	2.0991	2.0472	2.0231	2.0114	14
				2.7582	2.2639	2.1160	2.0547	2.0268	2.0131	15
					2.3138	2.1347	2.0631	2.0305	2.0154	16
					2.3724	2.1554	2.0720	2.0348	2.0170	17
					2.4415	2.1783	2.0817	2.0392	2.0192	18
					2.5243	2.2037	2.0920	2.0440	2.0215	19
					2.6259	2.2395	2.0950	2.0492	2.0240	20
					2.7529	2.2625	2.1155	2.0546	2.0266	21
						2.2970	2.1285	2.0603	2.0293	22
						2.3349	2.1428	2.0664	2.0321	23
						2.3787	2.1572	2.0729	2.0352	24
						2.4269	2.1734	2.0797	2.0383	25
						2.4817	2.1908	2.0869	2.0416	26
						2.5444	2.2093	2.0945	2.0451	27
						2.6170	2.2292	2.1025	2.0487	28
						2.7021	2.2507	2.1107	2.0527	29
						2.8042	2.2737	2.1198	2.0565	30

	k=-4\3	k=-5\4	k=-6\5	k=-7\6	k=-8\7
1	2.4818	2.1217	2.0490	2.0211	2.0104
2				2.4485	2.1810
3					2.7462

and the error in using B_N^* as an estimate of B_∞ will be $O(N^{-2})$ instead of $O(N^{-1})$ that occurs in (3.20). Higher order Richardson extrapolates are available, but rather dangerous to use if the sequence $\{B_n\}$ oscillates (as indeed it tends to do as α increases). So we contented ourselves with (5.2) and took B_N^* as our estimate of B_∞ in calculating Δ. This seemed satisfactory for $k \geq -18$; but below $k = -18$ even B_N^* became untrustworthy for the values of N that were feasible on the computer.

A better method was possible in [2] for the special case $\alpha = 0$. In that case

$$B_\infty = \tfrac{1}{2}\pi^{\frac{1}{2}} \prod_{n=1}^{\infty} \frac{(2n+1)(2n+2)}{(2n+1)(2n+2)+\mu(1-\rho_n)}\ , \quad \text{where } 1-\rho_n = 2a_n/b_n\ . \qquad (5.5)$$

It was found empirically that the sequence $(1-\rho_n)$ converged to a limit much faster than the sequence $\{B_n\}$, and this permitted accurate estimates of $\Delta(k,0)$ down to $k = -25$: details of the procedure appear in [2]. We had hoped that a similar procedure would work for $\alpha > 0$; but unfortunately the sequence $\{1-\rho_n\}$ does not converge so well when $\alpha > 0$, and even if it were better behaved it is not clear how it could be accommodated in the appropriate generalisation of (5.5). The problem of discovering a fast extrapolatory technique for $\alpha > 0$ therefore deserves further research.

6. CALCULATION AND STRUCTURE OF THE EIGENVALUES

For fixed k, let $0 < \alpha_1(k) < \alpha_2(k) < \ldots$ denote the successive eigenvalues of α that ensure $\Delta = 0$. When using STIFF, the procedure for locating an eigenvalue was to calculate $\Delta(k,\alpha)$ for $\alpha = 0.00(0.01)0.98$ looking for a change in the sign of Δ. When a pair of successive values of α was detected, the computer, automatically divided the difference between successive values of α by a factor of 10 and recalculated Δ for the new α–spacing between the given pair, and so on with successive divisions by a factor of 10 until two values of α differing by 10^{-5} were found that caused a change of sign in Δ. First and second differences of Δ with respect to this finest α–spacing were then printed out by the computer to indicate whether or not there was a discontinuity at the located eigenvalue α; and if there was no discontinuity the eigenvalue was accepted as correct to 5 places of decimals, or rejected in the case of a discontinuity. Eigenvalues were also found by STIFF by fixing α and looking for changes of sign in Δ for variable k, using a similar refining process whose output is illustrated in Table 5 for the fixed value $\alpha = 0.61$. All STIFF calculations of eigenvalues were restricted to the range $k > -7$.

For $k \leq -8$, DEND was necessary; and in one sense it became easier to search for eigenvalues, but in another sense much harder. It was easier because DEND, when successful, automatically guaranteed that the solution was monotonic and so there was no longer any need to check for discontinuities. It was harder because the eigenvalues became more numerous, more closely packed, and smaller as ϵ decreased; and hence an initially coarse mesh search contains too many changes of sign in Δ to be workable. Moreover DEND takes too long a time to calculate Δ accurately, if many trial values of α are required. To surmount these difficulties, we noticed empirically that the sign of Δ, as opposed to an accurate estimate of its magnitude, required far fewer coefficients b_n: in computing Δ, the coefficients a_n and b_n are required for $1 \leq n \leq N$ and the time taken by the calculation is proportional to N^2. When $N = 17000$, as is typically required for an accurate estimate of the magnitude of Δ, the computing time is about 1 hour. On the other hand, the sign of b_n and hence the sign of Δ often remains constant for $n \geq 1000$ or $n \geq 2000$. Thus, in a search confined to checking merely the sign of Δ, it suffices to take $N = 1000$ or 2000, say, given a watch on the computer screen to ensure that the sign of b_n has finally settled down. This reduces the computing time for each trial value of α to 1 minute or less. The next point is to know where to search, that is to say to know what trial value of α will bracket each wanted eigenvalue without bracketing any other

Table 8.3 Vertical Differences

k=-9	k=-10	k=-11	k=-12	k=-13	k=-14	k=-15	k=-16	k=-17	k=-18	
4.8363	4.7990	4.7808	4.7717	4.7673	4.7650	4.7639	4.7633	4.7631	4.7629	1\2
2.7643	2.6880	2.6525	2.6355	2.6271	2.6228	2.6209	2.6198	2.6193	2.6191	2\3
2.0468	1.9110	1.8543	1.8281	1.8155	1.8095	1.8063	1.8048	1.8040	1.8037	3\4
1.7790	1.5302	1.4448	1.4081	1.3909	1.3828	1.3787	1.3766	1.3757	1.3752	4\5
	1.3308	1.2026	1.1532	1.1316	1.1203	1.1156	1.1132	1.1118	1.1112	5\6
	1.2514	1.0499	0.9842	0.9563	0.9441	0.9375	0.9345	0.9330	0.9323	6\7
		0.9534	0.8659	0.8318	0.8165	0.8092	0.8057	0.8039	0.8030	7\8
		0.8997	0.7807	0.7381	0.7207	0.7124	0.7083	0.7063	0.7053	8\9
		0.8876	0.7186	0.6683	0.6467	0.6367	0.6323	0.6300	0.6289	9\10
			0.6742	0.6117	0.5872	0.5766	0.5712	0.5687	0.5674	10\11
			0.6449	0.5673	0.5393	0.5269	0.5212	0.5184	0.5170	11\12
			0.6297	0.5318	0.4995	0.4858	0.4794	0.4763	0.4748	12\13
			0.6308	0.5036	0.4663	0.4510	0.4441	0.4407	0.4391	13\14
			0.6547	0.4814	0.4383	0.4214	0.4139	0.4101	0.4084	14\15
				0.4645	0.4145	0.3958	0.3874	0.3837	0.3814	15\16
				0.4528	0.3943	0.3736	0.3647	0.3604	0.3588	16\17
				0.4461	0.3770	0.3542	0.3445	0.3401	0.3378	17\18
				0.4451	0.3623	0.3369	0.3266	0.3218	0.3195	18\19
				0.4511	0.3495	0.3137	0.3106	0.3055	0.3031	19\20
				0.4668	0.3398	0.3168	0.2963	0.2909	0.2883	20\21
					0.3309	0.2964	0.2835	0.2777	0.2750	21\22
					0.3241	0.2861	0.2718	0.2657	0.2628	22\23
					0.3193	0.2756	0.2612	0.2548	0.2517	23\24
					0.3161	0.2678	0.2516	0.2447	0.2416	24\25
					0.3149	0.2602	0.2428	0.2356	0.2322	25\26
					0.3160	0.2532	0.2347	0.2271	0.2236	26\27
					0.3198	0.2472	0.2273	0.2193	0.2157	27\28
					0.3270	0.2419	0.2205	0.2122	0.2083	28\29
					0.3393	0.2373	0.2142	0.2052	0.2014	29\30

	k=-6	k=-7	k=-8
1\2	5.5121	5.0847	4.9141
2\3		3.5104	2.9452
3\4			2.4857

unwanted eigenvalues. The key to this was the observation in figure 1 that $-\log_2\alpha_1(k)$ was roughly a linear function of k; and this suggested that we might take, for each fixed value of j, $[\alpha_j(k+1)]^2/\alpha_j(k+2)$ as a rough predictor of $\alpha_j(k)$; and that similar ratio predictors might be used for trial bracketing values of α. This empirical procedure of tracking $\alpha_j(k)$ as a function of k, for each fixed j, worked well. It involved an initial search for $\alpha_j(k)$ when it first appeared below DEND's threshold $\alpha < 0.14$, where the eigenvalues are not so closely packed together and rough positions of sign change could be guessed from Tables 4.1 and 4.2. Thereafter ratio estimates were used to guide the building of Table 7 from left to right for decreasing values of k, for each fixed j. Instead of the automatic search procedure used by STIFF, we embedded the basic programme for DEND in an interactive search procedure with manual on–line control allowing various options:

(i) manual setting of N, and the ability to vary N in the light of sequences $\{B_n\}$ together with their second–order Richardson extrapolations appearing on the computer screen;
(ii) search for a sign change in Δ using the manual setting of an initial α and a manual setting of a step size in α until a sign change in Δ was detected;
(iii) ratio prediction of $\alpha_j(k)$, and tracking of $\alpha_j(k)$ for decreasing k once the first two values of k had been found by (i) and (ii);
(iv) manual trial settings of α, and/or interval halving, and/or automatic interpolation/extrapolation in revisions of α in the light of previously estimated values of Δ.

After a little experience of learning the emergent pattern of eigenvalues, we were able to determine each eigenvalue to 4 or 5 significant decimal digits at a rate of roughly 5 minutes per eigenvalue. The results for the first 30 eigenvalues appear in Table 7. The structure of these eigenvalues is most readily appreciated from Table 8.1, which gives $-\log_2\alpha_j(k)$ for $j = 1,2,\ldots,30$ and $k = -3,4,\ldots,-18$, and Tables 8.2 and 8.3 which give the horizontal and vertical differences of Tables 8,1. These horizontal differences tend to 2 as $k \to -\infty$; which implies that

$$\alpha_j \sim \Psi_j \epsilon^2 \text{ as } \epsilon \to 0 , \tag{6.1}$$

where $\Psi_1 = 4.99$, $\Psi_2 = 135$, $\Psi_3 = 831,\ldots$ It would be interesting to find a more precise asymptotic formula in place of (6.1), which might be possible from scrutiny of Tables 8.2 and 8.3. In [1], Kruskal and Segur (allowing for their different notation of ϵ), expected the spacing between successive eigenvalues for fixed k to be proportional to $\epsilon^{1/2}$. This contrasts with (6.1), where the spacing is proportional to ϵ^2. However (6.1) assumes that j is fixed; and the situation might be different if j is allowed to be a function of ϵ. Figure 2 displays curves for the first 30 eigenvalues.

ACKNOWLEDGEMENTS

We thank Professor Segur for communicating the prepublication version of [1]. We are also indebted to Dr. C.R. Prior for checking some of our STIFF results with the alternative NAG routine DO2CBF on the IBM 3090 at the Rutherford Laboratory, Chilton.

REFERENCES

[1] M.D.KRUSKAL and H. SEGUR, "Asymptotics beyond all orders in a model of crystal growth," Stud. App. Math. (to appear).

[2] J.M. HAMMERSLEY and G. MAZZARINO, "A differential equation connected with the dendritic growth of crystals," IMA J. Appl. Math. (1989) **42**, 43–75.

[3] J.M. HAMMERSLEY and G. MAZZARINO, "Computational aspects of some autonomous differential equations." Proc. Roy. Soc. (1989) A**242**, 19–37.

DENDRITIC CRYSTAL GROWTH: OVERVIEW

Herbert Levine

Institute for Nonlinear Science, 0402
University of California, San Diego
La Jolla, California 92093

I. Introduction

The phenomenon of dendritic crystal growth is one of the earliest scientific problems, tackled first by Kepler[1] in his work on six-sided snowflake crystals. Nowadays, the dendrite system is most often tackled by metallurgists who have learned to relate mechanical properties of a solidified alloy to the microstructure patterns formed via the crystal growth process. Most recently, however, physicists and mathematicians have focused on this system as a good example of pattern selection and stability in non-equilibrium systems - it is this last perspective which we will try to explain in this overview.

For simplicity, let us imagine the growth of a solid, crystalline phase from a single seed placed in a super cooled melt[2]. At the atomic scale, atoms must preferentially attach to the growing lattice, due to local non-equilibrium forcing. Depending on the nature of the liquid-solid interface, this can either be easy or hard. It will be easy when the interface is rough, exhibiting large (10 Å–100 Å) deviations from a plane - this is typically the case for metallic alloys and plastic crystals, at normal solidification temperatures. If the interface is indeed rough, attachment kinetics will only play a minor role in determining the growth. Instead, macroscopic transport will be of primary importance.

Again for simplicity, let us consider the case where the only transport process is diffusion. Then, the most important growth consideration is the fact that the latent heat released by crystallization must diffuse away from the interface into the bulk liquid[2]. For non-pure materials, diffusion of impurities rejected by the solid plays a similar role. Because growth is coupled to diffusion, simple shapes (spheres, planes...) will not remain simple due to the well-known Mullins-Sekerka instability[3]. Roughly, outwardly protruding bumps are more effective at radiating heat (or impurities) than the surrounding regions; hence they grow more quickly and the interface becomes quite complex.

Asymptotics beyond All Orders, Edited by H. Segur *et al.*
Plenum Press, New York, 1991

The single dendrite is an isolated structure which often emerges following the onset of the Mullins-Sekerka instability[4]. Here, a dendrite tip moves at constant speed while maintaining a parabolic shape for about five tip radii. Beyond that distance, a complicated sidebranching pattern emerges and the pattern can no longer be described as steady in some moving frame. Quite a bit of work has studied the dendrite tip problem and this will be reviewed below. A more detailed exposition of dendrite phenomenology will be given by Jerry Gollub in his paper.

It is important to realize that the free dendrite problem is one of a variety of related problems which arise in this system. In directional solidification[2], one places the crystallization process in an externally determined gradient. This leads to periodic cell patterns[5] and interesting dynamical mechanisms of wavelength selection. When we deal with alloy systems, we can often arrange the concentration in the liquid to be such that the two solid phases will grow simultaneously; this is called eutectic growth[6]. These systems will be the subject of the paper of Heinar Muller-Krumbhaar.

II. Basic Model

As we have already mentioned, the basic physics our model deals with is the diffusion of heat away from the growing crystal. Thus, the temperature obeys

$$D\nabla^2 T = \frac{\partial T}{\partial t}; \qquad T(\infty) \sim T_M - \Delta. \tag{1}$$

Here D is the diffusion constant, T_M the melting temperature and Δ the undercooling. We will measure all temperatures in units of L/C_p; L latent heat and C_p the specific heat. This equation fails to allow for bulk fluid motion which can sometimes be quantitatively important.

At the actual liquid-solid interface, we impose two boundary conditions. The first is the classical Stefan condition

$$[D\hat{n} \cdot \nabla T]_{s-l} = v_n \tag{2}$$

where $\hat{n}$ is the interface normal, $v_n \equiv \vec{v} \cdot \hat{n}$, and $[]_{s-l}$ means discontinuity as we cross from solid to liquid. This is just the statement of conservation of energy. The second condition we take is that of local thermodynamic equilibrium; this essentially states the local attachment process could occur much more quickly than the diffusion-limited growth actually requires. This yields

$$T = T_M(1 - d_0\kappa) \tag{3}$$

where κ is the (mean) curvature and the capillary length $d_0 = \frac{\gamma(\theta)}{L}$ for surface tension γ. We have included the functional dependence on $\theta \equiv \cos^{-1} \hat{n} \cdot \hat{y}$ because the surface tension is actually anisotropic due to the underlying crystal structure. To be precise, the above relation is only valid in two dimensions with a more complicated generalization[7] to the real three dimensional situation.

In comparing predictions from this model to experimented data, one must keep in mind that there are two distinct sources of error. The first, which we will loosely characterize as wrong mathematics, is due to faulty analyses of the predictions of the model equations. This was the case in say, the work of Nash and Glicksman[8] who wrongly concluded that these equations admit a continuous family of steady-state solutions (see the next section). Another possible source of wrong mathematics is the use of an irrelevant solution, perhaps by the wrong choice of boundary conditions for the physical problem; this might be the case for dendritic growth where we look for infinitely long steady-state shapes even though the actual dendrite exhibits sidebranching as we move away from the tip. Wrong mathematics can be detected by consistency checks such as comparing analytical and computational results, long before any serious attempt is made to deal with experimental data. Conversely, it makes no sense to validate models with experimental data if the aforementioned tests indicate that the mathematics is incorrect.

The second type of error is wrong or more likely incomplete physics. In many experiments, the aforementioned model will be insufficient due to kinetic effects, for example. Minimizing this source of disagreement, while a serious challenge to both the theorist and the experimenter, is a well-defined interactive process once the basic mechanisms (i.e. the mathematics) have been understood. Of course this process requires that the original model does indeed model the relevant phenomenon; we believe this to be true of equations (1)–(3) via a combination of heuristic arguments and computer simulation evidence. We may of course be wrong; we will of course provisionally assume that we are not.

Before proceeding to explain current theories of dendritic growth based on the model presented here, we would like to provide a brief historical aside. The idea of looking for steady-state solutions originated with Ivantsov[9] in 1947. Since he did not include the effects of surface tension, the velocity of his solution could be set arbitrarily. Velocity selection ideas emerged via the circuitous route of postulating local equations[10] which would be easier to analyze; these have been reviewed by Harvey Segur. These local model ideas led simultaneously to numerical demonstrations[11] of velocity selection for our basic model and novel analytic schemes[12] (asymptotics beyond all orders, naturally) for demonstrating the selection. These trends converged to the final works of Ben-Amar and Pomeau and (later) Brener and Meln'kov[13] who completely solved the problem analytically and to Saito, Goldbeck-Wood and Muller-Krumbhaar[14] who verified agreement with computer simulation results.

III. Microscopic Solvability

In this section, we will sketch the current analysis of dendritic tips in our basic model. Our treatment will follow that of Ben-Amar and Pomeau[13] and will make use of the Kruskal-Segur procedure of finding a local equation near a singular point in the complex plane. At the end, we will briefly survey the picture that has emerged from the application of these ideas to dendritic growth[15].

The first step is the assumption that one should look for a steady-state "needle crystal" and neglect sidebranching for the moment. This approach is supported by the fact that the tip region is (approximately) uniformly translating and via the fact that predictions based on this premise agree with dynamical simulations even though the dendrites in the latter exhibit sidebranching. The resulting idea that one should impose boundary conditions at infinity as a way of selecting the correct needle crystal, even though the steady-state region does not extend to infinity, is highly nontrivial; this assumption can only be "proven" by a full analysis of the time-dependent problem. Since this has not been accomplished to date even for simpler systems (such as the Saffman-Taylor finger[15,16]; see talk by Saleh Tanveer), we must realize that our approach may be incomplete. That is, it may turn out that dendritic growth is an intrinsically time-dependent state and that the following method gives only approximate answers.

Given this assumption, we can transform the evolution system equations (1)–(3) into an integro-differential shape equation. The procedure for doing this is straightforward and we merely quote the result:

$$\Delta - \tilde{v}\kappa(1 - \epsilon\cos 4\theta) = \frac{1}{\pi}\int_{-\infty}^{\infty} dx'\, e^{y(x')-y(x)}\, K_0\left(\sqrt{(x-x')^2 + (y(x)-y(x'))^2}\right). \quad (4)$$

Here $\tilde{v}$ is the dimensionless velocity $v\bar{d}_0/2D$ and we have taken the specific form of a four-fold anisotropy $d_0 = \bar{d}_0(1 - \epsilon\cos 4\theta)$. The unknowns in (4) are the shape $y(x)$ in units of $\frac{2D}{v}$ and the velocity $\tilde{v}$; lengths here are measured in units of $\frac{2D}{v}$.

The first solution to these equations was obtained by Ivantsov by setting $\bar{d}_0 = 0$. It is easy to check that assuming $y_0(x) = -x^2/2p$ makes the integral term independent of x, i.e. a constant. We must therefore choose

$$\Delta = \frac{1}{\pi}\int_{-\infty}^{\infty} e^{\frac{-x^2}{2p}} K_0\left(\sqrt{x^2 + \frac{x^4}{4p^2}}\right) dx \quad (5)$$

which is the Ivantsov relationship between dimensionless tip radius (Peclet number) and undercooling. Obviously, v is arbitrary in this limit since it does not appear in the equations at all.

The important issue is then whether this continuous (in v) solution branch persists at non-zero $\bar{d}_0$. A perturbative calculation demonstrates that any problems must be "beyond all orders" in $\tilde{v}$. But exactly such effects were found numerically! So, we must analyze this system carefully, looking for singularities in the complex x plane which may yield non-perturbative terms. Exactly such a singularity takes place at the point $x = ip$ where the Ivantsov solution curvature is singular.

The Kruskal-Segur approach as developed for this problem by Ben-Amar and Pomeau, and Brener and Meln'kov, consists of deriving a local equation near $x = ip$, starting from (4). The left hand side is already a local function of $y(x) = y_0(x) + \delta(x)$. For the right hand side, we are allowed to expand around the Ivantsov solution $y(x) = y_0(x) + \delta(x)$ to obtain

$$\frac{1}{\pi}P\int_{-\infty}^{\infty} dx\, \frac{\partial}{\partial y}\left(K_0(\sqrt{(x-x')^2 + (y(x)-y(x'))^2}e^{y(x')-y(x)}\right)\Big|_{y(x)=y_0(x)} \times (\delta(x) - \delta(x')) . \quad (6)$$

The derivative $\frac{\partial}{\partial y}$ is defined to be the derivative of the Green's function with respect to the coordinate y assumed to be independent of x, x' and y'; only after the differentiation are the coordinates y and y' evaluated on the curve.

We have formally introduced the principal value prescription which is irrelevant since there is actually no singularity. We then break this integral into the $\delta(x)$ term, which again is local, and the term

$$\frac{1}{\pi} P \int dx' \frac{\partial}{\partial y} \left(K_0 \left(\sqrt{(x-x')^2 + (y(x) - y(x'))^2} \right) e^{y(x')-y(x)} \right) \Big|_{y=y_0(x)} \delta(x'). \quad (7)$$

It is then possible to show that the leading term in this integral near $x = ip$ is obtained from the singularity at $x = x'$ and hence is also local. Combining these pieces, we derive[17]

$$\frac{-\tilde{v} y''}{(1+y'^2)^{3/2}} \left(1 - \epsilon \cos(4 \tan^{-1} y')\right) = \delta(x) \frac{1 + i y_0'}{1 + y_0'^2}. \quad (8)$$

Not accidentally, the linearized version of this equation

$$\frac{-\tilde{v} \delta''}{(1+y_0'^2)^{3/2}} \left(1 - \epsilon \cos(4 \tan^{-1} y_0')\right) = \delta(x) \frac{1 + i y_0'}{1 + y_0'^2} \quad (9)$$

(where we have neglected $\tilde{v}$ multiplied by lower derivatives) is exactly the WKB formula for zero growth perturbation of the slowly varying background shape[18]. This equation admits exponentially growing perturbations at $+\infty$ which must be suppressed by proper choice of $\tilde{v}$. Since the $\tilde{v}$ term is "relevant" only near $y_0' = i$, the coefficient of this bad term will be "beyond all orders" in $\tilde{v}$. When the analysis is carried through, the following results emerge[15]:

a) If $\epsilon = 0$, there are no solutions of the desired form.

b) The local equation can be rescaled in the singular region via $\tilde{v} = \epsilon^{\frac{7}{4}} p^2 \sigma$ (for small p).

c) At fixed ϵ, there are an infinite set of discrete roots σ_n.

d) The largest root σ_0, and hence largest velocity solution, is the only linearly stable solution[19].

So, given the above, there is a unique prediction for the shape and speed of the dendrite tip. This prediction is then tested numerically and found to be correct.

IV. Extensions and Issues

By now, the aforementioned solvability theory has been extended in a variety of directions. First, the effects of varying the Peclet number p have been addressed[20]. Also, models which include kinetic coefficients have been analyzed. The effects of finite size (growth in a channel) have also been included[21]. None of these calculations have revealed any basic flaws in this approach.

An interesting generalization of velocity selection was recently addressed by E. Brener and myself[22]. We studied systems in which the lattice structure was not reflection symmetric; this would be the case for example, for the growth of solid domains of chiral molecules in monolayers[23]. The lack of symmetry leads to a lack of reflection symmetry of the local equation in the complex x plane, and suppressing the bad behavior requires the adjustment of two numbers. So, in addition to velocity selection, we also dynamically select a growth angle with respect to the underlying lattice - for reflection-symmetric cubic systems, growth could only occur at 0°.

Finally, however, we must face up to the tough questions which remain unanswered. Everyone has their own priority list of what needs to be done, but here is mine:

a) More complicated interface physics - How do we combine our macroscopic approach with more realistic microscopic models and allow for kinetic roughening, dendrites with facets etc. Some work is progressing along these lines.

b) Universality - Closely connected to (a) is the question of how accurate must the full model be before we can make quantitatively reliable predictions. This is troubling because our selection mechanism is controlled by singularity structure in the complex x plane which can be changed dramatically by minor changes in, say, the anisotropy for real angles. I personally am somewhat less pessimistic and point to work on Saffman-Taylor fingers where the real world (films and three dimensionality) usually makes only small changes to the idealized two-dimensional model predictions.

c) Extension to non-axisymmetric patterns - As will be discussed at length in the paper of David Kessler, serious questions still remain about the details of the solvability mechanism for non-axisymmetric anisotropy.

Hopefully, some of those of you reading this will take up the challenge and push forward the boundaries of our current knowledge of this system.

V. References

1. J. Kepler, "De Nive Sexangula" published in Frankfurt am Main (1611).
2. For a survey, see D. P. Woodruff, "The Solid-Liquid Interface", Cambridge Univ. Press (1973) and J. S. Langer, Rev. Mod. Phys. 52, 1 (1980).
3. W. W. Mullins and R. F. Sekerka, J. Applied Phys. 34, 323 (1963); J. Applied Phys. 35, 444 (1964).
4. For experimental data on dendritic growth, see M. Glicksman, Mat. Sci. and Eng. 65, 45 (1984); A. Dougherty and J. Gollub, Phys. Rev. A38, 3043 (1988); H. Chou and H. Cummins, Phys. Rev. Lett. 61, 173 (1988).
5. V. Seetharaman, M. A. Eschelman and R. Trivedi, Acta Metall. 36, 1165 (1988); S. de Cheveigne, C. Guthmann and M. M. Lebrun, J. de Physique 47, 2095 (1986).
6. K. A. Jackson and J. D. Hunt, Trans. Metall. Soc. 236, 1129 (1966); D. A. Kessler and H. Levine, J. Crystal Growth 94, 871 (1989).
7. D. A. Kessler and H. Levine, Acta. Metall. 36, 2693 (1988).
8. G. E. Nash and M. E. Glickman, Acta. Metall. 22, 1283 (1974).

9. G. P. Ivantsov, Dokl Akad. Nauk SSR 58, 567 (1947); G. Horvay and J. Cahn, Acta Metall. 29, 717 (1961).

10. R. Brower, D. Kessler, J. Koplik and H. Levine, Phys. Rev. Lett. 51, 1111 (1983); Phys. Rev. A29, 1335 (1984); E. Ben-Jacob, N. Goldenfeld, J. S. Langer and G. Schon, Phys. Rev. Lett. 51, 1930 (1983); Phys. Rev. A29, 330 (1984).

11. D. Kessler, J. Koplik and H. Levine, Phys. Rev. 33, 3352 (1986); D. Meiron, Phys. Rev. A33, 2704 (1986); M. Ben-Amar and B. Moussallam, Physica 25D, 155 (1987).

12. D. Kessler, J. Koplik and H. Levine in "Patterns, Defects and Microstructures", D. Walgraef ed. NATO ASI Series E, Nijhoff (1986); A. Barbieri, D. C. Hong and J. S. Langer, Phys. Rev. A35, 1802 (1986; B. Caroli, C. Caroli and B. Roulet, J. de Physique 48, 547 (1987).

13. M. Ben-Amar and Y. Pomeau, Europhys. Lett. 2, 307 (1986); M. Ben-Amar, Physica 31D, 409 (1988). E. Brener and V.I. Meln'kov, Adavances in Physics, to appear; S. Tanveer, Phys. Rev. A40, 4756 (1989).

14. Y. Saito, G. Goldbeck-Wood and H. Muller-Krumbhaar, Phys. Rev. Lett. 58, 1541 (1987); Phys. Rev. 38, 2148 (1988).

15. For more complete reviews, see J. S. Langer, "Chance and Matter", ed. J. Souletie, North-Holland, Amsterdam (1987); D. A. Kessler, J. Koplik and H. Levine, Adv. in Phys. 37, 255 (1988); P. Pelce. "Dynamics of Curved Fronts", Academic (1988).

16. P. G. Saffman and G. I. Taylor, Proc. Roy. Soc. London 245, 312 (1958).

17. For large Peclet numbers, this expression needs to modified; see E. Brener and V. I. Meln'kov, ref 13.

18. D. Kessler and H. Levine, Europhys. Lett. 4, 215 (1987).

19. D. Kessler and H. Levine, Phys. Rev. Lett. 57, 3069 (1986).

20. E. Brener and V. I. Melnikov, J. de Physique 51, 157 (1990).

21. D. Kessler, J. Koplik and H. Levine, Phys. Rev. 34, 4980 (1986).

22. E. Brener and H. Levine, Phys. Rev. A, to appear.

23. R. Weis and H. M. McConnell, Nature 310, 47 (1984).

AN EXPERIMENTAL ASSESSMENT OF CONTINUUM MODELS OF DENDRITIC GROWTH

J.P. Gollub

Physics Dept., Haverford College, Haverford, PA 19041 USA and
Physics Dept., University of Pennsylvania, Philadelphia, PA 19104 USA

Abstract

Experimental evidence pertinent to theories of needle crystals (dendrites) based on continuum models is reviewed and assessed critically. Some predictions, such as the dependence of the growth state on crystalline anisotropy, have not been convincingly demonstrated, and the models may not be appropriate in all cases, for example when kinetic effects are important. On the other hand, the continuum models provide an internally consistent explanation for many of the observations, including some related to sidebranching.

Introduction

The problem of explaining the needle crystals or dendritic growth patterns that occur during solidification has attracted much interest and effort for several distinct reasons. From a mathematical point of view, the dendrite is significant as a steady and stable solution of a nonlinear pattern-forming system. Furthermore, the solution exhibits scaling behavior, in the sense that dendrites for different specified boundary conditions (undercooling) are the same except for a change of scale. Another remarkable property is the fact that the macroscopic form of a dendrite is determined by microscopic parameters such as the length characterizing capillarity, which is of the order of Angstroms for typical materials. From a physical point of view, dendrites are significant because they are ubiquitous in nature, and because they represent an ordered non-equilibrium state. Dendrites are of particular interest to metallurgists in part because cast metals are often composed of them.

Recent theories of pattern selection in dendritic growth that have been summarized in reviews by Kessler et al. [1], Langer [2], and Pelcé [3] provide a mathematically consistent approach to predicting the shape and velocity of dendrites using continuum model equations. In brief, the theory shows that if the growth is diffusion-controlled, and if the crystal is symmetric about the growth axis, then the macroscopic transport equations allow only a single stable solution when anisotropic surface tension is taken into account. The term "microscopic solvability" is often used to describe the theory because the capillary length scale characteristic of surface tension is typically only a few Angstroms. The purpose of this article is to assess the experimental evidence for (or against) the applicability of this approach to real dendritic crystals.

Systems Displaying Dendritic Growth

Dendritic crystal growth occurs in a great variety of materials during solidification.[4] It has been studied in two quite different situations: growth into the melt (or solution) at constant undercooling (supersaturation); and directional solidification with an imposed temperature gradient. Examples of materials displaying dendritic growth include many alloys, transparent organic materials such as CBr_4 and succinonitrile (either pure or with added solute), inorganic salts such as NH_4Br and NH_4Cl, and rare gas solids such as He, Xe, and Kr. In fact, the dendritic growth form is a ubiquitous non-equilibrium stationary state whenever the interfaces are microscopically rough (not faceted). Rough interfaces can occur either as a result of thermal roughening, or as a result of the growth process itself.

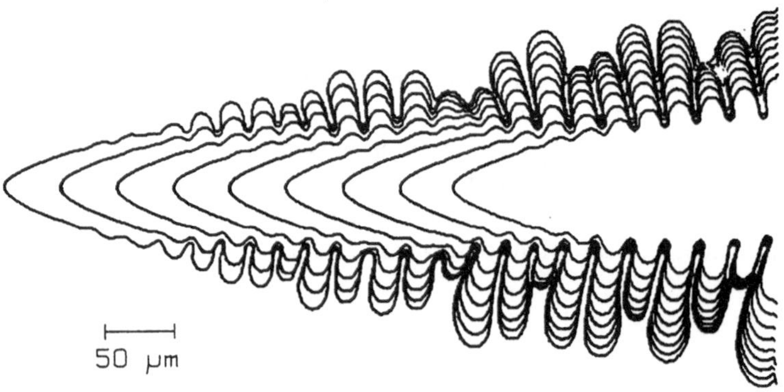

Fig. 1. Contours of an ammonium bromide dendrite growing from supersaturated solution, from Ref. [5]. The interval between contours is 20 s. Sidebranches also grow perpendicular to the plane of the diagram.

An experimental example of free dendritic growth from supersaturated solution, as studied by Dougherty et al.[5] is shown in Fig. 1. The main features worth noting are: constant growth speed; a fixed approximately parabolic shape near the tip; and sidebranches that grow progressively with distance back from the tip. It is often useful to describe the growth using a frame of reference moving with the dendrite; in this frame the branches appear to be waves propagating backward at constant speed. All of these features were documented much earlier in a classic study by Huang and Glicksman.[6]

Dendritic growth also occurs in quite different circumstances that are apparently unconnected with solidification. For example, dendrites are found in the electrolytic deposition of metals in solution, as shown in Fig. 2, which is taken from Ref. [7]. On the other hand, modification of the electrolyte concentration leads instead to disordered growth that has been compared quantitatively to diffusion-limited aggregation.[8] Another unconventional example of dendritic growth occurs in the transition between certain liquid crystal phases.[9]

Mullins-Sekerka Instability

It is worthwhile to note briefly the basic instability that leads to the dendritic state. Interfaces whose rate of growth is controlled by diffusion were shown to be unstable with respect to sinusoidal perturbations by Mullins and Sekerka.[10] The growth rate of the instability rises with wavenumber k, reaches a maximum, and falls at high k due to surface tension. A lovely quantitative study of this process was performed by Chou and Cummins,[11] and an example from their work is displayed in Fig. 3. These authors showed that the early growth process is irregular, and can be adequately described by a stochastic equation of the following form:

$$dA_k/dt = \sigma(k)A_k + \text{Gaussian noise} \quad . \tag{1}$$

In this equation, A_k is the amplitude of the Fourier mode with wavenumber k, and $\sigma(k)$ is its growth rate. Whether the noise is due to thermal fluctuations or some other source is unclear.

The sensitivity of the early growth process to noise has been nicely illustrated in a numerical simulation of the unstable growth of a spherical nucleus by Pines et al.[12] The growth was constrained to be axisymmetric and random noise was provided as part of the initial condition. The radius is shown as a function of the polar angle in Fig. 4 with an exaggerated vertical scale. The dominant features can be traced back to very early times. Once started, these features persist and compete with each other for material.

D. Predictions for Steady State Dendritic Growth

The central theoretical prediction for axisymmetric dendritic growth is that surface tension enters the problem as a singular perturbation, and solutions for stable needle crystals depend on a small anisotropy ε of the capillary length d_o (proportional to the surface energy). This effect is usually modeled (for cubic crystals) by a functional dependence of the form

$$d_o(q) = d_o(1 + \varepsilon\cos 4\theta) \quad , \tag{2}$$

where θ is the angle between the local normal to the surface and the (100) direction. The selected state of the dendrite is given for small undercooling by a parameter

Fig. 2. Dendritic growth in the electrodeposition of zinc from solution, from Ref. [7].

$$\sigma^* = 2D/v\rho^2 \ ,$$

where D is the diffusion constant (thermal or molecular) that controls the growth, ρ is the radius of curvature of the tip of the dendrite, and v is the speed of propagation of the interface. In addition, the product ρv is a function only of the undercooling (or supercooling) Δ for diffusion-controlled growth, so a unique state is predicted for each value of Δ. (The radius decreases while the speed increases as Δ is increased.) For axisymmetric growth, σ^* has been predicted to vary as $\varepsilon^{7/4}$ as $\varepsilon \to 0$. However, for the physically more relevant range $\varepsilon \sim 0.01$, the prediction [13][14] is closer to $\sigma \sim \varepsilon^1$. This prediction has been checked by two-dimensional numerical simulations. [15]

However, the theoretical situation is still apparently unresolved for non-axisymmetric three-dimensional growth [16]. In fact, the tip shape deviates significantly from a paraboloid somewhat before the sidebranches become prominent, so it is important to look at the experimental situation.

There is a long history of theoretical models of axisymmetric dendritic growth. Some references to the older approaches, such as marginal stability, are given in [17]. While these models give predictions that are in some cases not far from experimental results for steady state growth, their theoretical justification and mathematical consistency has been undermined by the microscopic solvability approach.

Measurements of Steady State Dendritic Growth from the Melt

Studies of dendritic growth radii and velocities have been undertaken by a number of investigators. However, a test of the theory also requires a quantitative measurement of the small parameter ε, which is quite difficult to perform accurately for small ε.

For the growth of highly purified material from the melt, Glicksman et al. [17] reported that succinonitrile and pure pivalic acid have nearly identical values of σ^* (0.0195 for SCN and 0.022 for PVA), while their anisotropies appear to differ substantially (0.005 for SCN and 0.05 for pure PVA). The anisotropies were determined from interfacial angular discontinuities at grain boundaries.

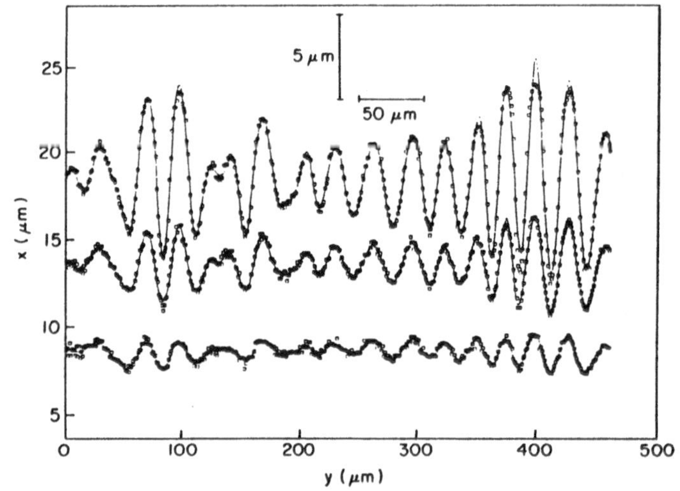

Fig. 3. Experimental measurement of the early stages of the Mullins-Sekerka instability, from Ref. [11].

Data on anisotropies and growth constants for these two materials is summarized in Table 1, along with corresponding data for solution growth to be discussed in the following section.

Studies of the dendritic solidification of krypton and zenon by Bilgram et al. [18] are particularly interesting because of the simplicity of rare gas solids. These authors found σ^* to be a function of supersaturation, something that would not be expected for small undercooling unless kinetic or impurity effects, or convection were important. The results are sufficiently surprising that further tests may be warranted.

Measurements of Steady State Dendritic Growth from Solution

The situation for growth from solution is somewhat different. It is perhaps worth indicating some of the experimental compromises that are involved in such measurements. It is desirable to use a growth cell larger than the diffusion length 2D/v to avoid finite size effects. On the other hand, the dendrites are small for solution growth, and adequate magnification can only be achieved with short focal length lenses. This unfortunately limits the cell thickness, but has the advantage of suppressing convection. Other difficulties are the achievement of spatially uniform solute concentration and adequate temperature control. Finally, one has to worry about possible optical distortions resulting from diffraction and the concentration gradients around the dendrite. These various constraints can be met in practice to a reasonable degree. Analogous problems affect studies of growth from the melt, but in different proportions.

Careful measurements for solution growth were undertaken by Dougherty and Gollub [19] for NH_4Br and later by Dougherty [20] for pivalic acid, using digital microscopy. The parameter ε was determined from the equilibrium radius $R(\theta)$ of a small spherical crystal (with mean radius roughly 40 μm) as a function of angle θ. This quantity has a fourfold component that gives ε directly. To make these measurements, it was necessary to hold the small crystal

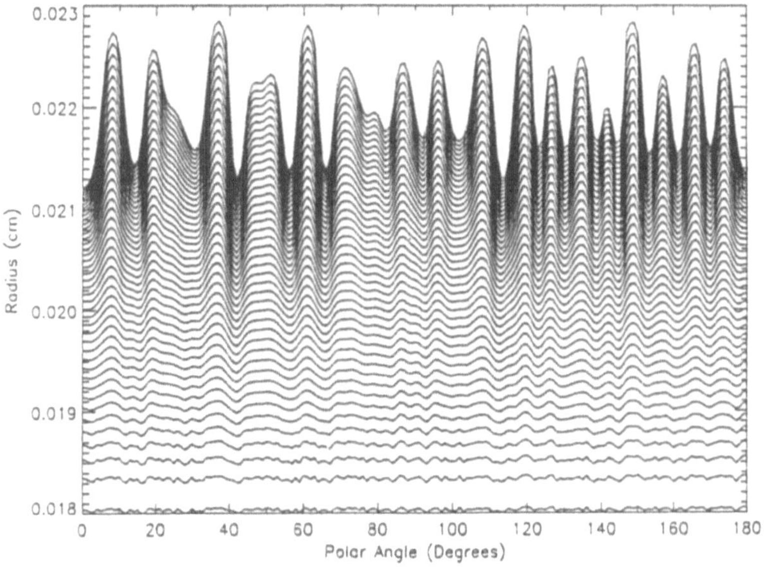

Fig. 4. Numerical computation of the axisymmetric growth of a spherical nucleus, illustrating the sensitivity of the structure to noise at early times, from Ref. [12].

Table 1. Comparison of predictions for the dendritic growth constant σ* based on microscopic solvability (assuming axisymmetric growth) with various experiments. The data was taken from Ref. [17] for growth from the melt and Refs. [9], [19], and [20] for growth from solution. PVA is pivalic acid; HET is hexaoctyloxytriphenylene, a discotic liquid crystal; SCN is succinonitrile. These materials have fourfold anisotropy, except for HET, which is hexagonal. See the text for discussion.

	ε	(σ*) from expt.	(σ*) from theory
SCN (melt)	0.005	0.0195	0.01
PVA (melt)	0.05	0.022	~0.1
PVA (soln.)	0.006±0.002	0.05±0.02	0.022±0.007
NH_4Br (soln.)	0.016±0.004	0.081±0.02	0.083±0.025
HET (soln.)	0.003±0.001	0.038	0.040

in equilibrium for long periods of time to allow equilibration. However, the crystal is unstable, tending either to grow or shrink. To keep the mean crystal size fixed, Dougherty developed a novel temperature control system with optical feedback. An example of R(θ) from Ref. [20] for pivalic acid in ethanol solution is given in Fig. 5. The measured anisotropy (averaged over several crystals) is ε=0.006±0.002 for this material when grown from 1% ethanol solution.

A comparison of theory and experiment for the growth from solution of ammonium bromide and pivalic acid is contained in Table 1. The theoretical predictions for σ* pertain to axisymmetric growth for the measured ε, and the errors quoted on the theoretical values stem from the uncertainty in ε. The experimental results on these two materials are consistent with the predictions, but the statistical errors are too large (or the anisotropies too similar) to allow a clear dependence on ε to be detected.

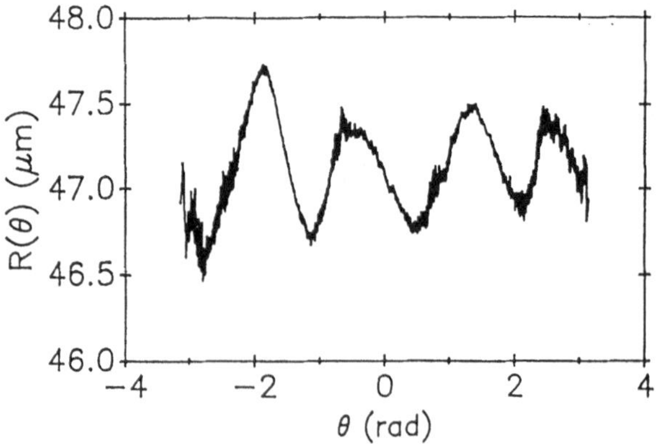

Fig. 5. Angular dependence of the equilibrium radius R(θ) of a small crystal of pivalic acid, from Ref. [20]. The fourfold component is proportional to the anisotropy ε.

Maurer et al. [21] have reported a somewhat higher value ($\sigma^*=0.12\pm0.01$) for the growth of NH_4Br in a gel, where possible convective effects should be strongly suppressed. This latter measurement is somewhat more precise than those reported in the table, but one must hope that the gel does not modify the growth process.

Finally, we note that Oswald [9] studied interfacial growth (in a very thin layer) at the boundary between two liquid crystal phases: a columnar hexagonal liquid crystal mesophase, and the isotropic phase of the same material. In this case, the anisotropy is 6-fold and was measured to be about 0.003, while $\sigma^*=0.038$. The corresponding theoretical value [22] for two-dimensional 6-fold growth (adjusted for the different diffusivities for impurities in the two phases) is 0.040. There are many uncertainties in the comparison, but qualitatively at least, the experiment seems not inconsistent with the predictions.

Now examining all the data in Table 1, we note that the case of pivalic acid grown from the melt is the only one that appears to contradict the theory. On the other hand, the values of

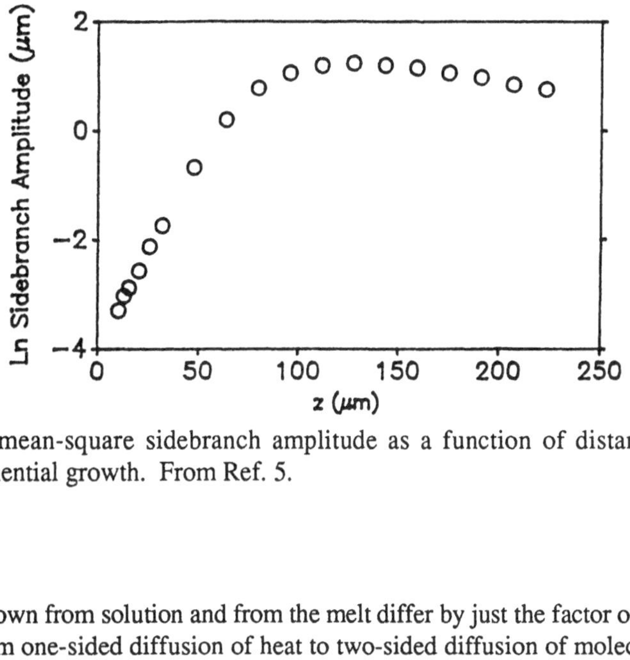

Fig. 6. Root-mean-square sidebranch amplitude as a function of distance from the tip, showing exponential growth. From Ref. 5.

σ^* for PVA grown from solution and from the melt differ by just the factor of two expected for the change from one-sided diffusion of heat to two-sided diffusion of molecules, without any additional change that might be attributed to a modification of the surface tension anisotropy upon addition of the ethanol "solvent".[20] The internal consistency of the data thus suggests that the measurements of the growth constants are correct. It would therefore seem worthwhile to remeasure ε for pure PVA using digital imaging methods.

G. Early Sidebranching

According to our present theoretical understanding, dendritic tips are stable with respect to variations of the radius of curvature at the tip. However, they are believed to be "convectively unstable" with respect to sidebranching waves. This means that a small perturbation can grow while propagating away from the tip (in the tip frame of reference), while leaving the tip unchanged. If this understanding is correct, one expects [32] that noise in the tip region would produce small wavepackets of mean wavelength

$$\lambda \sim 2\pi\rho(\sigma^*)^{1/2} \tag{4}$$

and amplitude A(s) as a function of arc length s given by

$$A(s) \sim \exp\{(\sigma^*)^{-1/2}(s/\rho)^{1/4}\} \tag{5}$$

Any source of noise might have this effect, including thermal noise or microscopic irregularities arising from atomic discreteness or defects in the crystal. The dendrite serves as a frequency-selective (or wavelength-selective) amplifier.

Something like this in fact seems to occur. For example, measurements of the logarithm of the root-mean-square sidebranch amplitude as a function of distance from the tip (from Ref. [5]) are shown in Fig. 6. An approximately exponential growth (which would be indistinguishable from the form given in (5) is seen to occur.

A more direct test has been made by Qian and Cummins,[23] who studied the response to a localized pulse of radiant energy sent through an optical fiber aimed at the tip. They indeed observed the expected wavepacket and have compared its form with the theoretical predictions.

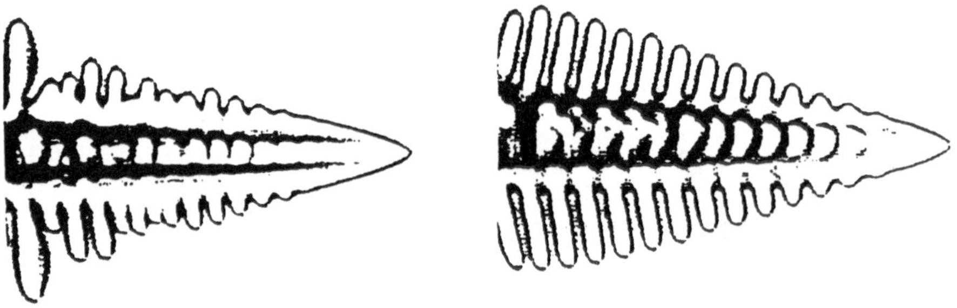

Fig. 7. Resonant response of a dendrite to a periodic perturbation, from Ref. [24]. Left, unforced; right, forced.

The concept that the dendritic tip is a frequency-selective amplifier was verified by Bouissou et al. [24]. These authors subjected the growing dendrite to a small oscillatory flow. They found that the sidebranch structure appears to resonate when forced weakly at a particular frequency, at which very large and regular branches are found, as shown in Fig. 7.

The mean sidebranch spacing has been predicted [1][14] based on the concept of noise-induced convective instability in the tip region. Though the predictions are somewhat model-dependent, the spacing should scale as $(\sigma^*)^{1/2}$. For ammonium bromide grown from solution, it is expected that λ/ρ should be roughly 3.7, while the experimental result [19] is somewhat larger, 5.2±0.8. For pure succinonitrile, both the prediction and the experimental observation [6] are lower than these values by a factor of 1.8, so the quantitative discrepancy is again about 30% of the experimental value. For pivalic acid grown from solution, the situation is worse, with the prediction and experimental value being about 3 and 6, respectively.[20] It is possible that the lack of axisymmetry of the real crystals plays a role in this quantitative disagreement.

The irregularity of the sidebranching pattern, as seen visually in Fig. 1, is affected eventually by local competition of adjacent branches for material. However, the irregularity is *also* seen quite close to the tip,[5] where the branches are barely visible in Fig. 1. This feature would also be a natural consequence of noise-induced convective instability in the tip region. The effect is qualitatively similar to the amplification of noise by the Mullins-Sekerka instability, as documented in Figs. 3 and 4. In a sense, the Mullins-Sekerka instability continues to operate, even in the steady state.

H. Kinetic Effects

One of the critical assumptions often made in describing dendritic growth is that departures from local equilibrium (or kinetic effects) can be neglected. However, experiments and theories have increasingly made efforts to deal with kinetic effects. For example, in the continuum theory of dendritic growth, one can allow the non-dimensional concentration at the interface to contain a term $-\beta(\theta)v$ that is linear in the velocity but depends on the angle of the local normal to a particular crystal axis. Such a model can explain the transition from (100) dendrites to (111) dendrites that was observed by Chan et al.[25] at high velocity long ago. For a recent example of the incorporation of kinetic effects into continuum growth models, see for example the discussion by Brener and Levine in Ref. [26].

Dendrites are sometimes partially faceted [27], and in this case kinetic effects must surely be taken into account. Theoretical predictions for such faceted dendrites have been proposed by Ben Amar and Pomeau.[28] Raz and Lipson [29] have noted that the dendritic growth of NH_4Cl shows a nonlinear dependence of the growth velocity on supersaturation, a likely hallmark of kinetic effects. In a later study, Raz et al. [30] discovered an interesting superlattice structure for dendritic growth in a very thin layer, which they attributed in part to kinetic effects.

In some cases, a crystal that is faceted in equilibrium will be roughened when grown even slowly. (For example, see Ref. [19].) This so-called "kinetic roughening" process is fairly complex. Phenomena of this type have been studied numerically by Xiao et al.[31] These authors used a modified Monte Carlo diffusion-limited aggregation model that includes fairly realistic boundary conditions. Surface diffusion prior to attachment, lattice anisotropy, and thermal fluctuations were included. At low temperatures, where the bond strength is much larger than the thermal energy kT, the faceted shape is preserved. However, at higher temperatures a clear transition from faceted to dendritic shapes is found in the numerical computations. Such simulations may be the only adequate way to treat these complex situations where kinetic effects are important.

J. Conclusion

A considerable body of experimental evidence now exists concerning the problem of pattern selection in dendritic growth. To what extent does it support the continuum models and their solution based on microscopic solvability? This is not an easy question to answer, and it is especially difficult to provide a summary that researchers in all of the relevant research disciplines (metallurgy and materials science, physics, and mathematics) will find equally acceptable, since they have divergent views as to the relevance of different types of evidence. The following is an effort, surely imperfect, to provide a balanced assessment.

First, there seems to be little doubt that the continuum models have been solved correctly and in a mathematically consistent way. The results have been checked by several independent investigators, and have also been compared to numerical computations.

Furthermore, the same methods have been found to produce correct predictions when applied to mathematically similar hydrodynamic situations, such as the viscous fingering problem.[1] As a solution to a mathematical problem of pattern formation within the framework of continuum models, the theory is impressive.

The real issue is that the models used to date are only approximate descriptions of growing crystals. First, the lack of axisymmetry of most three-dimensional dendritic crystals is obviously important, but still a subject of theoretical controversy [16] Second, kinetic effects are complex, and the simple addition of a linear term ($-\beta v$) to the boundary condition may not provide a sufficient remedy. Third, microscopic effects, including both molecular discreteness and crystal defects, may in some cases be significant, especially for sidebranching phenomena.

Still, it is important to ask to what extent the continuum models of axisymmetric dendritic growth adequately describe the basic phenomena. The following elements are important in such a comparison:

-- The theory provides a mathematically consistent way to understand the existence of a unique radius and velocity at given undercooling, as is found experimentally.

-- The theory provides a natural way to understand sidebranches that grow continuously from the tip, are sensitive to noise in that region, and resonate with an external perturbation at a particular frequency.

-- The evidence concerning the dependence of the growth constant σ^* on anisotropy is inconclusive at present. For growth from solution, the trend seems correct but a sufficient range of ε is unavailable. Further work on the rare gas solids also seems desirable to settle the question of the constancy of σ^* for these materials. For pure systems, several cases are at least roughly consistent with the theory (including the liquid crystal example). However, the one material (pure pivalic acid) for which a large value of ε has been reported does not show the predicted large growth constant. It would be worthwhile to remeasure ε for this material using different methods, given its importance to testing the theory.

-- The fact that dendrites grow along crystal axes and do not generally grow in amorphous materials certainly indicates that anisotropy is of some importance. Also, in the mathematically analogous case of hydrodynamic viscous fingering, dendrites (needle-like fingers with sidebranches) can be produced by introducing anisotropy into the boundary conditions, and do not occur in isotropic situations. This connection to hydrodynamics shows that the mathematics does produce qualitatively reasonable predictions in a situation where anisotropy can be controlled more easily than it can in solidification.

-- Even if the theory is eventually found to be reasonably consistent with experiment, there remain other important aspects of interfacial growth, such as the statistical properties of the coarsening process in the region of strong competition far from the tip, that are still incompletely understood.

Acknowledgements

This work was supported primarily by the National Science Foundation Low Temperature Physics Program (DMR-8901869). Some facilities were provided by the W.M. Keck Foundation and by the University Research Initiative Program (DARPA/ONR-N00014-85-K0759). I am indebted to A. Dougherty (of Lafayette College), some of whose work is reviewed in this article, and also to H. Levine and M. Glicksman for helpful comments. The hospitality of the Institute for Nonlinear Science at The University of California at San Diego, where this paper was written, is much appreciated.

This paper is reprinted from *Nonlinear Phenomena Related to Growth and Form,* ed. by M. Ben Amar, P. Pelce, and P. Tabeling (Plenum, 1991), by permission of the publisher.

References

[1] D.A. Kessler, J. Koplik, and H. Levine, Adv. Phys. **37**, 255 (1988).

[2] J.S. Langer, in *Chance and Matter*, ed. by J. Souletie, J. Vannimenus, and R. Stora (Elsevier, New York, 1987), p. 629; and Science **243**, 1150 (1989).

[3] P. Pelcé, *Dynamics of Curved Fronts* (Academic Press, New York, 1988).

[4] For a general review from a materials science perspective see W. Kurz and R. Trivedi, Acta Metall. Mater. **38**, 1 (1990).

[5] A. Dougherty, P.D. Kaplan, and J.P. Gollub, Phys. Rev. Lett. **58**, 1652 (1987).

[6] S.-C. Huang and M.E. Glicksman, Acta Metall. **29**, 701 and 717 (1981).

[7] Y. Sawada, A. Dougherty, and J.P. Gollub, Phys. Rev. Lett. **56**, 1260 (1986).

[8] F. Argoul, A. Arneodo, and G. Grasseau, Phys. Rev. Lett. **61**, 2558 (1988).

[9] P. Oswald, J. Phys. France **49**, 1083 (1988).

[10] W.W. Mullins and R.F. Sekerka, J. Appl. Phys. **34**, 323 (1963) and **35**, 444 (1964).

[11] H. Chou and H.Z. Cummins, Phys. Rev. Lett. **61**, 173 (1988).

[12] V. Pines, M. Zlatkowski, and A. Chait, Phys. Rev. A **42**, 6129 and 6137 (1990).

[13] D.A. Kessler and H. Levine, Phys. Rev. A **33**, 3352 (1986).

[14] A. Barbieri and J.S. Langer, Phys. Rev. A **39**, 5314 (1989); J.S. Langer, Phys. Rev. A **36**, 3350 (1987).

[15] Y. Saito, G. Goldbeck-Wood, and H. Müller-Krumbhaar, Phys. Rev. Lett. **58**, 1541 (1987) and Phys. Rev. A **38**, 2148 (1988).

[16] D. Kessler, in *Proceedings of the Conference on Asymptotics Beyond All Orders*, ed. by S. Tanveer (Plenum, 1991), to appear. A different approach to the effect of non-axisymmetry on state selection has been given by Y. Miata, S.H. Tirmizi, and M.E. Glicksman, "Dendritic growth with interfacial energy anisotropy," J. Cryst. Growth (1991), to appear.

[17] M.E. Glicksman and N.B. Singh, J. Cryst. Growth **98**, 277 (1989).

[18] J.H. Bilgram, M. Firmann, and E. Hürlimann, J. Cryst. Growth **96**, 175 (1989).

[19] A. Dougherty and J.P. Gollub, Phys. Rev. A **38**, 3043 (1988).

[20] A. Dougherty, Surface tension anisotropy and the dendritic growth of pivalic acid, J. Cryst. Growth (1991), in press.

[21] J. Maurer, B. Perrin, and P.Tabeling, Three dimensional structure of NH_4Br dendrites growing out of a gel (1990), preprint.

[22] M. Ben Amar, private communication.

[23] X.W. Qian and H.Z. Cummins, Phys. Rev. Lett. **64**, 3038 (1990).

[24] P. Bouissou, A. Chiffaudel, B. Perin, and P. Tabeling, Europhys. Lett. **13**, 89 (1990).

[25] S.-K. Chan, H.-H. Reimer, and M. Kahlweit, J. Cryst. Growth **32**, 303 (1976).

[26] E. Brener and H. Levine, Phys. Rev. A **43**, 883 (1991).

[27] J. Maurer, P. Bouissou, B. Perrin, and P. Tabeling, Europhys. Lett. **8**, 67 (1988).

[28] M. Ben Amar and Y. Pomeau, Europhys. Lett. **6**, 609 (1988).

[29] E. Raz, S.G. Lipson, and E. Polturak, Phys. Rev. A **40**, 1088 (1989).

[30] E. Raz, S.G. Lipson, and E. Ben-Jacob, Meta-ordering observed during dendritic growth of ammonium chloride crystals in thin layers (1990), preprint.

[31] R.-F. Xiao, J.I.D. Alexander, and F. Rosenberger, Phys. Rev. A **38**, 2447 (1988); J. Cryst. Growth **100**, 313 (1990).

[32] R. Pieters and J.S. Langer, Phys. Rev. Lett. **56**, 1948 (1986); M. Barber, A. Barbieri, and J.S. Langer, Phys. Rev. A **36**, 3340 (1987); D.A. Kessler and H. Levine, Europhys. Lett. **4**, 215 (1987); B. Caroli, C. Caroli, and B. Roulet, J. Phys. Paris **48**, 1423 (1987).

A NEW FORMULATION FOR DENDRITIC CRYSTAL GROWTH IN TWO DIMENSIONS

E. A. Coutsias* and H. Segur

*Department of Mathematics and Statistics
University of New Mexico
Albuquerque, New Mexico, 87131-1141

University of Colorado
Department of Mathematics
Boulder, Colorado 80309-0526

ABSTRACT

The objective of this paper is to study the growth of dendritic crystals in two spatial dimensions plus time. The paper makes three contributions. (i) We propose a new dynamic criterion to test physical mechanisms that might produce the velocity selection that is observed experimentally. We have not yet determined how the results of this criterion compare with those obtained by other criteria, such as microscopic solvability. (ii) The (known) equations of motion are restated in terms of orthogonal parabolic coordinates, a natural coordinate system in which to study perturbations of a parabolic (Ivantsov) interface. Among its other advantages, this formulation permits a larger class of behaviours far from the tip of the crystal than is allowed in the usual representation. (iii) On an initially parabolic interface, the analogue of the Mullins-Sekerka instability is more delicate than previously had been assumed. In particular, we find numerically that the range of unstable "wavenumbers" is bounded away from zero; i.e., sufficiently low wavenumbers are stable. Moreover, our preliminary calculations show parabolae, characterised by Peclet numbers of order 1, for which the linear instability is completely suppressed by enough surface tension. Such suppression is impossible on a flat interface.

INTRODUCTION

Of all of the physical problems discussed at this workshop, perhaps none has been more influential in distilling the fundamental concepts of the subject of Asymptotics Beyond All Orders than the problem of growing dendritic crystals. Within these Proceedings, all of the papers on the geometric model indirectly address the problem of dendritic crystals. Moreover, the papers by Gollub [6], Kessler [9], and Levine [15] (also in these Proceedings) discuss aspects of the hypothesis of "microscopic solvability", which asserts that the tip of a dendritic crystal grows with a constant velocity determined by a transcendentally small quantity related to the surface tension.

Asymptotics beyond All Orders, Edited by H. Segur *et al.*
Plenum Press, New York, 1991

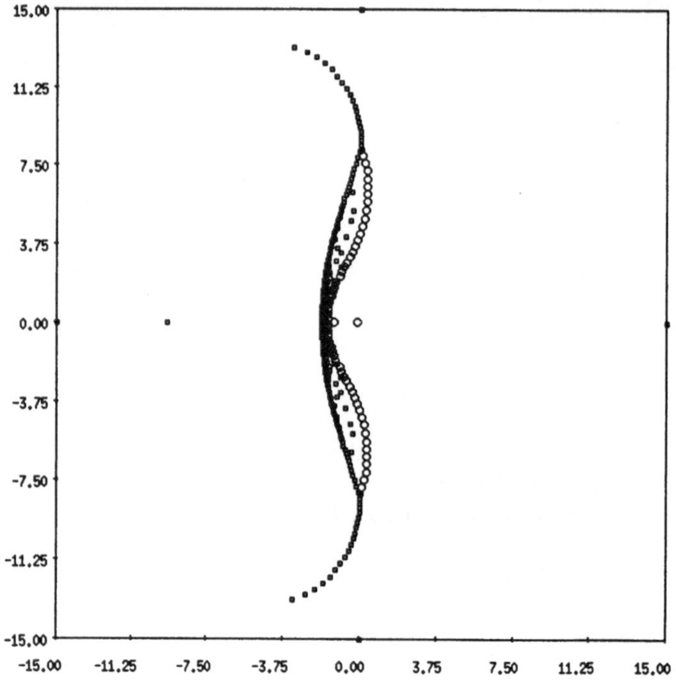

Figure 1. The spectrum for $Pe = 4$, $M = 512$ (even), $\epsilon = .005$. Eigenvalues on the stabilized arc are represented by circles, spurious eigenvalues by squares.

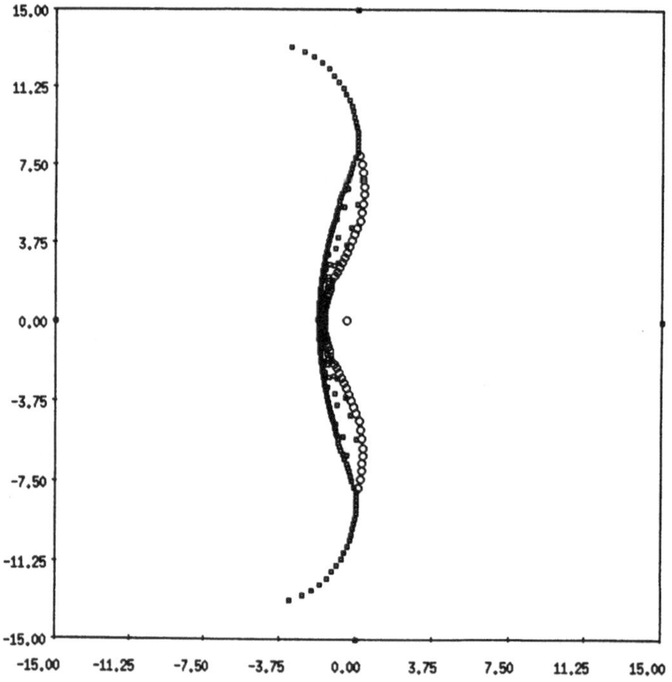

Figure 2. Same as in Fig. 1, for odd truncation.

A complete statement of the problem can be found in the papers just cited. Briefly, experiments show that when a dendritic crystal grows into a slightly supercooled melt, the shape of the moving interface between the (liquid) melt and the (solid) crystal is inherently unsteady and complicated, but also that the tip of the crystal moves with a nearly constant velocity, and with a nearly constant shape. (For a demonstration, see Figure 1 of Gollub, in these Proceedings, or [4], [5].) The steady-state model of Ivantsov [8] predicts that the temperature in the melt far from the crystal ("the undercooling") determines a dimensionless product of the tip radius and the tip velocity ("the Peclet number"), but not their separate values. Thus Ivantsov's model is too simple, in the sense that some physical effect omitted from Ivantsov's model selects a particular tip speed. The question is:
Which physical effects should be added to Ivantsov's model, to create a more complicated model in which both the tip speed and the tip radius are selected by the undercooling?
Candidates include: surface tension along the interface, with or without crystalline anisotropy, interfacial kinetics, and others.

In the presence of surface tension (or any of the other candidates mentioned above), Ivantsov's parabolic crystals no longer solve the equations of (steadily growing) motion. The hypothesis of microscopic solvability asserts that the physical mechanism that selects the tip velocity is the one for which the equations of motion admit a steadily growing, nearly parabolic interface, and the (nearly) parabolic tip that is observed is the one that grows steadily. If the equations admit more than one steadily growing interface shape, then the observed shape is the one that is dynamically stable. This hypothesis has led to the conclusion that surface tension with crystalline anisotropy provides the selection mechanism observed experimentally (cf. Barbieri et al. [1], BenAmar and Pomeau [2], Caroli et al. [3], Kessler and Levine [11], Meiron [16], Saito et al. [20]). Within these proceedings, see Levine [15] for a more complete statement of this approach, and see Gollub [6] for a comparison of the predictions of this model with experimental observations.

While this approach has been succesful in several respects, we feel that it suffers from two logical weaknesses.

1) The question of whether a steadily growing solution exists or not turns out to depend fundamentally, and very delicately, on boundary conditions imposed on the shape of the interface far from the tip of the crystal. However, real crystals are inherently unsteady (and very messy) far from their tips, so it is hard to imagine that steady boundary conditions imposed there ought to play an important role in the selection of the tip velocity.

2) The notions of a steadily growing solution, and of the dynamic stability of such a solution, both are relevant in the limit $t \to \infty$. However, real crystals apparently select their velocities rather quickly (e.g., after encountering an impurity), so an appropriate theory also ought to provide a mechanism to select a velocity on a fairly short time scale.

The primary objective of the research reported here is to formulate an alternative model to microscopic solvability, one that is free of the logical difficulties mentioned above. Our alternative hypothesis is stated in Section 2. The first step in implementing the hypothesis is to recast the equations of motion in parabolic coordinates; this reformulation is given in Section 3. In Section 4 we linearize these equations about an Ivantsov parabola, to obtain approximate equations governing the growth of an interface on a short time scale. (Our hypothesis is that the tip velocity is selected on this short time scale.) The homogeneous solutions of this linearized problem provide the analogue, on a parabolic interface, of the celebrated instability found by Mullins and Sekerka [19] on a flat interface. This instability is analyzed and discussed in Section 5. Our analysis indicates that the parabolic geometry changes the instability in fundamental ways, which apparently had not been noticed before.

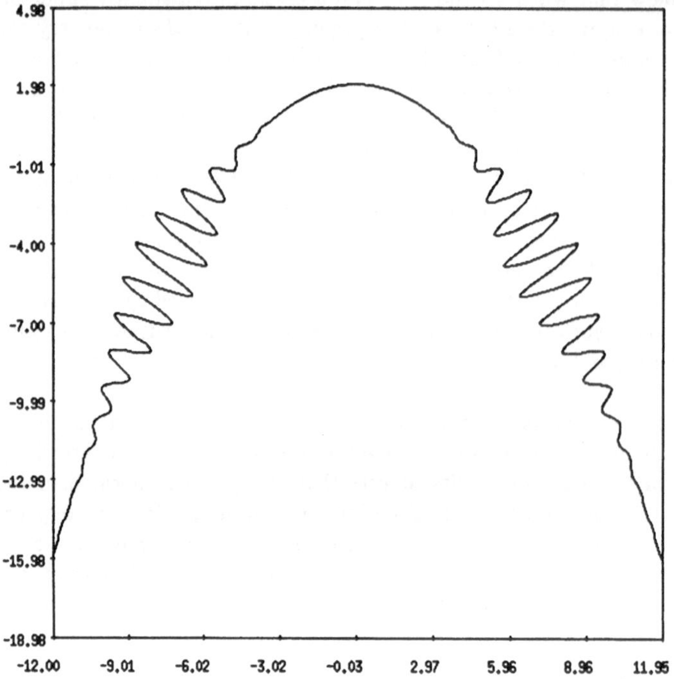

Figure 3. The real part of a localized eigenmode, corresponding to the eigenvalue $\lambda = 3.926 \pm i7.534$ for $\mathcal{P} = 4$, $\epsilon = .001$.

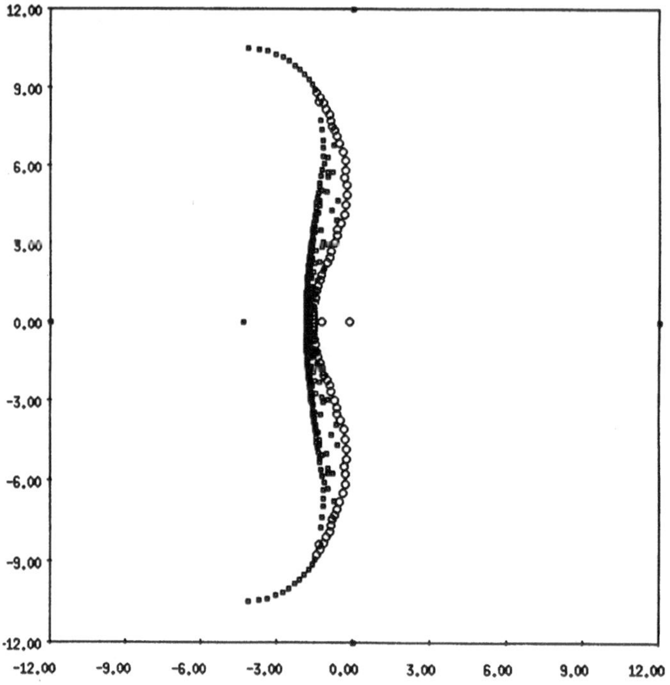

Figure 4. Same as in Fig. 1, with $\epsilon = .007$. All computed eigenvalues have negative real parts.

Our criterion is based on three assumptions:

(i) A growing dendritic crystal is inherently unsteady, except in the immediate neighborhood of the tip.
(ii) Selection of the tip velocity is done on a relatively short time scale.
(iii) The tip velocity is selected by some physical effect that is a small perturbation to the balance inherent in Ivantsov's model (in which latent heat, created at the interface, is carried away by diffusion).

In this paper, we demonstrate our proposed criterion by deriving the equations to determine whether surface tension, with or without crystalline anisotropy, can select a tip velocity in two spatial dimensions. We emphasize that the approach is not restricted to these physical effects.

Let us conduct a thought experiment, in which a crystal is growing steadily into an undercooled melt in two dimensions. For $t < 0$, there is no surface tension, and the undercooling would permit any one of an entire family of Ivantsov's parabolic crystals to grow. Pick one parabola from the family (i.e., select a velocity).

For $t \geq 0$, we turn on a small amount of surface tension. Now the parabolic interface is no longer an equilibrium solution, and the interface will deform. For short times, the size of the deformation will be proportional to the surface tension, and one can linearize the equations about the parabolic shape. We assume that the tip velocity is selected on this short time scale, so these linearized equations contain the selection mechanism, if one exists.

The time-dependent deformation of the interface away from its initially parabolic shape is essentially the Mullins-Sekerka [19] instability, but on a parabolic front. As one might expect, different wavelengths have different growth-rates. We apply the following selection criterion:
Given the initially parabolic shape of the interface, with its own tip velocity, does the resulting (Mullins-Sekerka) instability appear in the neighborhood of the tip? If so, then this particular parabola is not selected by surface tension. We say that a particular parabola is selected by surface tension only if the observed instability leaves quiescent a neighborhood of the tip. If no parabola is selected in this way, then we conclude that surface tension does not provide a selection mechanism.

The equations required to implement this criterion, derived in Sections 3 and 4, are similar to those previously obtained by Langer and Müller-Krumbhaar [13], with two significant differences.
(i) They were interested in questions of stability (as $t \to \infty$) of the parabolic front, which led them into delicate questions about boundary conditions far from the tip, and which eventually spawned the hypothesis of microscopic solvability. In the model proposed here, stability is irrelevant because the tip velocity is selected on a short time scale, and delicate questions about boundary conditions never arise.
(ii) Our linearized equations are obtained by expanding in the (small) surface tension parameter. However, this parameter multiplies the curvature (i.e., the highest derivative) in the Gibbs-Thomson condition, so the expansion becomes disordered for perturbations of sufficiently short wavelength. To overcome this difficulty, we include in our leading-order perturbation equations a higher-derivative term proportional to the surface tension. Our expansion is a singular-perturbation expansion in this sense.

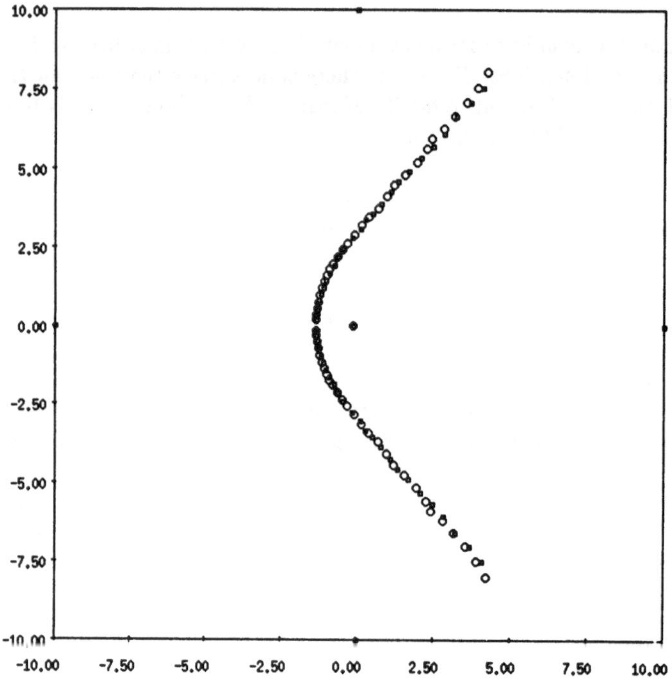

Figure 5. The spectrum for $\mathcal{P} = 4$, with $\epsilon = .001$. Only the proper eigenvalues are shown for truncations $M = 444$ (squares) and $M = 512$ (circles).

FORMULATION OF THE PROBLEM IN PARABOLIC COORDINATES

Consider a crystal, of initially parabolic shape, growing into an undercooled melt in two dimensions. Far from the solid-melt interface, the melt is assumed to be at temperature $T_\infty < T_M$, with T_M the melting temperature, while the solid is assumed to approach the temperature $\bar{T}$. Assume that the solid phase grows into the liquid in the positive z direction at a constant velocity V and that the solid and melt have molar specific heat and diffusivity, respectively, $C_p^\pm$, $D^\pm$, where + (−) refers to the liquid (solid) phases. Let L be the molar specific heat of solidification.

We introduce a coordinate system that is moving in the positive z direction with speed V with origin fixed at the tip. The location of the interface is expressed by the equation

$$\Phi(z,x,t) \equiv z - \zeta(x,t) = 0. \tag{1}$$

Defining the diffusive length scale $\ell = 2D^+/V$, replacing all lengths by dimensionless quantities in terms of ℓ, and scaling time by $2\ell/V$, we are led to the dimensionless system of equations for the temperature field

$$U^+ \to \Delta \quad , \quad z > \zeta(x,t), \begin{cases} z \to \infty \\ \text{or } |x| \to \infty \text{ , z fixed} \end{cases} \tag{2}$$

$$\Delta U^+ + 2\frac{\partial U^+}{\partial z} = \frac{\partial U^+}{\partial t} \quad , \quad z > \zeta(x,t) \tag{3}$$

$$\delta \, \Delta \, U^- + 2\frac{\partial U^-}{\partial z} = \frac{\partial U^-}{\partial t} \quad , \quad z < \zeta(x,t) \tag{4}$$

$$U^- \to \bar{\Delta} \quad , \quad z \to -\infty, \text{x fixed}, z < \zeta(x,t) \tag{5}$$

$$U^\pm = -\epsilon(1 - \alpha\cos 4\theta)\mathcal{K} \qquad \text{on } z = \zeta(x,t) \tag{6}$$

$$\left[\vec{V}_\perp - \beta\nabla U^- + \nabla U^+\right] \cdot \vec{n} = 0 \qquad \text{on } z = \zeta(x,t). \tag{7}$$

Here we have introduced the quantities

$$U = \frac{C_p^+}{L}(T - T_M), \quad \beta = \frac{D^- C_p^-}{D^+ C_p^+}, \qquad \epsilon = \frac{\gamma T_M C_p^+}{\ell L^2},$$
$$\Delta = \frac{C_p^+}{L}(T_\infty - T_M), \quad \bar{\Delta} = \frac{C_p^+}{L}(\bar{T} - T_M), \quad \delta = \frac{D^-}{D^+} \quad .$$

The ratio of capillary to diffusive lengths, ϵ, is a natural small parameter for our problem.

Of the two boundary conditions at the interface, (7) describes the energy balance there: the interface advances at a rate $(\mathbf{V}_\perp \cdot \mathbf{n})$ so that latent heat released is carried away by diffusion into both the solid and liquid phases. The other condition, (6), is the Gibbs–Thomson condition which asserts that the temperature at the interface is suppressed below T_M by capillary effects, by an amount proportional to the local curvature, $\mathcal{K} = \zeta_{xx}/\left[1 + (\zeta_x)^2\right]^{\frac{3}{2}}$. The other factor in (6), $(1 - \alpha\cos 4\theta)$, with θ the angle between the local normal and the z-direction, models the effects of crystalline anisotropy in the interfacial energy. Here a four-fold symmetry of the crystal is assumed [10], [1].

The shape of the crystal is assumed to be initially that of an Ivantsov parabola [8]. Following Horvay and Cahn [7], we introduce parabolic coordinates ξ and η defined by :

$$x = \xi\eta \quad , \quad z = \frac{\xi^2 - \eta^2}{2}. \tag{8}$$

Defining $\rho^2 = \xi^2 + \eta^2 = 2\sqrt{x^2+z^2}$ we have

$$\nabla = \frac{1}{\rho}\left(\vec{e_\xi}\frac{\partial}{\partial\xi} + \vec{e_\eta}\frac{\partial}{\partial\eta}\right) \tag{9}$$

while the unit vectors in the two systems are related by

$$\vec{e_x} = \frac{1}{\rho}(\eta\vec{e_\xi} + \xi\vec{e_\eta}) \quad , \quad \vec{e_z} = \frac{1}{\rho}(\xi\vec{e_\xi} - \eta\vec{e_\eta}) \tag{10}$$

and

$$\vec{e_\xi} = \frac{1}{\rho}(\eta\vec{e_x} + \xi\vec{e_z}) \quad , \quad \vec{e_\eta} = \frac{1}{\rho}(\xi\vec{e_x} - \eta\vec{e_z}) \ . \tag{11}$$

Throughout this discussion we have fixed a branch for our (double-valued) coordinate transformation by adopting $\xi \geq 0, \eta \cdot x \geq 0$. The fronts that we consider are nearly parabolic (typically of the form $\xi = A + \epsilon S(\eta, t, \epsilon)$). We assume that each can be described by

$$\xi = \Xi(\eta, t). \tag{12}$$

Then its unit normal $\vec{n}$ and curvature $\mathcal{K}$ are given by

$$\vec{n} = \frac{\vec{e_\xi} - \Xi_\eta\vec{e_\eta}}{\sqrt{1+\Xi_\eta^2}}, \tag{13}$$

$$\mathcal{K}[\eta, \Xi(\eta,t)] = \frac{(\Xi - \eta\Xi_\eta)(1+\Xi_\eta^2) - (\eta^2+\Xi^2)\Xi_{\eta\eta}}{\left[(\eta^2+\Xi^2)(1+\Xi_\eta^2)\right]^{\frac{3}{2}}}, \tag{14}$$

while its normal velocity is

$$V_\perp = \vec{n}\cdot(2\vec{e_z} + \rho\Xi_t\vec{e_\xi})\,. \tag{15}$$

The system (2-7), rewritten in parabolic coordinates, becomes

$$U^+ \to \Delta \quad , \quad \xi \to \infty \ , \ \eta \text{ fixed} \tag{16}$$

$$\left(\frac{\partial^2}{\partial\xi^2} + \frac{\partial^2}{\partial\eta^2}\right)U^+ + 2\left(\xi\frac{\partial}{\partial\xi} - \eta\frac{\partial}{\partial\eta}\right)U^+ = \rho^2\frac{\partial U^+}{\partial t} \quad , \quad \xi > \Xi(\eta,t) \tag{17}$$

$$\delta\left(\frac{\partial^2}{\partial\xi^2} + \frac{\partial^2}{\partial\eta^2}\right)U^- + 2\left(\xi\frac{\partial}{\partial\xi} - \eta\frac{\partial}{\partial\eta}\right)U^- = \rho^2\frac{\partial U^-}{\partial t} \quad , \quad \xi < \Xi(\eta,t) \tag{18}$$

$$U^- \to \bar{\Delta} \quad , \quad |\eta| \to \infty, \xi = 0 \tag{19}$$

$$2(\Xi + \eta\Xi_\eta) + (\eta^2+\Xi^2)\Xi_t = \left(\frac{\partial}{\partial\xi} - \Xi_\eta\frac{\partial}{\partial\eta}\right)[\beta U^- - U^+] \quad , \quad \text{on } \xi = \Xi(\eta) \tag{20}$$

$$U^\pm = -\epsilon\left(1 - \alpha + \frac{8\alpha\left(\frac{-\eta+\Xi\Xi_\eta}{\Xi+\eta\Xi_\eta}\right)^2}{\left[1+\left(\frac{-\eta+\Xi\Xi_\eta}{\Xi+\eta\Xi_\eta}\right)^2\right]^2}\right)\mathcal{K}[\eta,\Xi(\eta)] \quad , \quad \text{on } \xi = \Xi(\eta), \tag{21}$$

with $\mathcal{K}$ given by (14).

In the sequel we assume that the time scale governing the growth of the crystal is much slower than the time required by the temperature field to reach equilibrium (quasi-static approximation) so that the time derivatives in (17, 18) are neglected and the only time dependence in the problem enters through the energy balance (20).

LINEARIZATION FOR SHORT TIMES, WITH SMALL SURFACE TENSION

In this section we derive the equations for the evolution, under the influence of weak surface tension, of an interface that is initially close to an Ivantsov parabola. For $\epsilon << 1$, we assume that for short times, the interface and temperature fields have the form:

$$\Xi = A + \epsilon S(\eta, t; \epsilon) \tag{22}$$
$$U^{\pm} = U_I^{\pm}(\eta, \xi) + \epsilon u^{\pm}(\eta, \xi, t; \epsilon) \tag{23}$$

where $\xi = A$ and $U_I^{\pm}$ are the Ivantsov interface and temperature field respectively. Consider expansions of the various equations in powers of ϵ. At leading order, the Gibbs-Thomson relation (21) gives

$$U_I^{\pm} = 0 \text{ at } \xi = A, \tag{24}$$

and at higher order

$$\begin{aligned}\left(S\frac{\partial U_I^{\pm}}{\partial \xi}(\eta, A) + u^{\pm}(\eta, A, t; \epsilon)\right) + \epsilon S\left\{\frac{S}{2}\frac{\partial^2 U_I^{\pm}}{\partial \xi^2}(\eta, A) + \frac{\partial u^{\pm}}{\partial \xi}(\eta, A, t; \epsilon)\right\} + \mathcal{O}(\epsilon^2) \\ = \frac{1}{(\eta^2 + A^2)^{3/2}}\left[A + \epsilon\left\{S - \frac{3SA^2}{\eta^2 + A^2} - \eta S_\eta - (\eta^2 + A^2)S_{\eta\eta}\right\}\right] \\ \left[1 - \alpha + \frac{8\alpha A^2\eta^2}{(A^2 + \eta^2)^2} - \frac{16\epsilon\alpha A\eta(A^2 - \eta^2)}{(A^2 + \eta^2)^3}\left\{(A^2 + \eta^2)S_\eta + \eta S\right\}\right] + \mathcal{O}(\epsilon^2)\end{aligned} \tag{25}$$

We have retained higher order terms in (25) because we expect perturbations to develop high spatial derivatives and we specifically want to study their leading effects.

The equation of energy balance at the interface (20) becomes

$$\begin{aligned}2A + \epsilon\left[2(S + \eta S_\eta) + (\eta^2 + A^2)S_t + \epsilon 2ASS_t + \epsilon^2 S^2 S_t\right] \\ = \left(\frac{\partial}{\partial \xi} - \epsilon S_\eta \frac{\partial}{\partial \eta}\right)\left[\left(\beta U_I^- - U_I^+\right) + \epsilon\left(\beta u^- - u^+\right)\right]\Big|_{\xi = A + \epsilon S}.\end{aligned} \tag{26}$$

Expanding functions in Taylor series around the interface position we get at leading order

$$2A = \beta\frac{\partial U_I^-}{\partial \xi}(\eta, A) - \frac{\partial U_I^+}{\partial \xi}(\eta, A) \tag{27}$$

and at higher order

$$\begin{aligned}2(S + \eta S_\eta) + (\eta^2 + A^2)S_t + \epsilon\left(2ASS_t + \epsilon S^2 S_t\right) = \\ \left[S\frac{\partial^2}{\partial \xi^2} - S_\eta\frac{\partial}{\partial \eta} + \epsilon\left(\frac{S^2}{2}\frac{\partial^3}{\partial \xi^3} - SS_\eta\frac{\partial^2}{\partial \xi \partial \eta}\right)\right]\left(\beta U_I^- - U_I^+\right)\Big|_{\xi = A} \\ + \left[\frac{\partial}{\partial \xi} + \epsilon\left(S\frac{\partial^2}{\partial \xi^2} - S_\eta\frac{\partial}{\partial \eta}\right)\right]\left(\beta u^- - u^+\right)\Big|_{\xi = A} + \mathcal{O}(\epsilon^2).\end{aligned} \tag{28}$$

We proceed now with the derivation of the zeroth-order (Ivantsov) solution. Collecting all expressions, we have that to $\mathcal{O}(1)$ the temperature satisfies:

$$U_I^+ \to \Delta < 0 \quad , \quad \xi \to \infty \tag{29}$$

$$\left(\frac{\partial^2}{\partial \xi^2} + \frac{\partial^2}{\partial \eta^2}\right) U^+ + 2\left(\xi \frac{\partial}{\partial \xi} - \eta \frac{\partial}{\partial \eta}\right) U^+ = 0 \, , \, \xi > A \tag{30}$$

$$\delta \left(\frac{\partial^2}{\partial \xi^2} + \frac{\partial^2}{\partial \eta^2}\right) U^- + 2\left(\xi \frac{\partial}{\partial \xi} - \eta \frac{\partial}{\partial \eta}\right) U^- = 0 \, , \, \xi < A \tag{31}$$

$$U_I^- \to \bar{\Delta} \quad , \quad |\eta| \to \infty, \xi = 0 \tag{32}$$

$$U_I^\pm = 0 \quad \text{on } \xi = A \tag{33}$$

$$2A = \beta \frac{\partial U_I^-}{\partial \xi} - \frac{\partial U_I^+}{\partial \xi} \quad \text{on } \xi = A. \tag{34}$$

The solution is found by separating variables to be:

$$U_I^+(\xi) = \Delta - C \int_\xi^\infty e^{-s^2} ds, \tag{35}$$

$$U_I^-(\xi) = \bar{\Delta} - \bar{C} \int_0^\xi e^{-s^2/\delta} ds, \tag{36}$$

where the constants $C, \bar{C}$ are determined from the boundary condition (33) to be

$$C = \frac{\Delta}{\int_A^\infty e^{-t^2} dt} \quad , \quad \bar{C} = \frac{\bar{\Delta}}{\int_0^A e^{-t^2/\delta} dt}. \tag{37}$$

Define the Peclet number to be

$$\mathcal{P} = A^2 \tag{38}$$

where A^2 is the dimensionless tip radius. Then (34) results in the relation:

$$\sqrt{\pi \mathcal{P}} e^{\mathcal{P}} erfc(\sqrt{\mathcal{P}}) = -\Delta - \frac{\beta \bar{\Delta} e^{\mathcal{P}(1-1/\delta)} erfc(\sqrt{\mathcal{P}})}{\sqrt{\delta} erf(\sqrt{\mathcal{P}/\delta})} \tag{39}$$

in which the error function, $erf(z)$, and complementary error function, $erfc(z)$, are defined as usual:

$$erf(z) = \frac{2}{\sqrt{\pi}} \int_0^z e^{-t^2} dt, \quad erfc(z) = \frac{2}{\sqrt{\pi}} \int_z^\infty e^{-t^2} dt \ .$$

Relation (39) was first derived by Ivantsov ([8]) with $\bar{\Delta} = 0$.

At this point we assume that the various thermal coefficients for the solid and melt phases are the same and that the solid is kept at the freezing temperature at ∞, so that

$$\delta = 1 \quad , \quad \bar{\Delta} = 0 \quad , \quad \beta = 1.$$

Under these assumptions

$$\left.\frac{\partial U_I^+}{\partial \xi}\right|_A = -2\sqrt{\mathcal{P}} \, , \quad \left.\frac{\partial^2 U_I^+}{\partial \xi^2}\right|_A = 4\mathcal{P} \, , \quad \left.\frac{\partial U_I^-}{\partial \xi}\right|_A = \left.\frac{\partial^2 U_I^-}{\partial \xi^2}\right|_A = 0. \tag{40}$$

We now obtain the equations for the perturbation ($u^\pm$, S) at leading order as $\epsilon \to 0$. Using (40) we have

$$\left(\frac{\partial^2}{\partial\xi^2}+\frac{\partial^2}{\partial\eta^2}\right)u^+ + 2\left(\xi\frac{\partial}{\partial\xi}-\eta\frac{\partial}{\partial\eta}\right)u^+=0\ ,\quad \xi<A \tag{41}$$

$$\left(\frac{\partial^2}{\partial\xi^2}+\frac{\partial^2}{\partial\eta^2}\right)u^- + 2\left(\xi\frac{\partial}{\partial\xi}-\eta\frac{\partial}{\partial\eta}\right)u^-=0\ ,\quad \xi>A \tag{42}$$

$$\begin{aligned} u^+(\eta,\sqrt{\mathcal{P}},t)-2\sqrt{\mathcal{P}}S &= u^-(\eta,\sqrt{\mathcal{P}},t) \qquad (43)\\ &= \frac{\sqrt{\mathcal{P}}-\epsilon(\eta^2+\mathcal{P})S_{\eta\eta}}{(\eta^2+\mathcal{P})^{3/2}}\left[1-\alpha+\frac{8\alpha\mathcal{P}\eta^2}{(\mathcal{P}+\eta^2)^2}\right]+\mathcal{O}(\epsilon^2)\\ 2(S+\eta S_\eta)+(\eta^2+\mathcal{P})S_t &= -4\mathcal{P}S+\epsilon\sqrt{\mathcal{P}}(4\mathcal{P}-2)S^2 \qquad (44)\\ &+\left(\frac{\partial}{\partial\xi}+\epsilon S\frac{\partial^2}{\partial\xi^2}-\epsilon S_\eta\frac{\partial}{\partial\eta}\right)(u^- - u^+)|_{\xi=\sqrt{\mathcal{P}}}+\mathcal{O}(\epsilon^2). \end{aligned}$$

For solutions of (41–44) that are $\mathcal{O}(1)$ and whose derivatives are also $\mathcal{O}(1)$, leading order effects can be obtained simply by setting $\epsilon = 0$ in (43, 44). However, (41–44) also admit highly oscillatory solutions in which $S = \mathcal{O}(1)$, $\epsilon S_{\eta\eta} = \mathcal{O}(1)$, even though $\epsilon << 1$. For those highly oscillatory solutions we must retain the highest derivative term in (43), $\epsilon S_{\eta\eta}$, even for $\epsilon \to 0$. Thus the small ϵ expansion is a singular perturbation expansion, as had been anticipated by Langer and Müller–Krumbhaar [13]. It follows from (41, 42) that if $\frac{\partial}{\partial\eta} = \mathcal{O}(\epsilon^{-1/2})$ then $\frac{\partial}{\partial\xi} = \mathcal{O}(\epsilon^{-1/2})$ as well. Then estimates of the nominally small terms at (44) indicate that they all remain small in the limit in which the extra term in (43) becomes important: $\partial_\eta = \mathcal{O}(\epsilon^{-1/2})$, $\partial_\xi = \mathcal{O}(\epsilon^{-1/2})$ as $\epsilon \to 0$. To summarize, the small ϵ limit of (41–44) is obtained by retaining the term $\epsilon S_{\eta\eta}$ in (43) and letting $\epsilon \to 0$ elsewhere.

The Laplace equation (41, 42) is separable in parabolic coordinates [18]. Assuming for the temperature fields and interface deformation the expansions

$$u^\pm = \sum_{j=0}^{\infty} W_j^\pm(t)H_j(\eta)F_j^\pm(\xi), \tag{45}$$

$$S(\eta,t;\epsilon) = \sum_{j=0}^{\infty} S_j(t)H_j(\eta). \tag{46}$$

(where the coefficients S_j, W_j also depend on $\mathcal{P}$, ϵ and α), we find

$$H_j''-2\eta H_j'+2jH_j=0, \tag{47}$$

$$F_j''+2\xi F_j'-2jF_j=0. \tag{48}$$

The first of these is the Hermite equation, in which requiring H_j to be real and to grow no worse than algebraically at infinity results in j being a nonnegative integer ([14]):

$$H_j(\eta) = (-1)^j e^{\eta^2}\frac{d^j e^{-\eta^2}}{d\eta^j},\ j=0,1,2,\ldots \tag{49}$$

$$\text{with } \int_{-\infty}^{\infty} e^{-x^2}H_nH_m dx = \delta_{mn}2^n n!\sqrt{\pi}\equiv\delta_{mn}c_n. \tag{50}$$

Also, requiring that F_j^+ grow no worse than algebraically at infinity and that F_j^- and $F_j^{-\prime}$ be continuous at $\xi = 0$ results in the following expressions for $F_j^\pm$:

$$F_j^-(\xi) = \begin{cases} \sum_{k=0}^m a_{2k}^j \xi^{2k}, & j = 2m \\ \sum_{k=0}^m a_{2k+1}^j \xi^{2k+1}, & j = 2m+1 \end{cases} \tag{51}$$

$$\text{with } a_{k+2}^j = \frac{2(j-k)}{(k+2)(k+1)} a_k^j \quad , \quad a_1^{2j+1} = a_0^{2j} \quad , \quad a_0^{2j+2} = \frac{a_0^{2j}}{4(j+1)}$$

$$F_j^+(\xi) = \int_\xi^\infty e^{-s^2} \frac{(s-\xi)^j}{j!} ds. \tag{52}$$

For future reference, it is convenient to introduce

$$F_{-1}^- \equiv 0, \; F_{-1}^+(A) \equiv -e^{-A^2}, \tag{53}$$

and to note that the functions $F_j^\pm$, H_j satisfy the recursions:

$$F_j^{\pm\prime} = \mp F_{j-1}^\pm \tag{54}$$

$$F_{j-2}^\pm \mp 2\xi F_{j-1}^\pm - 2j F_j^\pm = 0 \tag{55}$$

$$H_j' = 2j H_{j-1} \tag{56}$$

$$2j H_{j-1} - 2\eta H_j + H_{j+1} = 0 \; . \tag{57}$$

It is well-known [14] that series of the form (45, 46) can be used to represent functions with quite general growth behaviour as $\eta \to \infty$. In fact, for a function $S(\eta)$ on the infinite interval $(-\infty, \infty)$ which is piecewise smooth on any finite subinterval and such that

$$\int_{-\infty}^\infty e^{-\eta^2} S^2(\eta) d\eta < \infty, \tag{58}$$

the expansion (46) with coefficients

$$S_j = \frac{1}{c_j} \int_{-\infty}^\infty e^{-\eta^2} H_j(\eta) S(\eta) d\eta$$

converges at each point to the value $1/2(S(\eta+) + S(\eta-))$. The broad class of interfacial shapes allowed by (58) is one of the main advantages of using a parabolic coordinate system.

Turning now to the perturbed Gibbs–Thomson relation (43), we define two sets of coefficients, $\sigma_{2k}(\mathcal{P}, \alpha)$ and $f_j(\mathcal{P}, \alpha)$ by

$$\frac{1}{(\eta^2 + \mathcal{P})^{3/2}} \left[1 - \alpha + \frac{8\alpha\mathcal{P}\eta^2}{(\mathcal{P} + \eta^2)^2} \right] \stackrel{\text{def}}{=} \sum_{k=0}^\infty \sigma_{2k}(\mathcal{P}; \alpha) H_{2k}(\eta) \tag{59}$$

and

$$\frac{S_{\eta\eta}}{(\eta^2 + \mathcal{P})^{1/2}} \left[1 - \alpha + \frac{8\alpha\mathcal{P}\eta^2}{(\mathcal{P} + \eta^2)^2} \right] \equiv \sum_{j=0}^\infty f_j(\mathcal{P}, \alpha) H_j. \tag{60}$$

Note that $\sigma_{2k+1} \equiv 0, k = 0, 1, \cdots$ as we are expanding an even function. Substituting the above expansions to (43) yields

$$\begin{aligned} \sum_{j=0}^\infty [\sqrt{\mathcal{P}} \sigma_j(\mathcal{P}; \alpha)] H_j - \epsilon \sum_{j=0}^\infty f_j H_j &= \sum_{j=0}^\infty [W_j^+ F_j^+(A)] H_j - 2\sqrt{\mathcal{P}} \sum_{j=0}^\infty S_j H_j \\ &= \sum_{j=0}^\infty [W_j^- F_j^-(A)] H_j \end{aligned}$$

and solving for $W_j^\pm$ we find:

$$W_j^+ = \left[\frac{\sqrt{\mathcal{P}}\sigma_j(\mathcal{P};\alpha)}{F_j^+(A)}\right] + \left[\frac{2\sqrt{\mathcal{P}}}{F_j^+(A)}\right] S_j - \left[\frac{\epsilon}{F_j^+(A)}\right] f_j(\mathcal{P},\alpha),$$

$$W_j^- = \left[\frac{\sqrt{\mathcal{P}}\sigma_j(\mathcal{P};\alpha)}{F_j^-(A)}\right] - \left[\frac{\epsilon}{F_j^-(A)}\right] f_j(\mathcal{P},\alpha).$$

Turning now to (44) we use the following identities:

$$\eta S = \sum_{j=0}^{\infty} S_j[\frac{1}{2}H_{j+1} + jH_{j-1}] \quad = \quad \sum_{j=0}^{\infty}[\frac{1}{2}S_{j-1} + (j+1)S_{j+1}]H_j$$
$$S_{-1} \equiv \; S_{-2} \equiv 0$$

so that

$$\eta S_\eta = \sum_{j=0}^{\infty}[jS_j]H_j + \sum_{j=0}^{\infty}[2(j+1)(j+2)S_{j+2}]H_j \tag{61}$$

and

$$(\eta^2 + \mathcal{P})S_t \;\; = \;\; \sum_{j=0}^{\infty}[\frac{1}{4}S'_{j-2} + \left(j + \mathcal{P} + \frac{1}{2}\right) S'_j + (j+2)(j+1)S'_{j+2}]H_j$$

with

$$S_t = \sum_{j=0}^{\infty} S'_j H_j.$$

Also,

$$\left.\frac{\partial}{\partial\xi}u^{\pm}\right|_{\sqrt{\mathcal{P}}} = \sum_{j=0}^{\infty} W_j^{\pm} F_j^{\pm\prime}(A) H_j$$

and using the relations (54) we have

$$\left.\frac{\partial}{\partial\xi}(u^- - u^+)\right|_{\sqrt{\mathcal{P}}} = \sum_{j=0}^{\infty}\{W_j^- F_{j-1}^-(A) + W_j^+ F_{j-1}^+(A)\}H_j$$

where we have used (53). We now substitute the expressions derived above into (44), to get

$$\begin{aligned}
&\sum_{j=0}^{\infty}(2S_j)H_j + 2\sum_{j=0}^{\infty}[jS_j + 2(j+1)(j+2)S_{j+2}]H_j \\
+ \quad & \sum_{j=0}^{\infty}\left[\frac{1}{4}S'_{j-2} + (j + \mathcal{P} + \frac{1}{2})S'_j + (j+1)(j+2)S'_{j+2}\right] H_j \\
= \quad & \sum_{j=0}^{\infty}\Big\{W_j^- F_{j-1}^-(A) + W_j^+ F_{j-1}^+(A)\Big\}H_j - 4\mathcal{P}\sum_{j=0}^{\infty} S_j H_j.
\end{aligned}$$

Here, the first two lines give the latent heat release from the basic motion of the parabolic front, due to the shape perturbations. The first term of the third line gives the flux of the perturbation field at the parabola, while the second term gives the flux of the leading field at the perturbed interface.

Collecting results, we finally get the system of equations, for $j = 0, 1, 2, \cdots$:

$$\begin{aligned}
&\frac{1}{4}S'_{j-2} + (j + \mathcal{P} + \frac{1}{2})S'_j + (j+1)(j+2)S'_{j+2} \\
+ \quad & 2(j + 1 + 2\mathcal{P}\tau_j^+(A))S_j + 4(j+1)(j+2)S_{j+2} \\
= \quad & 2\mathcal{P}(\tau_j^- - \tau_j^+)\sigma_j(\mathcal{P};\alpha) - 2\epsilon\sqrt{\mathcal{P}}(\tau_j^- - \tau_j^+)f_j(\mathcal{P},\alpha),
\end{aligned} \tag{62}$$

where $S_{-2} = S_{-1} = 0$, and the coefficients $\tau_j^{\pm}$ are defined by:

$$\tau_j^{\pm} = 1 \mp \frac{1}{2\sqrt{\mathcal{P}}} \frac{F_{j-1}^{\pm}(A)}{F_j^{\pm}(A)}. \tag{63}$$

Our analysis centers on system (62). It gives the evolution of perturbations to the parabolic shape of the front, in terms of the coefficients in the Hermite expansion of the shape perturbations. Our work is based on the hypothesis that if a selection mechanism exists, it must be possible to find it by studying the spectrum of this system as the controlling parameters ($\mathcal{P}$, ϵ, α) are varied, as well as the behavior of the unstable modes near the tip.

The $F_j^{\pm}$ are solutions of (48) but it is not necessary to solve such equations separately in order to determine the $\tau_j^{\pm}$. Indeed, by using the recursions (55) we see that

$$\tau_j^{\pm} = 1 + \frac{j}{2\mathcal{P}} \frac{1}{\tau_{j-1}^{\pm}} \tag{64}$$

which, together with the relations

$$\tau_0^{+} = 1 - \frac{e^{-\mathcal{P}}}{2\sqrt{\mathcal{P}} \int_{\sqrt{\mathcal{P}}}^{\infty} e^{-s^2}\, ds} \quad , \quad \tau_0^{-} = 1 \tag{65}$$

give complete recursion sequences that determines the coefficients $\tau_j^{\pm}$.

ANALOGUE OF THE MULLINS–SEKERKA INSTABILITY

The system of equations in (62) contains a great deal of information about the evolution in two dimensions of nearly parabolic fronts under the influence of weak surface tension, with or without crystalline anisotropy. Here are three kinds of information available from (62):
(a) Neglecting all time derivatives in the first line of (62) results in a set of nonhomogeneous, algebraic equations for the Hermite coefficients S_j of the steady-state shape of the perturbed interface, in the presence of nonzero surface tension. These steady shapes might or might not be dynamically stable.
(b) Neglecting the nonhomogeneous terms σ_j in the last line of (62) provides a linear, homogeneous system of differential equations for the shape of a growing interface. The family of solutions of these equations provides the analogue, on a parabolic interface, of the Mullins–Sekerka instability on a flat interface. This analogy is limited by the fact that the derivation of (62) required small surface tension ($\epsilon << 1$), whereas the original analysis of Mullins and Sekerka [19] was not restricted in this way.
(c) Having computed the time-dependent growth of the unstable modes of (62) for a particular Ivantsov parabola, one implements the dynamic selection criterion proposed in Section 2 by determining whether these growing modes leave quiescent a neighborhood of the tip of the growing crystal.

We develop now some notation for the study of system (62). We define $\mathbf{M}$ to be the operator of multiplication by η^2 and we let $\mathbf{N} = \mathbf{M} + \mathcal{P}\mathbf{I}$. The form of $\mathbf{N}$ is seen in the first line of (62). We also define the matrix $\mathbf{K}$ to be the Hermite representation of the operator $2 + 2\eta\frac{d}{d\eta} + 4\mathcal{P}\mathbf{T}^{+}$, given in the second line of (62) with $T_{ij}^{+} = \tau_j^{+}\delta_{ij}$. Finally we let $\mathbf{T}$ be the diagonal matrix with elements $T_j = \tau_j^{-} - \tau_j^{+}$, and $\mathbf{U}$ a matrix with only the second superdiagonal different from zero, and whose $(j, j+2)$ element is $U_{j,j+2} = 4(j+1)(j+2)$. Then, (62) can be written as

$$\mathbf{N}S' + \mathbf{K}S = 2\mathcal{P}\mathbf{T}\Sigma - 2\epsilon\sqrt{\mathcal{P}}\mathbf{T}F. \tag{66}$$

For the isotropic case, $\alpha = 0$. We introduce the symmetric matrix G_{mn} by writing

$$\begin{aligned} f_j(\mathcal{P},0) &= \frac{1}{c_j}\int_{-\infty}^{\infty} e^{-\eta^2}\frac{S_{\eta\eta}}{\sqrt{\eta^2+\mathcal{P}}}H_j(\eta)d\eta \\ &= \frac{4}{c_j}\sum_{l=0}^{\infty}(l+1)(l+2)S_{l+2}\int_{-\infty}^{\infty} e^{-\eta^2}\frac{H_l(\eta)H_j(\eta)}{\sqrt{\eta^2+\mathcal{P}}}d\eta \\ &= \frac{4}{c_j}\sum_{l=0}^{\infty}(l+1)(l+2)S_{l+2}G_{jl} = \frac{1}{c_j}\sum_{l,m=0}^{\infty} G_{jl}U_{lm}S_m, \end{aligned} \tag{67}$$

with the c_j defined in (50). The coefficients f_j, G_{mn} (as well as the $\tau_j^{\pm}$ of the previous section) can be found through recursions. These computations turn out to be quite involved numerically (due to the presence of undesired dominant solutions of the recursions) but a suitable use of asymptotics allows the determination of these coefficients to arbitrary accuracy. This computation will be presented in detail elsewhere.

For $\alpha \neq 0$ it follows from (60) that $\mathbf{N}^2F = ((1-\alpha)\mathbf{N}^2 + 8\mathcal{P}\alpha\mathbf{M})\,\mathbf{C}^{-1}\mathbf{GU}S$, so that (66) becomes

$$\mathbf{N}^2\mathbf{T}^{-1}(\mathbf{N}S' + \mathbf{K}S) - 2\mathcal{P}\mathbf{N}^2\Sigma = -2\epsilon\sqrt{\mathcal{P}}\left((1-\alpha)\mathbf{N}^2 + 8\mathcal{P}\alpha\mathbf{M}\right)\mathbf{C}^{-1}\mathbf{GU}S, \tag{68}$$

with $C_{ij} = c_i\delta_{ij}$. The latter form allows a computation for the anisotropic case to be performed with no additional complexity compared to the isotropic case.

Now we restrict attention to the isotropic case, $\alpha = 0$. For our actual computations, it proved beneficial to introduce a new scaling, in terms of orthonormal Hermite polynomials. Thus, we rescale the Hermite coefficients S_j as $\hat{S} = \mathbf{\Lambda}S$ with $\Lambda_{ij} = \delta_{ij}\sqrt{2^j j!}$. Then, equation (68) becomes

$$\begin{aligned} &\mathbf{\Lambda N \Lambda}^{-1}\mathbf{\Lambda}S' + \mathbf{\Lambda K\Lambda}^{-1}\mathbf{\Lambda}S - 2\mathcal{P}\mathbf{T\Lambda}\Sigma \\ &= -2\epsilon\sqrt{\mathcal{P}}\mathbf{TC}^{-1}\mathbf{\Lambda G\Lambda}^{-1}\mathbf{\Lambda U\Lambda}^{-1}\mathbf{\Lambda}S. \end{aligned} \tag{69}$$

We note that the matrix $\hat{\mathbf{N}} = \mathbf{\Lambda N\Lambda}^{-1}$ is symmetric, with main diagonal same as that of $\mathbf{N}$ (i.e. $N_j = \hat{N}_{jj} = (j + \mathcal{P} + 1/2)$), and with second super- and subdiagonals equal to ($\hat{N}_{j,j+2} = \hat{N}_{j+2,j} =$)$(1/2)\sqrt{(j+1)(j+2)}$. Similarly the matrix $\hat{\mathbf{K}} = \mathbf{\Lambda K\Lambda}^{-1}$ has nonzero elements $\hat{K}_{jj} = K_{jj} = 2(j+1+2\mathcal{P}\tau_j^+)$ and $\hat{K}_{j,j+2} = 2\sqrt{(j+1)(j+2)}$.

Also, $\hat{\mathbf{G}} = \mathbf{C}^{-1}\mathbf{\Lambda G\Lambda}^{-1}$ is given by

$$(\hat{G})_{ij} = \frac{\sqrt{2^i i!}}{c_i}G_{ij}\frac{1}{\sqrt{2^j j!}} = \frac{1}{\sqrt{\pi}}\frac{G_{ij}}{\sqrt{2^{i+j}i!j!}}$$

while $\hat{\mathbf{U}} = \mathbf{\Lambda U\Lambda}^{-1}$ is a matrix of zeroes except for the $(j, j+2)$ diagonal which contains

$$(\hat{U})_{j,j+2} = 2\sqrt{(j+1)(j+2)}.$$

Now we focus on the homogeneous problem associated with (66), in order to study the analogue of the Mullins-Sekerka instability on a parabolic interface. The homogeneous problem can be formulated as a generalized eigenvalue problem: assuming that $\hat{S}_j(t)$ has the form $\hat{S}_j(t) = s_j e^{\lambda t}$ we have the eigensystem

$$\lambda \hat{\mathbf{N}} s + \hat{\mathbf{K}} s = -2\epsilon\sqrt{\mathcal{P}}\mathbf{T}\hat{\mathbf{G}}\hat{\mathbf{U}} s$$

or

$$\lambda \hat{\mathbf{N}} s = -\left(\hat{\mathbf{K}} + 4\epsilon\sqrt{\mathcal{P}}\mathbf{B}\right) s, \tag{70}$$

where we have introduced $\mathbf{B} = \frac{1}{2}\mathbf{T}\hat{\mathbf{G}}\hat{\mathbf{U}}$. The matrix elements of $\mathbf{B}$ are

$$\begin{aligned} B_{ij} &= \frac{1}{2}T_{il}\hat{G}_{ln}\hat{U}_{nj} \quad \text{(sum over repeated indices)} \\ &= \frac{1}{2}T_{ii}\hat{G}_{i,j-2}\hat{U}_{j-2,j} \quad \text{(no sums)} \\ &= (\tau_i^- - \tau_i^+)\sqrt{j(j-1)}\hat{G}_{i,j-2} \quad , \quad j = 2,3,\cdots \\ &\quad \text{with } B_{i,0} = B_{i,1} = 0 \quad , \quad i = 1,2,\cdots \end{aligned}$$

Note that in this system even and odd modes uncouple, effectively doubling the size of any truncation. Thus all our computations were performed on purely even or odd mode truncations, and the term "order of truncation" is used to denote the number of even or odd modes actually used.

We present now some preliminary results that were obtained from our computations of the spectrum of the system (70) with small, but nonzero, ϵ. An approach that we found satisfactory in computing the spectrum utilizes the fact that (70) is in the standard form for the QZ–factorization algorithm of Moler and Stewart [17] to be applicable. We used the EISPACK generalized eigenvalue solver which allowed us to compute eigenvalues for truncations including up to 1600 even or odd modes. It was found by various comparisons that above truncations of order 500 – 700, roundoff contamination was appreciable. In our computations with $\epsilon = 0$ we found that we could readily identify eigenvalues that stabilize, i.e. remain unchanged to a given tolerance over several truncations and discard the rest. For given $\mathcal{P}$, after increasing the truncation above a certain limit, roundoff prevented any further stabilized eigenvalues to appear.

For any ϵ in the range $0 \leq \epsilon \leq .01$ for which we have carried out computations, most of the computed spectrum of the truncated system was composed of eigenvalues lying, roughly, on two arcs (Figs. 1, 2). The outer arc, horse-shoe shaped for $\epsilon = 0$, becomes wider as the truncation increases and is apparently composed entirely of "spurious" eigenvalues, or eigenvalues of the truncated system that do not correspond to eigenvalues of the full system. The second, bell-shaped arc spans the horseshoe and is composed of eigenvalues that, once they appear, quickly settle to values that change very little as the order of truncation is increased.

However, without surface tension, ($\epsilon = 0$), system (70) is ill-posed with the growth rate of the instability being an increasing function of the wavenumber. In this limit of vanishing surface tension, the only correction to eq. (62) enters through the nonhomogeneous term, modifying smoothly the steady shape of the Ivantsov parabola, proportionally to the surface tension. Since instability ensues for any $\mathcal{P}$, this nearly parabolic profile is always unstable.

For any ϵ arbitrarily small but finite there exists a sufficiently high mode number for which the term

$$2\epsilon\sqrt{\mathcal{P}}(\tau_j^- - \tau_j^+)f_j(A)$$

becomes important. This term makes our eigenvalue problem well-posed for $\epsilon > 0$, introducing an upper frequency cutoff. We performed calculations for $\mathcal{P} = 2,4$ and for truncations spanning the range 300 to 1600. For values of $\epsilon < .007$, only a part of the spectrum lies in the positive real half-plane, with nonzero lower wavenumber/frequency cutoffs (as well as an upper cutoffs for any $\epsilon > 0$, see Figs. 1, 2). The eigenfunctions corresponding to eigenvalues with positive real part appear as waves of constant phase velocity, traveling down the sides of the parabola. Some of the unstable eigenmodes are strongly localized away from the tip. As an example of such a localized eigenmode, for $\mathcal{P} = 4$, $\epsilon = .001$, the real part of the mode corresponding to the eigenvalue $\lambda = 3.926 \pm i7.534$ is shown in Fig. 3. Clearly this localized mode has an effective wavelength. We can use these effective wavelengths to compute phase velocities and dispersion curves.

We found that if the surface tension were chosen sufficiently high, the real parts of all the computed eigenvalues invariably became negative and the instability was suppressed. For $\mathcal{P} = 4$ and for $\epsilon > .007$ the spectrum lies entirely on the left half plane and no instability was found, as shown in Fig. 4.

For $\epsilon \neq 0$ however small, both arcs become bow-shaped and eventually they bend back to recross the imaginary axis. In (Figs. 1, 2) we give the spectra corresponding to even and odd eigenmodes, respectively, for $\mathcal{P} = 4$ and $\epsilon = .005$ for a truncation of order 512. As mentioned before, the "outer" arc (squares) does not settle but keeps changing shape as the order of truncation increase, while the "inner" arc (circles) simply elongates as truncation is increased and new eigenvalues appear, without changing shape. It appears that the computation for $\epsilon \neq 0$ becomes much more sensitive to roundoff errors as compared to the case $\epsilon = 0$, so that for the truncations we considered and with 16 decimal digit arithmetic, individual eigenvalues did not settle quite satisfactorily. However the arc they define did settle, as can be seen in Fig. 5 where the computed spectrum for even eigenmodes at $\mathcal{P} = 4$, $\epsilon = .001$ is shown for trucations $M = 444$ (squares) and $M = 512$ (circles), superimposed for comparison, with the spurious part of the spectrum omitted. In Fig. 4, (with $\epsilon = .007$), as well as in Figs. 1, 2, a number of values shown that lie on neither arc should be attributed to roundoff contamination. The results were also found to be quite sensitive to the accuracy by which various coefficients in (70) were known. Considerable care was required in employing techniques for the computation of these coefficients that gave them to the maximum accuracy of 16 digits employed in the rest of the computations.

In conclusion, we present a comparison of our preliminary findings with previous studies of dendritic growth. The analysis of Mullins and Sekerka [19] for the instability of a flat solidification front shows both similarities and differences from our anlysis of a parabolic front, due to inherent physical differences in the two processes. The main points are:

1. The flat front corresponds to $\mathcal{P} \to \infty$, while a parabolic front can be found for every positive finite value of $\mathcal{P}$.

2. On a parabolic front, we find that each of the unstable modes, like the one shown in Fig. 3, has an effective wavenumber ($k \geq 0$), that $|\Im\lambda|$ increases as k increases, and $|\Im\lambda| \to 0$ as $k \to 0$. Therefore, a comparison is possible between the findings for the spectrum of the parabolic case and that for the flat case which has fourier modes, and for which typical stability diagrams are given as relations between k and $\Re\lambda$ ([12]).

3. Both cases show a cutoff for large k; that is, disturbances with sufficiently small wavelengths decay in either geometry.

4. For k small ($\Im\lambda$ small), $\Re\lambda < 0$, which differs from the flat interface case.

5. At least for $\mathcal{P} = 2$ and $\mathcal{P} = 4$, the instability is apparently suppressed for sufficiently large surface tension.

6. It is inappropriate to apply the known results from a flat interface to the sides of a parabola (where the interfacial curvature approaches zero). Along the sides of a parabola we must take into account the tangential motion of the liquid phase, which is not present at the flat interface.

In order to compare our findings with the results of other approaches to the dendritic growth problem, such as those of microscopic solvability, we still need to analyze the behavior of the steady correction to the front as $\eta \to \infty$. This analysis will be reported elsewhere.

ACKNOWLEDGEMENTS

E.C. wishes to acknowledge the hospitality and the use of the computing facilities at Risø National Laboratory, Denmark. H.S. was partially supported by NSF grant # DMS-9096156. Charles Troup (UNM) helped with the graphics. Bruce Fast (CU) provided logistics and systems help. Some of the preliminary computations were done at the NCSA's Cray-YMP. Funding for the final calculations on the Cray2 at Kirtland Air Force base, was provided by AFOSR.

References

[1] A. Barbieri, D.C. Hong and J. Langer, Phys. Rev. A 35, (1988), 1801.

[2] M. Ben-Amar and Y. Pomeau, Europhys. Lett., 2, 307, (1986).

[3] B. Caroli, C. Caroli, B. Roulet and C. Misbah, J. Physique, 48, 547, (1987).

[4] A. Dougherty, P. D. Kaplan and J. P. Gollub, Phys. Rev. Lett. 58, (1987), 1652-1655.

[5] M. E. Glicksman, in *Crystal growth of electronic materials*, E. Kaldis, ed., pp 57-69, Elsevier Sci. Pub. B. V., 1985.

[6] J.P. Gollub, these Proceedings.

[7] G. Horvay and J. W. Cahn, Acta Metal. 9, (1961), 695-705.

[8] G. P. Ivantsov, Dokl. Acad. Nauk SSSR 58, (1947), 567.

[9] D.A. Kessler, these Proceedings.

[10] D.A. Kessler, J. Koplik and H. Levine, Adv. in Phys. 37, (1988), 255–336.

[11] D.A. Kessler and H. Levine, Phys. Rev. B 33, 7687, (1986).

[12] J. S. Langer, Rev. Mod. Phys. 52, (1980), 1-28.

[13] J.S. Langer and H. Müller-Krumbhaar, Acta Metall. 26, 1681–1687, (1978); (Ibid), 1689-1695; (Ibid), 1697–1708.

[14] N.N. Lebedev, *Special Functions and their Applications*, Dover, NY (1972).

[15] H. Levine, these Proceedings.

[16] D. Meiron, Phys. Rev. A 33, 2704, (1986).

[17] C. Moler and Stewart, SIAM J. Num. Anal., 10, 241–256, (1973).

[18] P.M. Morse and H. Feshbach, Methods of Theoretical Physics, Vol. 1, McGraw-Hill, NY 1953.

[19] W.W. Mullins and R.F. Sekerka, J. Appl. Phys. 34, 444 (1964). 58, (1987),1652-1655.

[20] Y. Saito, G. Goldbeck-Wood and H. Müller-Krumbhaar, Phys. Rev. A 38, (1988), 2148-2157.

DIRECTIONAL GROWTH OF DILUTE MIXTURES AND LAMELLAR EUTECTICS

K. Kassner,[a] C. Misbah,[b] H. Müller-Krumbhaar,[a]

Y. Saito[c] and D.E. Temkin[d]

[a] Inst. f. Festkörperforschung, Forschungszentrum Jülich, Germany

[b] GPS, associé au CNRS, Place Jussieu, Paris, France

[c] Physics Department, Keio University, Yokohama, Japan

[d] I.P. Bardin Institute for Ferrous Metals, Moscow, USSR

Pattern formation in directional growth of dilute mixtures and eutectics is studied. In the cellular and dendritic problems, we compare our results on tip selection with the theory of a needle-shaped crystal in a channel. The selection of a dendrite is based either on surface tension or kinetics anisotropy. The possibility for the interface to undergo chaotic motion at large growth speeds is discussed. For lamellar eutectics we present recent results on a parity-breaking transition and similarity laws.

I. INTRODUCTION

Our understanding of pattern formation in crystal growth has been greatly enhanced by the solution of the velocity selection problem for a free dendrite.[1] The key to this solution was the observation that surface tension is a singular perturbation. Another topic of intensive investigations is pattern formation in directional solidi-

Asymptotics beyond All Orders, Edited by H. Segur *et al.*
Plenum Press, New York, 1991

fication (pulling the sample at a constant speed V in an external thermal gradient G). There we distinguish between, mainly, two situations. (i) The growth of a dilute binary alloy where a planar front becomes unstable at a critical pulling speed and changes into a parallel array of cellular shapes that bifurcate into dendrites at larger speeds (essentially when the diffusion length falls below the wavelength). At even higher velocities (when the diffusion length becomes smaller than the capillary length) the planar front regains stability. This restabilization is probably preceded by a chaotic regime.[2] The farther the system is from the onset of instability, the more complex it behaves until the planar front restabilizes. (ii) The growth of a solid at finite concentration. Such a situation often produces *eutectic* growth.[3] The liquid-solid interface forms a parallel array of the two coexisting solid phases. Under some circumstances,[4] some initially axisymmetric lamellae undergo a parity-breaking transition: there appears a domain of a few tilted lamellae, with a well defined tilt angle, escorted by two symmetric domains. This tilted domain seems to play the role of a wavelength selector. A similar phenomenon was previously discovered during directional growth of a nematic crystal[5] from the isotropic phase. The importance of mass diffusion in the nematic phase together with the relatively large value of the partition coefficient in this system permit the exploration of regions of the parameter space that would not be easily accessible with an ordinary material. We should mention that the observation of tilted growth in eutectics can be traced back at least two decades;[6] its existence was often attributed to crystalline anisotropy. We know now that this idea is not generally correct: tilted patterns exist even in the fully isotropic model. The eutectic system has the advantage that the wavelength of the pattern is much smaller than the diffusion length. This situation offers a nice possibility to extract from the full growth equation similarity laws for the pattern. Below we will describe some features of pattern formation in directional solidification. We send the reader to other references for more details.

II. DIRECTIONAL GROWTH OF DILUTE SYSTEMS

1. *Basic equations*

The front dynamics is governed, in the one sided model considered here, by a closed nonlinear integrodifferential equation for the front profile $\zeta(x,t)$ (we consider one dimensional deformations only). In the quasi-steady limit, this equation takes the form[7]

$$\int d\Gamma' g(\vec{r},\vec{r}\,')\frac{\partial u}{\partial n'} = \int d\Gamma' h(\vec{r},\vec{r}\,')u(\vec{r}\,') \tag{1}$$

where the integration is performed along the liquid-solid boundary and where $g(\vec{r},\vec{r}\,')$

is the Green's function

$$g(\vec{r},\vec{r}') = \frac{1}{2\pi} e^{-(z-z')/l} K_0(\|\vec{r}-\vec{r}'\|/l) \tag{2a}$$

and

$$h(\vec{r},\vec{r}') = \frac{\partial g}{\partial n'} - 2l^{-1} n'_z g - c(\vec{r}')\delta(\vec{r}-\vec{r}'), \tag{2b}$$

where K_0 is the modified Bessel function, $l = 2D/V$ is the diffusion length, D being the mass diffusion coefficient and V the pulling velocity. The coefficient $c(\vec{r}')$ that enters eq. (2b) originates[7] from the conversion of a surface integral on a 2D δ function (which is not unity since $\vec{r}'$ lies on the boundary Γ) into a contour integral on the 1D δ function of eq. (2). u denotes the dimensionless concentration of impurities and is defined as $u \equiv (c - c_\infty)/\Delta c$, where c is the physical concentration, c_∞ its value far ahead of the front and $\Delta c = (1-k)c_\infty/k$ is the equilibrium miscibility gap, while k is the partition coefficient. At the liquid-solid interface ($z = \zeta(x,t)$), the normal derivative $\partial u/\partial n$ and u obey the continuity and the Gibbs-Thomson conditions, respectively

$$\{1 + (u-1)(1-k)\}v_n = -D\frac{\partial u}{\partial n}, \tag{3a}$$

$$u = 1 - d\kappa - \frac{\zeta}{l_T} - B_{kin}v_n \tag{3b}$$

where v_n is the normal velocity, $d = \frac{\gamma T_m}{L m_l \Delta c}$ is the capillary length, with γ the surface tension, T_m the melting temperature of the pure substance, m_l the liquidus slope and L the latent heat of fusion per unit volume, $l_T = \frac{G m_l}{\Delta c}$ is the thermal length, where G is the applied thermal gradient. B_{kin} denotes a phenomenological kinetic coefficient. In general both d and B_{kin} are orientation-dependent quantities. For the four-fold crystalline anisotropy considered here, B_{kin} and d read

$$B_{kin} = b_{kin}\{1 - \alpha_{4kin}\cos(4\theta)\} \tag{4a}$$

$$d = d_0\{1 - \alpha_4\cos(4\theta)\}, \tag{4b}$$

where θ is the angle between the normal to the interface and the z axis, b_{kin} and d_0 are constant factors and the $\alpha's$ represent the strength of the crystalline anisotropy (we assume that the $\alpha's$ account for both the force and torque contributions). Equation (1) has a planar front solution moving at a constant speed V. Mullins and Sekerka[8] were the first to perform systematic linear stability analyses. They showed that for a given thermal length l_T the planar front becomes unstable at a critical velocity V_c, approximately given by $V_c \simeq D/l_T$. Above V_c, all infinitesimal perturbations with a wavelength λ grow exponentially in the course of time until nonlinear effects intervene to saturate the linear growth; the interface develops a finite deformation. If the bifurcation from the planar front is supercritical, one can use, in principle, a standard Landau-Ginzburg amplitude expansion. This procedure is, however, inadequate, due to the flatness of the Mullins-Sekerka[8] neutral curve. We should therefore have recourse to other methods even when dealing with structures close to the threshold. The steady numerical calculation of Ungar and Brown[9] has demonstrated the complex nature of the bifurcation diagram. They have shown that the bifurcating state at $V = V_c$ ceases to exist slightly above V_c (saddle-node bifurcation) and merges with solutions at half the wavelength. More recently a forward time-dependent calculation has allowed us[10] to deal, in particular, with the dendritic regime. In the following we briefly review some of our results and give a few new ones.

2. *Cells and dendrites*

Here we restrict our attention to intervals of about one wavelength. This means that we do not address the question of wavelength adjustment via soft mode processes. The wavelength λ together with the prescribed pulling speed V parametrize the phase diagram shown in Fig. 1. The asterisks denote the points of numerical investigation, (a) is the critical curve as obtained from linear stability analysis for a plane front, (b) is the linearly most dangerous mode and (c) corresponds to $l = \lambda$. The material parameters here correspond to a typical steel-alloy. Kinetics is for the moment disregarded. We find that[10] surface tension anisotropy is necessary to stabilize the tip of the dendrite against tip splitting modes.

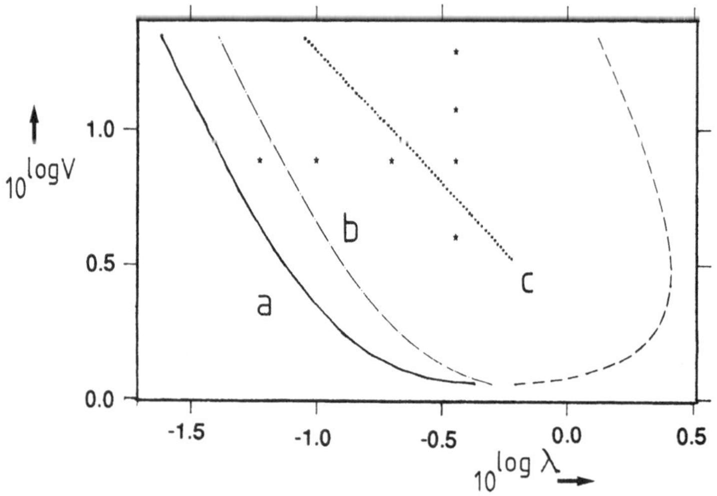

Fig. 1. Pattern phase diagram

For small velocities below the line (c) where the diffusion length is larger than the wavelength λ we obtain cellular structures without sidebranches. Close to the small wavelength limit of linear stability[8] we find quantitative agreement with scaling results calculated for the Saffman-Taylor problem[11] (see below). As we keep the wavelength fixed and increase the velocity, dendritic growth becomes apparent when the diffusion length is on the order of, or smaller than, the primary spacing (above line c in Fig. 1). We found there[10] quantitative agreement with the scaling law for the tip radius obtained for an isolated dendrite.

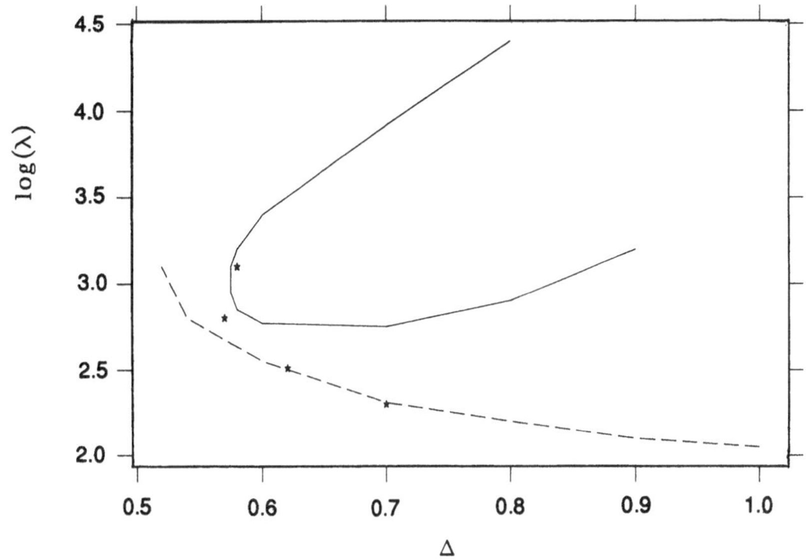

Fig. 2. λ vs Δ (see text)

We have compared our results with those obtained for growth in a channel. [12] Fig. 2 shows the comparison in the plane undercooling-periodicity. The undercooling Δ is defined as $\Delta = u(tip)$. If the profile is assumed to be exactly that of Saffman-Taylor the relation between λ and Δ for a given velocity is given by the dashed line in Fig. 2. Taking into account the fact that the profile, even for a small Péclet number (defined here as $P = \lambda/l$), differs from a Saffman-Taylor one, Brener *et al.*[12] have shown that the curve $\lambda(\Delta)$ is double-valued, i.e., it has two branches represented by the full line. Note that the Saffman-Taylor branch coincides with the lower branch obtained by Brener et *al.*[12] only in the limit $\Delta \to 1/2$, or in other words if the finger fills half the channel. The upper branch is the one which describes the transition to the growth of an isolated dendrite in the limit of infinite channel width. The asterisks in Fig. 2 are the results of numerical investigation. The Péclet number goes there from about $P = 0.05$ to $P = 1.5$. In the small Péclet number limit our results fall on the Saffman-Taylor branch. The last point, however, which corresponds to $P = 1.5$ lies close to the upper branch found by Brener et *al.*[12] In this limit the cells operate quasi-independently in the diffusion field; each cell is virtually free. This result is corroborated by a direct comparison of the scaling for the tip radius with that obtained for the free growth case.[10] Two remarks are in order. (i) Directional growth provides patterns with a minimum undercooling occuring at a cross-over from the Saffman-Taylor limit to the dendritic one. (ii) Both branches are stable with respect to hard mode fluctuations (as tip splitting) since they follow from a dynamical calculation. In an interesting paper, Pelcé[13] found the lower branch (Fig. 2) to be unstable, without resorting to the quasi-stationary approximation. His result seems to contradict ours. This may be due to the fact that we have used the quasi-stationary approximation, which possibly could supress the instability. However, this remains to be shown. Moreover, it should be kept in mind that the calculation referring to growth in a channel, on which the Pelcé result is based, misses the upper branch found by Brener *et al.*[12] Indeed the presence of an arbitrarily small correction, due to the growth, in the diffusion equation (advective term) completely destroys the mathematical structure because, according to Brener *et al.*,[12] it results in the appearance of an upper branch. Since this feature is not present in the Pelcé calculation, it can also be a reason for the difference from our result. Finally, we would like to recall that our growth problem is solved in the usual directional solidification set up (in the presence of an external gradient), and the extension of the stability result of Ref. 13 to that case remains to be shown as well.

3. Dendritic selection based on kinetics anisotropy

In the presence of kinetics it is legitimate to think of selection on the basis of kinetics anisotropy. For isotropic surface tension we find that an anisotropic kinetic coefficient is sufficient to stabilize the tip of dendrites. Fig. 3 shows two front morphologies. Fig. 3a corresponds to the kinetics-free case and an anisotropic surface tension. Fig. 3b corresponds to a situation where surface energy is isotropic while allowance is made for anisotropic kinetics. There $b_{kin} = 10^{-2}$ and the Péclet number is about 3 for both cases.

It is obvious that the overall shape of dendrites selected by kinetics anisotropy is significantly different from that selected by surface tension anisotropy. The tip in Fig. 3b still remains parabolic but slightly behind it the front develops quasi-planar sections. Moreover, the sidebranching activity is not efficient. These features are similar to those observed[14] on impure CBr_4. We are tempted to conjecture that dendrites in that system are selected by kinetics anisotropy. We have made a

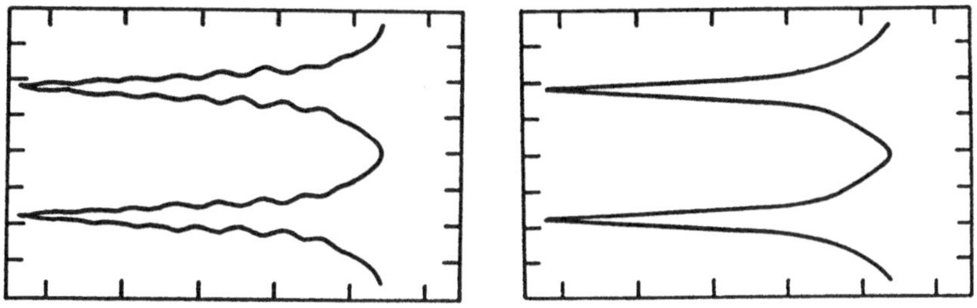

Fig. 3. Dendritic growth. (left) $b_{kin} = 0$. *(right)* $b_{kin} = 10^{-2}$.

systematic investigation with the aim to compare our results with those obtained in the free growth situation. Here we also find good quantitative agreement[15] with results obtained for an isolated dendrite.[16]

We should call to mind that little is known experimentally on the value of the kinetic coefficient. Thermal inertia seems to mask the manifestation of a front recession due to kinetics, and it is hard to imagine how a direct measurement of the interface supersaturation could be avoided in any experiment attempting to determine kinetic coefficients. Thermocouple techniques will probably be faced with obstacles such as the size effect and also run into difficulties because the interface moves with a finite velocity. An infra-red camera or interferometric techniques[17] might constitute promising tools to measure the interface supersaturation, which would be of great importance to promote further theoretical investigations.

4. Generic scenarios until restabilization

With increasing growth velocity the initially planar front becomes unstable and develops cellular structures. On further increase of the velocity cells bifurcate into dendrites. If the kinetic coefficient is not too large, selection should be ensured by surface energy anisotropy. As V gets larger and larger kinetics should become more and more important. We then expect a 'transition' to kinetics controlled dendrites. For some materials such as the nematic crystal (where mass diffusion is quasi-symmetric) the interface may undergo secondary bifurcations that are, generically, accompanied by the loss of symmetries. This is the case of parity-breaking transitions (see below). When V becomes so large that the diffusion length falls below the capillary length the planar front becomes stable again. Close to restabilization the front dynamics are governed by an equation of the Kuramoto-Sivashinsky type[2]

$$F_t = -F_{xx} - \frac{1}{2}F_{xxxx} + F_x^2 \tag{5}$$

where F is proportional to the front position. In the presence of an external thermal gradient we should add to the *r.h.s.* a term proportional to F, that plays the role of a 'damping' effect. This equation generates, besides order, temporal chaos as well as 'turbulence'.[18] For the growth of a nematic crystal where the large speed limit is reached, experimental evidence[19] for the transition to a chaotic regime close to the planar front restabilization is beginning to emerge.

III. DIRECTIONAL GROWTH OF LAMELLAR EUTECTICS

The eutectic problem is formulated in the same manner as for dilute systems. The front is still formally governed by eq. (1), where now the integration is performed along the liquid-α-phase + liquid-β-phase boundaries. The concentration field u is defined here as $u \equiv (c - c_\infty)/(c_\beta - c_\alpha)$, where c_α and c_β are the concentration of the α and β phases respectively, measured at the eutectic temperature. u and its normal derivative $\partial u/\partial n$ are given on the α-liquid and β-liquid boundaries by[20]

$$\frac{\partial u}{\partial n} = -(2 + \dot{\zeta})\{(1 - k_\alpha)(u - u_e) + \delta\}n_z \tag{6a}$$

$$\frac{\partial u}{\partial n} = -(2 + \dot{\zeta})\{(1 - k_\beta)(u - u_e) + \delta - 1\}n_z \tag{6b}$$

$$u = u_e - \frac{\zeta}{l_T^\alpha} - d_0^\alpha \kappa \tag{7a}$$

$$u = u_e + \frac{\zeta}{l_T^\beta} + d_0^\beta \kappa \tag{7b}$$

where $u_e \equiv (c_e - c_\infty)/(c_\beta - c_\alpha)$, $\delta = (c_e - c_\alpha)/(c_\beta - c_\alpha)$, where c_e is the liquid concentration at the eutectic temperature, k_i is the partition coefficient for the liquid-i-solid ($i = \alpha, \beta$) coexistence, d_0^i and l_T^i are the usual (chemical) capillary and thermal lengths for the i phase. In eqs. (7) – (8) we have reduced lengths and time by l and l^2/D, respectively. Eq. (1) should be supplemented by the mechanical equilibrium conditions where the three phases intersect (triple point)

$$\vec{\gamma}_{\alpha\beta} + \vec{\gamma}_{\alpha L} + \vec{\gamma}_{\beta L} = 0, \tag{8}$$

where $\vec{\gamma}_{ij}$ designates the surface tension between phase i and phase j. If the magnitudes of surface tensions are known, then condition (8) determines uniquely the contact angles θ_α and θ_β. If the pattern is tilted by an angle ϕ then the pinning angles at the two ends of a given lamella have different values. For example if θ_α and θ'_α denote these two angles (with, for example, $\theta_\alpha > \theta'_\alpha$), they are related by

$$\theta_\alpha = \theta'_\alpha + 2\phi\,, \tag{9}$$

and a similar relation holds for θ_β and θ'_β. Relation (9) follows frow a simple geometrical argument. The theory of axisymmetric growth was described in an important article by Jackson and Hunt.[20] In the following we will briefly review tilted growth and discuss some results on similarity.

1. Parity-breaking and the nature of the transition

Since the discovery by Simon et *al.*[5] of 'solitary' modes (which are asymmetric states moving along the growth front), similar phenomena were identified in other systems.[4,21] In directional solidification of the $CBr_4 - C_2Cl_6$ transparent eutectic Faivre et *al.*[4] observed that, under some circumstances, some initially axisymmetric lamellae become suddenly tilted with a well defined tilt angle (about 25°). As time elapses, the tilted domain travels transversely to the growth front. In fact metallurgists have been familiar with this type of phenomenon, and it was mentioned

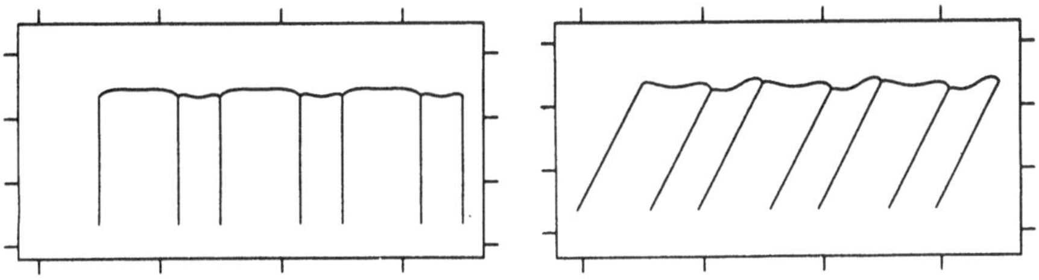

Fig. 4. Eutectic growth. (left) Axisymmetric. (right) Asymmetric.

at least 20 years ago.[6] Contrary to previous beliefs, the phenomenon itself is not a consequence of crystalline anisotropy. 'Solitary' modes,[5] tilted waves,[4] etc. have a common feature: they are asymmetric states. Coullet et *al.*[22] suggested that such travelling domains are localized inclusions of a new antisymmetric state. As soon as one accepts this idea, one expects that there exist tilted solutions filling the whole space. The tilted state should therefore emerge as a result of of a secondary bifurcation. In general, bifurcations are accompanied by the loss of some symmetry, which in the present case is the parity symmetry. We have shown[23] from a simple counting argument that the tilt angle is a well defined quantity, which belongs to a discrete set of solutions. Figs. 4a,b show a symmetric solution and a tilted one, respectively. We have used the material parameters for $CBr_4 - C_2Cl_6$.

The untilted solution shown in Fig. 4 is computed close to the critical velocity for the appearence of tilted solutions. Fig. 5 shows the bifurcation diagram; that is the 'order parameter' ϕ as a function of the growth velocity. This order parameter, when it is nonzero, induces an antisymmetric part in the front profile. Two remarks are in order. (i) The parity-breaking bifurcation is, for the set of parameters used so far, supercritical. The curve $\phi(V)$ is very stiff. This is taken to mean that even slightly above threshold (a few %) the tilt angle is about 20 to 30°. This indicates that a lowest order amplitude theory for the secondary bifurcation is probably inadequate. (ii) At $V = V_c$, the wavelength λ_{min} that provides the minimum undercooling for axisymmetric growth is approximately half the fixed wavelength λ for which Fig. 5 is computed. This means, with the proviso that axisymmetric growth operates, as suspected, at the minimum undercooling, that parity-breaking should occur, at least for the parameters used here, when the wavelength of the symmetric pattern is doubled. This result agrees with experiments.[4]

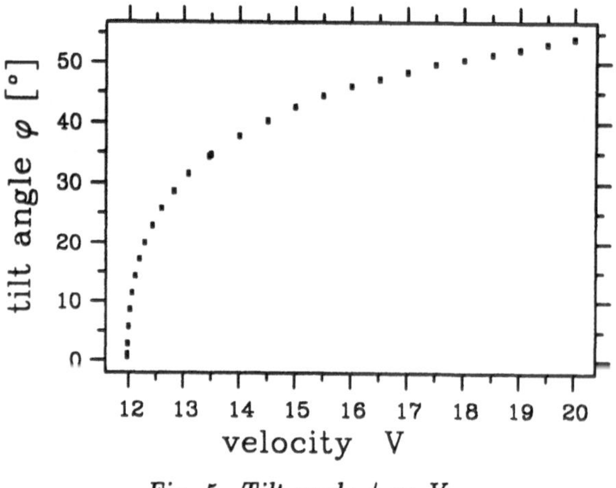

Fig. 5. Tilt angle ϕ vs V

2. Similarity laws

We found[23] that the critical velocity V_c at which parity symmetry breaks scales with the wavelength λ as $V_c \sim \lambda^{-2}$. A further result is that the tilt angle ϕ depends (for not too small $V's$) on $\lambda^2 V$ only. Moreover, many experiments reported that the selected wavelength scales with the growth velocity as $\lambda \sim V^{-1/2}$. The same scaling is obeyed by the wavelength at the minimum undercooling in the Jackson and Hunt[20] theory. However, many other experiments[24] claim that the relation $\lambda \sim V^{-1/2}$ is observed only at not too small $V's$; at small $V's$ the exponent is smaller than 1/2. It is therefore important to see whether these scalings can be understood on the basis of general considerations and to specify their degree of validity.

The eutectic problem contains, for periodic structures, four independent lengths: d_0, l, λ and l_T. One of these lengths may serve as a length unit so that the number of independent dimensionless quantities is equal to three. Or more generally, we are at liberty to choose any three independent dimensionless combinations to characterize the problem. We take for example $\sigma \equiv d_0 l/\lambda^2$, $\chi \equiv l/l_T$ and $P \equiv \lambda/l$ (the Péclet number). From dimensional considerations, the physical quantites should depend on these three combinations only. To reduce the number of degrees of freedom, details of the governing equations and the order of magnitude of different contributions must be considered. In standard experiments the diffusion length is much bigger than the wavelength λ; that is $P \sim 10^{-2}$. We have taken advantage of this fact[25] to show that P scales out of the integral equation so that the only remaining parameters are χ and σ. The small Péclet number limit is not uniform and care must be taken when performing it. Moreover, the kernel of the integral equation diverges logarithmically. We have explained elsewhere[25] how to deal with these difficulties. The resulting similarity equation tells us that the pattern is geometrically similar to other patterns obtained by multiplication of the wavelength λ by an arbitrary positive factor α and of both G and V by α^{-2}. Numerical integration of the original equation has confirmed our expectations: dimensionless physical quantities depend on σ and χ only. A simple example is the tilt angle: $\phi = \phi(\sigma, \chi)$. For small χ (which we call the large velocity domain) we find, as mentioned above, that the dependence on χ is so smooth that ϕ seems to depend on σ only. As the velocity decreases, however, that is when χ increases, we observe a strong dependence on χ (see Ref. 25 for more details).

An important consequence of the similarity laws is that the wavelength of the pattern scales as

$$\lambda \sim V^{-1/2} f(\chi), \tag{10}$$

where f is a function of χ only. Eq. (10) shows that $\lambda \sim V^{-1/2}$ (only) if $f \simeq$ *constant*. We have computed f for the minimum undercooling point. The condition of minimum undercooling has conventionally been assumed in the metallurgical literature to locate the operating point; and the resulting predictions seem to be in good agreement with experiments.

Fig. 6 displays the behaviour of f. At large $V's$ f is a smooth function and can hardly be distinguished from a constant. At small V f starts to decrease significantly.

This entails that $\lambda^2 V$ deviates from a constant. More precisely, at small velocities λ is smaller than what is predicted from the JH theory.

Our results agree well with a large number of experiments.[24] Careful investigations[24] have demonstrated that the law $\lambda^2 \sim V^{-1}$ holds only at large enough growth velocities. At small velocities, deviations from that law are observed. These deviations have the same tendency as the one reflected by the 'scaling' function (Fig. 6). We are now dealing with the full stability analysis of the pattern. First, the

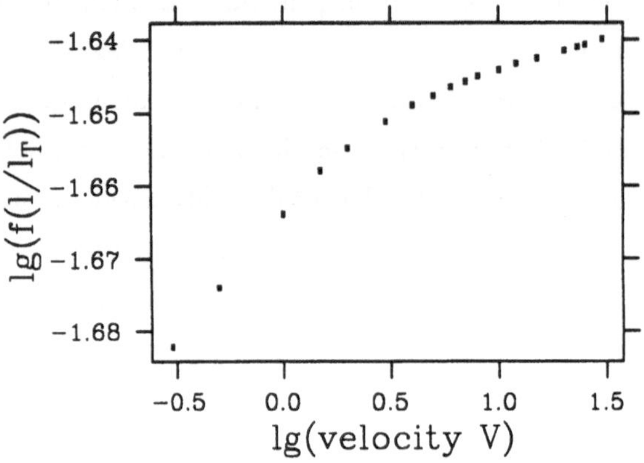

Fig. 6. The scaling function

phase diffusion treatment[13] can be generalized to deal with the eutectic problem. This will allow us to determine the limits of soft instability of phase type. This treatment is more efficient than a direct stability analysis; the price to be paid is the analytical effort needed to extract the relevant phase motion only. To study other instabilities, such as 'optical' modes, we have to resort, however, to a full stability calculation in the relevant region of the reciprocal space. In particular, soft instabilities should then appear as Bloch-like modes in the center of the first Brillouin zone.

REFERENCES

1. J.S. Langer, *in Proceedings of the Les Houches Summer School, Session 46,* edited by J. Souletie, J. Vannimenus, and R. Stora (Elsevier, Amsterdam, 1987).
2. C. Misbah, H. Müller-Krumbhaar and D.E. Temkin, J. Physique I **1**, 585 (1991).
3. J.D. Hunt and K.A. Jackson , *Trans. Metall. Soc. AIME* **236**, 843 (1966);

V. Seetharaman and R. Trivedi, *Metall. Trans.* **19A.**, 2955 (1988).

4. G. Faivre, S. de Cheveigné, C. Guthmann, and P. Kurowski, *Europhys. Lett.* **9**, 779 (1989).
5. A.J. Simon, J. Bechoefer and A. Libchaber, *Phys. Rev. Lett.* **61**, 2574 (1988).
6. R. Racek, *Thèse d'Université Nancy I*, (1973); see also H.E. Cline, *Mat. Sci. Engr.* **65**, 93 (1984).
7. Y. Saito, G. Goldbeck-Wood, and H. Müller-Krumbhaar, *Phys. Rev. A* **38**, 2148 (1988).
8. W.W. Mullins and R.F. Sekerka, *J. Appl. Phys.* **35**, 444 (1964).
9. L.H. Ungar and R.A. Brown, *Phys. Rev. B* **29**, 1367 (1984); **30**, 3993 (1984); **31**, 5923 (1985); **31**, 5931 (1985).
10. Y. Saito, C. Misbah and H. Müller-Krumbhaar, *Phys. Rev. Lett.* **63**, 2377 (1989).
11. P.G. Saffman and G.I. Taylor, *Proc. R. Soc.* **A 245**, 312 (1958). The analogy with directional growth was pointed out by P. Pelcé and A. Pumir, *J. Cryst. Growth* **73**, 337 (1985).
12. E.A. Brener, M.B. Geilikman and D.E. Temkin, *Sov. Phys. JETP* **67**, 1002 (1988). The problem of growth in a channel was considered earlier by D. Kessler, J. Koplik and H. Levine, *Phys. Rev. A* **34**, 4980 (1986). Brener *et al.* were the first to find the second upper branch (Fig. 2) which is relevant for directional growth.
13. P. Pelcé, *Europhys. Lett.* **7**, 453 (1988).
14. P. Kurowski, *Thèse d'Université, Paris 7*, 1990; P. Kurowski, C. Guthmann and S. de Cheveigné (to be published).
15. A. Classen, C. Misbah, H. Müller-Krumbhaar and Y. Saito, *Directional solidification with interface dissipation*, to appear in *Phys. Rev. A* **43**, 6920 (1991). Some aspects of the transition to kinetically controlled dendrites were discussed by E. Ben-Jacob, P. Garik and D. Grier, *Superlattices and Microstructures* **3**, 599 (1987).
16. E.A. Brener and V.I. Mel'nikov, Adv. in Physics, **40**, 53 (1991).
17. E. Raz, S.G. Lipson, and E. Polturak, *Phys. Rev. A* **40**, 1088 (1989).
18. See for example P. Manneville, *Dissipative structures and weak turbulence*, (Academic Press, New York 1990).
19. J.M. Flesselles, A.J. Simon and A.J. Libchaber, *Dynamics of one-dimensional interfaces: An experimentalist's view*, preprint 1990, to appear in Advances in Physics.
20. K.A. Jackson and J.D. Hunt, *Trans. Mettal. Soc. AIME* **236**, 1129 (1966).

21. M. Rabaud, S. Michalland and Y. Couder, *Phys. Rev. Lett.* **64**, 184 (1990).
22. P. Coullet, R.E. Goldstein and G.H. Gunaratne, *Phys. Rev. Lett.* **63**, 1954 (1989).
23. K. Kassner and C. Misbah, *Phys. Rev. Lett.* **65**, 1458 (1990); erratum *Phys. Rev. Lett.* **66**, (1991). For the counting argument and the parity-breaking problem in directional growth of liquid crystals, see also H. Levine and W. Rappel, *Phys. Rev. A* **42**, 7475 (1990).
24. For a review see G. Lesoult, *Ann. Chim. Fr.* **5**, 154 (1980).
25. K. Kassner and C. Misbah, *Phys. Rev. Lett.* **66**, 445 (1991).

A FLAT INTERFACE AND ITS UNFOLDING BIFURCATIONS

A. Libchaber, A. J. Simon and J.-M. Flesselles

The James Franck and Enrico Fermi Institutes
The University of Chicago
5640 South Ellis Avenue, Chicago, Illinois 60637, U.S.A.

1 THE PRIMARY INSTABILITY

The experimental setup used is a classical directional solidification one where a movable sample holder is displaced at a controled velocity along a temperature gradient provided by two thermally regulated copper blocks. The sample consists of a thin layer of liquid crystal (here the 4,4'-*n*-octylcyanobiphenyl, 8CB) sandwiched between two glass plates. The temperatures are adjusted such that the liquid crystal undergoes a phase transition above the gap between the two blocks. The 8CB is isotropic down to 40.5°C where it becomes nematic through a weakly first order transition. The sample is properly moved along the gradient so that the amount of the more ordered phase (here the nematic one) is increased with respect to the less ordered one (the isotropic one); hence the name *directional ordering experiment.* The experimental setup is mounted on an optical microscope to observe the interface and eventually record it for later analysis (Oswald *et al.*, 1987; Bechhoefer, 1988; Simon *et al.*, 1988; Simon and Libchaber, 1990).

Instead of using a liquid crystal, as we are doing, any material would be suitable, as long as it undergoes a first order transition. Usually (and historically) metals or plastic crystals are used; in that case the terms nematic and isotropic should be replaced by solid and liquid. The underlying mechanisms leading to the primary instability are identical.

First of all, because of the temperature gradient, there is diffusion of heat. We will assume that it is not perturbed neither by the interface nor by the motion of the sample. This leads to a linear gradient across the gap. According to the binary phase diagram, there is a jump in the concentration of impurities between the isotropic phase and the nematic one at the interface: during the directional ordering, the ordered phase expells impurities toward the disordered phase. Hence one has also to take into account the diffusion of impurities arising from this phenomenon. Finally the variations of the transition temperature with the interface curvature have to be considered. The balance between these mechanisms can lead to a destabilisation of the interface in the

Asymptotics beyond All Orders, Edited by H. Segur *et al.*
Plenum Press, New York, 1991

appropriate range of temperature gradient and moving velocity. This is known as the *Mullins-Sekerka instability* (Mullins and Sekerka, 1964). The interface, which is initially flat, becomes wavy, with a well-defined wavelength. The line separating the domain in the temperature gradient/velocity plane where the interface is stable from the one where it is unstable is called the *marginal stability curve.*

The three physical effects taken into account in this model—the diffusion of impurities, the presence of a coexistence region due to impurities and the depression of the transition temperature caused by curvature—are characterized by three lengths: *diffusion, thermal and capillary length.* The diffusion length l_D is inversely proportional to the velocity whereas the thermal length l_T is inversely proportional to the gradient and the capillary length l_C depends only on the material. These lengths have clear physical meaning: l_D is the falloff of the concentration profile in front of the advancing interface. Qualitatively, it is the wavelength of the pattern. It is also the length over which the time of advection of the front equals the diffusion time of impurities. l_T gives the spatial extent of the coexistence region within the temperature gradient for a given concentration. l_C is the smallest radius of curvature allowed by the system, *i.e.* for radii of curvature smaller than this length, capillary forces locally restabilize the interface.

Caroli *et al.*(1982) were the first to point out that the scale of the marginal stability curve can be significantly reduced by considering transitions from the isotropic to a more ordered liquid crystal phase (smectic or nematic). This would allow a convenient observation of the restabilization which is difficult in true directional solidification. In addition, the ratio η between the chemical diffusivities in the two phases is close to 1/2 (compared to $\eta \sim 10^{-5}$ for solid/liquid transitions), preventing the appearance of grooves. In other words, segregation is weak and the interface is permeable to impurities: since their concentration is nearly similar on both sides, the shape is nearly symmetric when looking from one phase or the other.

2 THE DYNAMICAL INSTABILITIES: AN OVERVIEW

During a typical run, the gradient is fixed and the velocity is varied. If the gradient is not too high (typically 20°C/cm in our experiment), for low enough velocities, the interface is stable. When the velocity reaches a critical value where the thermal length becomes of the order of the diffusion length (typically 10μmm/s), the interface destabilizes. As the velocity is increased, the amplitude of the pattern increases, then saturates, then decreases before finally restabilizing when the diffusion length becomes of the order of the capillary one. There also exists a critical gradient above which the interface is always stable whatever the velocity: the thermal length is then of order of the capillary one. Simultaneously, secondary instabilities appear along the background pattern.

Thinner samples exhibit an enhanced richness of secondary instabilities: empirically, the closer to the primary destabilization a dynamical bifurcation appears, the thinner the sample should be for it to be seen. In addition, thick samples reveal a "parasitic" hydrodynamic instability. Hence we describe the whole set of bifurcations observed in the thinnest samples (6 μm), thus encompassing all those seen in the thicker ones.

The first dynamical bifurcation encountered manifests itself as an *intermittency* of the pattern, whose amplitude even becomes temporarily flat. For these values of the control parameters, dynamical selection processes are few and inefficient: tip-splitting and cell collapse lead to local rearrangements over only two or three cell regions. Hence, wavelengths along the interface fluctuate within about 10%.

For higher velocities, ($\varepsilon \equiv (v - v_c)/v_c \sim 0.1$ where v_c is the critical velocity where the interface destabilizes) another dynamical state appears. Some cells, called *sources*, tip-split periodically in time, whereas others, called *sinks*, collapse an adjacent cell, leading to *travelling waves* between them. Now, the pattern resembles a sine wave with higher harmonics. This process of travelling waves leads to a well defined wavelength. As the velocity is increased, the amplitude of the pattern grows. Grooves develop between the cells which evolve toward more a square-like shape. Travelling waves then disappear and *solitary modes* appear for ε near two to three, depending on gradient. They are groups of adjacent tilted cells, moving along the interface, of longer wavelength than the surrounding pattern. If there existed a dispersion in wavelengths at this velocity (after a quench from a lower velocity for instance), these solitary modes would select the wavelength left behind them. They appear to be a more efficient and accurate dynamic selection mechanism than travelling waves.

At yet higher velocities, when the grooves become more pronounced and leave small channels of isotropic material between the nematic cells, *optical modes* are often observed: wavelengths of neighboring cells oscillate in phase opposition. Here, the pattern reaches its maximum amplitude and a further increase in velocity leads to droplet formation off the grooves. From therein to the restabilization, the amplitude continues to decrease. Before the droplet regime ends, the wave turbulence regime, also called *spatio-temporal chaos* begins, characterized by a rapid and disordered collapsing and splitting of cells. Close to the restabilization, the system enters a different regime where elastic properties of the interface dominate. And it finally restabilizes.

A central problem is and remains the wavelength selection mechanism. Though it is not yet solved, it appears to be an intrinsically dynamical process. As well as the fundamental characteristic of a phase transition is the spontaneous breaking of a spatial symmetry of the system, a bifurcation is a spontaneous breaking of a *spatio-temporal* symmetry. In these (quasi) one-dimensional systems such as the nematic/isotropic interface, the topmost importance of the parity breaking bifurcation has now been recognized: by spontaneous tilting of cells, the interface is not invariant any more by reversing x in $-x$. This leads to different parity broken states such as travelling waves and solitary modes. Since a parity broken state cannot be motionless, these patterns move along the interface and select the wavelength in their wakes.

These properties are not limited to the very nematic/isotropic experiment but appear instead in a wide variety of one-dimensional non linearly driven interfaces: directional solidification in plastic crystals (Cladis *et al.*, 1990), in eutectic mixtures (Faivre *et al.*, 1989), directional viscous fingering (Rabaud *et al.*, 1990), Taylor-Dean experiments (Mutabazi *et al.*, 1990), etc.

3 TRAVELLING WAVES

Although we observe a critical velocity, the nature of the transition from a steady state pattern to travelling waves is unresolved. Moreover the number of sources and sinks along the interface varies widely from one sample to the other. As a general rule, the lower the front velocity, the fewer sources and sinks, and the slower the tip-splitting and cell collapse. Whatever the parameters, the period remains of the order of 5 to 20 seconds for a front velocity of order 20μm/s. The aspect ratio also seems to influence

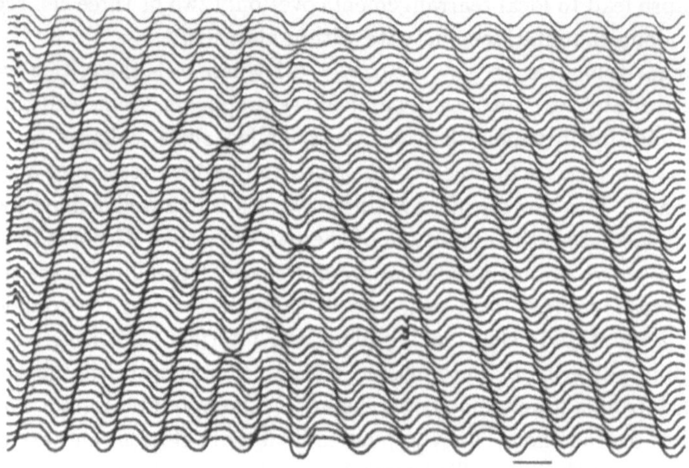

Figure 1. A sink. Space-time portrait sampled at 2.4 Hz. Undoped 8CB in a 12 μm thick sample, $v = 50\mu$m/s, and $G = 6.3$°C/cm. Bar in the right corner is 20μm long.

their density. In the capillary samples, there is at most a single source and a single sink. In a 2 mm wide sample, we have once observed a single source, the wires playing the role of sinks. In an extended 15 mm wide sample, we have counted from 2 sources at a velocity of 30 μm/s to 11 at 50 μm/s simultaneously present along the interface, with of course a sink between every two sources. Their density is also gradient dependent: close to the maximum gradient, none of them have been seen. Whatever the experimental conditions were, sources and sinks have not been observed in thicker samples (25 μm and above).

When observed between crossed polarizers, sources and sinks are easily distinguished from normal cells because of their characteristic signature in the meniscus. Both appear as dynamically stable localized defects. They usually emerge from point defects of the nematic director field within the meniscus. According to some recent experiments (Oswald, 1991), these defects occur from a twisting of the director field and induce an internal breaking of parity.

Although sources and sinks are complementary objects, the time reversed behavior of sources does not reproduce that of sinks, nor vice-versa. First of all sources are asymmetric: a given source always emits in the same direction and drifts in the opposite, whereas sinks have a less predictable behavior, absorbing either to the left or to the right, making them more mobile. However a perfectly symmetric sink has once been observed, which absorbed 22 cells in strict left/right alternation. The mechanism by which a source splits does not resemble the inverted mechanism by which a sink absorbs an incoming cell. A source cell elongates and slightly flattens until its width becomes 40 to 60 % larger than the selected wavelength. The flat section then progressively tip-splits into two cells. One of them is the source, which starts again the same process, leading to the characteristic herringbone like shape of travelling waves in a space-time representation. The region where two trains of cells emitted by two different sources collide corresponds to a sink. Sinking cells keep their shape but shrink until their width is about 30 %

smaller than the selected wavelength and then suddenly collapse within a few tenths of a second. From this behavior, since sources do not drift at the same velocity than the cells they emit, the relevant velocity is $v_{left} + v_{right}$, the sum of the velocity of left and right going cells. On the space time portraits, these velocities are obtained by measuring the angle of a front with respect to the interface. Because of the periodicity and the conservation of total number of cells, a simple geometric relation holds, which allows experimental consistency checks:

$$\lambda_{sel} = (v_{left} + v_{right})\tau \, , \tag{1}$$

where λ_{sel} is the wavelength of the selected pattern and τ, the characteristic time for a periodic source or sink. Typical travelling wave velocities are of order 1 to 5 μm/s which is about a tenth of the front velocity. Apart from a single observation, sources and sinks do not annihilate but tend to form bound states.

4 SOLITARY MODES

A solitary mode consists of a group of anywhere from 2 to about 30 elongated and tilted cells moving along the pattern of untilted cells which form the interface. Since this "bubble" is moving at constant speed, time and space translations are broken, unless they are coupled through the transformation $(x,t) \to (x - v_g\tau, t + \tau)$ where v_g is the group velocity of the solitary mode and τ an arbitrary time lag. Unlike travelling waves, the global spatial parity is broken in this regime. Since cells can be tilted to the right or to the left, solitary modes can propagate in either direction (Simon *et al.*, 1988).

The bifurcation is subcritical and shows a much clearer critical velocity than travelling waves. Solitary modes are robust and have been observed in all samples, regardless of secondary parameters. Though they share similar characteristics to travelling waves, both states are never seen simultaneously along the interface. The similarity and the difference between them are most clearly emphasized when viewed between crossed polarizers. As the velocity is increased above the threshold, the defects in the meniscus—black lines perpendicular to the interface—which are the signature of sources and sinks, suddenly emerge along the whole interface, creating in the meniscus a pattern of twice the spatial periodicity of the cellular one. Sources and sinks disappear and solitary modes appear.

Rarely solitary modes are created by dust particles crossing the interface but more often appear spontaneously, probably initiated by some intrinsic noise. What determines their length is still unknown: on any given interface, long and short solitary modes drifting in both directions are commonly observed. Once they exist, solitary modes are robust to even strong perturbations like dust particles passing through the interface.

Many parameters appear needed to characterize solitary modes and the relevant ones are still to be extracted. This is partly due to their complicated motion and their intriguing property of selecting the wavelength left in their wake. A solitary mode involves two different velocities: the group velocity v_g, characterizing the velocity of the bubble with respect to the fixed pattern, and the phase velocity v_ϕ of the tilted cells inside the bubble. The group velocity is obtained by measuring the angle of the edge of the bubble with respect to the average position of the interface, the phase velocity by the angle of the trajectory of a tilted cell. v_ϕ is always about one third of v_g, but large dispersion is observed. Since v_ϕ and v_g are always of opposite sign, the tilted cells

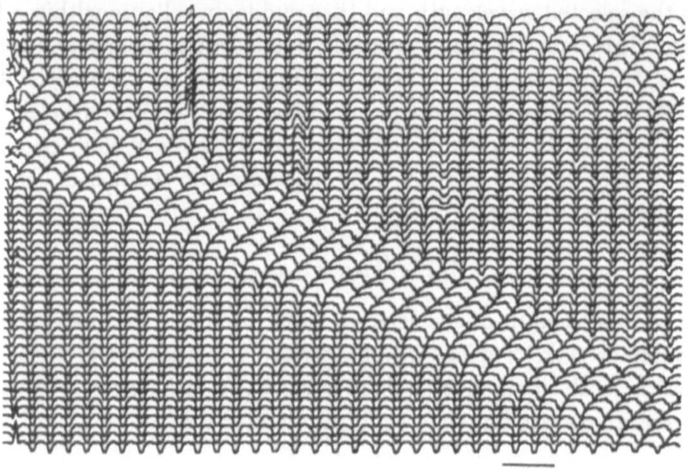

Figure 2. A solitary mode. Space-time portrait sampled at 2.0 Hz. Undoped 8CB in a 6 μm thick sample, $v = 40\mu$m/s, and $G = 10.2°$C/cm. Bar in the right corner is 40μm long.

within the bubble move backwards with respect to the solitary mode as a whole. As mentioned earlier, tilted cells are also elongated with respect to the background pattern, which defines an internal wavelength λ_{SM}.

However the situation is more complex if there is no initial wavelength selection. This can be achieved, for instance, by a quick change of the interface velocity from zero to a velocity where solitary modes are observed. After the passage of one solitary mode, the distribution of wavelengths is much more peaked; two or three solitary modes later, no improvement can be detected, eventually leaving the selected wavelength λ_{sel}. As a solitary mode moves through the background cells, it packs the trailing cells to a more uniform wavelength smaller or equal to the preceeding average one. The phase slip accumulates within the solitary mode which progressively elongates, and eventually turns into a new tilted cell, or into an additional normal cell. Hence, *solitary modes are wavelength selectors.* During the selection process, the leading edge of the solitary mode always moves faster than the rear: the bubble spreads. Once the wavelength is selected, since the total number of cell is conserved, another simple geometric relation holds:

$$\lambda_{SM} = \lambda_{sel}(1 + \frac{v_\phi}{v_g}) . \qquad (2)$$

No clear correlation has been found between v_g and the other parameters, apart from the loose equality of v_g and v_s, the velocity of the front close to the onset of solitary modes.

Solitary modes are dissipative structures and do not behave like solitons, although they are localized nonlinear propagative objects. When two solitary modes of opposite direction collide, a solitary mode emerges of length approximatively equal to the subtraction of the two previous lengths, in the direction of the longer one. In the rare

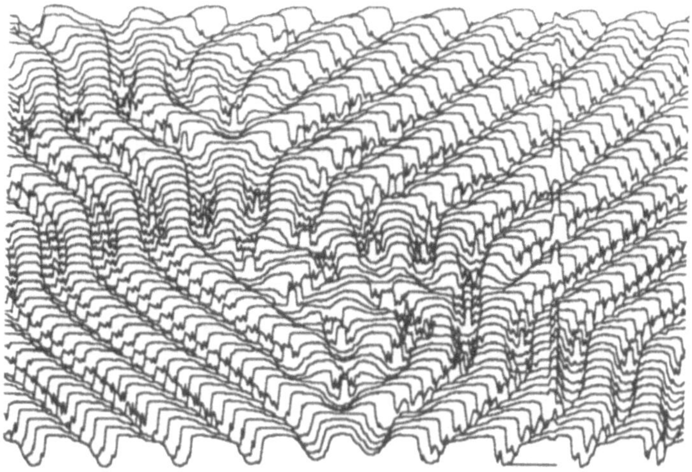

Figure 3. Interaction of many solitary modes. Space-time portrait sampled at 1.2 Hz. Undoped 8CB in a 25 μm thick sample, $v = 50\mu$m/s, and $G = 5.9$°C/cm. Bar in the right corner is 50μm long.

case when two identical solitary modes collide, they annihilate. However, since solitary modes are formed of elongated cells, any collision process creates normal cells.

A phenomenological model. By considering a solitary mode as an inclusion of an antisymmetric pattern in a symmetric one, Coullet *et al.*(1989) have suggested a phenomenological model, later extended by Goldstein *et al.*(1990), which accounts for most properties observed in travelling waves and solitary modes. A one-dimensional pattern $U(x,t)$, which represents the interface profile, is resolved into its symmetric and antisymmetric components, U_S and U_A, respectively:

$$U(x,t) = S(x,t)U_S(x + \phi(x,t)) + A(x,t)U_A(x + \phi(x,t)) \,, \qquad (3)$$

where S and A are real amplitudes, $\phi(x,t)$ is the phase of the pattern and with $U_S(z) = U_S(-z)$ and $U_A(z) = -U_A(-z)$. The actual functional forms of U_S and U_A are irrelevant in this model which only considers the envelope of the pattern: S, A and ϕ are slowly varying quantities. The amplitude A of the antisymmetric component serves as the order parameter of the bifurcation.

The lowest order equations satisfying the invariance requested by the system are:

$$\begin{aligned} A_t &= A_{xx} + f(A) + \gamma A A_x + \varepsilon \phi_x A + \cdots, && (4) \\ \phi_t &= \phi_{xx} + \omega A + \cdots, && (5) \end{aligned}$$

where γ, ε, ω are phenomenological constants, whose signs will have to be properly adjusted, and where $f(A)$ is an odd polynomial in A. Other terms of same order could be added, but would not change qualitatively the properties described here. For a proper description of the experimentally observed subcritical bifurcation, an appropriate form for f is:

$$f(A) = \mu A + \alpha A^3 - A^5 . \tag{6}$$

Once the bifurcation to a broken parity state is postulated, which implies a proper tuning of the phenomenological constants, this model properly describes the motion of "bubbles" of parity broken states, *i.e.* of segments of tilted cells on a non-tilted background. Another interesting feature of this model is the prediction of wavelength selection by successive passages of parity broken bubbles (now identified as solitary modes) in agreement with the experimental observation.

Nevertheless, this model is unable to describe collisions of solitary modes because it does not allow the creation of new cells. To circumvent this flaw and extend the model, the amplitude of the symmetric component S, previously kept fixed, has to be able to vanish: when the phase is undefined, it can vary by 2π giving birth to a new cell. By considering the symmetric component U_S as the real part of a complex amplitude $B \equiv S\exp(i\phi)$, and from the same symmetry considerations, Goldstein *et al.*(1990) have derived new equations similar to eq. (4) and (5), which only involve A and B. These new equations which recover properly the collisions as well as the approximate subtraction rule, provide a complete satisfactory description of solitary modes.

A mode coupling interpretation. Instead of starting from general symmetry considerations, a simple mode coupling approach can also be considered (Nozières, 1989; Fauve *et al.*, 1990). One first assumes that the fundamental mode and its first harmonic are enough to describe the interfacial pattern. Hence we write:

$$U(x,t) = a_1(t)e^{iqx} + a_2(t)e^{i2qx} + c.c.. \tag{7}$$

For a static pattern, by choosing an origin, one can take a_1 real and thus write:

$$U(x) = 2a_1 \cos qx + 2a_{2s} \cos 2qx + 2a_{2a} \sin 2qx . \tag{8}$$

The term in $\cos 2qx$ has the same parity as the fundamental, whereas $\sin 2qx$ breaks parity. More generally, a_1 and a_2 are complex numbers with phase and amplitude:

$$a_1 = A_1 e^{i\phi_1}, \quad a_2 = A_2 e^{i\phi_2} . \tag{9}$$

A broken parity state is characterized by the order parameter $\theta \equiv \phi_2 - 2\phi_1$, which measures the phase difference between the two modes.

The lowest order equations describing their coupled time evolution yield:

$$\dot{a}_1 = \alpha_1 a_1 + \beta_1 |a_1|^2 a_1 + \gamma_1 a_2 a_1^* + \delta_1 |a_2|^2 a_1 \tag{10}$$
$$\dot{a}_2 = \alpha_2 a_2 + \beta_2 |a_2|^2 a_2 + \gamma_2 a_1^2 + \delta_2 |a_1|^2 a_2 , \tag{11}$$

where $\alpha_{1,2}$, $\beta_{1,2}$, $\gamma_{1,2}$ and $\delta_{1,2}$ are real coefficients. The imaginary part gives the evolution equation for θ:

$$\dot{\theta} = -[\gamma_2 \frac{A_1^2}{A_2} + \gamma_1 A_2] \sin\theta . \tag{12}$$

If γ_1 and γ_2 are of opposite signs, the most common case, there exists a non zero θ solution, provided the amplitude ratio between the q and $2q$ modes is fixed so that the coefficient of $\sin\theta$ vanishes. Otherwise the only steady state solution arising from eq. (12) is the trivial fixed point $\theta = 0, \pi$ corresponding to a symmetric solution. To

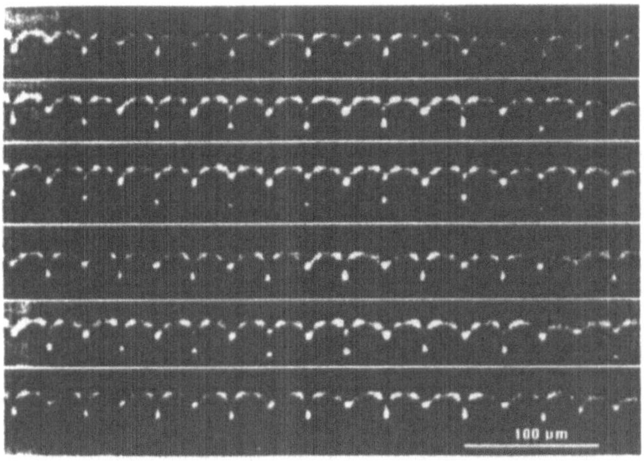

Figure 4. Correlated droplet emission. Images sampled from top to bottom at t=0, 0.20, 0.30, 0.64, 1.00 sec. Undoped 8CB in a 12 μm thick sample, $v = 80\mu$m/s, and $G = 4.8°$C/cm.

obtain a steady parity broken state, the expansion in equations (10) and (11) must then be pushed to at least the fifth order. The evolution equation (12) becomes:

$$\dot{\theta} = H(A_1, A_2, \theta) \sin\theta \,, \tag{13}$$

where H is a function depending on the coefficients α, β, γ, etc. Here, the steady state equation $\dot{\theta} = 0$ allows a parity broken state corresponding to an intermediate value $\theta = \theta^*$ which satisfies $H = 0$.

The drift velocity of the pattern is given by $\dot{\phi}_1$. It is zero for the trivial case of $\theta = 0$, but, as can be checked easily, is proportional to $A_2 f(\theta^*) \sin\theta^*$ for the parity broken solution, where f is a function. Hence the motion of the tilted cells arises only from the broken parity.

5 OTHER INSTABILITIES

5.1 Droplet formation

When increasing the velocity at a given gradient, once the solitary modes regime is passed, another kind of dynamic instability occurs: the ends of the grooves pinch off to produce droplets of impurity rich isotropic phase, which progressively disappear inside the nematic phase. Although this phenomenon is common to most directional solidification experiments, it is hardly to be considered as a dynamic state of a one-dimensional interface, because the notion of interface itself looses its meaning when there is no more connectivity. No analysis of the broken symmetry is possible here.

Nevertheless, droplet formation, is still a dynamic bifurcation occuring in any directional ordering experiment, which can be studied in itself. In the liquid crystal experiment, close to the onset of this instability, droplet emission is an ordered process with strong correlations in space as well as in time. Odd numbered cells emit simultaneously their drops in a perfect alternance with those of the even ones, generating a spatial period doubling with respect to the cellular pattern and inducing a temporal period of order one second. As the velocity is increased, these correlations weaken and extend over decreasing length until drops form independently from neighbors.

5.2 Spatio-temporal chaotic regime

Before the droplet emission process finishes, the interface enters a complicated regime called wave turbulence or spatio-temporal chaos. These terms are essentially words to mask our ignorance about this state on which no quantitative study has yet been done in the liquid crystal experiment. The correlation between cells is small and does not extend further than its two nearest neighbors. They constantly oscillate, appearing and disappearing with no obvious periodicity neither in time nor in space. The total number of cells is not conserved anymore. It rather seems to be constant on average. Moreover, at a given time, cells are hardly regular in shape and exhibit a broad distribution of wavelengths. The ratio between the longest one to the smallest detectable cells is of order five to ten. Since cell lifetime is of order a few tenths of a second, a proper experimental study of this state is complicated by the requirement of fast image processing.

REFERENCES

BECHHOEFER J, 1988, *Directional solidification at the nematic-isotropic interface* Ph. D. Thesis, The University of Chicago, unpublished.

BECHHOEFER J, SIMON A J, LIBCHABER A AND OSWALD P, 1989, Phys. Rev. A **40**, 2042–2056, *Destabilization of a flat nematic-isotropic interface.*

CAROLI B, CAROLI C AND ROULET B, 1982, J. Phys. France **43**, 1767–1780, *On the emergence of one-dimensional front instabilities in directional solidification and fusion of binary mixtures.*

CLADIS P E, GLEESON J T AND FINN P L, 1990, To be published in Phys. Rev. Lett., *Propagation of gaussian grooves from a boundary in directional solidification.*

COULLET P, GOLDSTEIN R AND GUNARATNE G, 1989, Phys. Rev. Lett. **63**, 1954–1957, *Parity-breaking transitions of modulated patterns in hydrodynamics systems.*

FAIVRE G, DE CHEVEIGNÉ S, GUTHMANN C AND KUROWSKI P, 1989, Europhys. Lett. **9**, 779–784, *Solitary tilt waves in thin lamellar eutectics.*

FAUVE S, DOUADY S AND THUAL O, 1990, Phys. Rev. Lett. **65**, 385, *Comment on "Parity-breaking transitions of modulated patterns in hydrodynamic systems."*

GOLDSTEIN R , GUNARATNE G, AND GIL L, 1990a, Phys. Rev. A **41**, 5731–5734, *Defects and traveling wave states in nonequilibrium patterns with broken parity.*

MULLINS W W AND SEKERKA R F, 1964, J. Appl. Phys. **35**, 444–451, *Stability of a planar interface during solidification of a dilute binary alloy.*

Mutabazi I, Hegseth J, Andereck C D and Wesfreid J, 1990, Phys. Rev. Lett. **64**, 1729–1732, *Spatiotemporal modulations in the Taylor-Dean system.*

Nozières P, 1989, *Instabilités interfaciales II* Cours au Collège de France 1989–1990, unpublished.

Oswald P, 1991, Preprint, *Role of elastic effects in the secondary instabilities of the nematic-isotropic interface.*

Oswald P, Bechhoefer J and Libchaber A, 1987a, Phys. Rev. Lett. **58**, 2318–2321, *Instabilities of a moving nematic-isotropic interface.*

Rabaud M, Michalland S and Couder Y, 1990, Phys. Rev. Lett. **64**, 184–187, *Dynamical regimes of directional viscous fingering: spatiotemporal chaos and wave propagation.*

Simon A J, Bechhoefer J and Libchaber A, 1988, Phys. Rev. Lett. **61**, 2574–2577, *Solitary modes and the Eckhaus instability in directional solidification.*

Simon A J and Libchaber A, 1990, Phys. Rev. A **41**, 7090–7093, *Moving interface: the stability tongue and phenomena within.*

VISCOUS DISPLACEMENT IN A HELE-SHAW CELL

S. Tanveer

Department of Mathematics
Ohio State University
Columbus, OH 43210

I. INTRODUCTION

A Hele-Shaw cell is a pair of parallel plates that are separated by a small gap b. The motion of a less viscous fluid displacing a more viscous fluid in this gap under the action of some imposed pressure gradient or gravity or fluid injection has been the study of intensive research over the last decade. (See Saffman[1], Bensimon et al[2], Homsy[3], Pelce[4] and Kessler,Koplik & Levine[5] for reviews from a range of perspectives). This has been spurred by the newly discovered mathematical analogies between this flow and dendritic crystal growth, directional solidification and diffusion limited aggregation (see references 4,5), though Darcian flow through a porous medium was the original motivation[6]. In most Hele-Shaw cell studies to date, the geometry consists of a long rectilinear channel where the width of the cell is $2a$, with $b << a$ (Figure 1). In this case, the interfacial motion is caused by an imposed pressure gradient which causes the more viscous fluid at infinity to be displaced with velocity V. Alternately, gravity can effect the interfacial displacement.

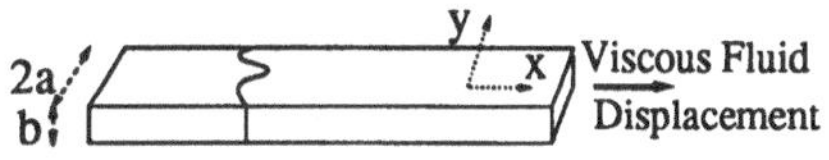

Figure 1. Rectilinear Hele-Shaw flow

Hele-Shaw flow has also been studied in a radial geometry with no side walls where the interface is driven by the injection of less viscous fluid through a hole on one of the parallel plates. The interfacial shape in the plane of the parallel plates (lateral plane) is shown in Fig. 2. A variation of the radial geometry is where sidewalls form a wedge[7] in the lateral plane with an opening angle of $2\theta_0$ $(-\pi \leq \theta_0 \leq \pi)$. Here, negative θ_0 corresponds to a convergent flow where the more viscous fluid is withdrawn at the intersection of the two side walls. Exact self-similar solutions in the wedge geometry in the absence of surface tension have been presented[8,9] at this conference generalizing earlier work[7] and including discussion of the effect of nonzero surface tension. It is to be noted that in the radial or wedge geometry, there are no other geometric length scales other than b. The Hele-Shaw equations discussed here will then be valid for $b << a$, where a in such cases is a typical interfacial length scale related to the shape of the interface in the lateral plane (Fig. 2).

Asymptotics beyond All Orders, Edited by H. Segur *et al.*
Plenum Press, New York, 1991

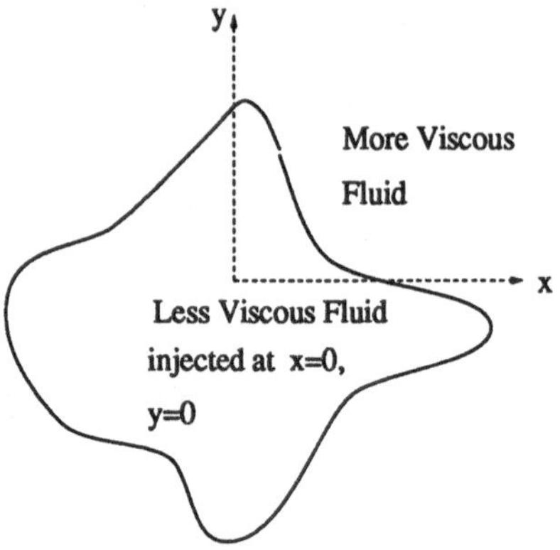

Figure 2. Radial Hele-Shaw flow

In conformity with the theme of this conference, we will limit ourslelves to discussion of the small surface tension problem, where asymptotics beyond all orders play an important role in the steady state selection and linear stability. Here is the general outline of this paper. In section 2, we describe some aspects of the experimentally observed features. In section 3, we describe the mathematical equations governing the Hele-Shaw flow and the boundary conditions that are derived under some drastic simplifying assumptions (We refer to these as the McLean-Saffman (MS) boundary conditions). In section 4, we present the findings of different investigators on steady states and their linear stability based on MS boundary conditions. In section 5, we summarize a procedure for extracting the asymptotics beyond all orders that gives rise to steady state finger selection. In section 6, we present an explicit example of a simple linear second order inhomogeneous differential equation to illustrate role of the inhomogenous term in determining the coefficients of transcendentally small terms. This clarifies a crucial step in the procedure described in section 5. Further, this example casts doubts on recent claims about existence of new kinds of steady state finger solutions[10] (so called 'Globally Neutral States'). In section 7, we discuss the initial value problem (with MS boundary conditions) and summarize some of our recent findings.

2. EXPERIMENTALLY OBSERVED PHENOMENA

It has been known since Hill[11], Chuoke et al[12] and Saffman & Taylor[6] that the planar interface between the viscous fluid of viscosity μ displaced by a less viscous fluid of viscosity μ_1 , $\mu_1 < \mu$ is unstable. Saffman & Taylor[6] studied the finite stages of this instability in a rectilinear Hele-Shaw cell. It was found that a small disturbance grows; and after a transient stage of competition between fingers, a single stable steadily travelling finger with speed U forms (see Fig. 3) except when the capillary effects are too small or equivalently the imposed pressure gradient (and therefore V) too large.

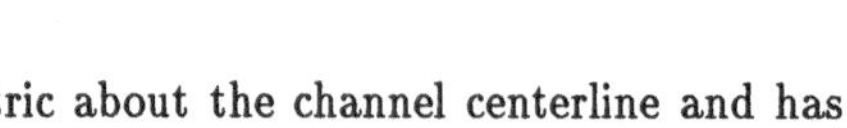

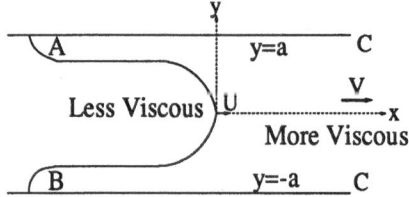

Figure 3. Final stages in a steady finger formation

This finger is symmetric about the channel centerline and has an asymptotic width of approximately one half the channel width for relatively large values of $Ca = \frac{\mu U}{T}$ (T : surface tension) or small values of $\frac{b}{a}$. This experiment[6] has since been repeated by other investigators[13–15] and the dependence of the eventual finger width on the different control parameters documented more precisely. It has been found that the actual width of the steady finger relative to the channel width can be somewhat less than a half and to a significant extent depends on two independent control parameters that may be taken as the gap to width ratio $\epsilon \equiv \frac{b}{2a}$ and the capillary number Ca . Further, it is found that the interface is unstable when $\frac{\epsilon^2}{Ca}$ is smaller than some small critical value[15]. This critical value seems to depend on the noise level in the experiment as well as on ϵ . The instability at very small values of $\frac{\epsilon^2}{Ca}$ leads to an intrinsically time dependent[15-17] pattern of tip splitting and side branching that can lead to the development of an apparently fractal pattern over a range of length scales.

Another important experimental observation is the extreme sensitivity to local flow inhomogeneities such as a little bubble[18] near the tip of a finger or a needle[19] piercing the interface or some anisotropy[20] introduced by etching the glass plates. Such a local perturbation dramatically alters the width of the observed fingers, these being a lot narrower and more stable than the regular ones. When instability does set in for these class of perturbed fingers, it is dendritic[21] rather than the usual tip splitting. To a remarkable degree, these perturbed fingers resemble a needle crystal.

For a circular or a divergent wedge geometry (i.e. $\theta_0 > 0$), the experiments of Bataille[22], Patterson[23] and Thome et al[7] show that a circular shape is unstable when the radius exceeds a surface tension dependent critical value. In a wedge geometry[7], for relatively small values of the wedge opening angle θ_0 ($\theta_0 < 20^o$), for a considerable period of time, the interface evolves into fingers that are self similar in time. However, in the divergent case, as the nose of such fingers fattens, they become susceptible to tip splitting. For larger values of the wedge angle in a divergent channel ($\theta_0 > 20^o$), or for the radial geometry, self similarity of any kind is not detectable as they appear to be sensitive to instabilities. In the range where capillary effects are small, λ , which is defined as the ratio of the angle subtended at the origin by these self-similar fingers to the wedge angle $2\theta_0$, varies roughly linearly with θ_0 for small θ_0 with λ close to a half for θ_0 approaching 0.

As for the rectilinear channel, small perturbations in the flow field dramatically alter the fingering pattern. Also, in this case, the self similarity is destroyed by such perturbation. For a convergent wedge flow, i.e. $\theta_0 < 0$, the self similar solutions are found to be quite stable though dendritic instability is still observed at sufficiently small surface tension values.

3. GOVERNING EQUATIONS AND BOUNDARY CONDITIONS

Here we pose the Hele-Shaw flow problem mathematically. Following McLean & Saffman[24] (MS), we make a number of simplifying assumptions. First, we ignore viscosity of the less viscous fluid and gravity effects. Each of these (with gravity parallel to the channel) have been included in previous steady state[6] and linear stability analysis[25] when other effects are neglected and it was found that there is no dramatic alteration of the qualitative features. However, gravity acting at an angle with respect to the channel axis, can result in a new effects[26] that will not be considered here. Further, for the sake of simplicity, we neglect any three dimensionality in the pressure field. The three dimensionality can arise near (distance of $O(b)$) the interface either because the interface touches the parallel plates at some contact angle other than 90^0 or because a thin film of the more viscous fluid is left behind next to the parallel plates. Effective two dimensional boundary conditions incorporating each of the localized three dimensional contact angle[27] and thin film effects[28,29] near the interface have been derived for steady flows. Calculations[27,30,31,32,33], based on these more complicated boundary conditions show important quantitative as well as qualitative effects[27,33] over some range of control parameters. However, analytical work[33] for a steady finger suggests that the mathematical analysis of the equations with thin film complications is, in a general sense, similar to the much simpler equations obtained here by neglecting such complications. In this conference where there is more interest in the mathematical structure of the equations rather than their physical implications and experimental comparisons, we will limit ourselves to the simpler theory with the MS boundary conditions, both for the steady state and the initial value problem.

With the simplifying assumptions discussed in the last paragraph, the Stokes equation for low Reynolds number flow through a narrow gap leads to a two dimensional pressure field[6]. The gap averaged fluid velocity in the laboratory frame can be written as the gradient of some velocity potential $\phi(x,y)$ that is proportional to the pressure. From incompressibility of the fluid flow, it follows that

$$\nabla^2 \phi = 0 \tag{1}$$

For a rectilinear geometry, the fluid far ahead of the finger moves at a constant velocity V in the x direction that is determined by the imposed pressure gradient. So, as $x \rightarrow +\infty$,

$$(u,v) \sim V\,\hat{x} + O(1) \tag{2}$$

where $\hat{x}$ is a unit vector along the x -axis (Fig. 1). In a circular geometry or a divergent wedge geometry, with fluid injected at the center at a rate Q , the fluid flow at ∞ is like a sink; i.e. as $x^2 + y^2 \rightarrow \infty$,

$$(u, v) \sim -\frac{Q}{2\pi}\nabla \, log \; r \tag{3}$$

where $r = \sqrt{x^2 + y^2}$. On the cell side walls (not relevant for the circular geometry)

$$(u, v) \cdot \vec{\hat{n}_w} = 0 \tag{4}$$

where $\vec{\hat{n}_w}$ denotes unit normal vector to the wall. Since this is a free boundary problem, there are two boundary conditions at the interface $F(x, y, t) = 0$:

$$\frac{\partial}{\partial t} F + u \frac{\partial}{\partial x} F + v \frac{\partial}{\partial y} F = 0 \tag{5}$$

and

$$\phi = \frac{b^2 T}{12\mu} \frac{1}{R}. \tag{6}$$

Here T is the interfacial surface tension, μ is the viscosity of the viscous fluid and R is the radius of curvature of the interface in the $x - y$ (lateral) plane. Equation (5) follows from the kinematic condition that a boundary element on the interface must have the same normal velocity as that of the fluid at that point; while (6) is a statement of the difference of pressure on the two sides of the interface and will be referred to as the dynamic boundary conditions. Equation (1) with boundary conditions (2) or (3), (4) (when applicable), (5) and (6) completely determine the evolution of the the fluid flow and the free boundary in the region occupied by the viscous fluid. We label the set of interfacial boundary conditions (6) and (7) as the McLean-Saffman[24] boundary conditions (MS).

Without any loss of generality, in the channel case, we set $V = 1$ and $a = 1$ as this corresponds to measuring length in units of a and time in units of $\frac{a}{V}$. Similarly, in the radial or wedge case, $Q = 2\pi$, and a , which in this case is some lateral length scale, can be set to unity.

Thus, with the MS boundary conditions, there is a single control parameter $\mathcal{B}$ characterizing the flow, where

$$\mathcal{B} \equiv \frac{b^2 T}{12\mu V a^2} \tag{7}$$

in the channel geometry, and for the circular or wedge case,

$$\mathcal{B} \equiv \frac{b^2 T \pi}{6\mu Q a} \tag{8}$$

Unfortunately for the reader, that there is no universally accepted choice of control parameters; it seems there are as many different definitions of the nondimensional surface tension parameter as the number of workers in the area!

Chuoke et al[12] and Saffman & Taylor[6] studied the linear stability of a planar front in a rectlinear geometry to find that it was unstable with growth rate

$$\sigma = |k|(1 - \mathcal{B}k^2) \tag{9}$$

where k is the wave number of the disturbance. Bataille[22] and Patterson[23] observed that the same relation (9) holds for the fractional growth rate of an expanding circular interface provided $\mathcal{B}$ in relation (9) is redefined as

$$\mathcal{B} \equiv \frac{b^2 T \pi}{6\mu Q R(t)}$$

where $R(t)$ is the radius of the interface at any time. This means that for sufficiently small $R(t)$, $\mathcal{B}$ is sufficiently large so that the circular interface is stable since the wave number k in this case is constrained to be larger than $\frac{1}{R(t)}$. With the injection of more fluid, $R(t)$ becomes larger and $\mathcal{B}$ smaller until the circular interface is destabilized.

Note that when $\mathcal{B} = 0$, the growth rate of the disturbance is proportional to the wave number; a feature that is in common with the other interfacial problems (Kelvin-Helmholtz instability of a planar interface between neighboring fluids moving with different velocities) or the Mullins-Sererka instability that of a planar interface in crystal growth). The initial value problem, in the absence of surface tension, is ill posed and that is reflected in the growth rate relation (9). Further discussion of the time evolving flow will be deferred till section 7.

4. STEADY SOLUTIONS AND ITS STABILITY

With the MS boundary conditions, for a steady state, the kinematic boundary condition (5) reduces to

$$\frac{\partial \phi}{\partial n} = U \cos \theta \qquad (10)$$

where U is the velocity of advance of the steady semi-infinite finger (Fig. 4) and θ is the angle between the interface normal and the positive x axis. Integrating $\nabla^2 \phi$ over the flow domain and using (4) and (10) one finds that

$$\lambda = \frac{V}{U} \qquad (11)$$

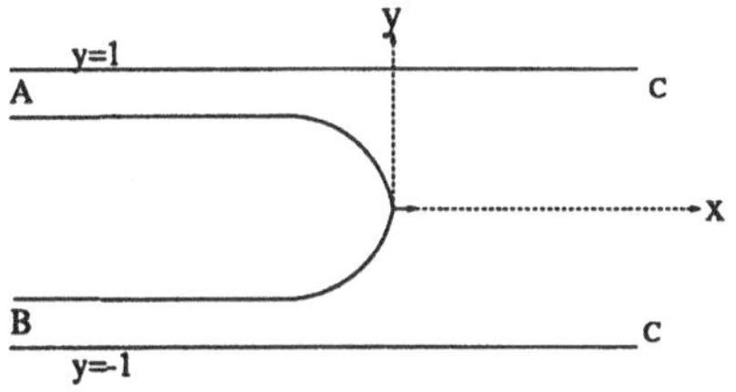

Figure 4. The physical $z = x + i\,y$ domain

Further, when $\mathcal{B} = 0$, the MS boundary conditions simplify as the nonlinear curvature term $\frac{1}{R}$ in (6) drops out. For these simplified boundary conditions, Saffman & Taylor[6] and Taylor & Saffman[34] found a two parameter family of steady semi-infinite fingers in closed form. The parameters can be taken to be the relative finger width λ and the distance of the tip from the channel centerline β . In the case when $\beta = 0$, the finger is symmetric about the channel centerline (Fig. 4) and the shape is described[6] by:

$$F(x,y,t) = x - \frac{t}{\lambda} - 2\frac{(1-\lambda)}{\pi}\, log\ cos\left(\frac{\pi y}{2\lambda}\right) = 0 \qquad (12)$$

λ is arbitrary in the interval (0, 1). Note that the time dependence disappears in (12) in the frame of the steady finger, i.e. if x is replaced by $x - \frac{1}{\lambda}t$. As mentioned in section 2, the experiment suggests that under normal conditions (without any needle or other tip perturbation), only a symmetric finger ($\beta = 0$) with specific λ is observed for a given set of experimental conditions. Despite this fundamental qualitative disagreement with experiment, the shape given by (12) matches quite well (except at very small Ca) with the experimentally observed finger[6] once the parameter λ in (12) is chosen in accordance with the experiment. Thus the main problem with the $\mathcal{B} = 0$ Saffman-Taylor theory[6] is a lack of unique steady state selection.

For nonzero $\mathcal{B}$, the mathematical problem is much more difficult owing to the presence of the nonlinear curvature term in (6). No closed form solutions exist though some have been found for other geometries[35,36]. McLean & Saffman[24] studied the steady finger for $\mathcal{B} \neq 0$ assuming that the finger was symmetric about the channel centerline. Using a hodograph representation in the frame of the steadily moving finger, equation (1) and boundary conditions (6) and (10) are converted into an integro-differential equation on a finite interval. Numerical calculations on these equations suggested that for nonzero values of $\mathcal{B}$, λ was determined as a function of $\mathcal{B}$. Further, the numerics suggested that $\lim_{\mathcal{B}\to 0} \lambda = \frac{1}{2}$ and in this limit the corresponding finger shape converged to the Saffman-Taylor finger (given by (12)) with $\lambda = \frac{1}{2}$. However, they also noted an apparent contradiction between their numerics and a formal perturbation series calculation. Subsequent numerical calculations by Romero[37] and VandenBroeck[38] confirmed the numerical findings of McLean & Saffman[24] and in addition led to the discovery of new branches of solutions. VandenBroeck[38] introduced a systematic procedure to find all branches of solutions. For the symmetric fingers, this procedure consisted of allowing for a finite angled cusp at the finger tip, i.e $\theta \neq 0$ at the tip. The finite angled cusp solutions are not physically relevant; however, they provide a convenient intermediate state between two physically relevant solutions (those with a continuous tangent). This provides a nice way of numerically continuing from one physical branch to another. In the McLean-Saffman integro-differential equation for a symmetric finger, allowance for a finite cusp angle at the tip can be conveniently made by taking λ as a known quantity along with $\mathcal{B}$, and by relaxing the continuous tangent boundary condition at the tip, which corresponds to one of the end point boundary conditions. When $\mathcal{B} = 0$, the continuity of the tip slope is automatically assured for the Saffman-Taylor solutions without any constraint on λ. Numerical evidence[38] suggested that this is not the case for nonzero $\mathcal{B}$. For given $\mathcal{B}$, by continuously increasing λ from 0 upwards, VandenBroeck[38] found that the cusp angle goes through zero for the first time when λ is equal to the value calculated by McLean & Saffman[24]. On further increase of λ, the cusp angle goes through oscillations about 0. For small $\mathcal{B}$, the number of zero crossings increases. Since each zero crossing corresponds to a tip with no discontinuity of the tangent, Vanden-Broeck[38] suggested that there was a discrete infinity of physically relevant branches of solution. As $\mathcal{B} \to 0$, Vanden-Broeck's numerical evidence suggested that λ on each and every branch asymptotes to $\frac{1}{2}$ as shown in figure 5. Romero[37] and VandenBroeck's[38] findings did not address the apparent contradiction between numerics and perturbation of McLean & Saffman[24]. On the contrary, it raised a new question about selection: If numerical calculations are correct, why is one branch of solution observed in experiment[6] ? As shall be seen later, the answer to this question has to do with stability.

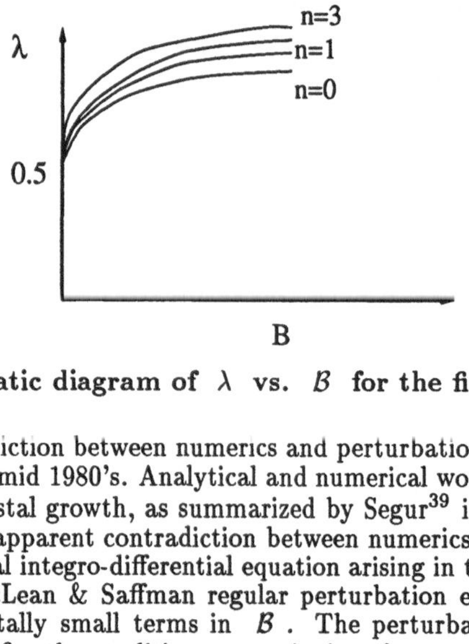

Figure 5. Schematic diagram of λ vs. $\mathcal{B}$ for the first few branches

The apparent contradiction between numerics and perturbation of McLean & Saffman[24] was not resolved until the mid 1980's. Analytical and numerical work on the simpler problem of geometric model of crystal growth, as summarized by Segur[39] in this proceedings, paved the way for resolving the apparent contradiction between numerics and perturbation for the more complicated non-local integro-differential equation arising in the Hele-Shaw flow. First, it was noted that the McLean & Saffman regular perturbation expansion was incomplete as it ignored transcendentally small terms in $\mathcal{B}$. The perturbation terms calculated by McLean & Saffman[24] satisfies the condition on continuity of tangent for any λ. It appeared that is the case for every term of the regular perturbation series of McLean & Saffman. On the other hand, the Vanden-Broeck calculations[38] suggests that the actual solution does not satisfy this condition for arbitrary λ. This suggests that the transcendentally small terms in $\mathcal{B}$ that were left out in the McLean-Saffman calculation can become important at the tip. Kessler & Levine[40] were the first to recognize this as a possible reason for the apparent contradiction between numerics and perturbation. They used the Vanden-Broeck formulation to calculate the cusp angle for arbitrary λ. Their numerical calculations[40] suggested that the cusp angle is a transcendentally small function of $\mathcal{B}$, as $\mathcal{B} \to 0$. Only for a discrete set of values of λ does the the cusp angle vanish. This was confirmed by formal calculations due to Combescot et al[41], Shraiman[42] and Hong & Langer[43]. Precise values of scaling constants have also been determined[44,45,46]. Besides selection of λ, the evidence suggests that only symmetric fingers are possible[44] for nonzero $\mathcal{B}$. Quite indepedendently of these analytical findings, the approximate equations for a small nearly circular Hele-Shaw bubble were analyzed[47] and it was found that a selection of bubble velocity occurs if one assumes that the interface is smooth and the fluid flow at infinity is free of any singularity. For arbitrary bubble velocity, unphysical singularities are present at ∞ that are transcendentally small for small surface tension. By requiring that this singularity be absent, a selection in bubble velocity occurred in the the Taylor-Saffman family of bubble solutions that are small and nearly circular. For bubbles of arbitrary size, numerical work[47,48] later supported by analytical evidence[49,50,51] suggests that surface tension causes a selection of bubble velocity and symmetry similar to that observed in the finger problem.

The linear stability of the steady finger has some surprising features. First, for zero surface tension, Saffman & Taylor[52] found a discrete set of tip breaking instability modes. McLean & Saffman[24] reported that there was no sign of stabilization even with surface tension. Subsequently, an error was found[53] in their stability equations nullifying their earlier conclusion. Kessler & Levine[54] found the $n = 0$ (McLean-Saffman) branch to be stable upto a small value of $\mathcal{B}$ below which reliable computations became difficult. They hypothesized that the McLean-Saffman branch of finger solution remains stable for any nonzero $\mathcal{B}$. This was supported by the work of Bensimon[55]. The Kessler-Levine conjecture about stability was later supported by analytical evidence[56] in the limit $\mathcal{B} \to 0$. It was found[56] that the spectrum of the linear stability operator with and without surface tension are fundamentally different. First, the $\mathcal{B} = 0$ discrete spectrum of Saffman & Taylor[52] is a special case of a more general continuous spectrum comprising the entire right half complex plane. However, analytical evidence[56] involving formal extraction of asymptotic terms beyond all orders suggest that almost all such eigen values and corresponding eigen modes do not continue to nonzero surface tension unless we relax the dynamic pressure condition at a point. Such a generalized eigenmode when superposed on the steady solution does not preserve the continuity of the tangent slope because of transcendentally small terms in surface tension. By requiring that the eigen modes introduce no discontinuity in the tangent slope, as is physically reasonable, a restriction on eigen values and corresponding eigen modes is imposed. Thus only very specific eigen values of the zero surface tension spectrum are possible limits as $\mathcal{B} \to 0$. A selection mechanism operates that is similar to the selection of λ for steady state fingers. For the McLean-Saffman finger branch, the only selected value of the spectrum in the limit of $\mathcal{B} \to 0$ is found to be zero. However, this analysis[56] does not include eigen modes that fail to decay at infinity as would be true for modes that change the width of the finger. The inclusion of such modes in the analysis remains an open problem since it is not clear what boundary conditions will be appropriate at infinity as an actual finger in an experiment is always of finite length. Further, subsequent numerical calculations[57,58] supported by analytical evidence[56,59] shows that all other branches ($n \geq 1$) are unstable with the n th branch unstable to n symmetric and n antisymmetric modes. On first glance, this result is rather surprising. All the branches tend to the same Saffman-Taylor finger, as shown in Fig. 5. So it might seem that for arbitrarily small surface tension, since the steady states corresponding to the different branches coalesce, so must the corresponding linear stability spectrum. This is not the case. The reason[56] is that selection of the spectrum in the limit $\mathcal{B} \to 0$ is caused by transcendentally small terms in $\mathcal{B}$. The precise form of this transcendental correction is sensitive to small changes in the steady state in the physical domain as the analytical continuation of these small changes do not stay small in the neighborhood of the critical point(s) in the unphysical complex plane. Yet, the form of the inner equation near the critical point(s), which is affected by this steady state differences, determine the coefficients of transcendentally small corrections to the zero surface tension eigen modes in the physical domain, and hence the constraints placed on the eigenvalues. This explains the difference in the linear stability spectrum between the $n = 0$ and $n \neq 0$ branches for arbitrarily small but nonzero $\mathcal{B}$. Numerical simulation of the initial value problem[2,60] suggests the instability of the $n = 0$ branch for sufficiently small $\mathcal{B}$. However, as Bensimon et al[2] noted, this is a nonlinear mechanism and does not contradict the linear stability results. In Fig. 5, all but one branch is unstable; yet when they coelesce in the limit $\mathcal{B} \to 0$, the amplitude of disturbance for nonlinear destabilization of the McLean-Saffman ($n = 0$) finger tends to zero.

We end this section by emphasizing that the results quoted thus far for nonzero $\mathcal{B}$ have not been proved; nor are they universally accepted. Based on numerical simulation of the initial value problem using the MS boundary conditions, Degregoria & Schwartz[61] suggest the possibility of branches of steady finger solutions whose relative widths are actually less than a half for small $\mathcal{B}$ in contradiction with the findings described in the previous paragraphs. However, numerical simulation by Meiburg & Homsy[62] casts doubt on these findings. Recently, Xu[10] analyzed the stability of a finger for small surface tension. He concludes that there are oscillatory instability modes. The determination of these nodes is carried out by analyzing the equations in the neighborhood of points in the unphysical complex plane. He also concludes that there are neutral modes (so called Globally Neutral States), when the finger widths have certain values less than a half and concludes that there exists selected steady states with $\lambda < \frac{1}{2}$. While linear instability modes affecting the tail of the finger does not contradict the previous analytical work[56] where these are disallowed, we are of the opinion that Xu's conclusion is questionable as they are based on a perturbing about the Saffman-Taylor finger given in (12) rather than the actual steady state. It is true that the true steady state is only a small deviation from the Saffman-Taylor finger; however, as noted in the previous paragraph, it is important to take this into account. Xu's analysis[10] does not discriminate between any of the branches shown in Fig. 5. For the same reason, the conclusion about the existence of neutrally stable modes (so called Globally Neutral Modes) and hence steady states (with $\lambda < \frac{1}{2}$) appears to be flawed. The condition[10] on λ found by requiring a neutrally stable mode, when surface tension effects are neglected in the steady state and included in the stability analysis, is generally incompatible to restriction on λ in an actual steady state calculation. In the first case[10], one requires that a homogeneous differential equation, which is the inner equation near a critical point in the complex plane, have nontrivial solutions that vanish as the inner independent variable goes to infinity in certain Stokes sectors. Only for specific value(s) of λ , the symmetry condition can be satisfied, giving rise to a selection. On the other hand, the correct determination of transcendental corrections to the Saffman-Taylor solution requires the analysis of an inner nonlinear equation[41,44,46]. Even if we were to linearize[46] this equation, we find that the inner equation is an inhomogeneous second order linear differential equation. Only the associated homogeneous equation[46] is equivalent to the inner equations and boundary conditions analyzed by Xu[10] near the critical point(s). In section 6, in a relatively simple and explicit example of an inhomogeneous linear ordinary differential equation, we illustrate that the determination of coefficients of transcendental corrections cannot be done by studying the associated homogeneous equation. This is true even though the form of the transcendental correction is determined by the associated homogeneous equation.

5. EXTRACTION OF EXPONENTIAL ASYMPTOTICS

We now elaborate the main steps in the mathematical determination of selection of steady finger in the limit of $\mathcal{B} \to 0$ via asymptotics beyond all orders. A generalization[56] of this procedure is applicable to the linear stability problem. The formulation used here is close to our previous work[44]. However, in hindsight, we have simplified some of the more complicated steps. This formulation is different from the integro-differential equation approach of McLean & Saffman[24] that has been the basis of most analytical work. We prefer our differing formulation because it is easily generalizable to the linear stability and the initial value problem. It is also generalizable to more complicated boundary conditions[30,33]. The general idea of some of the key steps presented here follows from the work of Combescot et al[41] and its subsequent comprehensive elaboration[46]. The details are different due to differences in formulation.

For the sake of simplicity of presentation, we assume a priori that the finger is symmetric about the channel centerline. The problem of determination of steady state selection for the class of symmetric fingers is equivalent[44] to determining an analytic function f in the unit upper half ζ semi-circle so that

$$Im\, f = 0 \tag{13}$$

on the real diameter and on the semi-circular boundary $\zeta = e^{i\nu}$, ν in $[0, \pi]$,

$$Re\, f = -\frac{\mathcal{G}}{|f_\zeta + h|}\, Re \left[1 + \zeta \frac{f_{\zeta\zeta} + h_\zeta}{f_\zeta + h} \right] \tag{14}$$

where

$$\mathcal{G} = \frac{1}{4}\pi^2 \lambda \mathcal{B}$$

is our new nondimensional surface tension parameter,

$$h = \frac{1 - p^2\zeta^2}{\zeta(\zeta^2 - 1)} \tag{15}$$

where

$$p^2 = 2\lambda - 1 \tag{16}$$

Note that p^2 is in the interval (-1,1) corresponding to λ in (0,1); so the zeros of h are outside the unit circle. Here f is defined by the expression:

$$z_\zeta(\zeta) = \frac{2}{\pi}[f_\zeta + h] \tag{15}$$

where $z(\zeta)$ is the conformal map from the unit semi-circle (Figure 6)

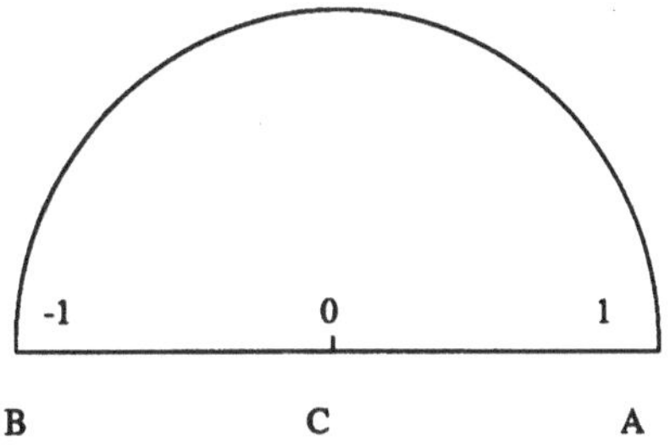

Figure 6. Unit ζ semi-circle

to the physical domain outside of the semi-infinite finger (Figure 4). As is physically reasonable, we will assume that the $f(\zeta)$ is continuous and piecewise analytic on the semi-circular arc $|\zeta| = 1$. We will also assume that any singularity of f' at $\zeta = \pm 1$ is better than a simple pole, i.e. $\lim_{\zeta \to \pm 1}(\zeta \mp 1)f' = 0$. This assumption ensures that the parameter λ appearing in (16) corresponds to the true finger width. Further, for a true physically relevant solution, $f'(e^{i\nu})$ is continuous for ν in the open interval $(0, \pi)$ for a finger boundary with a continuous tangent. However, in extracting asymptotics beyond all orders, this condition is relaxed for arbitrary λ and reimposed at the end to determine λ. A priori, it is also assumed that $z(\zeta) + \frac{2}{\pi} \ln \zeta$ is continuous at $\zeta = 0$. From this and the properties of a conformal map that maps a line segment to a line segment, it follows that f is continuous upto the real line for ζ in (-1,1). Schwartz's reflection principle holds implying that f is an analytic function for $|\zeta| < 1$. As a consequence, the boundary condition (14) holds on the lower half semi-circle as well. For $\mathcal{G} = 0$, (13) and (15) immediately imply that $f = 0$; this corresponds to the Saffman-Taylor solution[6]. On integrating $z_\zeta = \frac{2}{\pi} h$ on $\zeta = e^{i\nu}$, taking the real and imaginary parts and eliminating ν between them, we arrive at formula (12), noting that $z = x - \frac{1}{\lambda}t + i\,y$ in the finger frame. A regular perturbation expansion can be constructed in powers of $\mathcal{G}$:

$$f = \mathcal{G} f_1 + \mathcal{G}^2 f_2 + \ldots \tag{16}$$

as was first done by McLean & Saffman[24] in a differing but equivalent formulation. f_1 satisfies

$$Im\, f_1 = 0 \tag{17}$$

on the real diameter and on $\zeta = e^{i\nu}$,

$$Re\, f_1 = -\frac{1}{|h|} Re\left[1 + \zeta \frac{h'}{h}\right] \tag{18}$$

Clearly an expression for the analytic function f_1 can be found that satisfies (17) and (18) by using Poisson's integral formulae for a harmonic function and its conjugate in a circle, noting that from (17) and reflection principle, (18) holds on the lower half semi-circular arc as well. Since the right hand side of (18) is an analytic function of ν except at $\nu = 0$ and π, it follows that f_1 is an analytic function on $|\zeta| = 1$ except at $\zeta = \pm 1$. McLean & Saffman[24] and McLean[65] used a differing but equivalent formulation to determine the first few terms of (16). However, as they noted, (16) is not uniformly valid at $\zeta = \pm 1$, where the higher order perturbation terms are increasingly singular. These points correspond to the finger tails. While (17) and (18) suggest that near $\zeta = 1$,

$$f_1 = -\frac{4}{\pi}\,\frac{1+p^2}{(1-p^2)^2}\,(1-\zeta)\,ln(1-\zeta) + o(1-\zeta) \tag{19}$$

one finds that this behavior is inconsistent with (13) and (14). Indeed, through a local expansion, one finds a consistent behavior of the form:

$$f = a_1\,(1-\zeta)^{1-\tau} + a_2\,(1-\zeta) + o(1-\zeta) \tag{20}$$

where τ is determined by the transcendental equation:

$$tan(\frac{\pi\tau}{2}) = 4\,\mathcal{G}\,\frac{(1-\tau)^2}{(1-p^2)^2} \tag{21}$$

Thus for small $\mathcal{G}$, we find that the expression (16) is not uniformly valid in a neighborhood of $\zeta = \pm 1$. The coefficients a_1 and a_2 can be determined to first order in $\mathcal{G}$ by expanding (20) for small $\tau\, ln(1-\zeta)$ and matching with (19). McLean & Saffman[24] and McLean[65] carried out the inner-outer matching to fourth order perturbation term; yet were unable to find any condition on p^2 and hence λ. Further, the McLean-Saffman calculation[24,65] indicates that each term in the regular perturbation term in (16) is analytic on the semi-circular arc $\zeta = e^{i\nu}$ for ν in $(0,\pi)$ and so the condition on continuity of $f'(e^{i\nu})$ appears to hold automatically for any term in (16).

The failure of the regular perturbation expansion (16) in selecting a unique steady state is established more clearly in the related problem of velocity selection of a steady symmetric Hele-Shaw bubble[47]. In that case, one finds equations (13) and (14) holds (with a differing definition of $\mathcal{G}$), provided we also redefine

$$h = \frac{(1-p^2\,\zeta^2)}{(\zeta^2-\alpha^2)(1-\alpha^2\zeta^2)} \tag{21}$$

where α (bubble size parameter) and p are each in the interval (0, 1), with p related to the relative bubble velocity U through the relation:

$$p^2 = \frac{U(1+\alpha^2)-2}{U(1+\alpha^2)-2\alpha^2} \tag{22}$$

In this case, unlike the finger, the right hand side of (18) is a smooth 2π periodic function of the angular variable ν, where $\zeta = e^{i\nu}$ on the circle. From reflection principle, (14) holds on the entire circular boundary and from the smoothness of the boundary data, it immediately follows that f_1 will be analytic function of ζ for $|\zeta| \leq 1$. It has been shown[47] that any term f_n of the regular perturbation series (16) is analytic with the nearest singularity location (outside the unit circle) independent of n. Thus the formal expansion (16) is known to be consistent. Yet there is no constraint on the parameter p^2 arising from the restriction that the bubble boundary has a continuous tangent, i.e. f' is continuous on $|\zeta| = 1$. Numerical computation[47,48] on the other hand suggests a selection mechanism similar to that of a finger. However, as has been pointed out here quite a few times in the context of different problems, the fact that every term in the perturbation series satisfies a condition need not imply that f satisfies the same condition. Transcendentally small terms in $\mathcal{G}$ that are normally subdominant to every term in (16) can violate the condition that is desired. With respect to this condition, the terms that are normally beyond all orders can become dominant. Therefore, we need to extract such terms.

We now return to the finger problem and outline a procedure to extract the leading order transcendentally small correction to (16), as this does not satisfy the condition of continuity of tangent automatically. The constraint on λ found in this leading order analysis is expected to be modified slightly by inclusion of higher order transcendental asymptotics. However, comparison[44-46] with direct numerical calculations show agreement to a remarkable degree even when $\mathcal{G}$ is not too small.

Recall that for an harmonic function inside the unit circle

$$\phi(r,\theta) = \frac{1}{2\pi}\int_{-\pi}^{\pi} \frac{(1-r^2)}{1 + r^2 - 2r cos(\nu-\theta)}\,\phi(1,\nu)d\nu \tag{23}$$

Upto an additive constant, its harmonic conjugate is given by:

$$\psi(r,\theta) = = \frac{1}{2\pi}\int_{-\pi}^{\pi} \frac{2r sin(\theta-\nu)}{1 + r^2 - 2r cos(\nu-\theta)}\,\phi(1,\nu)\,d\nu \tag{24}$$

where (r,θ) are the standard polar coordinates. Further if the boundary data has the symmetry $\phi(1,\nu) = \phi(1,-\nu)$, as is true for $Re f(e^{i\nu})$ given by (14), then equations (23) and (24) can be written as

$$\phi(r,\theta) = \frac{1}{2\pi}\int_0^{\pi}\left[\frac{(1-r^2)}{1+r^2-2r cos(\nu-\theta)} + \frac{(1-r^2)}{1+r^2-2r cos(\nu+\theta)}\right]\phi(1,\nu)d\nu \tag{25}$$

$$\psi(r,\theta) = \frac{1}{2\pi}\int_0^{\pi}\left[\frac{2r sin(\theta-\nu)}{1+r^2-2r cos(\nu-\theta)} + \frac{2r sin(\theta+\nu)}{1+r^2-2r cos(\nu+\theta)}\right]\phi(1,\nu)d\nu \tag{26}$$

Identifying $\phi(r,\theta)$ with $Re\ f(re^{i\theta})$ and $\psi(r,\theta)$ with $Im\ f(re^{i\theta})$ in the above formulae and combining (25) and (26), we get

$$f(\zeta) = \frac{1}{2\pi i}\int_0^{\pi} d\nu\left[\frac{\zeta + e^{i\nu}}{e^{i\nu}-\zeta} + \frac{1+\zeta e^{i\nu}}{1 - e^{i\nu}\zeta}\right] Re\ f(e^{i\nu}) \tag{27}$$

Note that there is no possible additive constant for $Im\ f$ as it has been fixed by the condition $Im\ f(0) = 0$ that follows from (13). On introducing $\zeta' = e^{i\nu}$ and using (14), we get

$$f(\zeta) = \frac{\mathcal{G}}{2\pi i}\int_C \frac{d\zeta'}{\zeta'}\left[\frac{\zeta+\zeta'}{\zeta'-\zeta} + \frac{1+\zeta\zeta'}{1-\zeta\zeta'}\right]\kappa(\zeta') \tag{28}$$

where C is the upper half unit semi-circular contour going from +1 to -1 and $\kappa(\zeta)$ is defined to be a locally analytic function of ζ such that for $\zeta = e^{i\nu}$, ν in $(0,\pi)$,

$$\kappa(\zeta) = -\frac{1}{|f_\zeta + h|} Re\left[1 + \zeta\frac{f_{\zeta\zeta}+h_\zeta}{f_\zeta + h}\right] \tag{29}$$

On the semi-circular arc, κ is real and is actually the lateral curvature. Equation (29) is not valid off the unit upper semi-circular arc as the absolute value on the right hand side is not an analytic function. On the other hand, we note that owing to the property (13), on $\zeta = e^{i\nu}$,

$$f_\zeta^*(\zeta) = f_\zeta^*(e^{i\nu}) = f_\zeta(e^{-i\nu}) = f_\zeta(1/\zeta) \tag{30}$$

Similarly

$$h^*(\zeta) = h^*(e^{i\nu}) = h(e^{-i\nu}) = h(1/\zeta) \tag{31}$$

Thus on $\zeta = e^{i\nu}$,

$$|f_\zeta + h| = (f_\zeta(\zeta) + h(\zeta))^{1/2}(f_\zeta(1/\zeta) + h(1/\zeta))^{1/2} \tag{32}$$

where appropriate care need to be taken in the choice of branch cuts on the right hand side of (32). Similarly on $\zeta = e^{i\nu}$, we get

$$Re \left[1 + \zeta \frac{f_{\zeta\zeta} + h_\zeta}{f_\zeta + h}\right] = 1 + \frac{1}{2}\zeta \frac{f_{\zeta\zeta}(\zeta) + h_\zeta(\zeta)}{f_\zeta(\zeta) + h(\zeta)} + \frac{1}{2\zeta} \frac{f_{\zeta\zeta}(1/\zeta) + h_\zeta(1/\zeta)}{f_\zeta(1/\zeta) + h(1/\zeta)} \tag{33}$$

Then

$$\kappa(\zeta) = \frac{-1}{2(f_\zeta(\zeta)+h(\zeta))^{1/2}(f_\zeta(1/\zeta)+h(1/\zeta))^{1/2}} \left[2 + \zeta \frac{f_{\zeta\zeta}(\zeta)+h_\zeta(\zeta)}{f_\zeta(\zeta)+h(\zeta)} + \frac{1}{\zeta} \frac{f_{\zeta\zeta}(1/\zeta)+h_\zeta(1/\zeta)}{f_\zeta(1/\zeta)+h(1/\zeta)}\right] \tag{34}$$

Equation (34) holds off the semi-circular arc as well, as each side of (34) is an locally analytic function of ζ. Since we are interested in finding the form of equation in the unphysical region, we analytically continue (28) across the arc of the unit semi-circle to the region $|\zeta| > 1$ through contour deformation in the ζ' plane as illustrated in Fig. 7. On collecting the residue at $\zeta' = \zeta$,

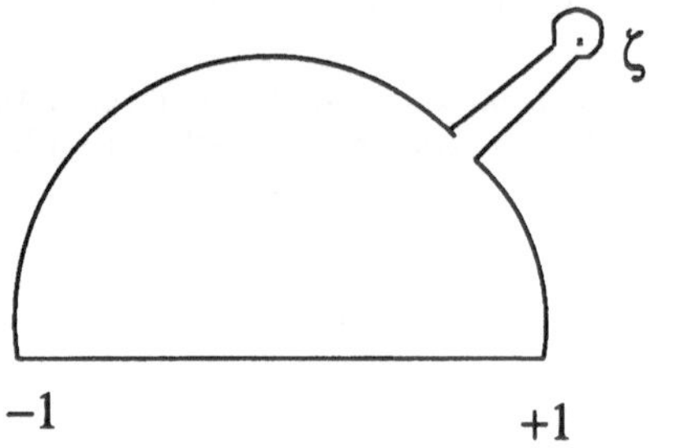

Figure 7. Contour deformation in the ζ' plane

it is clear that for $|\zeta| > 1$, the analytic continuation of (28) is

$$f(\zeta) = \mathcal{G}\, I(\zeta) + 2\,\mathcal{G}\, \kappa(\zeta) \tag{35}$$

where

$$I(\zeta) = \frac{1}{2\pi i} \int_C \frac{d\zeta'}{\zeta'} \left[\frac{\zeta + \zeta'}{\zeta' - \zeta} + \frac{1 + \zeta\zeta'}{1 - \zeta\zeta'}\right] \kappa(\zeta') \tag{36}$$

Substituting for κ from (34), equation (35) becomes an integro-differential equation for f since each of $f_\zeta(1/\zeta)$ and $f_{\zeta\zeta}(1/\zeta)$ can be expressed in terms of integrals of $Re\, f$ on $\zeta = e^{i\nu}$ through Poisson's integral formulae and its derivatives (as $\frac{1}{\zeta}$ is inside the unit circle). We can look for a solution with asymptotic series (16) for small $\mathcal{G}$. It is clear that

$$f_1 = I_0(\zeta) + 2\,\kappa_0 \tag{37}$$

where

$$\kappa_0(\zeta) = -\frac{1}{[h(\zeta)]^{1/2}[h(1/\zeta)]^{1/2}} \left[1 + \frac{\zeta}{2} \frac{h_\zeta(\zeta)}{h(\zeta)} + \frac{1}{2\zeta} \frac{h_\zeta(1/\zeta)}{h(1/\zeta)}\right] \tag{38}$$

is the curvature function corresponding to Saffman-Taylor solution $f = 0$, and

$$I_0(\zeta) = \frac{1}{2\pi i} \int_C \frac{d\zeta'}{\zeta'} \left[\frac{\zeta + \zeta'}{\zeta' - \zeta} + \frac{1 + \zeta\zeta'}{1 - \zeta\zeta'}\right] \kappa_0(\zeta') \tag{39}$$

Equation (37) can alternately be obtained by analytically continuing (18). Clearly, $I_0(\zeta)$ defines an analytic function everywhere in the region $|\zeta| > 1$ that is bounded everywhere as well. On the other hand the same is not true for the term κ_0 as it is singular at points where $h = 0$. This happens at critical points i.e. $\zeta = \pm\frac{1}{p}$, outside the unit circle. Therefore, the perturbation expansion (16) breaks down at these critical points.

This suggests the construction of an inner asymptotic expansion in the neighborhood of each of these points obtained by the introduction of new scaled dependent and independent variables into (35). For this problem, the relevant scalings for the dependent and independent variables are such that the the nonlocal term $\mathcal{G}I$, as well as $f_\zeta(1/\zeta)$ and $f_{\zeta\zeta}(1/\zeta)$ in the term κ (as given by 34), drop out of (35). This gives us a nonlinear ordinary differential equation. We will refer to this as the inner equation. From this point onwards, our description becomes more sketchy since the calculation procedure for exponentials arising in a nonlinear ordinary differential equation is essentially the one by Kruskal & Segur[67].

We must require that the solution to the inner equation match to the outer solution (16) in certain Stokes sectors that adjoin the physical domain boundary $|\zeta| = 1$, $Im\,\zeta > 0$. The Stokes sectors are bounded by Stokes lines that require determination of the WKB approximate solutions to an associated linear homogeneous equation found by linearizing (35) about $f = 0$. The non-local integral parts in (35), once again, do not play any part in determining the form of these leading order WKB solutions. Away from the critical points, their form, though not their coefficients, are determined by solutions to an associated second order homogeneous linear differential equation. The coefficients of these WKB corrections to (16), which are different in different Stokes sectors, are very important as they decide if in each sector on the finger boundary $\lim_{\mathcal{G}\to 0} f = 0$ (through elimination of exponentially large terms). They also decide if the above condition can be satisfied by relaxing the continuity of $f'(e^{i\nu})$ for ν in $(0, \pi)$. To find the coefficients the transcendental terms, it is necessary to carry out the inner-outer expansion near each critical point. Note that the two inner regions near the two critical points $\zeta = \pm\frac{1}{p}$ coalesce to one near ∞ for λ close to $\frac{1}{2}$.

The final conclusion from this procedure is that for arbitrary λ , it is not possible to find an analytic single valued solution f across the semi-circular arc (excluding $\zeta = \pm 1$) that tends to 0 as $\mathcal{G} \to 0$ (i.e. approaches a Saffman-Taylor solution) and is analytic everywhere in $|\zeta| < 1$. If we relax single-valuedness, solution exists corresponding to a single branch point on the semi-circular arc. This branch point corresponds to a cusp on the finger boundary where the tangent slope is discontinuous. For a symmetric finger, this cusp is located at the finger tip. On imposing the continuity of $f_\zeta(e^{i\nu})$ for ν in $(0, \pi)$, a discrete set of selected λ emerges such that on each branch in the small $\mathcal{G}$ limit, $\lambda \sim \frac{1}{2} + k_n\,\mathcal{G}^{2/3}$, where the positive numerical constant k_n is an increasing function of n (the positive integer characterizing the various branches (see Fig.5)). These results[41,44–46] support earlier numerical evidence[24,37,38,40].

We now comment on the relation of the procedure outlined in this section, which is basically a generalization of the Kruskal-Segur[67] procedure (see contribution by H.Segur) to other procedures used in determining steady state. One popular method, apparently due to Shraiman[42], is to linearize the equations about the Saffman-Taylor solution at the outset, retaining all the higher derivative terms, and cast the question of existence of solution in terms of Fredholm theory for a linear non self-adjoint operator. Approximate solution to the adjoint homogeneous equation is found in the WKB form. Fredholm alternative condition for existence of solution requires that certain constraint be satisfied which gives rise to selection of λ . In this simple and appealing formulation, there is apparently no need to analytically continue the equations to the unphysical plane. However, when it comes to actually evaluating the integral constraint for small $\mathcal{G}$, the steepest descent contour leads one to the neighborhood of the critical point(s) which gives most of the contribution. This gives an incorrect result for two reasons. First, the nonlinearity becomes important at the critical point (see contribution due to V. Hakim for comparison of linear and nonlinear method as applied to the geometric model of crystal growth). Second, even within the context of the linear equation, the WKB solution is invalid at the critical point. Despite these two errors, the method[42] gives the correct scaling result and qualitative conclusion. Another procedure used in the literature[10] in this and other mathematically related problems of crystal growth[63,64] is to use a multi-scale expansion. It relies on the fact that transcendentally small corrections (of the WKB form) to (16) vary over a fast scale in the physical domain compared to the terms in (16). One looks at the such fast varying solutions to the original partial differential equations (rather than the reduced integro-differential equation) with linearized boundary conditions. As can be expected, the form of the fast varying part of the solution (WKB-type) arise by studying the associated homogeneous equation and boundary conditions. A crucial ansatz is made at this stage: the slow varying inhomogeneous term in the equation has no effect with the fast varying part of the solution. This approach is appealing in that there is no need to use an integro-differential equation and therefore it would seem to be of value in more general problems that are not reducible to integro-differential equations on a line. Further, there seems to be no need to go to the complex plane except to resolve the Stokes phenomena near some complex critical points. However, as we illustrate through an explicit example in section 6, the ansatz about no interaction between the fast and slow varying parts of the solution is generally incorrect.

6. SIMPLE EXPLICIT EXAMPLE

The purpose of this section is to demonstrate that, in general, the coefficients of transcendentally small parts of a solution to a second order linear inhomogeneous differential equation cannot be deduced correctly by studying the associated homogeneous equation. We will also show that this incorrect resolution has immediate consequence on questions of existence of solution and the assumption that fast varying and slow varying parts of the solution do not interact as has been assumed in some work in the literature[10,63,64].

Consider the differential equation for $u(x)$:

$$\epsilon \frac{d^2u}{dx^2} - \frac{\epsilon}{x}\frac{du}{dx} - 4x^2 u = 1 \tag{42}$$

We require that as $x \to \infty$, with $Arg\, x$ in $[0, \frac{\pi}{2})$,

$$u \to 0 \tag{43}$$

and that on the positive real $x-$ axis

$$Im\, u = 0 \tag{44}$$

This problem, as posed above will be referred to as problem A.

First, at a formal level, through a repeated dominant balance procedure, a consistent asymptotic behavior of such a solution, assuming it exists, for small ϵ is given by

$$u(x) \sim -\frac{1}{4x^2} \sum_{n=0}^{\infty} (2n)! \frac{\epsilon^n}{x^{4n}} \tag{45}$$

Each and every term of the asymptotic series in (45) satisfies all the conditions of problem A including (44). However, to ensure that the actual solution satisfies (44), we have to extract transcendentally small terms in ϵ . These are generated by the solutions of the associated homogeneous equation:

$$\epsilon \frac{d^2u_H}{dx^2} - \frac{\epsilon}{x}\frac{du_H}{dx} - 4x^2 u_H = 0 \tag{46}$$

Generally, approximate solutions to such an equation is found by the well known WKB technique. In our simple example, the WKB solutions $e^{\epsilon^{-1/2}x^2}$ and $e^{-\epsilon^{-1/2}x^2}$. are actually exact solutions of (46). Clearly, the variation of the WKB solutions to (46) in a domain $\mathcal{D}$ that excludes a neighborhood of $x = 0$ is fast compared to any term in the regular perturbation expansion (45). Now following Xu[10], suppose we make the assumption that the fast varying parts of the solution (i.e. WKB parts) do not interact with the slowly varying parts (i.e. any term in (45)) and seek to superimpose appropriate solution u_H of (46) on the regular perturbation series (45). The general solution to (46) is

$$u_H = C_1 e^{\epsilon^{-1/2}x^2} + C_2 e^{-\epsilon^{-1/2}x^2} \tag{47}$$

Since any finite trunction of (45) satisfies condition (43) and (44), we must require that $u_H \to 0$ as $x \to \infty$ for $Arg\, x$ in $[0, \frac{\pi}{2})$. Since each of the exponentials in (47) grow for large x in some subrange of $Arg\, x$ in $[0, \frac{\pi}{2})$, it would then seem to follow that each of C_1 and C_2 will have to be zero in order to satisfy (43). Thus, it would seem that

$$u_H = 0 \tag{48}$$

Note that in this problem, there is no complication of Stokes phenomena related to the homogeneneous equation (46) itself, as the WKB solutions are exact. However, the conclusion (48) is generic even if (46) is replaced by a second order homogeneous differential equation for which (47) holds approximately in one Stokes sector. In that case, as we move from one Stokes sector, where (47) is valid approximately, to to another Stokes sector,

$$u_H \sim C_3\, e^{\epsilon^{-1/2}x^2} + C_4\, e^{-\epsilon^{-1/2}x^2} \tag{49}$$

The pair (C_1, C_4) would then be related to (C_2, C_3) through some 2x2 connection matrix. Generically, this connection matrix is nonsingular so that if $C_1 = 0$, $C_4 = 0$, as is for u_H satisfying (43), then (C_2, C_3) is also zero. Thus, the conclusion (48) would still be correct. In the exceptaional case that the connection matrix is singular then an infinite set of nonzero (C_2, C_3) is possible. In that case, it is possible to satisfy (44) with appropriate choice of $Arg\ C_2$. Thus, in all cases, this procedure of separating the slow varying and fast varying parts (WKB) of the solution, under the assumption that they do not interact, leads us to the specific conclusion that problem A has a solution.

Now, we explicitly show that the conclusions in the last paragraph are incorrect because it is not valid to separate out u_H from the so called slowly varying regular perturbation series (45). This should come as no big surprise to the experts. As has been pointed out at this conference (see presentation by M. Berry), a divergent asymptotic series such as (45) has encoded in it information on exponential terms. Through a routine use of variation of parameters, a solution to (42) satisfying (43) is given by

$$u(x) = \frac{1}{4\epsilon^{1/2}}\left[e^{\epsilon^{-1/2}x^2} \int_{\infty\, e^{i\,0}}^{x} \frac{dx'}{x'}\, e^{-\epsilon^{-1/2}x'^2} - e^{-\epsilon^{-1/2}x^2} \int_{i\,\infty}^{x} \frac{dx'}{x'}\, e^{\epsilon^{-1/2}x'^2} \right] \tag{50}$$

This can be seen intuitively in the following way. In Stokes sector I bounded by Stokes lines $Arg\ x\ \pm \frac{\pi}{4}$ that satisfy determined by $Re\ x^2 = 0$ (Note that this is Anti-Stokes line by definition of Berry), exponentially large contribution outside the first integral is balanced by exponentially small contribution from the integral itself (owing to the choice of lower limit) to leave terms that decay algebraically to the leading order. Again, for the second integral in sector I, exponentially large contribution of the integral itself is being multiplied by exponentially small contribution outside the integral to leave us with algebraically decaying factors. In Stokes sector II, which is bounded by the Stokes lines $Arg\ x = \frac{\pi}{4}$ and $Arg\ x = \frac{3\pi}{4}$, the exponentially large contribution from the first integral is balanced by the exponentially multiple outside the first integral. Again, in this sector, the exponentially small contribution of the second integral (owing to the choice of lower limit) is balanced by the transcendentally large multiple outside the second integral. Thus in Stokes sectors I and II, which include the region of our interest $Arg\ x$ in $[0, \pi/2)$, it is possible to find solutions to (42) that satisfy condition (43). Further, through integration by parts in (50), we can easily establish that $u \sim -\frac{1}{4x^2}$ as $x \to \infty$ for $Arg\ x$ in $[0, \frac{\pi}{2})$. On repeated integration by parts, we can verify the the formal result (45). Further, we cannot add any nonzero linear combination of $e^{\epsilon^{-1/2}x^2}$ and $e^{-\epsilon^{-1/2}x^2}$ to the solution (50) as condition (43) will then be violated.

By appropriate changes in integration variables in the first and second integral, for $Arg\ x$ in $(0, \frac{\pi}{2})$, we rewrite (50):

$$u(x) = \frac{1}{8\epsilon^{1/2}}\left[\int_0^{\infty} dy\, e^{-\epsilon^{-1/2}y} \left\{ \frac{1}{y - x^2} - \frac{1}{y + x^2} \right\} \right] \tag{51}$$

From Watson's Lemma, we can calculate the complete asymptotic expansion for $Arg\ x$ in $(0, \frac{\pi}{2})$, once again verifying (45). Note, in particular, that (45) is valid on the Stokes line $Arg\ x = \frac{\pi}{4}$, approaching it from either side. Thus coefficients of any possible exponentials on either side of the Stokes lines, which dominate the asymptotic expansion (45) on the Stokes lines, must be identically zero. However, this does not imply that that there are no exponentials in the solutions for other ranges of $Arg\ x$. Consider what happens as we cross the anti-Stokes line $Arg\ x = 0$ and move to the region where $Arg\ x$ is in $(-\frac{\pi}{2}, 0)$. We notice from the exact expression (51) that in doing so we will collect a residue from pole at $y = x^2$ and we will obtain

$$u(x) \sim -\frac{1}{4x^2} \sum_{n=0}^{\infty} (2n)! \frac{\epsilon^n}{x^{4n}} + \frac{2\pi i}{8\epsilon^{1/2}} e^{-\epsilon^{-1/2} x^2} \tag{52}$$

For $Arg\ x$ in $(-\frac{\pi}{4}, 0)$, the newly added transcendental term is exponentially small, however, it becomes dominant when $Arg\ x$ is in $(-\frac{3\pi}{4}, -\frac{\pi}{4})$. When $Arg\ x = 0$, i.e. on an anti-Stokes line, the coefficient of the exponentially small term is the average between the two sides. On including this exponentially small term,

$$u(x) \sim -\frac{1}{4x^2} \sum_{n=0}^{\infty} (2n)! \frac{\epsilon^n}{x^{4n}} + \frac{\pi i}{8\epsilon^{1/2}} e^{-\epsilon^{-1/2} x^2} \tag{53}$$

Note that (52) implies that on the real x axis

$$Im\ u(x) = + \frac{\pi}{8\epsilon^{1/2}} e^{-\epsilon^{-1/2} x^2} \tag{54}$$

which is nonzero. We note that transcendentally small terms that are usually subdominant now becomes dominant since every term of the regular perturbation expansion is identically zero. Thus, from (54), we conclude that there are no solutions to problem A.

Now consider the implications of our explicit solution for a general problem of the type

$$\epsilon\, \mathcal{L}\, u + g(x)\, u = f(x) \tag{55}$$

where $\mathcal{L}$ is some second order linear differential operator and ϵ is some small parameter. Through a formal procedure of dominant balance one obtains a possible asymptotic expansion of the form:

$$u \sim \frac{f(x)}{g(x)} - \frac{\epsilon}{g(x)} \mathcal{L} \left[\frac{f(x)}{g(x)}\right] + o(\epsilon) \tag{56}$$

One can continue the dominant balance procedure involving the nonhomogeneous term and obtain higher and obtain coefficients of higher powers of ϵ. Transcendental terms in ϵ are however generated by the WKB solutions to the homogeneous part of the equation

$$\epsilon\, \mathcal{L} u_H + g(x)\, u_H = 0 \tag{57}$$

Suppose the WKB solutions to (57) are of the form $e^{\pm\epsilon^{-1/2} W_0 + W_1}$. Suppose we require that (56) be valid in two adjoining Stokes sectors. This means that we want the coefficients of transcendentally large WKB solution in each of the two sectors to be zero. Thus

$$u \sim \frac{f(x)}{g(x)} - \frac{\epsilon}{g(x)} \mathcal{L} \left[\frac{f(x)}{g(x)}\right] + C_1\, e^{-\epsilon^{-1/2}\, W_0\, +\, W_1} \tag{58}$$

in one Stokes sector, where $Re\ W_0 > 0$; and in the adjoining Stokes sector where $Re\ W_0 < 0$, we require that

$$u \sim \frac{f(x)}{g(x)} - \frac{\epsilon}{g(x)} \mathcal{L} \left[\frac{f(x)}{g(x)}\right] + C_2\, e^{\epsilon^{-1/2}\, W_0\, +\, W_1} \tag{59}$$

Note that C_1 and C_2 are actually piecewise constants in any Stokes sector itself, since they take different values on different sides of an anti-Stokes line and on an anti-Stokes line itself (compare (45),(52) and (53) that are valid for different values of $Arg\ x$ in $(-\frac{\pi}{4}, \frac{\pi}{4})$). Our example shows that all such values of C_1 and C_2 are uniquely determined and that the righthand side of (55) plays an important role in determining these constants even though it is slowly varying and the form of the fast varying parts of the solution is determined by the homogeneous equation (57). The exception to this situation will occur if (55) admits a particular solution that is exact, either in closed form or in a convergent series representation.

We end this section by noting that in some of our own prior publications[33,44,56], we have not been careful to note that the values of C_1 and C_2, in the context of an equation of the form (55) (or its nonlinear version), can actually change within a Stokes sector as an anti-Stokes lines is crossed. However, our methods of calculation of constraints on λ relied on eliminating the growing exponential in each Stokes sector; and so this omission does not affect any of the prior results[33,44,56]

7. THE INITIAL VALUE PROBLEM

In this section, we will limit ourselves mainly to a description of our own recent work[66]. Consider the conformal map $z(\zeta,t)$ that maps the interior of the unit semi-circle in the ζ plane (Fig. 6) to the physical domain (Fig. 3) with point correspondences as shown in the figures. At any finite time t, the points A and B are finite points, unlike the case of a steady finger. We can clearly decompose

$$z(\zeta,t) = -\frac{2}{\pi} \ln \zeta + i + f(\zeta,t) \tag{60}$$

where $f(\zeta,t)$ is analytic inside the unit semi-circle. Further, on the real axis between -1 and +1, the geometric condition that the side walls are at $Im\, z = \pm 1$ correspond to the requirement

$$Im\, f = 0 \tag{61}$$

Assuming that the singularity of $z(\zeta,t)$ at $\zeta = 0$ is incorporated in the log term in (60) such that f is continuous at $\zeta = 0$, it follows from reflection principle that f is analytic for $|\zeta| < 1$. Further, for interfaces which are smooth, the finger is analytic on $|\zeta| = 1$ with the possible exception of $\zeta = \pm 1$. If the extended interface formed by reflection about each of the side walls is also assumed to be smooth, then analyticity follows at $\zeta = \pm 1$ as well. Thus, with these assumptions it is clear that analyticity of f extends to $|\zeta| < 1 + \epsilon$ for some $\epsilon > 0$. We decompose the complex velocity potential $W(\zeta,t)$ defined as $\phi + i\psi$ as:

$$W(\zeta,t) = -\frac{2}{\pi} + i + \omega(\zeta,t) \tag{62}$$

the fluid velocity at infinity is assumed to be unity without any loss of generality when the pressure gradient at infinity is time independent, as will be assumed here. It is clear that the condition of no flow through the walls implies

$$Im\, \omega = 0 \tag{63}$$

on the real diameter (-1, 1) of the unit semi-circle. As is physically reasonable, ω is assumed to be continuous up to the real diameter including $\zeta = 0$. Further, it will be assumed that ω is analytic on the semi-circular arc corresponding to the assumption of a smooth flow at the interface. From Scwartz reflection principle, (63) implies that ω is analytic in $|\zeta| \leq 1$. Thus the analyticity must extend to beyond the unit circle. As has been shown before[2,55,65], the pressure boundary condition (6) on $|\zeta| = 1$ corresponds to:

$$Re\, \omega = -\frac{\mathcal{B}}{|z_\zeta|} Re \left[1 + \zeta \frac{z_{\zeta\zeta}}{z_\zeta}\right] \tag{64}$$

The kinematic condition given in (5) can be written as:

$$Re \left[\frac{\zeta W_\zeta}{|z_\zeta|^2} - \frac{z_t}{\zeta z_\zeta}\right] = 0 \tag{65}$$

Equations (64) and (65) determine the evolution of the function f and ω as defined in (60) and (62). In the special case when $\mathcal{B} = 0$, it is clear from (63) and (64) that

$$\omega = 0 \tag{66}$$

In that case, it follows from (62) and (65) that

$$Re\left[\frac{z_t}{\zeta z_\zeta}\right] = -\frac{2}{\pi}\frac{1}{|z_\zeta|^2} \tag{67}$$

Although this equation has been derived only for the channel geometry, a small change in the decomposition (60) (putting a simple pole rather than a log term) and working with the full ζ circle allows us to consider the radial geometry problem as well.

Most of the analytical work[68-78] deals with the $\mathcal{B} = 0$ problem which amounts to finding solutions to (67), subject to the condition that z is of the form (60) with f analytic in $|\zeta| \leq 1$. Some of the techniques of finding exact solutions to the initial value problem appear to be first invented by the Russians[70-73] in the context of a theory of ground water movement, where the same equations arise. The review by Hohlov[74] summarizes much of this early work. Unaware of these techniques, classes of exact solutions were rediscovered in the west[68,69,75-79]. Recently, Howison[80], in a rather comprehensive review of the initial value problem for $\mathcal{B} = 0$, discusses the connection between the different works and the relation between the different techniques. Except for self similar solutions in time in a wedge geometry[7-9] which have a branch point at the origin, exact solutions to the initial value problem has been limited to initial conditions that are either combination of poles and/or polynomials for z_ζ that are located in the unphysical domain $|\zeta| > 1$ initially (except for the pole at the origin). From these works, it is known that the number of poles and zeros are preserved. Further, in terms of exact solutions, it has been shown that poles do not reach the boundary of the physical domain $|\zeta| = 1$ in finite time; whereas a zero of z_ζ can, leading to the formation of a cusped boundary. Further, Howison[68] showed that we can start with that two initial conditions that are arbitrarily close to each other in the physical domain $|\zeta| \leq 1$, one in the form of poles and the other a polynomial (except for the pole at the origin); yet at later times, there is no singularity in the physical domain for the first initial condition; in the second case, there is. Thus, in any norm measuring the slope of the interface, the deviation of the two solutions after some time cannot be controlled by making the difference of the initial data arbitrarily small in the physical domain. For the general initial value problem with arbitrary data, Gustaffson[81,82] has rigorously proved the existence of solution for finite time for analytic initial data on the real axis (the physical domain in his formulation).

For nonzero surface tension, Duchon & Robert[83] have proved the existence of solutions for short times. Numerical computations by Trygvasson & Aref[84,85], Schwartz & Degregoria[60], Bensimon et al[2] and Meiburg & Homsy[62] sheds light on the process of finger competition that eventually leads to a single steady finger and the onset of nonlinear instability when $\mathcal{B}$ is smaller than some critical value. However, an understanding of the complicated time dependent pattern beyond the onset stage is difficult using conventional simulation since widely disparate scales both in space and time arise in the limit of zero surface tension. This is not unexpected as one approaches an ill posed problem in this limit. Yet, the small surface tension limit appears to be the most interesting since from experiment complicated interfacial features are expected in that limit. On the theoretical side, applying a random walk simulation, Liang's[86] calculation suggests the formation of a fractal pattern. On the analytical side, Constantin & Kadanoff[87,88] consider surface tension effects on singularities when the interface (in the radial geometry) is approximately circular and singularities are very far from the physical domain. These assumptions allow an approximate localized partial differential equation. This analysis gives some idea about how singularities are affected by surface tension. However, it is limited by the assumption of singularities far from the physical domain.

Very recently, we[66] investigated the Hele-Shaw flow for small surface tension. With explicit initial conditions, we found that it was useful to study the evolution problem in the unphysical domain $|\zeta| > 1$. Through analytical continuation of boundary conditions (64) and (65), through a process similar to what is described in section 5, we find a nonlinear integro-differential equation:

$$z_t = q_1 z_\zeta + q_2 + \mathcal{B}q_3 + \mathcal{B}\frac{q_4}{z_\zeta^{1/2}} + \mathcal{B}\frac{q_5 z_{\zeta\zeta}}{z_\zeta^{3/2}} - \frac{3}{2}\mathcal{B}\frac{q_7 z_{\zeta\zeta}^2}{z_\zeta^{5/2}} + \mathcal{B}\frac{q_7 z_{\zeta\zeta\zeta}}{z_\zeta^{3/2}} \tag{68}$$

where each of q_1 through q_7 are analytic functions everywhere in $|\zeta| > 1$ and can be expressed as integrals of z_ζ and $z_{\zeta\zeta}$ in the physical domain. When $\mathcal{B} = 0$,

$$z_{0t} = q_{1_0} z_{0\zeta} + q_{2_0} \tag{69}$$

where subscript 0 referes to simplifications of the q_i due to the substituting $\mathcal{B} = 0$. Because of the analytic nature of q_1 and q_2 in $|\zeta| > 1$, each singularity $\zeta_0(t)$ of z present at the initial time moves with velocity determined by

$$\dot{\zeta}_0 = -q_1(\zeta_0(t), t) \tag{70}$$

There is an important property crucial property

$$Re \frac{q_{1_0}}{\zeta} > 0 \tag{71}$$

for $|\zeta| > 1$, which immediately implies from (70) that all information flows inwards towards the unit circle just as for a scalar hyperbolic partial differential equation with inward pointing set of characterestics. Using this property, a numerical scheme is suggested that allows calculation of solution to arbitrary initial condition. Recently G. R. Baker (private communication) has implemented this scheme successfully. Our evidence suggests that the problem is well posed in the domain $|\zeta| > 1$ in contrast to the original problem posed in $|\zeta| \leq 1$. In particular, all the singularities of z outside the unit circle move towards the unit circle. Analytical evidence was given that suggested that for branch point singularities of the type $z_\zeta \sim B_0(t)\,(\zeta - \zeta_s(t))^{-\beta}$, as $\zeta \to \zeta_s(t)$, the singularities approach but never reach the physical domain for $\beta \geq \frac{1}{2}$. For $0 < \beta < \frac{1}{2}$, the evidence suggests that the singularity hits the physical domain boundary in finite time. The same appeared to be true for $\beta < 0$, though it was based on certain assumption that was not possible to check. One implication of our findings can be illustrated through a specific example. Consider two conditions denoted by $z^1(\zeta, 0)$ and $z^2(\zeta, 0)$, so that

$$z^2_\zeta(\zeta, 0) = z^1_\zeta(\zeta, 0) + \epsilon^n\, \zeta\, (\zeta^2 + 1 + \epsilon)^{-1/3} \tag{72}$$

We assume that z^1_ζ has no singularity for $|\zeta| < 2$, except for the simple pole one at the origin (consistent with (60)). To ensure that $z^2_\zeta(\zeta, 0)$ is nonzero for $|\zeta| \leq 0$ (an unphysical initial condition otherwise), we will also assume that $z^1_\zeta \neq 0$ in $|\zeta| < 2$ and that z^1_ζ is an odd function of ζ on the real axis in the interval (-1,1) and that $Im\, z^1{}_\zeta(i,t) > 0$. Clearly other weaker conditions can be applied to ensure that z^2_ζ is nonzero in $|\zeta| \leq 1$ initially. Our findings[66] on finite time singularity implies that for $t = 0(\epsilon)$, the singularity of $z^2_\zeta(\zeta, t)$ hits the physical domain (actually with the symmetry, it will hit $\zeta = i$). At that time, $\max_{|\zeta|=1} |z^1_\zeta(\zeta, t) - z^2_\zeta(\zeta, t)|$ is $O(1)$ or larger even when their initial difference in the physical domain $|\zeta| \leq 1$ is made arbitrarily small (by increasing n). This suggests that the previous results[68,77] can be extended to ill-posedness in the sense of Hadamard for the physical domain ($(\zeta| \leq 1$) problem for any norm that measures the slope of the finger boundary. This ill-posedness is reflected in the growth rate relation in (9) when $\mathcal{B} = 0$. However, our evidence suggests that with an extended domain in $|\zeta| > 1$, the problem is well posed. Note that the example just cited does not qualify as a demonstration of ill-posedness in this domain as the initial conditions are not close. Their singularity locations are very different and so even at initial time the two functions z^1 and z^2 are not close. The ill posedness in the physical domain can then be understood in simple terms. Two analytic functions can be close in $|\zeta| \leq 1$, yet they need not be so outside the unit circle; they can have different singularity locations. Yet at a later time, because of the property (70) and (71), the singularities affect the time evolution. in $|\zeta| \leq 1$ (the physical domain). For this reason, a direct numerical simulation of solution to (67) will fail because of very fast growth of roundoff errors, unless, one follows a procedure of filtering out the large wave numbers, as has been successfully intoduced in another ill-posed problems[89].

Once, the zero surface tension problem has been imbedded into a well posed problem by extending the physical domain and studying exactly analytically continuable initial condition, one can formally construct a regular perturbation expansion

$$z = z_0(\zeta,t) + \mathcal{B}z_1(\zeta,t) + \mathcal{B}^2 z_2(\zeta,t) + .. \tag{73}$$

and it becomes clear that one can numerically calculate the evolution of higher order perturbation terms $z_1(\zeta,t)$, etc by using a variation of the numerical method[66] described to calculate z_0 . Analytical evidence suggests that this perturbation series is nonuniform and breaks down near the moving singularities and zeros of z_ζ^0 . However, for an initial branch point singularity, the leading order inner equation is the Harry Dym equation, as was originally realized by Constantin & Kadanoff[87,88], except when the singularity is very close to the physical domain. It is found[66] that the requisite inner-outer matching is possible over a broad class of singularities suggesting that the only effect of surface tension for small surface tension for $t << \frac{1}{\mathcal{B}}$ is to smear out these moving singularity over a surface tension dependent length scale. On the other hand, an initial zero of z_ζ spawns a singularity in the higher order perturbation z_1 , z_2 , etc, which are also smeared out over a surface tension dependent inner length scale. These daughter singularities (actually smeared out singularities) always move towards the physical plane just like the singularities of z_0 . The implication of all these findings is that even for small nonzero surface tension, we can start with slightly different initial conditions in the physical plane and at latter time end up with structures that are quite different. This difference is again traceable to the fact that the analytic continuation of the two initial conditions in the region $|\zeta| > 1$ will have different singularity and zero structure. As time evolves, these singularities and those born at each initial zero, while approaching the physical domain continuously, gets smeared out over a local surface tension dependent scale. However, it is possible (depending on the initial singularity nature) that z_ζ can scale as some inverse power of $\mathcal{B}$), corresponding to significant localized effects on the physical interface itself.

Clearly, the location and time when these structures are seen on the physical interface will depend on where the singularities and zeros were located initially. It appears that this explains what M. Berry termed as 'zero Reynolds number chaos'. Whether it is chaos in the sense of possessing a global attractor besides having extreme sensitivity to initial conditions is an open question. Further, as of now, it is unclear if and how the dynamics of singularities can lead to the formation of a fractal interface over some range of length scales.

REFERENCES

1. P.G. Saffman, Viscous fingering in a Hele-Shaw cell, J.FluidMech. 173:73 (1986).
2. D. Bensimon, L.P. Kadanoff, S. Liang, B.I. Shraiman, & C. Tang, Rev. Mod. Phys., 58, 977 (1986)
3. G.M. Homsy, Ann. Rev. Fluid Mech., 19, 271 (1987)
4. P. Pelce, 1988, "Dynamics of Curved Front", Academic Press.
5. D. Kessler, J. Koplik, & H. Levine, Patterned Selection in fingered growth phenomena, Advances in Physics, 37:255 (1988)
6. P.G. Saffman, & G.I. Taylor, The penetration of a fluid into a porous medium of Hele-Shaw cell containing a more viscous fluid, Proc. R. Soc. London Ser. A 245:312 (1958)
7. H. Thome, M. Rabaud, V. Hakim & Y. Couder, The Saffman Taylor Instability: From the linear to the circular geometry, Phys. Fluids A 1:224 (1989)
8 M. Ben Amar & R. Combescot, Saffman-Taylor viscous fingering in a wedge, to appear in This Conference Proceedings, (1991)
9. Y. Tu, Saffman-Taylor problem in sector geometry, to appear in This Conference Proceedings, (1991)
10. J.J.Xu, Globally unstable oscillatory modes in viscous fingering, European Journal of Applied Math., 2:105, (1991)
11. S. Hill, Channelling in packed columns, Chem. Engng. Sci 1:247 (1952)

12. R.L.Chuoke, P.Van Meurs & C.Van Der Poel, The instability of slow immiscible viscous liquid-liquid displacements in permeable media, Trans. AIME 216: 188 (1959)

13. E. Pitts, Penetration of fluid into a Hele-Shaw cell: the Saffman-Taylor experiment, J. Fluid Mech. 97: 53 (1980)

14. P. Tabeling & A. Libchaber, Film draining and the Saffman-Taylor problem, Phys. Rev. A, 33: 794 (1986)

15. P. Tabeling, G. Zocchi, & A. Libchaber, An experimental study of the Saffman-Taylor instability, J. Fluid Mech., 177: 67 (1987)

16. T. Maxworthy, The nonlinear growth of a gravitationally unstable interface in a Hele-Shaw cell, J. Fl. Mech., 177, 207 (1987)

17. A. Arneodo, Y. Couder, G. Grasseau, V. Hakim & M. Rabaud, Uncovering the Analytical Saffman-Taylor Finger in Unstable Viscous fingering and Diffusion Limited Aggregation, Phys. Rev. Lett., 63, 984 (1989)

18. Y. Couder, N. Gerard & M. Rabaud, Narrow fingers in the Saffman-Taylor instability, Phys. Rev. A, 34: 5175 (1986)

19. G. Zocchi, B. Shaw, A. Libchaber & L. Kadanoff, Finger narrowing under local perturbations in the Saffman-Taylor problem, Phys Rev. A, 36:1894 (1987)

20. E. Ben-Jacob, R. Godbey, N.D. Goldenfeld, J.Koplik, H.Levine, T.Mueller & L.M. Sander, Experimental demonstration of the role of anisotropy in interfacial pattern formation, Phys. Rev. Lett., 55:1315 (1985)

21. M. Rabaud, Y. Couder & N. Gerard, Dynamics and stability of anomalous Saffman-Taylor fingers, 1987, To be published in the J. Fl. Mech.

22. J. Bataille, Stabilite d'un deplacement radial non miscible Revue Inst. Pe'trole, **23**, 1349 (1968)

23. L. Paterson, Radial Fingering in a Hele-Shaw cell, J. Fl. Mech, **113**, 513 (1981)

24. J. W. McLean, & P.G. Saffman, The effect of surface tension on the shape of fingers in a Hele-Shaw cell, J. Fluid Mech. 102:455 (1981)

25. S. Tanveer & P.G. Saffman, The effect of finite viscosity ratio on the stability of fingers and bubbles in a Hele-Shaw cell, cell, Phys. Fluids, 31: 3188 (1988)

26. E. Brener, D.Kessler, H.Levene & W. Rappel, Europhys. Lett.,, 13:161 (1990)

27. S.J. Weinstein, E.B.Dussan & L.H.Ungar, A theoretical study of two phase flow through a narrow gap with a moving contact line: viscous fingering in a Hele-Shaw cell J. Fluid Mech., 221: 53 (1990)

28. C. W. Park, & G. M. Homsy, Two-phase displacement in Hele-Shaw cells: theory, J. Fluid Mech., 139: 291, (1985)

29. D. A. Reinelt, Interface conditions for two-phase displacement in Hele-Shaw cells, J. Fluid Mech, 183: 219 (1987)

30. D. A. Reinelt, The effect of thin film variations and transverse curvature on the shape of fingers in a Hele-Shaw cell, Phys. Fluids 30: 2617 (1987)

31. L.W. Schwartz, & A. J. Degregoria, Simulation of Hele-Shaw cell fingering with finite capillary number effects included, Phys. Rev. A, 35:276, (1987)

32. S. Sarkar, & D. Jasnow, 1987, Quantitative test of solvability theory for the Saffman-Taylor problem, Phys. Rev. A, 35:4900 (1987)

33. S.Tanveer, Analytic theory for the selection of Saffman-Taylor finger in the presence of thin-film effects, Proc. Roy. Soc. London, A 428:511 (1990)

34. G.I. Taylor & P.G. Saffman, A note on the motion of bubbles in a Hele-Shaw cell and porous medium, Q. Jour. Mech. Appl. Math 12:265 (1959)

35. L.P. Kadanoff, Phys. Rev. Lett., 65, 2986 (1990)

36. G.L.Vasconcelos & L.P.Kadanoff, Stationary solutions for the Saffman-Taylor problem with surface tension, To appear in Phys. Rev. A, (1991)

37. L. A. Romero, Ph.d thesis, Department of Applied Math, California Institute of Technology, 1982

38. J.M. Vanden-Broeck, Fingers in a Hele-Shaw cell with surface tension, Phys. Fluids, 26:2033 (1983)
39. H. Segur, The geometric model of crystal growth- an overview, in this proceeding (1991)
40. D. Kessler & H. Levine, The theory of Saffman-Taylor finger, Phys. Rev. A 32:1930 (1985)
41. R. Combescot, T. Dombre, V. Hakim, Y. Pomeau, & A. Pumir, Shape selection for Saffman-Taylor fingers, Phys. Rev Lett., 56:2036 (1986)
42. B.I. Shraiman, On velocity selection and the Saffman-Taylor problem, Phys. Rev. Lett., 56:2028 (1986)
43. D.C. Hong, & J.S. Langer, Analytic theory for the selection of Saffman-Taylor finger, Phys. Rev. Lett., 56:2032 (1986)
44. S. Tanveer, Analytic theory for the selection of symmetric Saffman-Taylor finger, Phys. Fluids 30:1589 (1987)
45. A.T. Dorsey & O. Martin, Saffman Taylor fingers with anisotropic surface tension, Phys. Rev. A 35: 3989 (1987)
46. R. Combescot, T. Dombre, V. Hakim, Y. Pomeau, & A. Pumir, Analytic theory of the Saffman-Taylor fingers, Phys. Rev. A 37: 1270 (1987)
47. S. Tanveer, The effect of surface tension on the shape of a Hele-Shaw cell bubble, Phys. Fluids, 29, 3537 (1986)
48. S. Tanveer, New solutions for steady bubbles in a Hele-Shaw cell, Phys. Fluids, 30, 651 (1987)
49. R. Combescot, & T. Dombre, Selection in the Saffman-Taylor bubble and assymmetrical finger problem, Phys. Rev. A, 38 (5), 2573 (1988)
50. D.C. Hong, & F. Family, Bubbles in the Hele-Shaw cell: Pattern selection & Tip perturbations, Phys. Rev. A, 37, 2724 (1988)
51. S. Tanveer, Analytic theory for the selection and stability of bubbles in a Hele-Shaw cell, Part I: Velocity Selection, J. of Th. & Comp. Fl. Mech, 1 (3), 135-164 (1989)
52. P.G. Saffman & G.I. Taylor, in Proc. 2nd Ann. Naval Symp. Hydrodynamics, p.277 (1959)
53. S.K.Sarkar, private communication to P.G. Saffman (1986)
54. D. Kessler & H. Levine, Stability of finger patterns in Hele-Shaw cells, Phys. Rev. A 33: 2632 (1986)
55. D. Bensimon, Stability of viscous fingering, Phys. Rev. A 33: 1302 (1986)
56. S. Tanveer, Analytic theory for the linear stability of Saffman-Taylor finger, Phys. Fluids 30: 2318 (1987)
57. D. Kessler & H. Levine, Phys. Fluids, (1987)
58. S.Tanveer & P.G. Saffman, Stability of bubbles in a Hele-Shaw cell, Phys. Fluids, 30:2624 (1987)
59. D. Bensimon, P. Pelce & B.I. Shraiman, Dynamics of Curved Front and pattern selection, J. Physique, 48, 2081 (1987)
60. A.J. Degregoria & L.W. Schwartz, A boundary-integral method for two-phase displacement in Hele-Shaw cells, J. Fluid Mech., 164: 383 (1986)
61. A.J. Degregoria & L.W. Schwartz, Saffman-Taylor finger width at low interfacial tension, Phys. Rev. Lett., 58:1742 (1987)
62. E. Meiburg & G.M. Homsy, Nonlinear unstable viscous fingers in Hele-Shaw flows. II Numerical simulation, Phys. Fluids, 31, 429, (1988)
63. J.J.Xu, Asymptotic theory of steady axi-symmetrical needle growth, Studies in Appl. Math., 82:71, (1990)
64. J.J.Xu, Interface wave theory of solidification: Dendritic pattern formation and selection of growth velocity, Phys. Rev. A, 43:930 (1991)

65. J.W.McLean, Ph.d Thesis, Department of Applied Math, California Institute of Technology, (1980)
66. S.Tanveer, Evolution of a Hele-Shaw interface for small surface tension, submitted to Phil. Trans R. Soc. London, 1991.
67. M. Kruskal & H. Segur, Asymptotics beyond all orders, Aeronautical Res. Associates of Princeton, Technical Memo, 85-25, 1986
68. S.D. Howison, Fingering in Hele-Shaw cells, J. Fl. Mech, **167**, 439 (1986)
69. B.I. Shraiman & D. Bensimon, Singularities in nonlocal interface dynamics, Phys. Rev. A, **30**, 2840 (1984)
70. L.A. Galin, Dokl. Akad. Nauk. S.S.S.R, 47, 246-249 (1945)
71. P.Ya Polubarinova-Kochina, Dokl. Akad. Nauk. S.S.S.R, 47:254 (1945)
72. P.Ya Polubarinova-Kochina Prikl. Mathem. Mech., 9:79 (1945)
73. P.P.Kufarev, Dokl. Akad. Nauk. S.S.S.R, 60:1333 (1948)
74. E.Yu. Hohlov, M.I.A.N.Preprint, no. 14, Steklov Institute, Moscow (1990)
75. P.G. Saffman, Exact solutions for the growth of fingers from a flat interface between two fluids in a porous medium or Hele-Shaw cell, Q. J. Mech. appl. Math, **12**, 146-150 (1959)
76. S.D.Howison, Proc. Roy. Soc. Edin., A102, 141-148 (1985)
77. S.D.Howison, Siam J. Appl. Math, 46:20 (1986)
78. A.A.Lacey, J. Australian Math. Soc., B24:171 (1982)
79. S. Richardson, Hele Shaw flows with a free boundary produced by the injection of fluid into a narrow channel, J. Fl. Mech, **56** (4), 609 (1972)
80. S.D. Howison, Complex variable methods in Hele-Shaw moving boundary problems, Submitted to the European J. Appl.Math. (1991)
81. B.Gustaffson, Nonlinear Analysis, T.M.A., 9:203 (1984)
82. B.Gustaffson, Arkiv. fur Mathematik, 25:231 (1987)
83. Duchon, J & Robert, R., Evolution d'une interface par capillarite' et diffusion de volume, Ann l'Inst. H. Poincare' , **1**, 361 (1984)
84. G. Trygvasson, & H. Aref, Numerical experiments on Hele-Shaw cell with a sharp interface, J. Fl. Mech., **136**, 1 (1983)
85. G. Trygvasson, & H. Aref, Finger-interaction mechanisms in stratified Hele-Shaw flow, J. Fl. Mech., **154**, 287 (1985)
86. S.Liang, Random walk simulations of flow in Hele-Shaw cells, Phys. Rev. A., 33:2663 (1986)
87. P.Constantin & L.P.Kadanoff, Physica D, 47:450 (1991)
88. P.Constantin & L.P.Kadanoff, Phil. Trans. R. Soc. Lond. A, 333:379 (1990)
89. R.Krasny, A study of singularity formation in a vortex sheet by the point-vortex approximation, J. Fluid Mech., 167:65 (1986)

SAFFMAN - TAYLOR VISCOUS FINGERING IN A WEDGE

Martine Ben Amar and Roland Combescot

Laboratoire de Physique statistique*

24 rue Lhomond, 75231 Paris Cedex 05, France

I. INTRODUCTION

Despite all the recent progress in the field[1], the Saffman - Taylor instability keeps raising a lot of interest. The reason is the wide variety of phenomena displayed despite the apparent simplicity of the physical situation. The Saffman - Taylor (ST) instability is perhaps the simplest example of pattern selection. In addition to standard fingers, it displays fractal structures[2] and even dendritic growth[3] under anisotropic conditions.These are remarkably similar to a DLA aggregate, which itself is a standard model of crystal growth.

The standard finger is by now well understood in the linear geometry[4] developed by Saffman and Taylor[5].The main advantage of traditional experiments in a linear channel is to produce steady fingers with a constant velocity. For a long time, the major question has been to understand the observed value for the relative width λ of the experimental finger[5] whereas Saffman and Taylor[5] had shown the existence of a possible continuum of symmetric solutions with every possible λ. It is now well established, by analytical[4] or numerical means[6] that isotropic surface tension selects a solution, which occupies half the channel width in the limit of zero surface tension. In fact, an infinite discrete set is selected but only the finger with the lowest λ value is linearly stable[7].

Viscous fingering in a wedge was introduced recently in a clever experiment by Thomé et al.[8] who decided to grow Saffman - Taylor fingers in Hele - Shaw cells with the shape of a disk sector, in order to bridge the gap between the well established linear geometry and the completely open circular geometry where the theory is much less advanced. The aim was to understand the dynamical evolution of viscous patterns in radial growth, i.e. what we call now the Bataille-Paterson geometry[9]. Although fairly common in nature, radial growth phenomena have been neglected, to some extent, by theorists in hydrodynamics. However, recently, this growth process has suggested several statistical approaches[2,10]. Several aspects remain unexplained at present like the destabilization of the

Asymptotics beyond All Orders, Edited by H. Segur *et al.*
Plenum Press, New York, 1991

pattern by tip-splitting or the cross-over between petal and dendritic shapes for crystal seeds. The sector geometry is an intermediate step between the two geometries.

In an open geometry, steady solutions with the finger moving at constant velocity, as in the linear case, no longer exist even if the imposed fluid flux is time-independent. Neththerless, Thomé et al.'s experiment[8] indicates that the interface develops fingers whose growth is found to be roughly self-similar in time at large velocities. This suggests, that after some typical time, a self-similar regime is reached, which is a nice starting point for any theoretical treatment. At large velocities, they found that the asymptotic value of the ratio λ of the angular width of these fingers to the sector angle $2\theta_0$ varies linearly with the sector angle. They observed an increasing sensitivity to tip splitting for larger angle which made it impossible to observe the asymptotic regime of large velocities beyond roughly $2\theta_0 = 20°$.

Thomé et al.[8] have shown that, in the presence of capillary effects, self-similar patterns can be solutions of the time-dependent equations only with a specific time dependent growth. In the absence of these effects, they found a continous set of solutions for the right angle geometry $2\theta_0 = 90°$. This set is labelled, as in the linear case, by λ which is estimated near the center (for divergent flow) or in the tail for convergent flow. Very recently by an extension of the hodograph method due to McLean and Saffman, we succeeded[11] in finding a continuous set of solutions for arbitrary sector angle, in the divergent or convergent flow regime.

In this paper, we consider theoretically the width selection of Saffman - Taylor fingers in this sector geometry. This problem has already been considered by Brener et al.[12] for the $2\theta_0 = 90°$ sector and in this proceeding[12] Yuhai Tu has also reported on this problem. We have investigated both divergent and convergent case. But we will concentrate on the divergent case since the convergent situation happens to be quite similar to the linear geometry. In the divergent case we find a qualitatively new phenomenon. In the linear geometry one has a discrete set of solutions all converging to a relative width $\lambda = 1/2$ for vanishing surface tension T. Here instead we find that the solutions disappear below some surface tension (or equivalently above some sector angle). This disappearance of the selected solution provides a natural and intrinsic explanation for the tip splitting instability observed experimentally, with a rough quantitative agreement (the threshold is difficult to locate experimentally with precision).

For a general sector angle $2\theta_0$ and surface tension T, our results have been obtained numerically by solving the problem on the finger interface. This has been done by a generalization of the approach followed by McLean and Saffman for the linear geometry. This generalization is not easy and is explained in details below. The resulting set of equations provide a convenient and fast algorithm to explore the selected fingers for a wide range of parameters. Due to the disappearance of the solutions by branch merging, the spectrum for the finger width displays curled branches (see Fig.3).

In order to understand the physical origin of this new behaviour, we have for small angle and surface tension performed an analytical study by a complex extension . The resulting differential equation indeed displays the branch merging phenomenon. The reason is found by a WKB analysis. Instead of a single singularity on the channel axis or two singularities on the channel sides, as for the linear geometry, we find always three singularities (for the divergent case). One of them corresponds to the axis singularity of the linear geometry, the other two are analogous to the channel side singularities.

These singular points are the sources of transcendental corrections to a regular perturbation series for small nonzero surface tension. To get a selection of λ to the leading order, the tip smoothness condition has to be imposed on the leading order of the transcendentally small corrections obtained through a WKB analysis, since all the regular perturbation terms satisfy the smoothness condition automatically. When the side singularities dominate (are "closer" to the finger), the situation is analogous to the one found in the linear channel geometry for $\lambda > 1/2$: smooth tip solutions are possible for a discrete set of λ. However at low T, the axis singularity always becomes dominant and, as for the linear channel when $\lambda < 1/2$, it is not possible to satisfy the smooth tip condition. Therefore the crossover of dominance from side-like to axis-like singularity leads to a disappearance of the solutions which is done effectively by branch merging. The location of this disappearance in the WKB analysis (and all the width spectrum) is in very good agreement with the direct solution of the differential equation in the vicinity of the singular points ("inner problem"). The latter one agrees very well itself for $2\theta_0= 20°$ with the numerical results on the finger (compare Fig.3 and Fig.7). Actually the scaling found in the inner problem for small angle agrees with numerics up to $2\theta_0= 70°$.

II. THE SELF- SIMILAR FREE BOUNDARY PROBLEM

A. The viscous fingering equation in the physical plane

We take the disk sector symmetrical with respect to the positive x axis. When the experiment takes place in a wedge, air can push oil either from the center O, which we take as the origin, or from the periphery. In the first case, the fluid motion is divergent from the center O, in the second it converges toward O. One can treat simultaneously these two situations by assigning an algebraic value to the sector angle: positive for divergent flow, negative for convergent one. We look for a complex potential $w = \phi + i\psi$, analytical function of the complex position $z=x+iy$, with complex velocity $u = u_x - iu_y = dw/dz$. The boundary conditions are $u \approx 1/\theta_0 z$ at infinity, and $\mathrm{Arg}(u)=\pm\theta_0$ on the cell sides. On the finger, the normal components of fluid and finger velocity must be equal (continuity equation); and $\phi=\alpha/R$ where R is the radius of curvature of the interface and $\alpha=Tb^2/12\mu$ with b the cell thickness and μ the fluid viscosity.The angular width of the finger $2\theta_1$ is defined as the finger angle at the origin and the relative width is $\lambda=\theta_1/\theta_0$.

In the physical plane, which we call the π plane, the shape of the inteface looks like a petal for divergent flows, whereas for convergent flows it has the more traditional finger shape. We expect to reach situations for arbitrary value of the sector angle so apriori $-2\pi \leq 2\theta_0 \leq 2\pi$. $2\theta_0 = \pm 2\pi$ represents a rather interesting experimental situation of a unique wall crossing an half -space. In the sector, the flow is governed by Darcy law for porous medium, which is valid if the two parallel plates of the Hele - Shaw cell are closely spaced:

$$\mathbf{v} = -\frac{b^2}{12\mu}\nabla p = \nabla\phi \qquad \text{(II-1)}$$

p is the pressure, μ the viscosity and b the distance between the two plates. We have introduced, as usual, the velocity potential ϕ which is Laplacian since the fluid is incompressible. Experiment[8] suggests that the interface evolves in times nearly without changing its shape so we will look for self-similar solutions:

$$\mathbf{r}(t) = A(t)\ \mathbf{r}(0) \qquad \text{(II-2)}$$

where $\mathbf{r}(t)$ is a point of the interface at time t. In this particular case, the continuity equation for the velocity, to be applied to the interface, gives:

$$\mathbf{n}.\nabla_{\mathbf{r}}\, \phi \;=\; R_0^2 A'(t)\; A(t)\; \mathbf{n}.\mathbf{r} \qquad \text{(II-3)}$$

where $\mathbf{n}$ is the normal to the interface at time t = 0. Here we have taken as unit of length the position R_0 of the nose of the finger at time t = 0, so that $\mathbf{r}$ refers to a dimensionless finger with length equal to unity. Here A'(t) means the first derivative of A(t) with respect to the natural variable t. Moreover the capillary effects fixes the value of the velocity potential for each point of the interface:

$$\phi - \phi_0 = -\frac{b^2}{12\mu}\,\frac{T}{A(t)\,R_0}\,\Omega \qquad \text{(II-4)}$$

since we have assumed that the pressure P_0 is constant inside the finger (we neglect air viscosity). Ω , once scaled by 1/A(t), is independent of time.

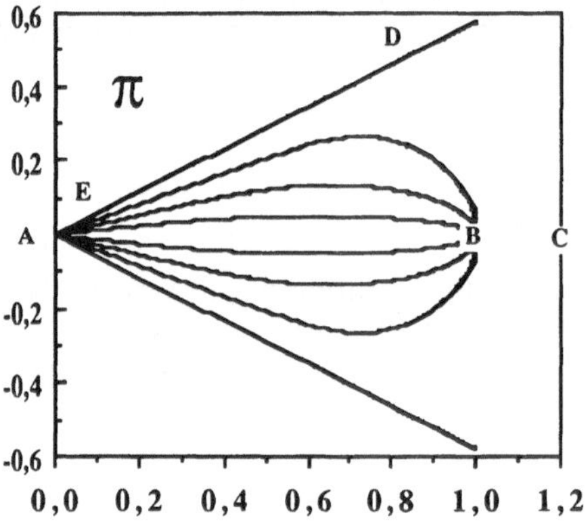

Figure1a. Self-similar shapes of relative width λ = 0.25, 0.5, 0.75 for the divergent flow ($2\theta_0$ =60°)

In the absence of capillary effects, T = 0 and self-similar patterns are dynamical solutions of this free-boundary problem. If we take the injection or extraction rate Q as constant, as this is essentially the case experimentally, we have:

$$2\;Q \;=\; \int dS_1\; \mathbf{v}.\mathbf{n} \;=\; R_0^2\; A'(t)\; A(t) \int ds_1\; \mathbf{r}.\mathbf{n} \qquad \text{(II-5)}$$

which implies, up to a normalization constant, A'(t)A(t) = ±1 and A(t) = $(1 \pm 2t\,)^{1/2}$ (the ± sign comes in because the flux rate is always positive while the integral can be either positive or negative depending on wether we have divergent or convergent fingers; since the origin of time is fixed arbitrarily, we assume that time has negative values for the convergent flow regime).

For a non zero surface tension, self - similarity implies from Eq.(II-4) either a time - dependent surface tension [12] with $T \sim A(t)$ or a time dependence for $\phi \sim A^{-1}(t)$. Naturally a time - dependent surface tension is not physical, but since for long times it changes slowly with time we can assume that, in experiment, the finger follows a kind of adiabatic evolution and has at any given time the shape corresponding to the instantaneous value of this surface tension. On the other hand we can keep a rigorous self - similarity if $\phi \sim A^{-1}(t)$, which leads[8] to $A'(t)A^2(t) = \pm 1$, and to $A(t) = (1 \pm 3t)^{1/3}$. However this leads to a time dependent injection rate $Q \sim A^{-1}(t)$ in contrast with experiment. Nevertheless we can again consider that, because the time evolution of this extraction rate is slow at long times, the experimental finger adjusts adiabatically to the self - similar shape corresponding to the instantaneous value of the injection rate. Either way we are lead to a time independent problem obtain by setting $A(t) = 1$ and $A'(t)A(t) = \pm 1$ in the above equations, which corresponds to take $t = 0$.

From now on, we focus on the selection mechanism of these self-similar fingers by a constant effective surface tension. Since the linear geometry is most familiar, we explain in the following section how to transform this free boundary problem in a wedge into a new one in a linear channel.

B. Transformation of the wedge into an infinite strip

In the past the Saffman-Taylor instability which takes place in an infinite strip has been extensively studied experimentally[5], numerically[6] and analytically[4]. For the physicist, this instability does not seem to hide any mystery anymore, at least in the steady-state regime. Even the deviations from ideal situations like three-dimensional effects[13] or inhomogeneity in the fluid[14] (presence of a bubble, a thread...) seem rather well understood. For this geometry, only time-dependent situations which occur at very low surface tension (or large pushing velocity) form the only remaining challenge.

In a wedge, experiments and the associated D.L.A simulations[2] indicate some continuity of the physical results with respect to the linear geometry, at least for values of the angle not too large. As an example, at small angles θ_0, the shape of the experimentally selected finger is not very different from the Saffman-Taylor solution with the relative width λ once modified by the following conformal map:

$$Z = \frac{1}{\theta_0} \mathrm{Log}(z) \qquad \text{(II-6)}$$

In the following we will call the infinite strip the Π plane and π the physical plane. Finally the width λ, selected either experimentally or by the D.L.A. simulations[2] deviates from the well-known 0.5 value approximately linearly with $2\theta_0$, at small surface tension. All this suggests a continuous behavior. This leads us to recover the linear geometry instead of the wedge one, via the conformal map (Eq.II-6).

Figs. (1a) and (1b) show the corresponding characteristic points in the two planes. The nose of the finger fixed at (1,0) in the Π plane becomes the origin in the π plane. The center of the sector is sent either at $+\infty$ for the convergent flow or $-\infty$ for the divergent one. The two walls are parallel, the distance between them is two length units[6]. In the π plane, the velocity potential remains Laplacian and the walls plus the centerline BC are always streamlines. Anyway one has to transform the boundary conditions Eq. (II-3) and (II-4) on the interface. This is made easily if one introduces first polar coordinates r and θ in the π plane. With these coordinates, the normal gradient is:

$$\mathbf{n}.\nabla\phi = [\, r\frac{d\theta}{ds_1}\ \frac{\partial\phi}{\partial r}\ -\ \frac{dr}{ds_1}\ \frac{1}{r}\ \frac{\partial\phi}{\partial\theta}\,]$$

with $ds_1 = (dr^2 + r^2 d\theta^2)^{1/2}$. Since the polar coordinates are simply related to Z via Eq. (II-6):

$$X = \frac{1}{\theta_0}\ \mathrm{Ln}\,(r) \qquad \text{and} \quad Y = \frac{\theta}{\theta_0}$$

we obtain from Eq.(II-3) the velocity continuity equation in the Π plane:

$$\mathbf{N}.\nabla_{\Pi}\phi = r\,|\theta_0|\,\mathbf{n}.\nabla\phi = |\theta_0|\,\exp(2\theta_0 X)\ \cos(\Theta) \qquad \text{(II-7)}$$

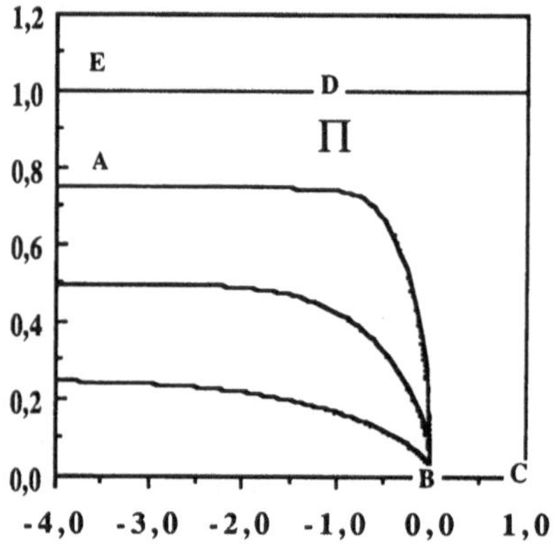

Figure 1b. Self-similar shapes of width $\lambda = 0.25, 0.50, 0.75$ for $2\theta_0 = 60°$ after conformal map

with Θ the angle in the Π plane between the normal to the interface and the X axis. As expected, this quantity is always positive since $-\pi/2 \leq \Theta \leq \pi/2$. The curvature in polar coordinates is given by:

$$\Omega = \frac{2r' - rr'' + r^2}{(r^2 + r'^2)^{3/2}}$$

r' (resp. r") means the first (resp. the second) derivative with respect to the polar angle θ. These derivatives can be written in terms of X and Y and we get:

$$\Omega = \frac{1}{|\theta_0|}\ \exp(-\theta_0 X)\ [\,\frac{Y''}{(1+Y'^2)^{3/2}} - \theta_0 \frac{|Y'|}{(1+Y'^2)^{1/2}}\,] = \frac{1}{|\theta_0|}\ \Omega_\theta \qquad \text{(II-8)}$$

Eq.(II-8) serves as a definition for Ω_θ . Henceforth we do not need the intermediate polar coordinates so Y' and Y'' means the derivatives with respect to X.We deduce the Laplace equation in the Π plane:

$$\phi - \phi_0 = -4\sigma \, |\theta_0| \, \Omega_\theta \tag{II-9}$$

with σ the effective surface tension parameter given by:

$$\sigma = \frac{b^2}{12\mu} \frac{T}{A(t)\, R_0{}^3} \frac{1}{(2\theta_0)^2} = \frac{b^2}{12\mu} \frac{T}{U(t)\, R_1(t)^2} \frac{1}{(2\theta_0)^2}$$

with $U(t)$ and $R_1(t)$ the instantaneous velocity and position of the finger nose in the physical plane π. Note that this quantity has been measured[8] during the growth experiment. We see also, by rescaling ϕ by a factor $|\theta_0|$, that the classical ST instability of the linear geometry appears just as a particular case of the more general class of instability processes which we are considering here. It merely corresponds to the limiting value $\theta_0 = 0$ or equivalently to asymptotic long times. If the fluid flux decreases in times like $t^{-1/3}$, σ is no longer time-dependent and is only related to the finger position at time t =0. In this case, the preceding reads:

$$\sigma = \frac{b^2}{12\mu} \frac{T}{R_0{}^3} \frac{1}{(2\theta_0)^2}$$

Now we know the equations in the Π plane. As it is usual now, let us transform the free boundary problem in the Π plane in an unique equation for the interface suitable for the numerics.

III. INTEGRO-DIFFERENTIAL EQUATIONS FOR THE INTERFACE: THE HODOGRAPH METHOD

Since the extension of this method to our case is not obvious, we recall here the strategy of this method applied first for the linear geometry and we show how it can be extended to any geometry. Note that this method especially requires Laplacian fields. For $\theta_0 = 0$, once rescaled, the equation for the normal velocity (II-7) is :

$$N.\nabla_\Pi \phi = \cos(\Theta) = \frac{\partial \psi}{\partial s} \tag{III-1}$$

Eq. (III-1) introduces the stream potential ψ which is the imaginary part of a generalized complex potential Φ. Both real (ϕ) and imaginary (ψ) part of Φ satisfy Laplace equation. We recall the boundary conditions on the walls and at infinity:

$$\begin{aligned} &\phi \approx 0 && \text{when} \quad X \to -\infty \quad \text{and} \quad \lambda < |Y| < 1 \\ &\phi \approx \lambda X && \text{when} \quad X \to \infty \quad \text{and} \quad -1 < Y < 1 \\ &\frac{\partial \phi}{\partial Y} = 0 && \text{on } Y = \pm 1 \end{aligned} \tag{III-2}$$

Here, these equations are dimensionless. McLean and Saffman (McS)[6] have introduced an analytical function whose imaginary part vanishes on the finger. In the linear geometry, this function is easily found:

$$\Phi^* = \frac{\Phi - Z}{1-\lambda} \qquad \text{(III-3)}$$

Note that the centre-line BC and the finger corresponds to a streamline ($\psi^*=0$). The walls $Y=\pm 1$ are also streamlines with $\psi^* = \pm 1$ (because of the normalization) . Φ^* has an obvious physical meaning : it is the velocity potential in the frame of the finger but, in the following, its main interest for us is to transform the finger into a streamline. Thereafter, the McS method consists in getting rid of the walls by transforming half the infinite strip into an upper half Σ space. Explicitely Σ is given by:

$$\Sigma = s + it = \exp[\,(\phi_0^* - \Phi^*)\pi] \qquad \text{(III-4)}$$

ϕ_0^* is the potential at the nose. Since the unknown finger has been transformed into the streamline $\psi^* = 0$, it is located on the interval $s \in [0,1]$. The center-line maps on $1<s<\infty$, and the upper wall ($\psi^* = -1$) on the negative s axis $-\infty < s \leq 0$. All the remaining of the treatment is a matter of algebra , the keypoint being the localization of the unknown interface in the Π plane on a segment of the real axis in the Σ plane. The extension of the McS method to an arbitrary geometry seems to present serious difficulties due to the continuity equation Eq.(II-7). This is only true in appearence. Let us construct an analytical function H(Z) of Z whose imaginary part on the unknown finger is given by:

$$\mathrm{Im}(H(Z)) = \int_0^S dS' \exp(2\theta_0 X) \cos(\Theta) = \int_0^S dS' \exp(2\theta_0 X) \frac{dY}{dS'} \qquad \text{(III-5)}$$

S is the arclength defined on the finger, from the nose. Moreover, we will impose that Im(H(Z)) vanishes on the centre-line and is a constant Q_0/λ on the wall. Here, $Q_0 = Q / (R_0^2 A'(t)\, A(t))$ represents the dimensionless half fluid flux across the interface, which was equal to λ for the linear cell. .

$$Q_0 = \int_0^\infty dS' \exp(2\theta_0 X) \quad \cos(\Theta) \qquad \text{(III-6)}$$

We also construct a generalized potential which transforms the finger into a stream-line:

$$\Phi^* = \frac{\Phi - H(Z)}{(1-\lambda)\, Q_0/\lambda} \qquad \text{(III-7)}$$

We recover the same situation as above Eq.(III-3) and we will follow exactly the same strategy : the conformal map from Π to Σ. The only difference comes from the absence of physical meaning for the generalized potential and of a closed analytical expression for H(Z) which makes the following algebra a little more complicated.

Let us first write the Laplace equation (II-8,II-9) in term of Φ^*:

$$\frac{\partial\phi^*}{\partial S} = -\frac{1}{(1-\lambda)Q_0/\lambda}\left[\sigma\frac{\partial\Omega_\theta}{\partial S} + \frac{\partial \mathrm{Real}(H)}{\partial S}\right] = \frac{q}{(1-\lambda)} \qquad \text{(III-8)}$$

where we have introduced the analytical function[6]:

$$\frac{d\Phi^*}{dZ} = \frac{q}{(1-\lambda)}\exp[-i(\Theta+\pi/2)]$$

In the linear geometry, it represents the complex fluid velocity. We calculate Real (H) using the Cauchy integral theorem, taking advantage that $(\mathrm{Im}(H(s)-\frac{Q_0}{\lambda}h(-s))$ vanishes for every real s-value outside [0,1] (h(-s) equal 1 if s is negative, zero otherwise). We deduce, without any difficulty that:

$$\frac{\partial \mathrm{Real}(H)}{\partial s} = \frac{Q_0}{\lambda\pi s} - \frac{(1-\lambda)}{\pi^2 s}\, P\int_0^1 dt\, \frac{\exp(2\theta_0 X)\cos(\Theta)}{q(t-s)} \qquad \text{(III-9)}$$

since $dS = -ds\,\frac{(1-\lambda)}{\pi q s}$. P denotes the principal value of the integral. Now, we define $\tau=\Theta-\pi/2$ in order to use the same notations as McLean and Saffman [6] and we derive from the Laplace equation the integro-differential equation of this free boundary problem.

$$\kappa q s\frac{\partial}{\partial s}\exp(-\theta_0 X)\left[qs\frac{\partial\tau}{\partial s}+\frac{\theta_0}{\pi}(1-\lambda)\sin(\tau)\right] - q = \frac{-q}{(1-\lambda)} - \frac{q}{\pi Q_0/\lambda}\int_0^1 dt\,\frac{\exp(2\theta_0 X)\sin(\tau)}{q(t-s)}$$

(III-10)

with $\kappa = \frac{4\pi^2\sigma}{(1-\lambda)^2 Q_0/\lambda}$ and the following boundary conditions: $\tau(0) = 0$, $\tau(1) = -\pi/2$, $q(0) = 1$ and $q(1) = 0$.

This equation relates three unknown functions q,τ and X. We need two other equations which are given in McS paper[6]. Since $q\exp(-i(\tau-\pi))$ is an analytical function of Z, the Cauchy integral equation relates Ln(q) and τ :

$$\mathrm{Ln}(q) = -\frac{s}{\pi}\int_0^1 dt\,\frac{\tau(t)}{t\,(t-s)} \qquad \text{(III-11)}$$

Here also we use the fact that τ(s) vanishes everywhere on the real s axis, except for $s\in[0,1]$. The other relation simply indicates that z is an analytical function of Φ^*:

$$X(s)+i\,Y(s) = -\frac{(1-\lambda)}{\pi}\int_s^1 dt\,\frac{\exp(i\tau)}{t\,q} \qquad \text{(III-12)}$$

The reader can check that we recover the McS equations for $\theta_0 = 0$. In this case, the set is reduced to two coupled equations. In the radial geometry, one has to solve three coupled equations: one which is integro-differential plus two integral equations, so the algorithm is a little more complicated. Nevertheless it remains rather effective as soon as one follows the numerical procedure of McLean and Saffman.Solutions at $\kappa = 0$ can be found analytically[11]:

$$x = \varepsilon(\theta_0)\ s^{\theta_0(1-\lambda)/\pi}\ F\left(\frac{\theta_0\,(2-\lambda)}{\pi},\ -\frac{\lambda\theta_0}{\pi},\ \frac{1}{2},\ 1-s\right)$$

$$y = A\ s^{\theta_0(1-\lambda)/\pi}\ (1-s)^{1/2}\ F\left(\frac{1}{2}+\frac{\theta_0\,(2-\lambda)}{\pi},\ \frac{1}{2}-\frac{\lambda\theta_0}{\pi},\ \frac{3}{2},\ 1-s\right)$$

with $$A = 2\ \tan(\lambda\,\varepsilon(\theta_0)\theta_0)\ \frac{\Gamma\left(1-\frac{\theta_0\,(2-\lambda)}{\pi}\right)\Gamma\left(1+\frac{\lambda\theta_0}{\pi}\right)}{\Gamma\left(\frac{1}{2}-\frac{\theta_0\,(2-\lambda)}{\pi}\right)\Gamma\left(\frac{1}{2}+\frac{\lambda\theta_0}{\pi}\right)} \qquad \text{(III-13)}$$

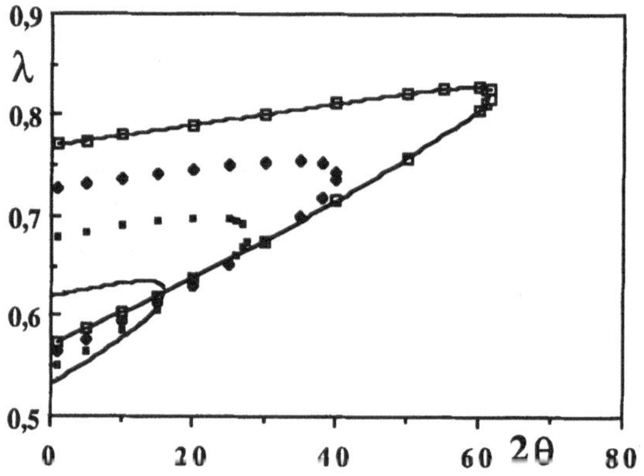

Figure 2. Relative widths λ_1 and λ_2 versus the sector angle $2\theta_0$, for various effective surface tension $\sigma = 0.001$ (——), $\sigma = 0.002$ (■ ■ ■), $\sigma = 0.003$ (◆ ◆ ◆), $\sigma = 0.004$ (-□-□-□-)

Γ is the Euler function, F is the hypergeometric function and $\varepsilon(\theta_0)$ the sign of θ_0.

As a consequence, one notices that there exists a lower bound for λ values as soon as $2\theta_0$ is greater than 90° : $\lambda_{min} = 2 - \frac{\pi}{2\theta_0}$ and an upper bound $\lambda_{max} = -\frac{1}{\theta_0}$ for $2\theta_0$ less than $-\pi$.

IV. NUMERICAL RESULTS FOR THE DIVERGENT FLOW REGIME AT LOW SURFACE TENSION

We focus here on the most striking feature of this analysis and explain the tip-splitting instability observed in radial growth process. The most amazing numerical result is the presence of a loop at fixed effective surface tension parameter σ in the graph $\lambda = f(2\theta_0)$, for divergent flow as shown in Fig.2 . This loop occurs at low σ and seems to disappear for $\sigma>0.005$, at least for $2\theta_0 \leq 90°$ (see Fig.2) . Results for larger values of $2\theta_0$ are very difficult to obtain, because of the very high values of the selected λ: $\lambda \geq 0.90$. This loop couples the first and second levels, if one refers to the ordinary S-T eigenvalue spectrum, while the third level seems not too much affected by the opening of the sector. It is shifted toward large angles when σ increases. As soon as σ is greater than 0.005, the plot $\lambda = f(2\theta_0)$ is approximately linear in θ_0. At fixed sector angle, the $\lambda = f(\sigma)$ spectrum recovers

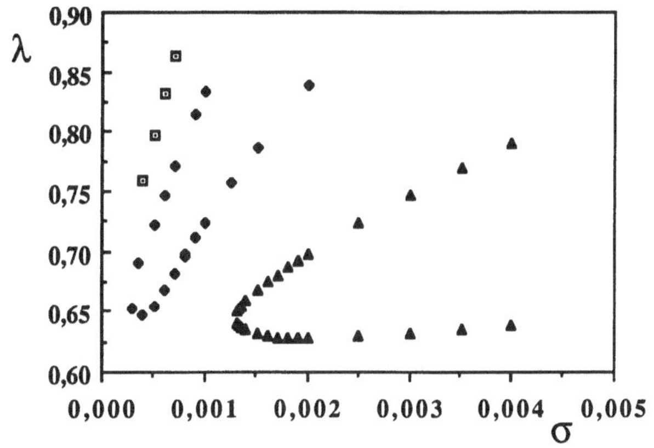

Figure 3. Eigenvalue spectrum versus σ for $2\theta_0 = 20°$

the same looping phenomena as shown by Fig.3 (for $2\theta_0=20°$) and Fig.4 (for $2\theta_0= 90°$). Regarding the right angle geometry (for $2\theta_0=90°$), a W.K.B analysis[12] gave $\lambda_{min}(2\theta_0) \approx 0.85$ in good agreement with Fig. 4 which indicates 0.858 for the loop $\lambda_{3,4}$.

Let us focus, for the moment, on the first and second eigenvalues which are perfectly defined at high surface tension (the right part of the Fig.3 and 4). There is a characteristic threshold of surface tension which depends of the sector angle $\sigma_{1,2}(2\theta_0)$. Below this threshold, the first and second eigenvalues λ_1 and λ_2 do not exist.Above it, they separate together from a common value $\lambda_{1,2}$ $(2\theta_0)$. So it is a kind of bifurcation with a non-vanishing surface tension threshold. A more careful numerical analysis at lowest surface tension "indicates" that the other eigenvalues show the same feature, that is the existence of a threshold $\sigma_{n,n+1}$ (n is odd). Below this threshold, the eigenvalues λ_n and λ_{n+1} do not exist ; at the threshold they have a common value $\lambda_{n,n+1}$;above it they split into two different levels n and n+1. Our intuition is that this phenomenon occurs for any divergent flow and any eigenvalue but it is not always easy to show numerically this scenario.

As shown in Fig 2, the threshold $\sigma_{1,2}$ is an increasing function of the sector angle, and we can extend this result to any threshold $\sigma_{n,n+1}$ as suggested by the comparison of two sector angles $2\theta_0 = 20°$ and 90°. Probably $\sigma_{n,n+1}$ is less than $\sigma_{m,m+1}$ if n is greater to m. The W.K.B analysis explained in the next section provides us scaling laws for these characteristic values $\sigma_{1,2}$ and $\lambda_{1,2}$ in good agreement with the numerics. Experimentally, it had also been observed[8] that the finger loses its stability at a limit value of σ, increasing with $2\theta_0$. If one assumes that only the first level is stable, this stable solution disappears at the loop, for $\sigma = \sigma_{1,2}$. So we can imagine the following scenario for the experimental dynamics where the finger grows at constant velocity. As time passes, the effective surface tension $\sigma(t)$ is decreasing because the channel width increases. The observed petal finger shows a relative width $\lambda(t)$ which slightly decreases following the stable branch λ_1 , as

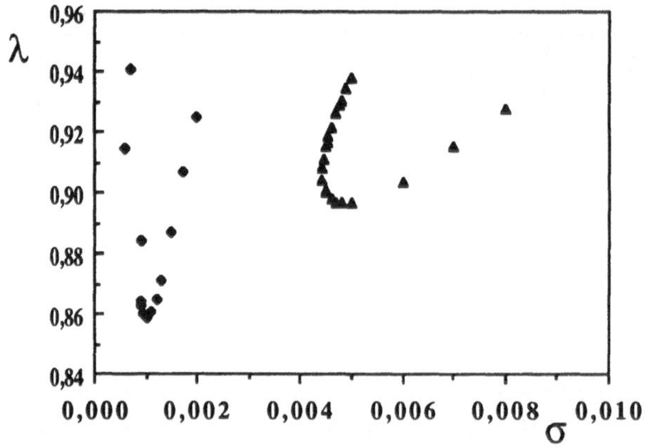

Figure 4: Eigenvalue spectrum versus σ for $2\theta_0 = 90°$

shown in Fig. 3 for $2\theta_0 = 20°$. When its shape has a λ very close to $\lambda_{1,2}(2\theta_0)$, it becomes unstable and prefers to split into two petal fingers. Doing that, after sometimes, it feels an effective sector angle, the value of which is half the experimental one. For this half sector value, the petal finger recovers stability : as the angle is smaller, its new characteristic σ is larger. Furthermore at this angle the instability threshold $\sigma_{1,2}$ is smaller. In other terms, tip-splitting reduces the radius of curvature at the nose, so increases the stabilizing effect of the capillarity. This scenario for the explanation of the tip-splitting instability can be compared to the experimental results.

On Fig. 5, we have plotted numerical and experimental results [8], for $2\theta_0 = 23°$. The agreement between predicted λ_1 and measured λ values is excellent when σ is greater than $\sigma_{1,2}$. So, as time passes, the σ parameter decreases and the time dependent relative width of the finger $\lambda(t)$ follows closely the first level which is really the relevant one for the experiment, because we expect that this branch only will be stable. On the other hand the global evolution of the instability threshold observed experimentally is well understood by the existence of loops and the finger appears to destabilize at $\sigma_{1,2}$. As the angle $2\theta_0$ increases so does the threshold.

So our conclusion is the following: in a radial growth process, the tip splits because there is no more possible self-similar shape for the finger. Since this occurs at low surface tension a W.K.B analysis can detect it and this is the subject of the next section.

V. ANALYTICAL INVESTIGATION FOR SMALL ANGLE AND SMALL SURFACE TENSION

We turn now to the analytical investigation[15] of selection in the limit of low surface tension T. The method can be applied in principle for any angle[15]. The first step is naturally to have a solution for the T=0 problem. However, whatever the more or less explicit form we take for this solution, one ends up with numerical calculations if one wants to obtain

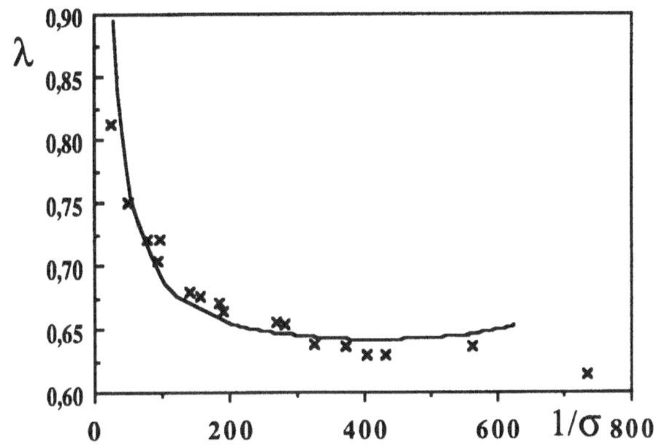

Figure 5: Comparison between predicted and measured relative width λ versus the experimental parameter σ^{-1} for $2\theta_0 = 23°$. x denotes experimental values.

completely explicit results for the selection problem. Therefore we will concentrate on sectors with small angles where essentially the theory can be carried out completely explicitely. We will see that this limit displays all the behaviour found numerically for a general angle, and that the physical reasons for it will appear quite clearly and simply ; this limit provides all the qualitative understanding in which we are interested. Since numerical calculations show that the behaviour is not different for a general angle, it is reasonable to expect that the qualitative reasons behind it are the same and that only the numbers are changed.

For small angle it can be shown[15] that, for the complex potential $w = \phi + i\psi$, the T=0 solution can be written as: :

$$\text{Ln } z - w\theta_0 = \frac{2\theta_0}{\pi} (1 - \lambda) \text{ Ln } (1 + e^{-\pi w}) \qquad \text{(IV-1)}$$

for proper time and length scale. This is actually the first term of a θ_0 expansion. This solution is in good agreement with the small θ_0 experimental results[8] (the finger shape is obtained by setting $\phi = 0$ in the complex potential).

As usual surface tension gives the boundary condition $\phi = \alpha/R$ on the finger. By taking its derivative with respect to arclength along the finger and combining with the relation between finger and fluid velocity, this condition is conveniently transformed into[16] a differential equation along the finger :

$$\varepsilon \frac{d^2f}{dz^2} + \frac{z}{f^2} = 2\lambda\, \theta_0\, u - z^* \qquad \text{(IV-2)}$$

where $\varepsilon = 2\alpha\lambda\theta_0$. Here z^* must be understood as the analytical continuation of $z^*(z)$ or $z^*(w)$. The complex velocity u can be obtained from Eq.(V-1) to lowest order in surface tension. Finally $f = e^{\,i\theta_n}$ with θ_n being the angle between the normal to the finger and the cell axis. In Eq.(V-2), the first term acts as a singular perturbation and the selection condition is obtained by requiring that no transcendental divergent term is produced by the perturbation.

This condition is most conveniently expressed by performing an analytical continuation of Eq.(V-2) in order to go to the singular points which generate the divergence. In the present case, this continuation merely amounts to extend Eq.(V-2) into the viscous fluid[17]. From this differential equation a perturbation expansion in ε can be generated for f around the $\varepsilon = 0$ solution $f_0 = (2\lambda\theta_0 u/z - z^*/z)^{-1/2}$. However this expansion breaks down around the singularities of f_0, where $\varepsilon\, d^2f_0/dz^2$ diverges. These singular points are the zeros of the r.h.s. of Eq.(V-2). For small θ_0 and $\lambda \approx 1/2$ (which is expected for small T by

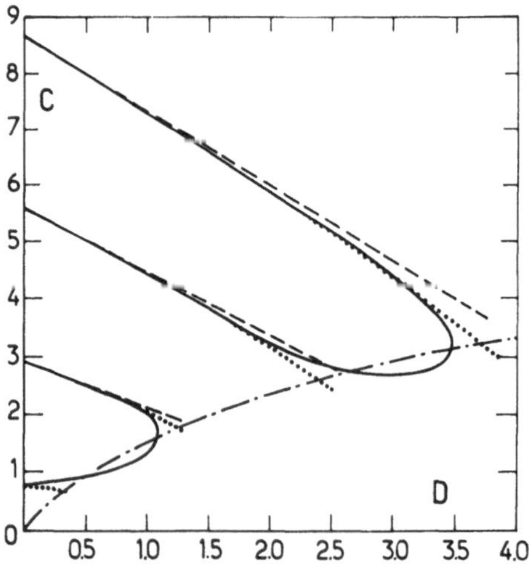

Figure 6 Non linear eigenvalue C as a function of D. Full curves: numerical integration of Eq.(V-3). Dotted curves: WKB results. Dashed curves: simple linear approximation to WKB. Dashed - dotted curve: WKB limit.

continuity from the $\theta_0 = 0$ case), these singularities are found for $z \approx 1$: they are located near the finger tip compared to the finger length. However they satisfy $e^{\pi w} \gg 1$ and are not located near the finger tip compared to the channel width $2\theta_0$ at the finger tip. In this range $2\lambda\theta_0 u \approx 1 + 2\eta + e^{-\pi w}$ with $\eta = \lambda - 1/2$, and $z \approx z^* \approx 1 + \theta_0 w$. It is then more convenient to use the rescaled variable $y = \sigma_0^{-1/3} e^{-\pi w/2}$ and the function $F = \sigma_0^{1/3} f$ with $\sigma_0 = \varepsilon (\pi/2\theta_0)^2$. This leads to the equation :

$$y \frac{d}{dy}\left(y \frac{dF}{dy}\right) + \frac{1}{F^2} = y^2 + C + D \,\mathrm{Ln}\, y \qquad \text{(IV-3)}$$

with (as a definition for the non linear eigenvalue C):

$$\frac{\eta}{\theta_0/\pi} - \frac{1}{2} \mathrm{Ln} \frac{\pi}{2\theta_0} = \frac{C}{D} + \frac{1}{2} \mathrm{Ln}\, D = R \qquad D = \frac{2\theta_0}{\pi} \sigma_0^{-2/3} \qquad \text{(IV-4)}$$

This non linear equation has to be solved with the boundary condition $F \approx 1/y$ for large $|y|$ and $|\mathrm{Arg}\, y| \leq \pi/2$. For $D \to 0$, i.e. $\theta_0 \to 0$ or large σ_0, the equation reduces to the one studied for the linear case[18]. We note that Eq.(V-4) provides a mapping between the problems for different surface tension and angle, but same D. This scaling law is a generalization of the relation $\eta \sim \varepsilon^{2/3}$ for linear fingers.

Eq.(V-3) can be solved numerically. From the linear case, we know that, for D = 0, Eq.(V-3) has solutions only for a discrete set of values for C, which produces finger width selection. These values are[18] $C_0 = 0.8158$, $C_1 = 2.950$, $C_2 = 5.63$ and $C_n = 2^{2/3}(n+4/7)^{4/3}$ for large n, as obtained from the WKB method (the precision for n = 2 is already very good). Therefore in the same way we expect for C a discrete set of values $C_n(D)$ for each value of D.One finds indeed for C(D) discrete branches, shown in Fig.6, which display for

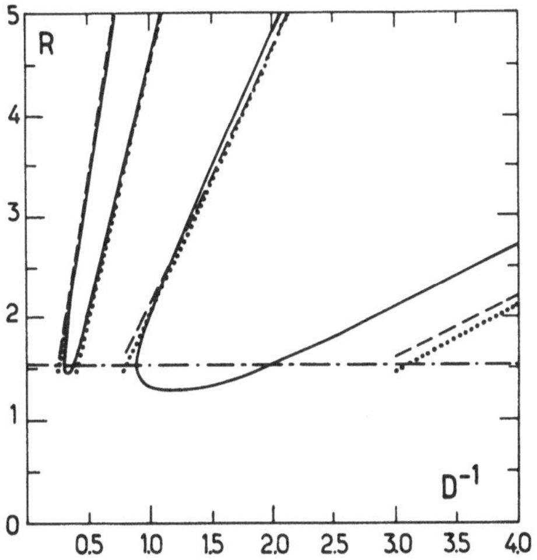

Figure 7 Finger width spectrum R as a function of the reduced viscosity D^{-1}. Full curves: numerical integration of Eq.(V-3). Dotted curves: WKB results. Dashed curves: simple linear approximation Eq.(V-14) to WKB.Dashed-dotted horizontal line: WKB limit R=1.52.

large D the branch merging or loop phenomenon. When translated for the finger width as a function of D^{-1}, which can be considered as a reduced surface tension, one finds the results shown in Fig.7 which agree very well with numerical results on the finger as it can be seen from the comparison with Fig.3. In particular the minimum width for $2\theta_0=20°$ is $\lambda=0.63$ which agrees with these results within numerical uncertainty.

In order to gain more insight into branch merging, we proceed to a WKB treatment of Eq.(V-3). Although valid in principle only for large C, this has proved to be quite sucessful qualitatively and quantitatively for linear fingers[18]. This is done by the new scaling $x = \gamma y$ and $F = \gamma G$ which gives :

$$\gamma^3 x \frac{d}{dx}\left(x \frac{dG}{dx}\right) + \frac{1}{G^2} = x^2 + 1 + \delta \operatorname{Ln} x \qquad \text{(IV-5)}$$

with $\gamma^2 = \delta/D$ and $R = 1/\delta + (1/2)\operatorname{Ln}\delta$. For small γ, Eq.(V-5) has the solution $G_0=(x^2+1+\delta \operatorname{Ln} x)^{-1/2}$ to lowest order. The small corrections h to this solution are obtained by linearizing Eq.(V-5) with $G= G_0+h$.The transcendental terms generated are solutions of the homogeneous part of this equation :

$$\gamma^3 x \frac{d}{dx}\left(x \frac{dh}{dx}\right) - 2h\, G_0^{-3} = 0 \qquad \text{(IV-6)}$$

and can be obtained from the WKB approximation as :

$$h\pm \approx G_0^{3/4} \exp\left[\pm \sqrt{2}\, \gamma^{-3/2} \int^{x} \frac{dx}{x} G_0^{-3/2}\right] \qquad \text{(IV-7)}$$

The approximation is not valid in the vicinity of the turning points of Eq.(V-6) which correspond in the present case to the zeros of the r.h.s. of Eq.(V-5). There is always a single zero x_0 on the real positive x axis, which is easily seen directly in the present case. But there are also always two complex conjugate zeros $x_\pm$ as can be seen more easily from the r.h.s. of Eq.(V-3) by setting $t = D^{-1/2}y = \delta^{-1/2}x$, which leads to the equation:

$$t^2 + \operatorname{Ln} t + R = 0 \qquad \text{(IV-8)}$$

with :

$$R = \frac{C}{D} + \frac{1}{2}\operatorname{Ln} D = \frac{\eta}{\theta_0/\pi} - \frac{1}{2}\operatorname{Ln}\left(\frac{\pi}{2\theta_0}\right) = \frac{1}{\delta} + \frac{1}{2}\operatorname{Ln}\delta \qquad \text{(IV-9)}$$

One finds that Eq.(V-8) has always three zeros by counting them in the complex t plane by the contour method. Or explicitely, with $t_\pm = \rho\, e^{i\varphi}$, Eq.(V-8) is equivalent to :

$$\rho^2 = -\frac{\varphi}{\sin(2\varphi)} \qquad R = -\rho^2 \cos(2\varphi) - \operatorname{Ln}\rho \qquad \text{(IV-10)}$$

which is always satisfy for $\pi/2 < |\varphi| < \pi$, since R as given by Eq.(V-10) goes from $-\infty$ to ∞ when $|\varphi|$ goes from π to $\pi/2$.

Therefore we find a situation which is qualitatively different from the one found for the linear finger. We have always three singularities, one being on the channel axis, the two other ones being complex conjugate. In contrast for the linear finger we have either a singularity on the axis or two singularities on the channel sides. Naturally the axis singularity for the sector is analogous to the axis singularity of the linear finger, and reduces to it for $\theta_0 = 0$ and $\eta < 0$. Similarly the two complex conjugate singularities are analogous to the channel side singularities of the linear finger and reduce to them for $\theta_0 = 0$ and $\eta > 0$. However , since they have $\pi/2 < |\varphi| < \pi$, they are no longer on the sector sides (which correspond to $|\varphi| = \pi/2$), but they are rather located in the analytic continuation beyond the sector sides. Therefore we see that the limit $\theta_0 \to 0$ is somewhat singular since there is always a qualitative difference between the situation for $\theta_0 \neq 0$, however small, and the linear case $\theta_0 = 0$.

Whenever the axis singularity x_0 "dominates", i.e. is "nearer" to the finger, we find the same situation as for the linear finger with an axis singularity and there is no solution. When the side singularities $x_\pm$ dominate, we find solutions as for the linear case. The transition between the two regimes occurs when x_0 and $x_\pm$ are on a same Stokes line :

$$\mathrm{Re} \int_{x_0}^{x_\pm} \frac{dx}{x} G_0^{-3/2} = 0 \tag{IV-11}$$

which gives the limiting values $\delta_0 = 0.55$ and $R_0 = 1.52$ for δ and R. The corresponding boundary in the (C,D) plane coincides very well with the place where branch merging occurs as seen in Fig.Z (numerical integration of Eq.(V-3) leads to an effective $R_0 = 1.25$ for the first "curl" and $R_0 = 1.4$ for the second one).The corresponding minimum width is given by:

$$\eta_{min} = \frac{\theta_0}{\pi} \left(R_0 + \frac{1}{2} \mathrm{Ln} \frac{\pi}{2\theta_0}\right) \tag{IV-12}$$

This result shows that η_{min} has a weak singularity for $\theta_0 = 0$, but the overall behaviour is roughly linear in agreement with experiment. Note on the other hand that η is linear with θ_0 at constant σ_0, since C is essentially constant on the lower branch.

In the regime $\delta \leq \delta_0$, we require that η is exponentially small for large $|x|$, $|\mathrm{Arg}\, x| \leq \pi/2$. The situation is quite analogous to the linear case[4,18] and leads in the same way to the selection condition :

$$I(\delta) = \frac{2}{\pi} \mathrm{Im} \int_{x_0}^{x} \frac{dx}{x} G_0^{-3/2} = (2\gamma^3)^{1/2} (n + 4/7) \tag{IV-13}$$

Actually we have neglected here a very small correction due to the fact that $x_\pm$ are no longer purely imaginary as in the linear case[15,16]. Once δ is found from Eq.(V-13), the finger width is obtained from $2\pi\eta/\theta_0 = 2/\delta + \mathrm{Ln}(\pi\delta/2\theta_0)$. For $\delta = 0$, $I(0) = 1$ and we recover the result for the linear finger. When I(δ) is calculated numerically, one obtains the results shown in Fig.2, which agree very well with the numerical integration, except naturally near branch merging. Since δ is never large, it is possible to expand I(δ) as $I(\delta) - 1 \approx (3\delta/8)(3\,\mathrm{Ln}2 - \pi/2) = 0.19\,\delta$. This gives a very good approximation to the WKB result as seen in Fig.2.The corresponding result for the finger width is:

$$\frac{\eta}{\theta_0/\pi} = C_n D^{-1} - \frac{1}{2}\,\mathrm{Ln}\,\frac{2\theta_0 C_n D^{-1}}{\pi} - 0.25 \qquad \text{(IV-14)}$$

Together with the WKB limit Eq.(V-12) it gives a quite good qualitative and quantitative description of the finger width picture (see Fig.1).

As a conclusion our WKB analysis demonstrate clearly that branch merging occurs physically when the axis singularity is no longer hidden by the side singularities and becomes dominant. Then the finger disappears exactly as it does in the linear case for λ<1/2. Let us point out that this axis singularity is a physical object which can be manipulated experimentally, as shown recently[17].Therefore we can safely predict that a bubble or a disk in front of the finger will produce narrow fingers as for the linear case.

We are quite grateful to V. Hakim and Y. Couder for very enlightening and stimulating discussions.

REFERENCES

1. For reviews, see D. A. Kessler, J. Koplik and H. Levine, Adv.Phys.**37**, 255 (1988); P. Pelcé, "Dynamics of curved fronts" Perspective in Physics, Ed.H.Araki, A.Libchaber, G. Parisi (Academic, Orlando,Florida,1988).

2. A. Arnéodo,Y. Couder, G. Grasseau, V. Hakim and M. Rabaud, Phys.Rev.Lett.**63,** 984 (1989).

3. Y. Couder, F. Argoul A. Arnéodo, J. Maurer and M. Rabaud, Phys.Rev.**A42**, 3499 (1990).

4. B. I. Shraiman, Phys.Rev.Lett.**56**, 2028 (1986); D. C. Hong and J. S. Langer, Phys.Rev.Lett.**56,** 2032 (1986); R. Combescot, T. Dombre, V.Hakim,Y. Pomeau and A. Pumir,Phys.Rev.Lett.**56**, 2036 (1986); S. Tanveer Phys. **30**,1589 (1987).

5. P.G. Saffman and G.I. Taylor Proc.Roy.Soc.A **245**,312 (1958);P.G. Saffman J.Fluid Mech. **173**,73 (1986); G.M. Homsy, Annu. Rev. FLuid Mech.

6. J.W. McLean and P.G. Saffman, J.Fluid Mech **102**, 455 (1981); J.M. Vanden-Broeck, Phys.Fluid **26**,2033 (1983).

7. D.A. Kessler and H. Levine Phys. Rev. **A 32**,1930 (1985) , D. Bensimon Phys.Rev.**A 33**, 1302 (1986); S. Tanveer Phys. Fluid **30**, 2318 (1987).

8. H. Thomé, M. Rabaud,V. Hakim and Y.Couder, Phys.Fluids.**A1**, 224 (1989).

9. J.Bataille, Rev.Inst.Pet.23,1349 (1968); L.Paterson, J.Fluid.Mech. **113**, 513 (1981).

10. S.Sarkar, Phys.Rev.Lett. **65**,2680 (1990); S.E. May and J.V. Maher, Phys.Rev.**A 40**,1723 (1989); P. Meakin in "Phase transitions and Critical Phenomena", edited by C. Domb and J.L. Lebowitz (Academic, Orlando, FL,1988),Vol.12, and references therein.

11. M. Ben Amar,to be published in Phys.Rev.A ; V. Hakim, M. Mashaal and Y. Couder, Phys. Fluid A. 3,1687 (1991).

12. E.A. Brener, D.A. Kessler, H. Levine and W-J. Rappel, Europhys.Lett.**13**,161 (1990), Yuhai Tu, this proceeding.

13. P. Tabeling, G. Zocchi and A. Libchaber, J. Fluid Mech.**177**, 67 (1987) ; P.G. Saffman and S. Tanveer Phys. Fluids **A1**,219 (1989), S. Tanveer in Proceedings of Nato-Asi "Growth and Forms: Nonlinear Aspects", Cargèse, July 1990, Ed. M. Ben Amar, P.Pelcé and P. Tabeling).

14. Y. Couder, N. Gerard and M. Rabaud , Phys.Rev.A **34**,5175 (1987), G. Zocchi, B.Shaw, A. Libchaber and L. Kadanoff, Phys.Rev.A **36**,1894 (1987), D.C. Hong and J.S. Langer, Phys.Rev. **A 36**,2325 (1987), R. Combescot and T. Dombre, Phys.Rev.**A 39**,3525 (1989).

15. For full details, see R.Combescot, to be published.

16. R. Combescot and T. Dombre, Phys.Rev.**A38**, 2573 (1988).

17. H. Thomé, R. Combescot and Y. Couder, Phys.Rev.**A41**, 5739 (1990).

18. R. Combescot, T. Dombre, V.Hakim,Y. Pomeau and A. Pumir, Phys. Rev. **A 56**, 1270 (1988).

SAFFMAN-TAYLOR PROBLEM IN SECTOR GEOMETRY

Yuhai Tu

Department of Physics
Institute for Nonlinear Science
University of California, San Diego
La Jolla, CA 92093

I. Introduction

Patterns formed by the instabilities in propagating interfaces between different phases have received much attention[1] in recent years. One of the well known examples is the Saffman-Taylor problem in a Hele-Shaw cell[2], where an unique finger pattern is observed when a viscous fluid is displaced by a less viscous fluid.

Recently, the same experiment was also done in different geometries[3], in particular, in a sector shaped cell with arbitrary opening angle. It was observed that just as in the linear geometry, at large velocity, a unique finger tends to occupy a well determined fraction of the cell angular width. This fraction is an increasing function of the opening angle and it goes to .5 as the angle approaches 0. The only theoretical attempt to explain the selected finger width in this type of geometry is by Brener, Kessler, Levine and Rappel(BKLR)[4]. They applied the same WKB method as in the linear geometry to a family of analytic solutions found in ref. 3 in the 90 degree cell. Their result of the selected finger width at small surface tension agrees with the experiment. The use of the WKB method (or any analytical method) is based on knowing the analytical zero-surface tension solution. But until now no analytical solution has been found for any other angle.

In the following sections, we develop a systematic method for deriving the zero surface tension solutions analytically at arbitrary opening angle. The solutions can be very easily extended to the complex plane. By using the WKB method, we can obtain the selected finger width at small surface tension. We also show that the structure of the selected solutions are different from that of the linear geometry due to the difference of distribution of poles and branch cut for the phase integral. We further discuss the stability of the selected solutions within the WKB scheme based on the special structure of the solutions.

II. Solutions Without Surface Tension

The evolution equations for the interface in the sector geometry are easy to derive

Asymptotics beyond All Orders, Edited by H. Segur *et al.*
Plenum Press, New York, 1991

(see fig. 1(a)). Inside the viscous fluid, the velocity potential ϕ satisfies the Laplace equation:

$$\nabla^2 \phi = 0 \tag{1}$$

with the boundary conditions

$$\phi = \gamma\kappa|_{AOB} \quad and \quad \partial\phi/\partial n|_{AC(BD)} = 0 \tag{2}$$

where γ is the surface tension parameter, and κ is the curvature of the interface. By normalizing the injection rate at infinity to 2π, ϕ has the behavior $\phi \sim \frac{2\pi}{\theta_0} \ln \sqrt{x^2 + y^2}$ as $x^2 + y^2 \to \infty$. The interface moves with the velocity:

$$v_n = \vec{n} \cdot \nabla\phi|_{AOB} \tag{3}$$

We are interested in obtaining the zero surface tension in this section, so we let $\gamma = 0$.

We consider the problem using hodograph method, i.e., considering $z = x + iy$ as an analytical function of the complex velocity potential $\omega = \phi + i\psi$, ψ is the stream function. The interface equation of motion eq.(3) is now:

$$\frac{\partial x}{\partial t}\frac{\partial y}{\partial \psi} - \frac{\partial y}{\partial t}\frac{\partial x}{\partial \psi} = 1 \quad on\ \phi = 0 \tag{4}$$

Because of the two dimensional nature of the geometry, instead of looking for a solution translating with constant velocity as in the linear channel, we are seeking for self similar solutions here, written as:

$$x(t, \phi = 0, \psi) = h(t)x(\psi) \quad ; \quad y(t, \phi = 0, \psi) = h(t)y(\psi) \tag{5}$$

Substituting this into eq.(4), we have (superscript "*" represent complex conjugate):

$$Im(z^* \frac{\partial z}{\partial \psi}) = 1 \tag{6}$$

while $h(t)$ satisfies:

$$h(t)\frac{dh(t)}{dt} = 1 \quad i.e. \quad h(t) = \sqrt{2(t - t_0)} \tag{7}$$

where t_0 is the initial time.

So the problem of finding the zero surface tension solutions reduces now to finding an analytical function z in the region $\phi \geq 0$, $-\pi < \psi < \pi$(see Fig. 1(b)) satisfying the following three conditions as stated in ref. 3: (i) $\arg(z) = \pm\theta_0/2$ on "AC" and "BD", i.e. the half lines $\psi = \pm\pi$. (ii)$z \sim \exp(\omega\theta_0/2\pi)$ when $\phi \to \infty$. (iii) z satisfy eq. (6) on "AOB", i.e., on $\phi = 0$.

Let us start by analyzing condition (iii) first. Differentiating both sides of equation (6) with respect to ψ, and because $\frac{\partial z^*}{\partial \psi}\frac{\partial z}{\partial \psi}$ is real, we get:

$$Im(z^* \frac{\partial^2 z}{\partial \psi^2}) = 0 \quad on\ \phi = 0 \tag{8}$$

so $\partial^2 z/\partial\psi^2$ should be equal to z times some real function of ψ, $f(\psi)$:

$$\partial^2 z/\partial\psi^2 = f(\psi)z \quad on\ \phi = 0 \tag{9}$$

To interpolate the equation from $\phi = 0$ to the whole region, we can replace ψ by ω/i, and the equation becomes:

$$-\partial^2 z/\partial\omega^2 = f(-i\omega)z \tag{10}$$

Next, we consider the second condition, namely, as $\omega \to \infty$, $z \sim e^{\theta_0\omega/2\pi}$. This determines the limiting behavior of $f(-i\omega)$: $f(-i\omega) \to -(\theta_0/2\pi)^2$ as $\omega \to \infty$. We subtract the constant part from $f(-i\omega)$ and define:

$$f(-i\omega) \equiv -((\theta_0/2\pi)^2 + f_1(\omega)) \tag{11}$$

We also make the corresponding transformation for z:

$$z = \exp(\theta_0\omega/2\pi)z_1 \tag{12}$$

The equation for z_1 is then:

$$d^2 z_1/d\omega^2 + \frac{\theta_0}{\pi}dz_1/d\omega = f_1(\omega)z_1 \tag{13}$$

with the conditions for f_1 that $f_1(\omega) \to 0$ as $\omega \to \infty$ and $f_1(\omega)$ must be real for pure imaginary ω. For future convenience, we change the region of interest to a finite region by following transformation (see Fig. 1(c)):

$$\tau = (-\exp(\omega))^{-1/2} \tag{14}$$

The equation for z_1 changes to:

$$dz_1^2/d\tau^2 + \frac{(1 - 2\theta_0/\pi)}{\tau}dz_1/d\tau = 4\tau^{-2}f_2(\tau)z_1 \tag{15}$$

where $f_2(\tau) = f_1(\ln(-\tau^{-2}))$. So the conditions for f_2 are: $f_2(\tau) \to 0$ as $\tau \to 0$ and $f_2(\tau)$ must be real on the unit circle, i.e., when $|\tau| = 1$.

Finally, we come to the last condition that requires $\arg(z) = \pm\theta_0/2$ on $\psi = \pm\pi$. Because of the transformation (12) and (14), this condition becomes that the solution z_1 of equation (15) has to be real and on the real τ axis in the range $-1 \leq \tau \leq 1$, and it has to be finite when $\tau \neq \pm 1$.

Now the problem has been reduced to finding an analytic function $f_2(\tau)$ which satisfies all the above conditions, and then solving the linear ordinary differential equation (15) to obtain the mapping $z_1(\tau)$ from the τ plane (Fig. 1(c)) to the real space (Fig. 1(a)).

We consider the behavior of z_1 around the singular point $\tau = 0$ first. Because $f_2(\tau) \to 0$ as $\tau \to 0$, the leading order behavior is determined by the two terms at the left hand side of eq. (15). We can express the two solutions in power series around $\tau = 0$:

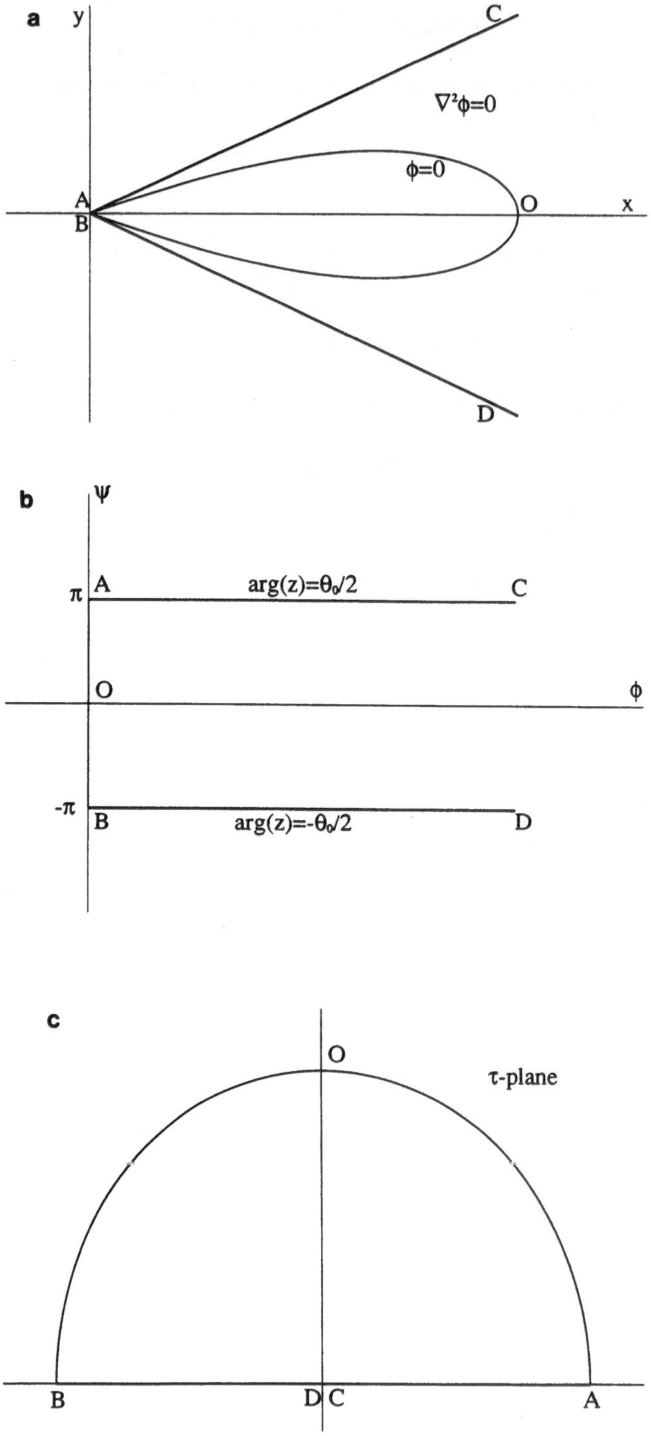

Figure 1. Schematic picture of the Saffman-Taylor problem in sector geometry ("AC" and "BD" are the boundaries of the sector; "AOB" is the interface.): (a) In physical space; (b) In ω plane; (c) In τ plane.

$$z_1^{(1)} = 1 + a_1\tau + a_2\tau^2 + \cdots$$
$$z_1^{(2)} = \tau^{2\theta_0/\pi}(1 + b_1\tau + b_2\tau^2 + \cdots) \qquad (16)$$

Where a's and b's are constants, and the general solution is $z_1 = Az_1^{(1)} + Bz_1^{(2)}$ with A and B arbitrary constants. In order to satisfy the condition that z_1 has to be real on the real line segment $-1 \leq \tau \leq 1$ and finite except for $|\tau| = 1$, we must choose $B = 0$ (except for the case that $\theta_0 = \pi/2$, π and $3\pi/2$ which will be discussed at the end of this section) and $A \neq 0$. Also $f_2(\tau)$ has to be a real function of τ.

Other singular points of the equation (15) will come from the poles of $f_2(\tau)$. Assume that $f_2(\tau)$ has a pole at $\tau = \tau_0$, then $z_1(\tau)$ will map τ_0 into $z_1 = 0$ which is on the interface, so τ_0 has to be on the semi-unit-circle, $|\tau_0| = 1$. This is also required by the previous consideration, because if $z_1(\tau)$ has a singular point at some τ inside the unit circle, solution $z_1^{(1)}$ can not be extended towards $\tau = \pm 1$. More strictly, because we are only interested in the physically meaningful case where the interface has only one bump[1], τ_0 has to be ± 1.

Let's analyze the local behavior of the mapping around $\tau = \pm 1$. Comparing the "A" corner in Fig. 1(a) and 1(c), we can see that the angle between the interface and the boundary wall changes in these two coordinate system. So the mapping around $\tau = \pm 1$ should have the form $z_1(\tau) \sim (\tau \mp 1)^{s_\pm}$, where $s_\pm (> 0)$ are constants determine the relative angular width of the finger to the sector. Substituting this behavior into equation (15), we can easily see that $f_2(\tau)$ has two second rank poles at $\tau = \pm 1$.

Now we can finally gather all the properties of function $f_2(\tau)$ and determine its form. First, $f_2(\tau)$ is a real function of τ, and it also has to be real on the unit circle, so:

$$f_2(e^{i\psi}) = (f_2(e^{i\psi}))^* = f_2^*(e^{-i\psi}) = f_2(e^{-i\psi}) \qquad (17)$$

Interpolating to the whole region, we have:

$$f_2(\tau) = f_2(1/\tau) \qquad (18)$$

So the function $f_2(\tau)$ is constructed by two building blocks: $\tau + 1/\tau$ and $(\tau - 1/\tau)^2$. Remembering that $f_2(\tau)$ has 2nd rank poles at $\tau = \pm 1$ and $f_2(\tau) \to 0$ as $\tau \to 0$, we immediately have the general form $f_2(\tau)$:

$$f_2(\tau) = (c_1 + c_2(\tau + \tau^{-1}))/(\tau - \tau^{-1})^2 \qquad (19)$$

where c_1 and c_2 are real constants. We make the following transformation:

$$c_1 = \frac{1}{4}\left(\frac{1}{2}(\lambda_1^2 + \lambda_2^2) - 1\right) \quad ; \quad c_2 = \frac{1}{16}(\lambda_1^2 - \lambda_2^2)$$

The reason for this transformation will be clear soon. Substituting the expression of $f_2(\tau)$ into equation (15), we have:

$$d^2z_1/d\tau^2 + \frac{(1 - 2\theta_0/\pi)}{\tau}dz_1/d\tau = \frac{1}{\tau(\tau^2 - 1)^2}\left(\left(\frac{\lambda_1^2 + \lambda_2^2}{2} - 1\right)\tau + \frac{\lambda_1^2 - \lambda_2^2}{4}(\tau^2 + 1)\right)z_1 \quad (20)$$

---

[1]Generally, this problem allows to have more than one bump, and this is accomplished by choosing $f_2(\tau)$ to have 2nd rank poles at points on the semi-unit-circle other than $\tau = \pm 1$. For example, if $f_2(\tau)$ generates a single bump solution in angle θ_0, then function $f_2^{new}(\tau) = f_2(\tau^n)$ will generate a n-bump solution in angle $n\theta_0$. It is clear that $f_2^{new}(\tau)$ has (n-1) other poles on the semi-unit-circle.

We have to choose the right solution of this equation: $z_1^{(1)}(\tau)$ which has finite derivatives to any order at $\tau = 0$. In fact, the solutions for the symmetric interface $\lambda_1 = \lambda_2 = \lambda$ can be expressed as hypergeometric functions:

$$z_1 = (1-\tau^2)^{(1+\lambda)/2} F\left(\frac{1+\lambda}{2}, \frac{1+\lambda}{2} - \frac{\theta_0}{\pi}, 1 - \frac{\theta_0}{\pi}, \tau^2\right) \tag{21}$$

This expression is first found by M. Benamar. For non-symmetric solutions, because the equation (21) has four singular points $\tau = 0, \pm 1, \infty$ and there is no symmetry between τ and $-\tau$, the solutions can not be expressed by any known function.

Let us check the behavior of $z_1(\tau)$ around $\tau = \pm 1$. At the neighborhood of $\tau = 1$, the leading order behavior of $z_1(\tau)$ is $(1-\tau)^{(1-\lambda_1)/2}$ (assuming $\lambda_{1(2)} \geq 0$). In order that $z_1(1) = 0$, we have $0 \leq \lambda_1 \leq 1$, the same condition for λ_2 also applies. So solving the linear ordinary differential equation (21), we obtain a two parameters family of solutions parameterized by λ_1 and λ_2, analogous to the two parameters Saffman-Taylor solutions found in the linear channel. The solutions with $\lambda_1 \neq \lambda_2$ will give asymmetric interfaces, while the solutions with $\lambda_1 = \lambda_2$ correspond to symmetric interfaces. The effective angular width λ_{eff} of the finger with respect to the opening angle of the sector can be expressed as:

$$\lambda_{eff} = 1 - \frac{\pi}{2\theta_0}\left(1 - \frac{1}{2}(\lambda_1 + \lambda_2)\right) \tag{22}$$

where λ_1 and λ_2 are restricted by the condition $\lambda_{eff} \geq 0$.

We finish this section by analyzing the equation (21) at the special value of opening angle $\theta_0 = m\pi/2$, with m=1,2,3. At these value of θ_0, there are essential differences of z_1 behavior around $\tau = 0$ from that of other angles. In general, there will be terms like $\tau^m \ln \tau$ present in $z_1^{(1)}$. For m=1 or 3, we can eliminate these terms by letting $\lambda_1 = \lambda_2$, they are also two-parameter family of solutions because $z_1^{(2)}(\tau)$ is also a valid solution . In fact, an closed form of these solutions can be obtained in the $m = 1$ case and has the form:

$$z_1 = (1-\tau)^{(1-\lambda_1)/2}(1+\tau)^{(1+\lambda_1)/2} + c(1+\tau)^{(1-\lambda_1)/2}(1-\tau)^{(1+\lambda_1)/2} \tag{23}$$

The solutions correspond to any real positive constant c and $0 \leq \lambda_1 \leq 1$. The solution found by Hakim in ref. 3 correspond to the case $c = 1$ which is the symmetric solutions. For m=2, the coefficients of the logarithmic terms in $z_1^{(1)}$ do not vanish for any physical λ_1 and λ_2 except for $\lambda_1 = \lambda_2 = 1$. So there is no one-bump solution for this angle except the trivial solution of the semi-circle. We can also see this from expression (21), where the hypergeometric function is not defined for $\theta_0 = \pi$.

III. Selections with Surface Tension

As is well known now, in the linear channel geometry, the inclusion of surface tension will break the continuous family of solutions into an infinite countable discrete set of solutions, and all the selected finger widths go to an unique value as the surface tension approaches zero[5,6]. In the channel geometry, this value is equal to .5. Although there exist more rigorous methods of analysing the selection problem, the WKB method is simpler, gives the correct qualitative results including the selected finger width, the correct power law. But it is quantitatively not quite right, for example, it predicts the wrong prefactor in front of the power law. However, in our case, where we are not interested in the detailed scaling behaviour, WKB method is sufficent, and we will show later that its results are in very good agreement with the numerical caculation[10].

Following BKLR, we assume a time dependent surface tension $\gamma_{phys} = \gamma_0 t^{\frac{1}{3}}$ in order to maintain the self-similarity of the solution in presence of the surface tension. Let the deviation from the zero-surface tension self-similar solution be $\delta(y)$, and linearize the equation around the known zero surface tension solution $x_0(y)$, we get the following equation for $\delta(y)$:

$$\frac{\gamma_0 \delta''(y)}{(1 + x_0'(y)^2)^{3/2}} + L[x_0(y)]\delta(y) = \gamma_0 \kappa_0(y) \qquad (24)$$

where L is some linear integral operator. The solvability condition is then [2]:

$$\int \kappa_0(y)\hat{\delta}_0(y)dy = 0 \qquad (25)$$

where $\hat{\delta}_0(y)$ is the zero mode of the adjoint linear operator. Using a WKB method and a local approximation of the integral operator, $\hat{\delta}_0(y)$ is expressed as:

$$\hat{\delta}_0(y) \sim \exp\left(\frac{i}{\sqrt{\gamma_0}} \int^{x'(y)} \frac{(1 - iz)^{3/4}(1 + iz)^{1/4}(x_0(z) + iy_0(z))^{1/2}}{g(z)} dz\right) \qquad (26)$$

where $g(z) = d^2x_0/dy_0^2$ with $z = dx_0/dy_0$.

It is easy to see that the integral (25) has a stationary phase point at $z = \pm i$, it is also obvious that the integral (25) will have branch cut at point $\tilde{z}$ with $g(\tilde{z}) = 0$.

In the sector geometry, there are two branch points in the upper half plane distributed symmetrically around the imaginary axis. So, for condition (25) to be satisfied, the contribution of integral around the branch point has to be large enough to cancel the contribution from the stationary phase point. The selected finger width at small surface tension limit is then determined by:

$$Im \int_{\tilde{z}}^{i} \frac{(1 - iz)^{3/4}(1 + iz)^{1/4}(x_0(z) + iy_0(z))^{1/2}}{g(z)} dz = 0 \qquad (27)$$

In ref. 4, this calculation was done for the 90° geometry case, and very good agreement was obtained between the theoretically derived selection width and the experimentally observed one.

We now extend this derivation to a sector geometry of arbitrary angle using the results obtained in the previous section. Because there is no asymmetric forcing in this problem, we assume the selected finger is symmetric around the center line of the sector[6,9], i.e., $\lambda_1 = \lambda_2 \equiv \lambda$.

Let us introduce a new variable φ: $\tau = \exp(i\varphi)$, the interface "AOB" corresponds to the segment on the real axis[3] : $0 \leq \varphi \leq \pi$. To extend the solution to complex plane, we substitute all the transformations back to equation (10) and obtain a rather simple equation for $z(\varphi)$:

$$d^2z/d\varphi^2 = -\left(\left(\frac{\theta_0}{\pi}\right)^2 + \frac{1 - \lambda^2}{4}\frac{1}{sin^2\varphi}\right)z \qquad (28)$$

The real and imaginary part of this equation can be extended from the real axis to the whole complex plane. The equations satisfied by $x(\varphi)$ and $y(\varphi)$, the complex extension

---

[2]**Due to different coordinate system, our x (or y) coordinate is the y (or x) coordinate in ref. 10**

[3]φ **is related to** ψ **by:** $\varphi = \frac{1}{2}(\pi - \psi)$

of x_0 and y_0 are (we omit the subscript 0 from now on):

$$d^2U/d\varphi^2 = -((\frac{\theta_0}{\pi})^2 + \frac{1-\lambda^2}{4}\frac{1}{sin^2\varphi})U \tag{29}$$

where U can be x or y.

From equation (29), we can see that both $d^2x/d\varphi^2$ and $d^2y/d\varphi^2$ and therefore d^2x/dy^2 are equal to zero when:

$$(\frac{\theta_0}{\pi})^2 + \frac{1-\lambda^2}{4}\frac{1}{\sin^2\varphi} = 0 \tag{30}$$

i.e.,

$$\varphi = \pm i\alpha, \quad \pi \pm i\alpha \quad with \ \sinh(\alpha) = \frac{\pi}{2\theta_0}\sqrt{1-\lambda^2} \tag{31}$$

It is easy to show that $dx/dy|_{\varphi=\pm i\alpha} = -(dx/dy|_{\varphi=\pi\pm i\alpha})^*$, so the branch points of the integral (25) are distributed symmetrically around the the imaginary axis. We also know from numerical calculation that $\varphi = -i\alpha, \pi - i\alpha$ correspond to the two branch points in the upper half plane.

It is convenient to calculate the integral of equation (27) in the φ plane. The location of the stationary points in φ-plane are $\varphi = \pi/2 \pm i\beta$, where β can be determined numerically. Now equation (25) becomes:

$$Im\int_{\pi/2-i\beta}^{-i\alpha}(1+i\frac{dx/d\varphi}{dy/d\varphi})^{3/4}(1-i\frac{dx/d\varphi}{dy/d\varphi})^{1/4}(x+iy)^{1/2}dy/d\varphi d\varphi = 0 \tag{32}$$

The results of selected λ, defined as λ_s, for different opening angles are shown in table I. The real finger width λ_{eff} is obtained according to the relation:

$$\lambda_{eff} = 1 - \frac{\pi}{2\theta_0}(1-\lambda) \tag{33}$$

The finger width determined by our calculation agrees with the experimental result within five percent of accuracy. Even more strikingly, the selected λ value stays almost as a constant in the range $30° \leq \theta_0 \leq 90°$, $\lambda \doteq .87$. So according to relation (33), λ_{eff} should have the behavior:

$$\lambda_{eff} \doteq 1 - 11.7°/\theta_0 \tag{34}$$

Experimentally the same behavior was observed in approximately the same range with the constant 10° instead of 11.7° in our expression.

We also list the value of real part of dx/dy at the branch point: $Re(\tilde{z})$ in table I. We show in the table that the branch points approach the imaginary axis as the opening angle decreases.

Another important issue of the problem is the linear stability of the selected solutions. Experimentally, it was found that at least for large opening angles the finger pattern is unstable against tip splitting after some time (or certain distance from the origin). Theoretically, in channel geometry, it was demonstrated numerically[7] and later analytically[8] using WKB method that the only stable solution is the one which has the

Table 1. Various quantities (defined in the text) vs opening angle

θ_0	λ_s	λ_{eff}	$Re(\tilde{z})$	$I(\lambda_s)$
0.175	0.952	0.572	0.175	0.028
0.248	0.937	0.602	0.214	0.031
0.322	0.923	0.624	0.248	0.052
0.395	0.911	0.646	0.280	0.072
0.468	0.901	0.668	0.319	0.092
0.542	0.892	0.686	0.374	0.113
0.615	0.884	0.705	0.418	0.135
0.689	0.879	0.724	0.459	0.155
0.762	0.874	0.741	0.496	0.194
0.836	0.870	0.756	0.543	0.211
0.909	0.869	0.773	0.577	0.246
0.983	0.867	0.787	0.619	0.283
1.056	0.863	0.796	0.656	0.327
1.130	0.861	0.807	0.694	0.367
1.203	0.859	0.816	0.747	0.394
1.277	0.856	0.822	0.794	0.444
1.350	0.856	0.832	0.838	0.488
1.424	0.856	0.842	0.885	0.533
1.497	0.858	0.851	0.934	0.581
1.571	0.861	0.861	0.980	0.647

the lowest width, and the nth solution will have (n-1) unstable modes. The analytical analysis by Bensimon, Pelce and Shraiman in ref. 8 was based on the local flatness of the interface compared with the capillary length and the use of the Mullins-Sekarka instability result. If we assume the structure of the solutions in sector geometry is similar to that of the channel geometry, this analysis would certainly be applicable to the sector geometry case. This would predict that the selected self-similar solution (the lowest branch) would be linearly stable, which is clearly in contradiction with the experiment.

In fact, numerical caculation[10] has shown that the structure of the selected width looks similar to that of the channel geometry for large surface tension; But when the surface tension becomes smaller, the structure of the solutions changes dramatically, the solutions dissapear in pairs. We are going to demonstrate below that this is a consequence of the branch cut distribution in the complex plane. As we showed before, the main contributions to the integral in (25) come from three parts, two from the path near the branch points $\varphi = -i\alpha$ *and* $\pi - i\alpha$ and one from the path around the stationary point $\varphi = -i\beta$. Neglecting all the slow varying functions of γ_0 at small γ_0, we can then write (25) approximately as :

$$\exp(i\hat{\delta}(\varphi = -i\alpha)/\sqrt{\gamma_0}) + \exp(i\hat{\delta}(\varphi = \pi - i\alpha)/\sqrt{\gamma_0}) + \exp(i\hat{\delta}(\varphi = \pi/2 - i\beta)/\sqrt{\gamma_0}) = 0 \tag{35}$$

Let's define:

$$\begin{aligned} R(\lambda) &= -Im(\hat{\delta}(\varphi = -i\alpha) - \hat{\delta}(\varphi = \pi/2 - i\beta)) \\ I(\lambda) &= Re(\hat{\delta}(\varphi = -i\alpha) - \hat{\delta}(\varphi = \pi/2 - i\beta)) \end{aligned}$$

Because $\hat{\delta}(\varphi = -i\alpha) - \hat{\delta}(\varphi = \pi/2 - i\beta) = (\hat{\delta}(\varphi = \pi - i\alpha) - \hat{\delta}(\varphi = \pi/2 - i\beta))^*$, we have:

$$1 + \exp(R(\lambda)/\sqrt{\gamma_0})\cos(I(\lambda)/\sqrt{\gamma_0}) = 0 \tag{36}$$

As shown in figure 2, solutions of the above equation exist in the region with $R(\lambda) > 0$. The limiting value of selected value of λ: λ_s as $\gamma_0 \to 0$ is determined by the equation $R(\lambda_s) = 0$. We check the behavior of $I(\lambda)$ around $\lambda = \lambda_s$ and we get numerically: $I(\lambda_s) > 0$ and $dI/d\lambda|_{\lambda=\lambda_s} > 0$ for all angles. Because of this behavior of $I(\lambda)$, it is easy to see that as γ_0 decreases, the lowest and the 2nd lowest solutions will approach λ_s, and eventually disappear in pair; as γ_0 decreases even more, the 3rd and the 4th lowest solutions will disappear in pair; and so on.

We also list the value of $I(\lambda_s)$ in table I. As we can see, $I(\lambda_s)$ goes to zero as the angle decreases to zero, which means that the solutions disappear at smaller γ_0 as we go to smaller angles. In fact, as is well known, the solutions never disappear at finite surface tension in the channel geometry.

Finally, we notice that for constant real surface tension γ_{phys}, γ_0 decrease with increasing time as $\gamma_0 = \gamma_{phys}t^{-1/2}$. So for real physical system where γ_{phys} is a constant, γ_0 will eventually decrease to the region where the only linearly stable solution is unavailable and the pattern will develop tip splitting instability. For smaller opening angles, it will take longer time for the tip splitting instability to occur.

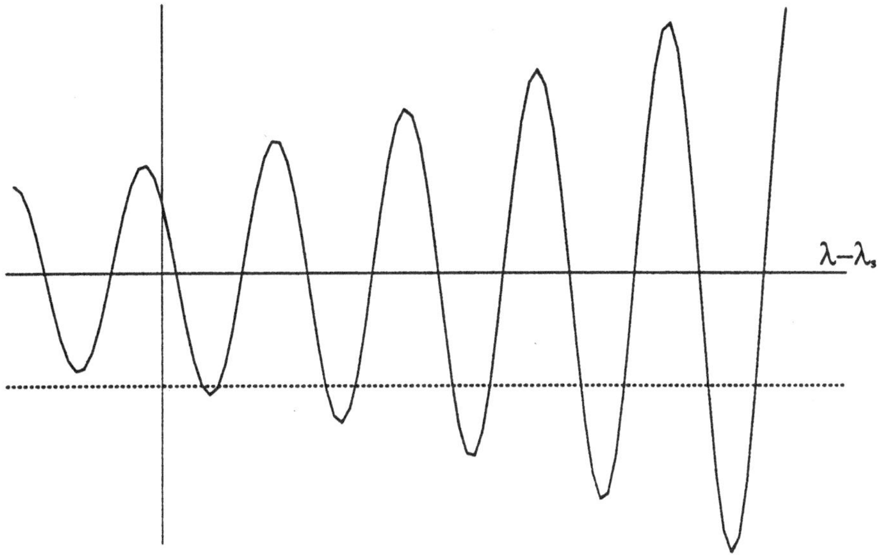

Figure 2. Illustration of the solutions of eq. (36).

IV. Summary

In the above sections, we have found an analytical method to find the self-similar solutions at zero surface tension. The solutions form a two parameter family, exactly analog to the two parameter family solutions found in channel geometry. In one special case $\theta_0 = \pi/2$, we get a closed form of the solution, and Hakim's solutions are just the symmetric subset of our general solutions. We extend the WKB calculation of BKLR of the 90° geometry to arbitrary opening angles, and an excellent agreement with the experiment is reached in the whole range of opening angles. We also demonstrated that the selected solutions will disappear in pairs due to the structure of the branch points and stationary point in the complex plane. Based on this result and the fact that the effective surface tension decreases with increasing time, we are able to explain the instability found in the experiment.

Acknowledgment

The author would like to thank Prof. H. Levine and W. Rappel for a lot of helpful discussions. This work was supported in part by the U. S. Defense Advanced Projects Administration under the University Research Initiative, Grant no. N00014-86-K-0758.

Note added: After finishing this work, we learnt that analytical self-similar solutions were also obtained for arbitary angle by M. Benamar in different coordinate system(see the previous paper), her solutions correspond to the symmetric case in our derivation where $\lambda_1 = \lambda_2$.

References

1. For a general review of interfacial pattern formation, see D. Kessler, J. Koplik and H. Levine, *Adv. in Phys.* **37**, 255(1988); J. S. Langer in *Chance and Matter*, J. Souletie ed., North-Holland (1987).

2. For more details about the Saffman-Taylor problem, see D. Bensimon, L. Kadanoff, S. Liang, B. I. Shraiman and L. Tang, *Reviews of Modern Physics* **58**, 977(1986); G. M. Homsy *Ann. Review of Fluid Mech.* **19**, 271(1987).

3. H. Thome, M. Rabaud, V. Hakim and Y. Couder, *Phys. Fluids* **A1**, 224(1989).

4. E. Berner, D. Kessler, H. Levine and W. Rappel, *Europhys. Lett.*, **13** (2), 161(1990).

5. J. M. Mclean and P. G. Saffman, *J. Fluid Mech.* **102**, 455(1981); L. Romero, Ph.D thesis, California Institute of Technology(1982); J. M. Vanden-Broeck, *Phys. Fluids* **26**, 8(1983).

6. B. I. Shraiman, *phys. Rev. Lett.* **56**, 2028(1986); D. Hong and J. S. Langer, *Phys. Rev. Lett.* **56**, 2032(1986); R. Combescot, T. Dombre, V. Hakim, Y. Pomeau and A. Pumir, *Phys. Rev. Lett.* **56**, 2036(1986) and *Phys. Rev.* **A37**, 1270(1988); S. Tanveer, *Phys. Fluids* 30, 1589(1987).

7. D. Kessler and H. Levine, *Phys. Rev.* **A33**, 2621, 2634(1986); D. Bensimon, *Phys. Rev.* **A33**, 1302(1986); D. Kessler and H. Levine, *Phys. Fluids* **30**, 1246(1987).

8. D. Bensimon, P. Pelce and B. I. Shraiman, *J. Phys. Fluids* **48**, 2081(1987); S. Tanveer *Phys. Fluids* **30**, 2318(1987).

9. The non-symmetric patterns can be selected when there exists asymmetric forcing in the problem. For some detail, see, "Non-symmetric Saffman-Taylor fingers", by E. Berner, H. Levine and Y. Tu, to appear in *J. Fluid Dynamics.*

10. The non-existence of the lowest branch of solutions at small surface tension was first observed in numerical calculation, M. Ben-Amar in the proceedings of NATO-Asi, *Pattern and Growth*, Cargese, 1990 (to appear); a qualitative explanation was given by V. Hakim, which is essentially the same as the one we discuss here.

EXPONENTIALLY SMALL ESTIMATES FOR SEPARATRIX SPLITTINGS

Jürgen SCHEURLE

Angewandte Mathematik
Universität Hamburg
Bundesstrasse 55
2000 Hamburg 13, Germany

Jerrold E. MARSDEN[1]

Department of Mathematics
University of California,
Berkeley, CA 94720 USA

Philip HOLMES

Departments of Mathematics and
Theoretical and Applied Mechanics
Cornell University
Ithaca, N.Y. 14853 USA

ABSTRACT This paper reviews our previous estimates and gives an example exhibiting a new phenomenon. In problems involving asymptotics beyond all orders in a perturbation parameter ε, it is a common *assumption* that the quantity being studied (such as a separatrix splitting distance or angle, a solitary wave mismatch, etc.) can be "estimated" by an expression of the form $a\varepsilon^b e^{-c/\varepsilon}$ as $\varepsilon \to 0$. Here, a, b and c are constants (where b can be negative and c is "sharp", often the distance from the real axis to a pole in the complex plane). The main purpose of our example is to show that this assumption can be wrong. The example, which concerns the splitting of separatrices in a rapidly forced system with a heteroclinic orbit shows that the even the estimate from above (using the sharp value of c) is incorrect. We argue that this situation is not isolated or particular, but happens rather generally. We especially note that in situations involving asymptotics beyond all orders, when an estimate of the form $a\varepsilon^b e^{-c/\varepsilon}$ is assumed, it needs to be justified.

[1]Research partially supported by NSF grant DMS 89-22704 and a Humboldt award during a visit to the Universität Hamburg

1 INTRODUCTION

In HOLMES , MARSDEN, and SCHEURLE [1988] and SCHEURLE [1989], both upper and lower exponentially small estimates for separatrix splitting in rapidly forced planar systems were obtained. Although the results are rather general for analytic reversible planar systems, we recall them for the pendulum:

UPPER ESTIMATE *Consider*

$$\ddot{\varphi} + \sin\varphi = \delta \sin(t/\varepsilon). \tag{1.1}$$

For any $\eta > 0$ *there is a* $\delta_0 > 0$ *and a constant* $C(\eta, \delta_0)$ *such that for all* ε *and* δ *satisfying* $0 < \varepsilon \leq 1, 0 < \delta \leq \delta_0$,*we have*

$$|\text{separatrix splitting}| \leq C(\eta, \delta_0) e^{-(\frac{\pi}{2} - \eta)/\varepsilon} \tag{1.2}$$

LOWER ESTIMATE AND SHARP UPPER ESTIMATE *Consider*

$$\ddot{\varphi} + \sin\varphi = \varepsilon^p \delta \sin(t/\varepsilon) \tag{1.3}$$

and assume that $p > 8$. *Then there is a* $\delta_0 > 0$ *and constants* C_1 *and* C_2 *such that for all* ε, δ *satisfying* $0 < \varepsilon \leq 1, 0 < \delta \leq \delta_0$, *we have*

$$C_2 \varepsilon^p \delta e^{-\pi/2\varepsilon} \leq |\text{separatrix splitting}| \leq C_1 \varepsilon^p \delta e^{-\pi/2\varepsilon} \tag{1.4}$$

Remarks

1. Estimates similar to our upper estimates were obtained by NIESHTADT [1984]. The coefficient in the exponent for our estimate, namely $\pi/2$, is the distance from the real axis to the closest pole of the homoclinic orbit in the complex t-plane. For NIESHTADT, the exponent is related to the width of a strip in the complex plane on which the angle variable in action-angle coordinates is analytic. The exact relation between these two approaches would be interesting to explore further. NIESHTADT also makes the interesting remark that a variant of KAM theory can be used to bound the whole stochastic layer between exponentially close KAM curves.

2. There are similar upper estimates for *mappings* in FONTICH and SIMO [1990].

3. The upper estimate (1.2) shows that the splitting distance is *beyond all orders in* ε (without any assumption on p, as in the second result). An analyticity argument shows that in (1.1), splitting *does occur* (with at most discrete exceptions) as $\varepsilon \to 0$, but it does not provide an estimate.

4. There are similar estimates for the splitting angle.

5. Based on the example below, we conjecture that for situations of most interest in KAM theory, perhaps even for $\ddot{\varphi} + \sin\varphi = \delta\varepsilon \sin(t/\varepsilon)$, there *is no estimate of the form*

$$C_2 \varepsilon^b \delta e^{-c/\varepsilon} \leq |\text{splitting distance}| \leq C_1 \varepsilon^b \delta e^{-c/\varepsilon}$$

with a sharp constant c/ε (like $\pi/2$ for the pendulum) for any constant b (positive or negative), for all small δ and ε. In fact, in problems like this, it would seem that *one cannot avoid essential singularities that develop in the estimate.* We will see this explicitly in the example.

6. On the other hand, what we *do* believe occurs in examples like the $p = 1$ pendulum and related maps like the ones in FONTICH and SIMO [1990], is a convergent expression of the form

$$|\text{splitting distance}| = \left(\cdots + a_2\varepsilon^2 + a_1\varepsilon + a_0 + \frac{a_{-1}}{\varepsilon} + \frac{a_{-2}}{\varepsilon^2} + \cdots\right) e^{-c/\varepsilon}$$

for some constants a_i. Which power of ε to put in front of $e^{-c/\varepsilon}$ depends on how small ε is. Papers giving numerical evidence for a *fixed* power of ε as $\varepsilon \to 0$ need to confront the example in this paper in which there is *no upper estimate of the form* $a\varepsilon^b e^{-\pi/2\varepsilon}$ as $\varepsilon \to 0$. However, it *suggests* that $a\varepsilon^b e^{-\pi/2\varepsilon}$ may give a good *approximation*, but that b may need to be adjusted as $\varepsilon \to 0$. Despite this extreme delicacy with the sharp exponent $-\pi/2\varepsilon$, the upper estimate, with $\pi/2$ replaced by a slightly smaller exponent, remains valid.

2 THE EXAMPLE

We consider the following family of planar systems.

$$\begin{aligned} \dot{x} &= 1 - x^2 \\ \dot{y} &= [2x - (\alpha + 2\beta x)(1 - x^2)]y + \delta\cos(t/\varepsilon) \end{aligned} \tag{2.1}$$

where $\alpha, \beta, \delta, \varepsilon$ are constants. For $\delta = 0$, this system has the heteroclinic orbit

$$\Gamma: \quad x = \tanh t, \quad y = 0$$

joining $(-1, 0)$ to $(1, 0)$. We are interested in the splitting of this orbit for small $\alpha, \beta, \delta, \varepsilon$ and $\beta, \delta, \varepsilon$ non-zero. Before proceeding with the example, we make a series of remarks.

Remarks

1. The system (2.1) is chosen so the *variables separate* (the first equation is independent of y). This enables one to perform explicit and *exact* calculations, but does not seem to be essential for the phenomenon we want to illustrate.

2. Many systems of interest in the splitting of separatrices are Hamiltonian (see HOLMES , MARSDEN, and SCHEURLE [1988] for instance), so we make some remarks on this structure. First, consider the Hamiltonian

$$H(q,p) = \frac{1}{2m}p^2 + V(q) + pf(q) \tag{2.2}$$

where V is a potential and $pf(q)$ is a *gyroscopic term*. The variables separate in the limit $m \to \infty$ and Hamilton's equations become

$$\begin{aligned} \dot{q} &= f(q) \\ \dot{p} &= -V'(q) - pf'(q) \end{aligned} \tag{2.3}$$

For $f(q) = 1 - q^2$, (2.3) has a heteroclinic orbit joining $(-1, -\frac{1}{2}V'(-1))$ to $(1, \frac{1}{2}V'(1))$. With the addition of $\delta\cos(t/\varepsilon)$ to the p-equation, one can readily compute the splitting and one finds that it is given *exactly* by

$$\text{splitting} = \int_{-\infty}^{\infty} \frac{\delta e^{it/\varepsilon}}{\cosh^2 t}\, dt = \frac{\delta\pi}{\varepsilon\sinh(\pi/2\varepsilon)} \tag{2.4}$$

and so the system (2.3) is too simple to illustrate what we want. That is, one *does* have, for this especially simple situation, a valid prefactor ε^b. Note that if $V = 0$ and $f(q) = 1 - q^2$, then (2.3) reduces to (2.1) with $\alpha = 0, \beta = 0$. However, (2.1) exhibits the behavior we want to illustrate only for α or β non-zero. If in (2.2), m is finite, it seems difficult to calculate the splitting directly; however, we suspect the same phenomena can happen as in the nearby system (2.1)—that is, we suspect that there is no prefactor ε^b one can use (along with the sharp exponential factor $e^{-\pi/2}$) to get an estimate.

3. The equations (2.1) are Hamiltonian when $\delta = 0$, with Hamiltonian function

$$H = e^{\alpha x + \beta x^2}[(1 - x^2)y]$$

and symplectic form

$$\Omega = e^{\alpha x + \beta x^2} dx \wedge dy$$

as is readily checked. For $\delta \neq 0$ the equations are still Hamiltonian in the non-autonomous sense.

4. A formal calculation of the Melnikov function for (2.1) gives the function

$$M_\varepsilon(t_0) = \delta \int_{-\infty}^{\infty} (1 - x^2) \cos((t + t_0)/\varepsilon) dt = \frac{\delta \pi \cos(t_0/\varepsilon)}{\varepsilon \sinh(\pi/2\varepsilon)}, \tag{2.5}$$

whose magnitude coincides with the exact result (2.4). This occurs because the evolution equation along the heteroclinic orbit for $\alpha = \beta = 0$ is linear, so that the iteration procedure of HOLMES, MARSDEN, and SCHEURLE [1988] terminates after the first term. Note, however, that the formal result (2.5) is obtained for *all* α and β. As we shall see, the ε dependence of this "leading" term is in general incorrect for $\beta = \varepsilon^p$.

5. Essentially the same equation (2.1) with $\alpha = \beta = 0$ with forcing, but not rapid forcing, was presented by BOUNTIS, PAPAGEORGIOU, and BIER [1987] as an example which is integrable in the sense that it is separable, so can be explicitly integrated, but which nonetheless exhibits separatrix splitting.

6. A key point that leads to the invalidity of the assumption of a prefactor of the form ε^b for (2.1) is the *essential singularity* in the resulting formula for the splitting distance in (3.7) below. It should be noted that we did not put in this essential singularity by hand–it arises naturally even though there are no obvious essential singularities in the given problem. From the proofs of the splitting estimates, one sees that one should expect this most of the time, even in simple problems.

3 THE SPLITTING DISTANCE FORMULA

An interesting feature of the equations (2.1) is that there is a relatively explicit formula for the exact splitting distance. From $\dot{x} = 1 - x^2$, we find that the first component of the stable and unstable manifolds near Γ are given by $x = \tanh t$. Substituting in the second equation, we get

$$\begin{aligned} \dot{y} &= [2x - (\alpha + 2\beta x)(1 - x^2)]y + \delta \cos(t/\varepsilon) \\ &= [2\tanh t - (\alpha + 2\beta \tanh t)(\operatorname{sech}^2 t))]y + \delta \cos(t/\varepsilon) \end{aligned} \tag{3.1}$$

Let $u = y/\cosh^2 t$ so that

$$\dot{u} = \frac{\dot{y}}{\cosh^2 t} - \frac{2y \sinh t}{\cosh^3 t} = \frac{\dot{y} - 2y \tanh t}{\cosh^2 t}$$

i.e.,

$$\dot{u} = -\frac{(\alpha + 2\beta\tanh t)}{\cosh^2 t}u + \frac{\delta\cos(t/\varepsilon)}{\cosh^2 t} \tag{3.2}$$

Since an indefinite integral of $\alpha\mathrm{sech}^2 t + 2\beta\mathrm{sech}^2 t\tanh t$ is $\alpha\tanh t - \beta\mathrm{sech}^2 t$, we get

$$u(t) = e^{-\alpha\tanh t + \beta\mathrm{sech}^2 t}\left\{e^{-\beta}u(0) + \int_0^t e^{\alpha\tanh s - \beta\mathrm{sech}^2 s}\frac{\delta\cos(s/\varepsilon)}{\cosh^2 s}ds\right\}$$

i.e.,

$$y(t) = \cosh^2 t\, e^{-\alpha\tanh t + \beta\mathrm{sech}^2 t}\left\{e^{-\beta}y(0) + \int_0^t e^{\alpha\tanh s - \beta\mathrm{sech}^2 s}\frac{\delta\cos(s/\varepsilon)}{\cosh^2 s}ds\right\} \tag{3.3}$$

The unstable manifold of the periodic point near $(-1,0)$ starting at $(0,y(0))$ at $t=0$ is characterized by choosing $y(0) = y_u(0)$ so that $y(t)$ is bounded as $t \to -\infty$. Then (3.3) gives

$$y_u(0) = \delta e^{\beta}\int_{-\infty}^{0} e^{\alpha\tanh s - \beta\mathrm{sech}^2 s}\frac{\cos(s/\varepsilon)}{\cosh^2 s}ds \tag{3.4}$$

Subtracting an analogous formula for $y_s(0)$ gives

$$y_u(0) - y_s(0) = \delta e^{\beta}\int_{-\infty}^{\infty} e^{\alpha\tanh s - \beta\mathrm{sech}^2 s}\frac{\cos(s/\varepsilon)}{\cosh^2 s}ds \tag{3.5}$$

The formula with a starting time t_0 and position $(0, y(t_0))$ similarly gives

$$y_u(t_0) - y_s(t_0) = \delta e^{\beta}\int_{-\infty}^{\infty} e^{\alpha\tanh s - \beta\mathrm{sech}^2 s}\frac{\cos(s+t_0)/\varepsilon}{\cosh^2 s}ds \tag{3.6}$$

Thus, the *splitting distance* is the absolute value

$$d = \left|\delta e^{\beta}\int_{-\infty}^{\infty} e^{\alpha\tanh s - \beta\mathrm{sech}^2 s}\frac{e^{is/\varepsilon}}{\cosh^2 s}ds\right| \tag{3.7}$$

The integrand has an essential singularity at $s = i\pi/2$. If we shift the contour to $s = i\pi + w$, we get

$$\delta e^{\beta}\left|\int_{i\pi-\infty}^{i\pi+\infty} e^{\alpha\tanh w - \beta\mathrm{sech}^2 w}\frac{e^{iw/\varepsilon}}{\cosh^2 w}dw\right| e^{-\pi/\varepsilon}$$

Thus,

$$d = \left|\frac{\delta e^{\beta}}{1 - e^{-\pi/\varepsilon}}\oint_{C'} e^{\alpha\tanh w - \beta\mathrm{sech}^2 w}\frac{e^{iw/\varepsilon}}{\cosh^2 w}dw\right|$$

where the integration contour C' encloses the point $i\pi/2$. Now make the change of variables $z = w - i\pi/2$ and use $\cosh(z + i\pi/2) = i\sinh z$ to get

$$d = \left|\frac{\delta e^{\beta}}{1 - e^{-\pi/\varepsilon}}\oint e^{\alpha\coth z + \beta/\sinh^2 z}\frac{e^{iz/\varepsilon}}{\sinh^2 z}dz\right| e^{-\pi/2\varepsilon} \tag{3.8}$$

where the integration contour encloses the origin (and none of the other singularities at $(2n+1)\pi i/2$). Formulas (3.7) and (3.8) are the splitting distance formulas we shall work with.

The main problem we now pose is this: If $\beta = -\varepsilon$ and $\alpha = 0$, can one estimate (3.8) above by an expression of the form $a\varepsilon^b e^{-\pi/2\varepsilon}$ for constants a and b (not depending on ε) as $\varepsilon \to 0$? As we shall see in the next section, the answer is NO.

Remarks

1. The main difficulty with (3.8) is the presence of the essential singularity of the function

$$g(z) = e^{\alpha \coth z + \beta/\sinh^2 z} \frac{e^{iz/\varepsilon}}{\sinh^2 z} \tag{3.9}$$

at $z = 0$. One can ask if essential singularities occur typically or are a peculiarity of this example. For the pendulum example, one sees from SCHEURLE [1989] that there is an uncontrollable accumulation of poles as the iteration procedure is carried out. This accumulation is the counterpart of the essential singularity of g, so one can expect a similar consequence. A similar phenomenon seems to occur in the bifurcation example in HOLMES, MARSDEN and SCHEURLE [1988].

2. If $\beta = C\varepsilon^p$ where $p \geq 3$, then the essential singularity can be controlled, as in the pendulum case, and in this case the upper and lower estimates are valid.

3. Equations (2.1) can be modified to

$$\begin{aligned} \dot{x} &= 1 - x^2 \\ \dot{y} &= [2x - (\alpha + 2\beta x)(1 - x^2)]y - \mu[2x - (\alpha + 2\beta x)(1 - x^2)\sin(t/\varepsilon) \\ &\quad + \left(\delta + \frac{\mu}{\varepsilon}\right)\cos(t/\varepsilon), \end{aligned} \tag{3.10}$$

which may be regarded as a perturbation of (2.1) with the new parameter μ. However, for $\delta = 0$ and all α, β and μ, this system is integrable, since it is transformed back to the autonomous Hamiltonian system (2.1) with $\delta = 0$ under the change of variables

$$x = \tilde{x}, \qquad y = \tilde{y} + \mu \sin(t/\varepsilon),$$

and so the splitting is zero. In this case, there is still an essential singularity in the splitting distance formula, but there is in fact no splitting. Examples like this show how delicate the splitting condition can be!

4 EXPONENTIALLY SMALL ESTIMATES

We now turn to estimates on d given by (3.7) and (3.8). We shall focus on the region in parameter space where $\alpha = 0, \beta = -\varepsilon$ and $\varepsilon > 0$. However, there are many possible ways of expanding the integral, so we leave β independent of ε for the moment.

Let

$$D = \int_{-\infty}^{\infty} e^{-\beta/\cosh^2 s} \frac{e^{is\varepsilon}}{\cosh^2 s} ds \tag{4.1}$$

As in (3.8), we get

$$D = \frac{-1}{2\sinh(\pi/2\varepsilon)} \oint e^{\beta/\sinh^2 z} \frac{e^{iz/\varepsilon}}{\sinh^2 z} dz \tag{4.2}$$

Thus, with $\alpha = 0$, we get $d = \delta e^{\beta} D$. Now we expand the first exponential in (4.1), giving

$$D = \sum_{n=0}^{\infty} \int_{-\infty}^{\infty} \frac{(-\beta)^n}{n!} \frac{e^{is/\varepsilon}}{\cosh^{2n+2} s} ds \tag{4.3}$$

Observe that the (inverse) Fourier transform of $1/\cosh^2 s$ gives

$$\int_{-\infty}^{\infty} \frac{e^{is\eta}}{\cosh^2 s} ds = \frac{\eta}{\sinh(\pi\eta/2)} \geq 0 \tag{4.4}$$

With $\eta = 1/\varepsilon$ in (4.4), we see that each term in the expansion (4.3) is a multiple convolution of the *non-negative* function $\eta/\sinh(\pi\eta/2)$ and therefore it is clear that these terms are postive. Therefore we get an inequality if we discard the tail of the series. Thus, we have proved the following:

Lemma 1 *If $\beta < 0$ and N is a positive integer, then*

$$D \geq \sum_{n=0}^{N} \frac{(-\beta)^n}{n!} \int_{-\infty}^{\infty} \frac{e^{is/\varepsilon}}{\cosh^{2n+2} s} ds \tag{4.5}$$

Next, rewrite (4.5), as we did in (4.2), to give

$$D \geq \frac{1}{2\sinh(\pi/2\varepsilon)} \sum_{n=0}^{N} \frac{(-\beta)^n}{n! i^{2n+2}} \oint \frac{e^{iz/\varepsilon}}{\sinh^{2n+2} z} dz. \tag{4.6}$$

Making the change of variables $\zeta = z/\varepsilon$, we get

$$\begin{aligned} D &\geq \frac{-\varepsilon}{2\sinh(\pi/2\varepsilon)} \sum_{n=0}^{N} \frac{\beta^n}{n!} \oint \frac{e^{i\zeta}}{\sinh^{2n+2}(\varepsilon\zeta)} d\zeta \\ &= \frac{-\varepsilon}{2\sinh(\pi/2\varepsilon)} \sum_{n=0}^{N} \frac{\beta^n}{n!} \oint \frac{1}{(\varepsilon\zeta)^{2n+2}} h_n(\varepsilon\zeta) e^{i\zeta} d\zeta \end{aligned} \tag{4.7}$$

where $h_n(w) = w^{2n+2}/\sinh^{2n+2} w$, which is analytic near zero. Therefore, by Cauchy's theorem for the derivatives, we get

$$D \geq \frac{-1}{2\sinh(\pi/2\varepsilon)} \sum_{n=0}^{N} \frac{\beta^n}{n!} \frac{2\pi i}{(2n+1)!} \frac{1}{\varepsilon^{2n+1}} \frac{d^{2n+1}}{d\zeta^{2n+1}} [h_n(\varepsilon\zeta) e^{i\zeta}]_{|\zeta=0}. \tag{4.8}$$

The power series expansion of $h_n(w)$ has the form

$$h_n(w) = 1 + a_{n,2} w^2 + a_{n,4} w^4 + \cdots$$

for constants $a_{n,m}$, and so

$$h_n(\varepsilon\zeta) e^{i\zeta} = \left(1 + a_{n,2}\varepsilon^2\zeta^2 + a_{n,4}\varepsilon^4\zeta^4 + \cdots\right)\left(1 + i\zeta + \frac{(i\zeta)^2}{2!} + \cdots\right)$$

The coefficient of ζ^{2n+1} has the form

$$\frac{i}{(2n+1)!}\left[(-1)^n + \varepsilon^2 b_{n,2} + \varepsilon^4 b_{n,4} + \cdots + \varepsilon^{2n} b_{n,2n}\right]$$

for real constants $b_{n,2}, \ldots, b_{n,2n}$. Thus, (4.8) gives

$$D \geq \frac{-\pi}{\sinh(\pi/2\varepsilon)} \sum_{n=0}^{N} \frac{\beta^n}{n!} \frac{-1}{(2n+1)!\varepsilon^{2n+1}} \left[(-1)^n + \varepsilon^2 b_{n,2} + \cdots + \varepsilon^{2n} b_{n,2n}\right] \tag{4.9}$$

For ε small, and since this is a *finite* sum, the expression (4.9) is dominated by the term from $n = N$; i.e., we have proved that

Lemma 2 *For ε sufficiently small and $\beta < 0$, we have*

$$D \geq \frac{1}{2} \frac{\pi}{\sinh(\pi/2\varepsilon)} \frac{|\beta|^N}{N!(2N+1)!} \frac{1}{\varepsilon^{2N+1}} \tag{4.10}$$

If $|\beta| = \varepsilon^p$, then as in our main lower estimate, for $p \geq 3$, this lower estimate is consistent with a splitting of the form $a\varepsilon^b e^{-\pi/2\varepsilon}$. However, with $\beta = -\varepsilon$, we get

$$D \geq \frac{\pi}{2\sinh(\pi/2\varepsilon)} \frac{1}{N!(2N+1)!} \frac{1}{\varepsilon^{N+1}} \geq \frac{C_N}{\varepsilon^{N+1}} e^{-\pi/2\varepsilon} \tag{4.11}$$

Putting this all together, we have established our main result:

Theorem 1 *For any integer N, there are constants $c_N > 0$ and $\varepsilon_N > 0$ such that for $\alpha = 0, \beta = -\varepsilon$ and $\delta = \varepsilon$, we have*

$$d \geq \frac{c_N}{\varepsilon^N} e^{-\pi/2\varepsilon} \tag{4.12}$$

for all $0 < \varepsilon < \varepsilon_N$

Remarks

1. This result shows, that *no sharp upper estimate of the form $a\varepsilon^b e^{-\pi/2\varepsilon}$ can exist.* We emphasize that our "rough" upper estimate still applies, giving $d \leq C_\eta e^{-(\frac{\pi}{2}-\eta)/\varepsilon}$ uniformly as $\varepsilon \to 0$ for any $\eta > 0$, which is possible, despite (4.12).

2. The above calculations show that with $\beta = -\varepsilon, \delta = \varepsilon, \alpha = 0$ and $\varepsilon > 0$, the splitting distance has the form

$$d = \left(\cdots + a_2\varepsilon^2 + a_1\varepsilon + a_0 + \frac{a_{-1}}{\varepsilon} + \frac{a_{-2}}{\varepsilon^2} + \cdots\right) e^{-\pi/2\varepsilon} \tag{4.13}$$

where the Laurent series is *convergent* and infinitely many of $a_{-1}, a_{-2} \ldots$ are non-zero. In other words,

$$d = \varphi(\varepsilon) e^{-\pi/2\varepsilon} \tag{4.14}$$

where φ has an *essential singularity* in ε at $\varepsilon = 0$. Of course in particular examples, like this one, one can compute, in principle, the coefficients a_k. We suspect the asymptotic form (4.13), (4.14) is valid rather generally and that φ will often have an essential singularity. Moreover, from (4.11) it is reasonable to suspect that the coefficients a_k get small quickly for large $|k|$, so for ε not *too* small, a truncation of (4.13) may yield a useful numerical approximation.

CONCLUSIONS

In this paper we have given an explicit example of a system in the plane of the form $\dot{u} = f_0(u) + \varepsilon f(u, t/\varepsilon)$, where $\dot{u} = f_0(u)$ has a heteroclinic connection, with pole at $i\pi/2$ in the complex t-plane, and there is *no* upper estimate of the form $a\varepsilon^b e^{-\pi/2\varepsilon}$ for *any* constants a, b uniformly as $\varepsilon \to 0$ for the separatrix splitting.

This example illustrates phenomena that we believe are generic, and not isolated. In particular, it shows that the assumption that the splitting is obtained by (*i.e.*, estimated above and below by) $a\varepsilon^b e^{-c/\varepsilon}$, where a, b and c are constants and the latter is sharp, is not, in general, correct; in fact, not even the upper estimate is correct. (Note that for reliable numerical work, one would ideally like error bounds corresponding to an estimate both above and below). Rather, it seems to us that the correct splitting is given by a prefactor that is an infinite expansion in powers of ε, both positive and negative. Certain ranges of powers may be useful for numerically estimating the splitting over corresponding ranges of ε, and this would be of interest to explore further.

The source of the basic difficulty that the example illustrates is the essential singularity, which builds up in the iteration process used to give the exact splitting distance; this essential singularity will typically be present in examples, even though the original problem may have only poles in the complex t plane and there is no obvious essential singularity in the given data. In this example, one is able to see the essential singularity explicitly in the exact formula (3.7). The essential singularity corresponds to one that the proof suggests will build up through a successive iteration process in most examples. There may be particular mechanisms that can be used to control the essential singularity, and this is the purpose of the powers ε^p in equation (1.3). This type of phenomenon also holds in our main example (2.1), with $\beta = C\varepsilon^p$, where the power of p needed is given by inspection in this case and in general by a proof analysis of what is required to control the growth of the pole order in the iteration proccess. Without such special assumptions, one should expect the type of behaviour in our example.

Acknowledgements-We thank Martin Kummer, Jim Ellison, Harvey Segur, and Saleh Tanveer for several helpful suggestions.

REFERENCES

T. Bountis, V. Papageorgiou, and M. Bier [1987] On the singularity analysis of intersecting separatrices in near-integrable dynamical systems *Physica D* **24**, 292–304

P. Holmes, J. Marsden and J. Scheurle [1988] Exponentially small splittings of separatrices with applications to KAM theory and degenerate bifurcations, *Cont. Math., AMS* **81**, 213–244

E. Fontich and C. Simo [1990] The splitting of separatrices for analytic diffeomorphisms, *Ergodic Th. and Dyn. Sys.* **10**, 295–318

A. I. Nieshtadt [1984] The separation of motions in systems with rapidly rotating phase, *PMM USSR* **48**, 133–139

J. Scheurle [1989] Chaos in a rapidly forced pendulum equation, *Cont. Math. AMS* **97**, 411–419

EXPONENTIALLY SMALL PHENOMENA IN THE RAPIDLY FORCED PENDULUM

Martin Kummer*, James A. Ellison** and A.W. Sáenz***

*Department of Mathematics, University of Toledo, Ohio 43606 USA
**Department of Mathematics, University of New Mexico, Albuquerque, New Mexico 87131 USA
***Naval Research Laboratory and Department of Physics, Catholic University Washington, D. C. 20064 USA

Abstract

The rapidly forced pendulum equation with forcing $\delta \sin \frac{t}{\varepsilon}$, where $\delta = \delta_0 \varepsilon^p$, p = 5, for δ_0, ε sufficiently small, is considered. We sketch our proof that stable and unstable manifolds split and that the splitting distance $d(t_0)$ in the $\dot{x}$ - t plane satisfies

$$d(t_0) = 2\pi \delta \sin \frac{1}{\varepsilon} t_0 \operatorname{sech} \frac{1}{2\varepsilon} \pi + O(\delta_0 \delta \exp(-\pi/2\varepsilon)),$$

and the angle of transversal intersection, ψ, in the t = 0 section satisfies

$$\tan \frac{1}{2} \psi = \frac{\delta}{2\varepsilon} \pi \operatorname{sech} \frac{1}{2\varepsilon} \pi + O(\delta_0 \frac{\delta}{\varepsilon} \exp(-\pi/2\varepsilon)).$$

It follows that the Melnikov term correctly predicts the exponentially small splitting and angle of transversality. Our method reduces the previous result of Holmes, Marsden and Scheuerle from p = 8 to p = 5. Our proof is elementary, self - contained including a stable manifold theorem and places an emphasis on the phase space geometry.

I. Introduction

We consider the rapidly forced pendulum equation

$$\ddot{x} = -\sin x + \delta \sin \frac{t}{\varepsilon}, \tag{1}$$

where $\delta = \delta_0 \varepsilon^p$, p = 5. For $\delta = 0$, Eq.(1) has a homoclinic orbit, connecting the equilibrium

**) Supported by NSF Grant DMR - 8704348.

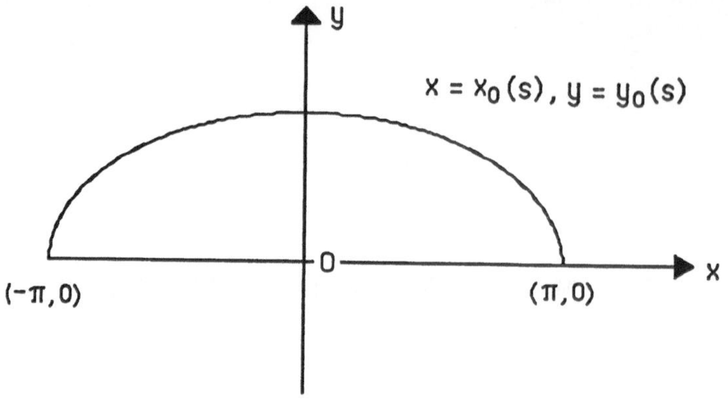

Fig. 1

points ($\pm\pi$,0) in the x - y phase plane ($y = \dot{x}$), given by

$$x_0(s) = 2\tan^{-1}(\sinh s),\ \ y_0(s) = x_0'(s) = 2\ \mathrm{sech}\ s, \tag{2}$$

$$s = t - t_0, \qquad -\infty < s < \infty ,$$

as shown in Fig. 1. For $\delta > 0$, the equilibrium points are replaced by nearby hyperbolic periodic orbits, $x = x_p(t)$. We sketch our proof that for δ_0 sufficiently small the stable and unstable manifolds of these periodic orbits split and become transversal with exponentially small angle. Detailed proofs will appear elsewhere [(1)]. Our work was inspired by the work of Holmes, Marsden and Scheuerle [(2-3)] and can be considered an outgrowth of their study. The workshop contribution of Marsden [(4)] and Ref.2 put this problem in historical perspective, discuss its contemporary significance and emphasizes that Eq.(1) with p = 1 is a model problem for separatrix splitting in certain Hamiltonian systems. In fact, our initial interest in this problem arose from a two - degree - of - freedom Hamiltonian system with quartic symmetry, occuring in the theory of particle channeling in crystals. This system [(5)] has a transition from order to chaos similar to the transition in the Henon - Heiles system, and appears to have an exponentially small splitting. Scheuerle,[(6)] in his workshop contribution, discusses the case $p \geq 8$ of Refs.[2-3] and in addition presents a model problem which is easier to analyse. Our contribution extends the $p \geq 8$ results to the $p \geq 5$ case with an emphasis on the geometry and an elementary self - contained proof. Informal discussions at the workshop included rumors of separatrix splitting for lower values of p.

In Section II, we discuss the geometry and formulate the problem. In Section III, we do a two - time embedding of (1), extend to the complex plane and derive an integral equation for the stable manifold. Section IV contains our stable manifold theorem and finally, in Section V, we outline our proof of the exponential smallness of the splitting distance and angle.

II. Geometry and Formulation of the Problem

We parametrize the stable manifold of $x_p(t)$ by solutions $x(t,t_0)$ of (1) (close to $x_0(t - t_0)$) satisfying $x(t_0,t_0) = 0$. With the abbreviation $y(t,t_0) := D_1x(t,t_0)$, (where generally D_i denotes the derivative of a function w. r. to the i - th argument), the requirement that $(x(t,t_0), y(t,t_0), t)$ lies on the stable manifold is expressed by the condition that its ω - limit set is the periodic solution $x_p(t)$ near $(\pi,0)$. The time reflection principle, namely the observation:

"*If* $x(t)$ *is a solution of* (1) *so is* $x^*(t) := -x(-t)$"

enables us to cut labor in half since it relates stable and unstable manifold in a simple way. It follows that $(x_u(t,t_0),y_u(t,t_0)) = (-x(-t,-t_0),y(-t,-t_0))$ are solutions characterizing the unstable manifold. Thus, the distance between the two manifolds in the y - t plane at $t = t_0$ (splitting distance) is given by the formula

$$d(t_0) := y(-t_0,-t_0) - y(t_0,t_0). \tag{3}$$

Note that $d(0) = 0$, so that the stable and unstable manifolds intersect at $x = 0$ in the $t_0 = 0$ section.

The stable manifold in the $t = 0$ section is given parametrically by $(x(0,t_0), y(0,t_0))$ for $-\infty < t_0 < \infty$ and is shown in Fig.2. Notice that $x > 0$ for $t_0 < 0$ and that $x(0,t_0) \to x_p(0)$ for $t_0 \to -\infty$. By the time reflection principle, the unstable manifold is just the reflection of the stable manifold about the y - axis. Thus we have the perfect symmetry shown in Fig.2 and it follows that $x_p(0) = \pi$. Furthermore, from the parametrization above, the expression for the slope of the stable manifold at $x = 0$ is

$$S_s = \frac{D_2y(0,0)}{D_2x(0,0)} = -\frac{\frac{d}{dt_0}y(t_0,t_0)|_{t_0=0}}{y(0,0)}, \tag{4}$$

and by the time reversal principle the angle of intersection of stable and unstable manifolds, ψ, satisfies $\tan\frac{\psi}{2} = S_s$.

The situation in the extended phase space is illustrated in Fig. 3. In addition to the intersection of the manifolds with the $t = 0$ plane, which is just the $t = 0$ section of Fig.2, we show (1) their intersection with the y - t plane, which illustrates the splitting distance, (2) a typical orbit on the stable manifold and (3) the periodic orbit $x_p(t)$ which emerges from the $\delta = 0$ equilibrium solution.

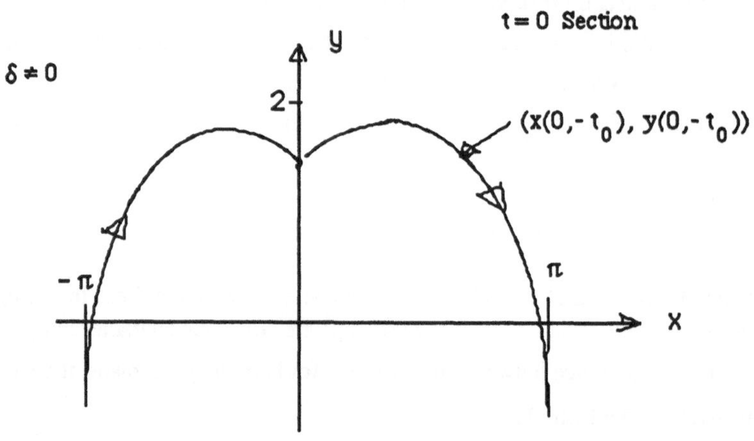
t = 0 Section
y
δ ≠ 0
2
(x(0,- t0), y(0,- t0))
- π
π
x

Fig. 2

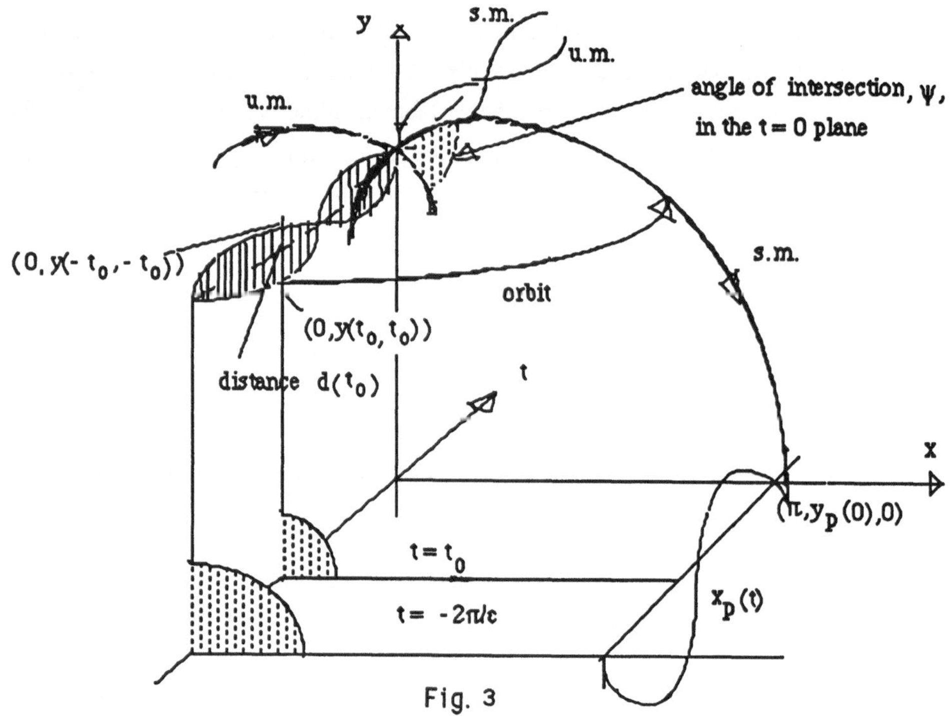
y
s.m.
u.m.
u.m.
angle of intersection, ψ,
in the t = 0 plane
(0, y(- t0, - t0))
s.m.
orbit
(0,y(t0, t0))
distance d(t0)
t
x
(π,yp (0),0)
t = t0
xp (t)
t = -2π/ε

Fig. 3

Our ultimate goal is to prove the

Main Theorem:

For $p = 5$ *and* δ_0, ε *sufficiently small,*

$$d(t_0) = 2\pi\,\delta\,\sin\frac{1}{\varepsilon}t_0\,\mathrm{sech}\frac{1}{2\varepsilon}\pi + O(\delta_0\delta\exp(-\pi/2\varepsilon))$$

$$S_s = \frac{\delta}{2\varepsilon}\pi\,\mathrm{sech}\frac{1}{2\varepsilon}\pi + O(\delta_0\frac{\delta}{\varepsilon}\exp(-\pi/2\varepsilon)).$$

Remarks:

(1) We know $d(0) = 0$, however, the theorem shows that splitting does occur and that the splitting distance is of order $\exp(-\pi/2\varepsilon)$.

(2) The angle of intersection, ψ, in the $t = 0$ section is non - zero and is of order $\exp(-\pi/2\varepsilon)$.

(3) δ_0 can be of order ε^q for $q > 0$, so our result includes the result of Ref.2-3.

III. Two - Time Embedding and Associated Stable Manifold Integral Equation

Next we construct the family of solutions $x(t,t_0)$, that is used to parametrize the stable manifold associated with the periodic solution $x_p(t)$. For this purpose we introduce coordinates (a,b) close to the homoclinic orbit of Eq.(2) by setting

$$x(t,t_0) = x_0(s) + \delta\,a(s,t_0), \quad y(t,t_0) = y_0(s) + \delta\,b(s,t_0), \tag{5}$$

where $b(s,t_0) := D_1 a(s,t_0)$. In order to separate, at least partially, the two time scales manifest in the problem, namely the "slow" drift of the solutions on the stable manifold to the attractor $x_p(t)$ and the "fast" periodic behavior inherent in $x_p(t)$ itself, we let the variable a depend on two different times s and z by

$$a(s,\varepsilon\phi) = \alpha(s,z,\phi), \qquad \text{where} \qquad z := \frac{s}{\varepsilon}, \qquad \phi := \frac{t_0}{\varepsilon}.$$

Here, we require that z and ϕ be treated independently and that α be periodic with period 2π in each. This is in contrast to Refs.2-3, where α depends only on z and ϕ through the variable $\theta := \phi + z$. This is because we want to impose the condition $\alpha(0,0,\phi) = 0$; furthermore, it allows us to take ϕ complex. Using the two times s and z and the equation for a which results from Eqs.(1) and (5), we obtain the partial differential equation:

$$\left(\frac{\partial}{\partial s} + \frac{1}{\varepsilon}\frac{\partial}{\partial z}\right)^2 \alpha + g(s)\,\alpha = F(\alpha,s) + \sin(z + \phi), \tag{6}$$

where

$$F(\alpha,s) := 2f(s)\frac{1-\cos\delta\alpha}{\delta} + g(s)\left(\alpha - \frac{\sin\delta\alpha}{\delta}\right),$$

with $\quad f(s) := \tanh s \operatorname{sech} s$, $g(s) := 2\operatorname{sech}^2 s - 1$.

In view of the properties of $x(t,t_0)$ the auxiliary conditions for α are as follows:

(i) $\alpha(0,0,\phi) = 0$,

(ii) $\alpha(s,z,\phi)$ approaches $\alpha_p(z+\phi)$ as $s \to \infty$.

In addition we impose the regularity condition:

(iii) α to be 2π periodic and C^1 in z, analytic in a strip S about the real axis in ϕ and analytic in s in a complex domain $\mathfrak{D}$ to be defined.

Remark: The way we have introduced the two time variables may or may not seem reasonable. The ultimate justification comes from our stable manifold theorem in Section IV and the proof of the exponential smallness discussed in Section V.

We expand α into a Fourier series in the variable z as

$$\alpha(s,z,\phi) = \sum_{k=-\infty}^{\infty} \alpha_k(s,\phi)\, e^{ikz}.$$

The initial condition (i), expressed in terms of the Fourier coefficients, becomes

$$\alpha_0(0,\phi) = -\sum_{k\neq 0} \alpha_k(0,\phi). \tag{7}$$

Moreover, the Fourier coefficients must satisfy the system of ODE's:

$$\left(\frac{\partial}{\partial s} + \frac{1}{\varepsilon} ik\right)^2 \alpha_k(s,\phi) + g(s)\, \alpha_k(s,\phi) = f_k(s,\phi), \tag{8}$$

where the right side represents the k - th Fourier coefficient of the function

$$f(s,z,\phi) := F(\alpha(s,z,\phi),s) + \sin(z+\phi).$$

The k-th Fourier coefficient of the first term alone will be denoted by $F_k(s,\phi)$. Since $\ell(s) := \operatorname{sech} s$ and $h(s) := \sinh s + s\,\ell(s)$ is a fundamental set of solutions of the variational equation

$$a'' + g(s)\, a = 0,$$

along the homoclinic orbit, it is easy to check that the bounded solutions of

$$a'' + g(s)\, a = f(s)$$

are given by $a(s) = a(0)\ell(s) - \ell(s)\int_0^s h(\sigma)f(\sigma)d\sigma - h(s)\int_s^\infty \ell(\sigma)f(\sigma)d\sigma$

and this can be used to find solutions of (8) for the Fourier coefficients $\alpha_k(s,\phi)$.
It turns out that the conditions on α listed above are all satisfied if we choose the following solutions of the ODE (8)

$$2\alpha_k(s,\phi) = \oint_s^\infty G(s,\sigma)\exp(\frac{ik}{\varepsilon}(\sigma - s))f_k(\sigma,\phi)d\sigma, \quad (k \neq 0), \tag{9}$$

where $G(s,\sigma) := \ell(s)\,h(\sigma) - h(s)\,\ell(\sigma)$ and the path of integration is defined as indicated in Fig.4.

That is, for any s in $\mathfrak{D}$ the integral in (9) is on the straight line through s with slope +1 if $k > 0$ and slope -1 if $k < 0$. The circle on the integral symbol in Eq. (9) will always denote this path. Figure 4 also depicts the s - domain $\mathfrak{D}$ in which our functions are analytic as a function of s. $\mathfrak{D}$ depends on the quantity η which is the distance from $\mathfrak{D}$ to $\pi/2$, along the imaginary axis, where many of our functions have a first order pole. Eventually, we will be forced to put $\eta = \varepsilon$. Since the $f_k(\sigma,\phi)$'s depend on all the Fourier coefficients of α, Eqs.(9) constitute integral equations for the α_k's. However, they must be supplemented by an equation for α_0. In view of (7) and (8), we obtain

$$-2\,\alpha_0(s,\phi) = \int_0^s G(s,\sigma)f_0(\sigma,\phi)d\sigma + h(s)\int_0^\infty \ell(\sigma)f_0(\sigma,\phi)d\sigma +$$

$$\ell(s)\sum_{k\neq 0}\oint_0^\infty h(\sigma)f_k(\sigma,\phi)e^{ik\sigma/\varepsilon}d\sigma. \tag{10}$$

The last term in (10) is required so that $\alpha(0,0,\phi) = 0$. This term is the reason that the (z,ϕ) dependence of the function $\alpha(s,z,\phi)$ can not be reduced to $z + \phi$ alone.

The relevant geometrical quantities, namely the distance and slope of Eqs. (3) and (4), are now expressed in terms of $\alpha(s,z,\phi)$ as follows. First we define

$$\beta(\phi) := -D_1\alpha(0,0,\phi) - \frac{1}{\varepsilon}D_2\alpha(0,0,\phi),$$

so that

$$y(t_0,t_0) = 2 - \delta\beta(\phi).$$

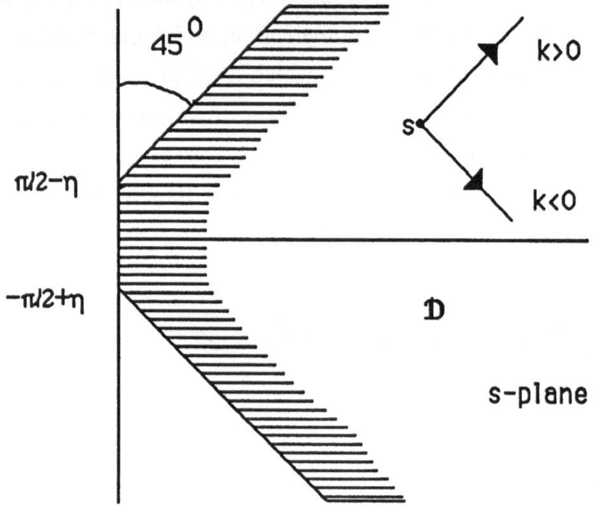

Fig. 4

Setting

$$\Delta(\phi) := \beta(\phi) - \beta(-\phi),$$

the distance and the slope become

$$d(t_0) = \delta\, \Delta(\phi), \tag{11}$$

$$S_s = \frac{\frac{\delta}{2\varepsilon}\Delta'(0)}{2 - \delta\, \beta(0)}. \tag{12}$$

Our strategy is to prove that $\Delta(\phi)$ is exponentially small but not identically zero. Since by construction, $\Delta(\phi)$ will be real analytic in ϕ in an $O(1)$ - neighborhood of $\phi = 0$, a Cauchy estimate will show that $\Delta'(\phi)$ (and therefore S_s) is also exponentially small but not zero. The crucial quantity $\Delta(\phi)$ can be written

$$\Delta(\phi) := Q(\phi) + 2\pi \sin\phi \operatorname{sech}\frac{\pi}{2\varepsilon}, \tag{13}$$

where
$$Q(\phi) := \sum_k Q_k(\phi), \tag{14}$$

and

$$Q_k(\phi) = \int_0^\infty \ell(\sigma)[F_k(\sigma,\phi)e^{ik\sigma/\varepsilon} - F_{-k}(\sigma,-\phi)e^{-ik\sigma/\varepsilon}]. \tag{15}$$

The second term on the rhs of (13) is the so - called Melnikov term. The reader who follows the calculation (13) - (15) may end up with -k replaced by k in the second term on the rhs of (15). The above form gives the same result by a rearrangement in (14) and is preferred because of the anticipated exponential smallness of the Q_k's in (15).

Our task now breaks into two parts:

(1) Show that the integral equations (9) and (10) for the Fourier coefficients of α have a unique solution in an appropriate Banach space, thus implying that $Q(\phi)$ in (13) is well - defined.

(2) Show that $Q(\phi)$ defined in (13) is exponentially small.

The former is the subject of Section IV and the latter of Section V.

IV. Stable Manifold Theorem

The integral equations (9) and (10) can be written

$$\mathfrak{G}\, \alpha = \alpha,$$

where

$$\mathfrak{G}\, \alpha = \mathfrak{L}\, (\mathfrak{F}\, \alpha + \mathfrak{S}).$$

Here, $\mathfrak{S}$ is the function $\sin(z+\phi)$, $\mathfrak{F}$ the nonlinear operator

$$(\mathfrak{F}\,\alpha)(s,z,\phi) = F(\alpha(s,z,\phi),s),$$

and $\mathfrak{L}$ the linear operator affiliated with the integral equation which maps the Fourier coefficients of f to those of α.

In order to define a suitable Banach space in which α is a fixed point of $\mathfrak{G}$, we first introduce the Banach space Y of bounded real analytic functions $\beta(s,\phi)$*) with continuous extension to the boundary equipped with the supremum norm

$$\|\beta\|_\infty := \sup_{\mathfrak{D}\times S} |\beta(s,\phi)|,$$

where $\mathfrak{D}$ is the domain introduced above and S is a strip about the real ϕ - axis. Since our functions should have limits for Re $s \to \infty$, we consider the subspace Y_0 of Y such that $\lim_{\mathrm{Re}\, s\to\infty} \beta(s,\phi)$ exists for each fixed value of Im $s \in \mathbb{R}$, uniformly in ϕ and such that the limit is independent of Im s. It is not difficult to show that Y_0 is also a Banach space. Our ultimate Banach space, X, will be the space of sequences of Fourier coefficients $\beta = \{\beta_k\}$, $\beta_k \in Y$, equipped with the norm

$$\|\beta\| = \sum_{k=-\infty}^{\infty} (|k|+1)\, \|\beta_k\|_\infty,$$

and X_0 is defined as the subspace of those elements $\beta \in X$ whose Fourier coefficients belong to Y_0.

The linear operator $\mathfrak{L}$ viewed as an operator over X is bounded and maps X_0 into itself. In general we find

$$\|\mathfrak{L}f\| \le B\eta^{-2}\|f\|;$$

however, if the zero-th Fourier coefficient of f is zero, we obtain the better estimate

$$\|\mathfrak{L}f\| \le \varepsilon\eta^{-2}\hat{B}\,\|f\|.$$

Here, B and $\hat{B}$ are constants and η is defined in Fig. 4. The nonlinear operator $\mathfrak{F}$ which is explicitly defined via

$$\mathfrak{F}\,\alpha = 2f(s)\sum_{n=1}^{\infty}\frac{(-1)^{n+1}}{(2n)!}\delta^{2n-1}(\alpha)^{*2n} + g(s)\sum_{n=1}^{\infty}\frac{(-1)^{n+1}}{(2n+1)!}\delta^{2n}(\alpha)^{*2n+1},$$

where $\alpha == \{\alpha_k\}$ and * stands for the convolution product in X, is also bounded on X and maps X_0 into itself. Its bound is determined by the formula

$$\|\mathfrak{F}\alpha\| \le D\|\alpha\|^2\delta \int_0^1 (1-t)\, e^{t\delta\|\alpha\|}dt,$$

*) In this section β will always denote a generic function.

where $D \leq \hat{D}\eta^{-2}$ with some positive constant $\hat{D}$. Using these bounds, we find for $\beta \in X_0$ and $\|\beta\| \leq R$ that

$$\|\mathfrak{G}\,\beta\| \leq B\hat{D}\eta^{-4}R^2\delta \int_0^1 (1-t)\, e^{t\delta R} dt + \hat{B}\,\|\mathbf{S}\|\varepsilon\,\eta^{-2}.$$

Thus for $\eta = \varepsilon$, we see that there exists an R_0 such that the ball with radius $R = \frac{R_0}{\varepsilon}$ (with R_0 a sufficiently large positive number) is mapped by $\mathfrak{G}$ into itself for ε and δ_0 sufficiently small. Furthermore, this is the first place we need $\delta = \delta_0\varepsilon^p$ with $p \geq 5$, for otherwise we cannot map the ball into itself. It is now easy to show that $\mathfrak{G}$ is a contraction mapping and we have the following fixed point

Theorem (Stable Manifold Theorem):
Eqs. (9) *and* (10) *have a unique fixed point in the ball in* X_0 *defined by* $\|\beta\| \leq \frac{R_0}{\varepsilon}$. *That is, there exists a unique* $\alpha(s,z,\phi)$ *which satisfies* (9) *and* (10), *and conditions* (i), (ii) *and* (iii) *listed after Eq.*(6). *In particular,* α *satisfies the* PDE *of Eq.* (6) *and approaches the periodic solution in the precise sense expressed in* (ii). *When z is replaced by* s/ε, *we obtain the parametrization of the stable manifold defined in Section* II.

V. Outline of Proof of Exponential Smallness

Here we describe our proof that $Q_0(\phi)$ and $\tilde{Q}(\phi) = \sum_{k \neq 0} Q_k(\phi)$, are exponentially small. By shifting the path of the integral in Eq.(15) for $k \neq 0$ away from the real axis and by an appropriate application of the mean value inequality, we obtain the following estimate for the non-zero frequency part

$$\tilde{Q}(\phi) = O\left(\frac{\delta}{\varepsilon}\eta^{-3}\|\gamma\|^0 e^{-r/\varepsilon}\right) + O(\delta\eta^{-3}\varepsilon^{-1}e^{-r/\varepsilon}), \tag{16}$$

where

$$\gamma(is,z,\phi) := \alpha(is,z,\phi) + \alpha(-is,-z,-\phi)$$

and $r := \frac{\pi}{2} - \eta$. For $k > 0$, the path is taken along the imaginary axis from 0 to r and then along the path $ir + (1+i)t$, $t \in [0,\infty)$ and for $k < 0$ the path is just the reflection of the $k > 0$ path about the real axis. It turns out that the major contribution to (16) (i.e.the first term) comes from the imaginary axis. The norm $\|\gamma\|^0$ is defined via the formula

$$\|\gamma\|^0 := \sum_{k=-\infty}^{\infty} (|k|+1)\|\gamma_k\|^0_\infty\,,$$

where we employ the same weighted sup - norm as in Refs. 2 - 3, namely

$$\|\gamma_k\|^0_\infty := \sup_{\substack{-r\le s\le r \\ \phi \in S}} |\gamma_k(is,\phi)| e^{(r-|s|)/\varepsilon} .$$

For $\eta = \varepsilon$ and $\delta = \delta_0\,\varepsilon^5$, δ_0, ε sufficiently small, $\tilde{Q}$ becomes

$$\tilde{Q}(\phi) = O(\delta_0\,\varepsilon\,\|\gamma\|^0 e^{-\pi/2\varepsilon}) + O(\delta_0\,\varepsilon\, e^{-\pi/2\varepsilon}). \qquad (17)$$

The exponential smallness of $\tilde{Q}(\phi)$ will be established if we can show that $\varepsilon\|\gamma\|^0$ is uniformly bounded in ε for $\varepsilon\downarrow 0$.

In order to obtain a similar estimate for Q_0, we have to resort to a new argument which has no counterpart in previous work. We first observe that our PDE can be written in the form

$$x_{ss} + \frac{2}{\varepsilon} x_{sz} + \frac{1}{\varepsilon^2} x_{zz} = -\sin x + \delta \sin(z+\phi),$$

where $x(s,z,\phi) = x_0(s) + \delta\,\alpha\,(s,z,\phi)$. Multiplying by x_s, averaging over z and integrating by parts yields

$$I(\phi; x) = \frac{1}{2}\left(\frac{1}{\varepsilon^2}\,\overline{x^2_z} - \overline{x^2_s}\right) + \overline{\cos x} + \delta\,\overline{x\sin(z+\phi)}$$

as independent of s, where bar denotes z - average. After some algebra, we end up with the conclusion that the functional of $\alpha(s,z,\phi)$ defined by

$$J(\alpha, s, z+\phi) = \frac{1}{2}\left(\frac{\delta}{\varepsilon^2}\,\alpha_z^{\,2} - \delta\,\alpha_s^{\,2}\right) - 2\,\ell(s)\,\alpha_s - g(s)\frac{1-\cos\delta\alpha}{\delta} - 2f(s)\frac{\sin\delta\alpha}{\delta}$$
$$+\,\delta\,\alpha\sin(z+\phi)$$

has average (i.e. 0 - th Fourier coefficient) which is independent of s. This means that its value on the periodic solution, namely

$$J(\alpha_p, z+\phi) = \frac{1}{2}\frac{\delta}{\varepsilon^2}\,\alpha_{pz}^{\,2} + \frac{1-\cos\delta\alpha_p}{\delta} + \delta\,\alpha_p \sin(z+\phi)$$

and its value at s = 0 agree, when both are averaged over z. The same holds for the evaluation of the functional J on the solutions

$$\alpha^*(s,z,\phi) := -\alpha\,(-s,-z,-\phi)$$

associated with the unstable manifold. Since there is really only one periodic orbit, we find that the 0 - th order Fourier coefficients of

$$J(\alpha(0,z,\phi), 0, z+\phi) \text{ and } J(-\alpha(0,-z,-\phi), 0, z+\phi)$$

agree. This gives an expression for $Q_0(\phi)$ which leads to the estimate

$$Q_0(\phi) = O(\delta_0\varepsilon^4\,\|\gamma\|^0 e^{-\pi/2\varepsilon}) + O(\delta_0\varepsilon^3 e^{-\pi/2\varepsilon}). \qquad (18)$$

The final step in our proof uses the integral equation $\alpha = \mathcal{L}\,(\mathcal{F}\,\alpha + \mathcal{S})$ to show that

$$\|\gamma\|^0 = O(\delta_0 \|\gamma\|^0) + O(\delta_0) + O(1/\varepsilon), \tag{19}$$

that is, for δ_0 sufficiently small $\varepsilon\|\gamma\|^0$ is uniformly bounded as $\varepsilon \downarrow 0$. We sketch the proof. The integral equation (9) for α_k, where s is replaced by i s, $s \in \mathbb{R}$, is

$$2\alpha_k(is,\phi) = \oint_{is}^{\infty} G(is,\sigma) e^{(ks + ik\sigma)/\varepsilon} f_k(\sigma,\phi)d\sigma,$$

where the integration proceeds along the boundary of $\mathbf{D}$, as discussed in conjunction with (16). Splitting the integral into a part from i s to i $r\kappa$, where $\kappa := \operatorname{sgn} k$ and $r = \pi/2 - \eta$, and into one from i $r\kappa$ to ∞, we obtain

$$2\alpha_k(is,\phi) = i \int_s^{r\kappa} G(is,i\sigma) e^{k(s-\sigma)/\varepsilon} f_k(i\sigma,\phi)d\sigma + O(\varepsilon\eta^{-2}\|f_k\|_\infty e^{(|s|-r)/\varepsilon}).$$

Making the substitutions $k \to -k$, $s \to -s$, $\phi \to -\phi$ in the last formula and adding the result yields

$$2\gamma_k(is,\phi) = i \int_s^{r\kappa} G(is,i\sigma) e^{(ks - k\sigma)/\varepsilon} [f_k(i\sigma,\phi) + f_{-k}(-i\sigma,-\phi)]d\sigma + O_k$$

with $\qquad O_k := O(\varepsilon\eta^{-2} e^{(|s|-r)/\varepsilon}(\|f_k\|_\infty + \|f_{-k}\|_\infty)).$

Applying the mean value theorem in the form

$$F(\alpha(is,z,\phi),is) + F(\alpha(-is,-z\,-\phi),-is) = \rho(is,z,\phi)\,\gamma(is,z,\phi)$$

to the last expression yields

$$2\,\gamma_k(is,\phi) = i \int_s^{r\kappa} G(is,i\sigma) e^{(ks - k\sigma)/\varepsilon} \sum_{h=-\infty}^{\infty} \rho_{k-h}(i\sigma,\phi)\,\gamma_h(i\sigma,\phi)d\sigma + O_k.$$

This leads to an estimate

$$\|\gamma_k\|_\infty^0 \le C_1 \int_{s\kappa}^{r} e^{(s\kappa - \sigma + |\sigma| - |s|)/\varepsilon} d\sigma \sum_{h=-\infty}^{\infty} \|\rho_{k-h}\|_\infty \|\gamma_h\|_\infty^0 + O(\varepsilon\eta^{-2}(\|f_k\|_\infty + \|f_{-k}\|_\infty)),$$

where C_1 is a positive constant. It turns out that the exponential is always less than unity and therefore can be dropped in our estimate yielding

$$\|\gamma_k\|_\infty^0 \le 2r\,C_1 \sum_{h=-\infty}^{\infty} \|\rho_{k-h}\|_\infty \|\gamma_h\|_\infty^0 + O(\varepsilon\eta^{-2}(\|f_k\|_\infty + \|f_{-k}\|_\infty)).$$

Using the fact that $\sum_{k=-\infty}^{\infty} (|k|+1)\|f_k\|_\infty \le \|\mathcal{F}\alpha\| + \|\mathcal{S}\| = \|\mathcal{S}\| + O(\delta_0\,\varepsilon)$ for the choice

$\delta = \delta_0 \varepsilon^5$ and $\eta = \varepsilon$, we arrive at the estimate

$$\|\tilde{\gamma}\|^0 = 2r\, C_1 \sum_{k=-\infty}^{\infty} (|k| + 1) \sum_{h=-\infty}^{\infty} \|\rho_{k-h}\|_\infty \|\gamma_h\|_\infty^0 + O(\delta_0) + O(1/\varepsilon),$$

where $\tilde{\gamma}$ is the non - zero frequency part of γ. Observing that the double sum on the right side is bounded by $\|\rho\|\, \|\tilde{\gamma}\|^0$ and that $\|\rho\| = O(\delta_0)$, we finally obtain (19) with $\|\tilde{\gamma}\|^0$ in place of $\|\gamma\|^0$ on the left side. The proof of (19) is completed by establishing the same type of estimate for $\|\gamma_0\|^0$ as for $\|\tilde{\gamma}\|^0$ (1).

Using (17) and (18) and (19), we obtain $Q(\phi) = O(\delta_0 \exp(-\pi/2\varepsilon))$, where r has been replaced by $\pi/2$ since $\eta = \varepsilon$. From (13) and (11), it is now apparent that the first statement of our main theorem is true. A Cauchy estimate yields the second statement.

VI Conclusion.

Our approach to the rapidly forced pendulum problem is fairly elementary in that the most sophisticated tool it uses is the fixed point theorem of a contraction mapping in a metric space. In addition to making use of many ideas from Refs. 2-3, it also differs in at least five key - points.

- ✖ We focus on the geometry as illustrated in Figs.2 and 3, namely the splitting distance in the y - t plane and angle of transversality in the $t = 0$ section. In order to compute this distance, we found it convenient to design our integral equation in such a way that its solutions strictly vanish at the initial time $t = t_0$. In this way, we are lead to consider functions that not only depend on the variable $s = t - t_0$, but also depend on two 2π - periodic variables $z = s/\varepsilon$ and $\phi = t_0/\varepsilon$, and not just on their sum $\theta := z + \phi$. Correspondingly, the Fourier coefficients of these functions in the variable z depend not only on s but on ϕ as well.

- ✖ In order to formulate the integral equation for $p = 5$, we found it necessary to consider a larger domain in the complex s - plane.

- ✖ The proof that the difference of the 0 - th Fourier coefficients of the y - coordinates of the solutions on the stable and unstable manifolds is exponentially small is based on the construction of a quantity J whose average over the fast variable z is independent of the slow variable s. See Eqs. (3), (11), (13) - (15) and (18).

- ✖ We immediately obtain exponentially small bounds for the angle, once they are known for the distance, using a Cauchy - type estimate.

✖ We show explicitly that the solution of our integral equation for $s \rightarrow \infty$ tends to the hyperbolic periodic solution $x_p(t)$ which emerges from the unstable equilibrium point $(\pi,0)$ for non-zero δ. Thus we also prove a stable manifold theorem.

Of course, the most obvious difference is the improvement that we achieve using our approach, that is, we are able to reduce p in ε^p from $p = 8$ to $p = 5$.

Acknowledgement: We thank the organizers of the workshop for inviting us to make this contribution and Harvey Segur for several helpful comments.

References

1. M. Kummer, J.A. Ellison and A.W. Saenz, to be submitted.
2. P. Holmes, J. Marsden and J. Scheuerle, "Exponentially Small Splittings of Separatrices with Applications to KAM Theory and Degenerate Bifurcations", Contemporary Mathematics 81 (1988) 213.
3. J. Scheuerle, "Chaos in A Rapidly Forced Pendulum Equation", Contemporary Mathematics 97 (1989) 411.
4. J. Marsden, this proceedings.
5. H.S. Dumas and J.A, Ellison in "Local and Global Methods of Nonlinear Dynamics", pp. 200-230 (A. W. Saenz, W.W. Zachary and R. Cawley, Eds.) Springer Verlag, New York, 1986.
6. J. Scheuerle, this proceedings.

PROOF OF AN ASYMPTOTIC SYMMETRY OF THE RAPIDLY FORCED PENDULUM

Yi-Hua Chang

Program in Applied Mathematics, University of Colorado
Boulder, CO 80309-0526
Department of Mathematics, State University of New York
Buffalo, NY 14214-3093

ABSTRACT

This paper concerns with the split of a separatrix of a nonlinear pendulum by a small but rapid sinsusoidal forcing. A partial answer is given, with main result, road map and proof of δ-asymptotic expansion.

1. BACKGROUND

We have been studying the rapidly forced pendulum, and the differential equation is:

$$x'' + \sin x = \delta \sin (t+t_0)/\varepsilon , \qquad 0 < \delta \ll 1, \qquad 0 < \varepsilon \ll 1. \tag{1.1}$$

If $\delta=0$ there is a separatrix,

$$x_0(t) = 2 \tan^{-1} [\sinh(t)]; \tag{1.2}$$

The motivation for studying this apparently simple problem is its connection to Poincare three body problem and the converse KAM theorem[2]. We want to know whether the complement of the region, in which tori are invariant under a small perturbation, is a chaotic region.

Our results are as follows:

(1) We find an asymptotic symmetry carrying all orders that is the first step to use M.Kruskal-H.Segur's method.[3]

(2) If the separatrix split, the splitting distance is exponentially small[1].

Asymptotics beyond All Orders, Edited by H. Segur *et al.*
Plenum Press, New York, 1991

2. MAIN RESULTS

Here we present the statement of some theorems that have been proved in [4]. We will only present the proof of asymptoticity of a perturbation expansion in powers of δ (Theorem 3), assuming the results of the previous theorems. The figure in the next page charts the general direction of our attack on this problem.

The differential equation is:

$$x'' + \sin x = \delta \sin[(t+t_0)/\varepsilon] , \quad \delta , \varepsilon \ll 1. \tag{2.1}$$

With $x_0(t)$ given by (1.2), set

$$x(t, t_0, \varepsilon, \delta) = x_0(t) + \delta\, y(t, t_0, \varepsilon, \delta), \tag{2.2}$$

so that y satisfies

$$y'' + (\cos x_0)y = f(y, t, \delta) + \sin[(t+t_0)/\varepsilon] \tag{2.3a}$$

where

$$f(y, t, \delta) = -\{(\sin x_0)[(\cos \delta y)-1] + (\cos x_0)[(\sin \delta y)-\delta y]\}/\delta. \tag{2.3b}$$

Let y^+ denote a solution of (2.3) that is bounded as $t \to +\infty$, and let $c^+ = y^+ |_{t=0}$.

Let y^- denote a solution of (2.3) that is bounded as $t \to -\infty$, and let $c^- = y^- |_{t=0}$.

Theorem 1: For $t \ge 0$, there is a $\delta_0>0$ such that a one-parameter (c^+) family of solutions, $y^+(t, t_0, \varepsilon, \delta, c^+)$, of (2.3) exists for $\delta \le \delta_0$, uniformly in ε for $\varepsilon > 0$. Each of these solutions is uniformly bounded for $t \ge 0$. For $t \le 0$, $y^-(t, t_0, \varepsilon, \delta, c^-)$ is similarly defined and bounded.

Now represent $y^\pm(t, t_0, \varepsilon, \delta, c^\pm)$ by formal series for $0 < \delta \ll 1$:

$$\delta\, y^\pm(t, t_0, \varepsilon, \delta, c^\pm) \sim \sum_1^\infty \delta^n x_n^\pm(t, t_0, \varepsilon, c_n^\pm). \tag{2.4}$$

Substituting (2.4) into (2.3) and equating powers of δ yields

$$(x_1^\pm)'' + (\cos x_0)\, x_1^\pm = \sin[(t+t_0)/\varepsilon]; \tag{2.5a}$$

and for n=2,3,...,

$$(x_n^\pm)'' + (\cos x_0)\, x_n^\pm = (\sin x_0)\bullet\Phi_n^\pm(x_1, x_2, \dots, x_{n-1}) + (\cos x_0)\bullet\Psi_n^\pm(x_1, x_2, \dots, x_{n-1}), \tag{2.5b}$$

where $\Phi_n^\pm$ and $\Psi_n^\pm$ are polynomials in $(x_1^\pm, x_2^\pm, \dots, x_{n-1}^\pm)$; see (3.4).

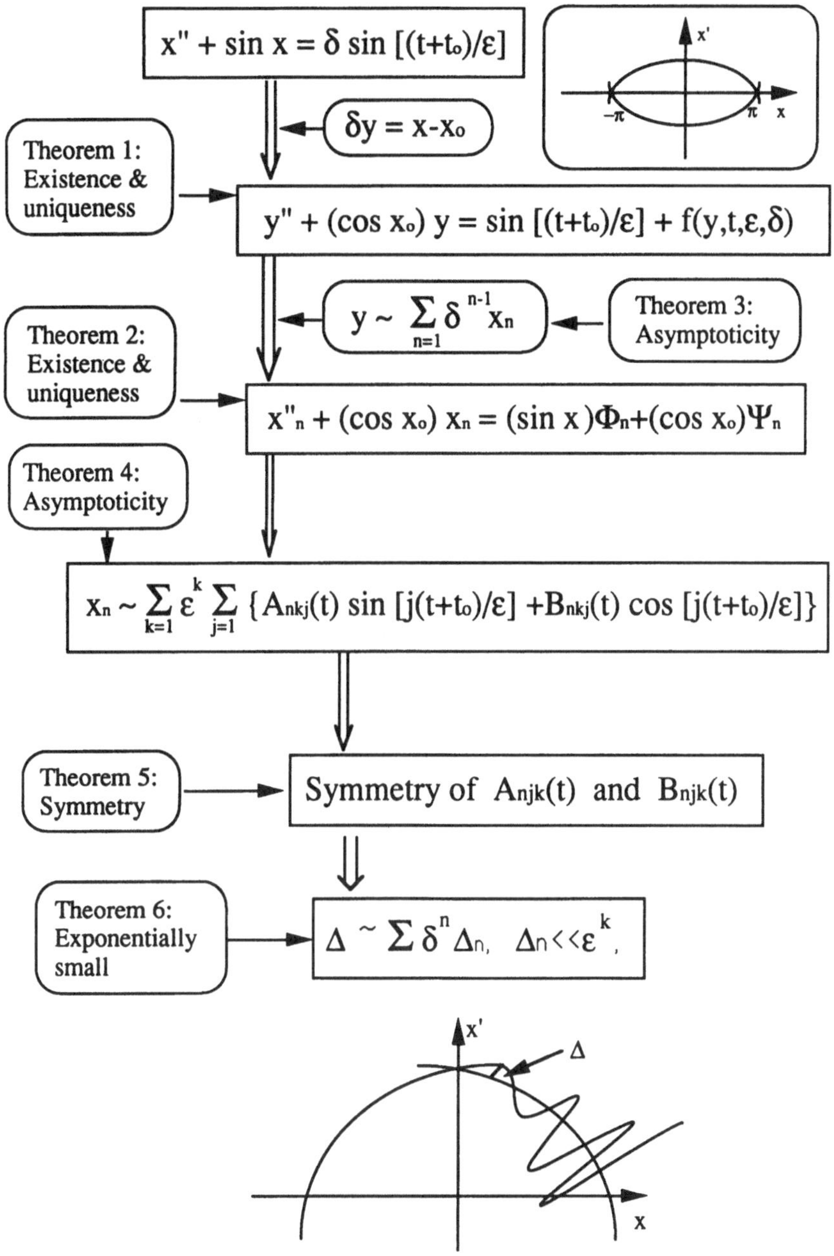

x" + sin x = δ sin [(t+t₀)/ε]
x'
−π
π
x
δy = x-x₀
Theorem 1: Existence & uniqueness
y" + (cos x₀) y = sin [(t+t₀)/ε] + f(y,t,ε,δ)
y ~ Σ_{n=1} δ^{n-1} x_n
Theorem 3: Asymptoticity
Theorem 2: Existence & uniqueness
x"_n + (cos x₀) x_n = (sin x)Φ_n+(cos x₀)Ψ_n
Theorem 4: Asymptoticity
x_n ~ Σ_{k=1} ε^k Σ_{j=1} {A_nkj(t) sin [j(t+t₀)/ε] +B_nkj(t) cos [j(t+t₀)/ε]}
Theorem 5: Symmetry
Symmetry of A_njk(t) and B_njk(t)
Theorem 6: Exponentially small
Δ ~ Σ δ^n Δ_n, Δ_n<<ε^k,
x'
Δ
x

Road Map

For $n = 1, 2, \ldots$, let x_n^+ denote a solution of (2.5) that is bounded as $t \to +\infty$, and let $c_n^+ = x_n^+|_{t=0}$. Let x_n^- denote a bounded solution as $t \to -\infty$, and let $c_n^- = x_n^-|_{t=0}$.

Theorem 2: For every $n \geq 1$, (2.5) has a one-parameter family of solutions, $x_n^+(t, t_0, \varepsilon, c_n^+)$, that are uniformly bounded for $t \geq 0$ and $\varepsilon > 0$. Similarly, every $x_n^-(t, t_0, \varepsilon, c_n^-)$ is uniformly bounded for $t \leq 0$ and $\varepsilon > 0$.

In order for (2.4) to hold at t=0, we require that

$$c^+ \sim \sum_1^\infty \delta^{n-1} c_n^+ \text{ and } c^- \sim \sum_1^\infty \delta^{n-1} c_n^- \qquad \text{as } \delta \to 0. \quad (2.6)$$

Theorem 3: Given (2.6), the series in (2.4) are asymptotic to $y^\pm(t, t_0, \varepsilon, \delta, c^\pm)$ as $\delta \to 0$, uniformly in ε and t for $\varepsilon > 0$, $t \geq 0$ ($t \leq 0$).

Now we let $\varepsilon \to 0$ as $\delta \to 0$. For $n=1,2,\ldots$, we seek expansions of the form:

$$x_n^\pm \sim \sum_{k=1}^\infty \varepsilon^k \sum_{j=0}^n \{A_{nkj}(t) \bullet \sin[j(t+t_0)/\varepsilon] + B_{nkj}(t) \bullet \cos[j(t+t_0)/\varepsilon]\}. \quad (2.7)$$

Theorem 4 asserts that such expansions exist for a particular choice of $c_n^\pm$, $n=1,2,\ldots$.

Theorem 4: One can choose $\{c_1^\pm, c_2^\pm, c_3^\pm, \ldots\}$ so that as $\varepsilon \to 0$, the series in (2.7) are asymptotic to $x_n^+(t, t_0, \varepsilon, c_n^+)$ uniformly in t for $t \geq 0$, and to $x_n^-(t, t_0, \varepsilon, c_n^-)$ uniformly for $t \leq 0$. Moreover for each (n, k, j), $A_{nkj}(t)$ and $B_{nkj}(t)$ in (2.7) and all of their derivatives are uniformly bounded in t.

This representation leads to our main result, that (2.7) exhibits a symmetry to all orders.

Theorem 5: For every $(n, j, k=1, 2, \ldots)$ in (2.7), $A_{nkj}(t)$ is an even function of t, and $B_{nkj}(t)$ is an odd function of t. If $(n \bullet j)$ is even and either n or j is odd, then $A_{nkj}(t)$ and $B_{nkj}(t)$ vanish for every k. In particular, if n is odd and $j=0$, then $B_{nkj}(t)$ vanishes.

Finally, the splitting distance $\Delta(t_0, \varepsilon, \delta)$ is defined as follows:

$$[\Delta(t_0, \varepsilon, \delta)]^2 := \delta^2 [y^+|_{t=0} - y^-|_{t=0}]^2 + \delta^2 [(y^+)'|_{t=0} - (y^-)'|_{t=0}]^2. \quad (2.8)$$

We can require $[y^+|_{t=0} = y^-|_{t=0}]$, so that (2.8) becomes:

$$\Delta(t_0, \varepsilon, \delta) = \delta \left| [(y^+)'|_{t=0} - (y^-)'|_{t=0}] \right|. \quad (2.9)$$

Let

$$\Delta_n(t_0, \varepsilon) = [(x_n^+)'|_{t=0} - (x_n^-)'|_{t=0}], \quad (2.10)$$

so that

$$\Delta(t_0, \varepsilon, \delta) \sim \sum_1 \delta^n |\Delta_n|. \qquad (2.11)$$

Theorem 6: As $\varepsilon \to 0$, for every $n \geq 1$, $\Delta_n \ll \varepsilon^k$ for every positive integer k.

This means that the splitting distance is at most transcendentally small in ε as $\varepsilon \to 0$. This result had previously been obtained by Holmes, Marsden and Scheurle, Scheurle, and Neishtadt, [2], by other means.

3. THE TECHNIQUE IN THE PROOF OF ASYMPTOTICITY (THEOREM 3):

Some people do an asymptotic expansion and never prove the asymptoticity. It is not very easy to prove an asymptoticity. The major difficulty is how to find the general representation in order to use the inductive method. We have found a technique to overcome this difficulty in the δ-expansion of the pendulum problem. Our approach is reminiscent of that of M.Kruskal [5], whose work came to our attention after this study was completed. The present paper describes the technique in more explicit detail than can be found in [5].

In the pendulum problem we want to prove

$$y^{\pm}(t, t_0, \varepsilon, \delta, c^{\pm}) \sim \sum_1^{\infty} \delta^{n-1} x_n^{\pm}(t, t_0, \varepsilon, c_n^{\pm}), \quad \delta \to 0. \qquad (3.1)$$

We will give the proof for $y^+(t, t_0, \varepsilon, \delta, c^+)$; since the proof of $y^-(t, t_0, \varepsilon, \delta, c^-)$ is similar, we omit it. Also, in the following we drop the "+" superscript. From (2.3) the $y(t, t_0, \varepsilon, \delta, c)$ satisfies the integral equation,

$$y(t, t_0, \varepsilon, \delta, c) = c\alpha_1(t) - \alpha_1(t)\int_0^t ds\alpha_2(s)\{f(y, t, \delta)+\sin[(t+t_0)/\varepsilon]\}$$
$$+ \alpha_2(t)\int_{\infty}^t ds\alpha_1(s)\{f(y, t, \delta)+\sin[(t+t_0)/\varepsilon]\} := c\alpha_1(t) + G\{y(t)\}, \qquad (3.2)$$

where $f(y, t, \delta)$ is given by (2.3b), c is independent of t, and

$$\alpha_1(t) = \operatorname{sech} t; \qquad \alpha_2(t) = (\sinh t + t \operatorname{sech} t)/2.$$

From (2.5), $x_n(t, t_0, \varepsilon, c_n)$ satisfies ,

$$x_n(t, t_0, \varepsilon, c_n) = c_n\alpha_1(t) - \alpha_1(t)\int_0^t ds\alpha_2(s)\{(\sin x_0)\cdot\Phi_n + (\cos x_0)\cdot\Psi_n\}$$
$$+ \alpha_2(t)\int_{\infty}^t ds\alpha_1(s)\{(\sin x_0)\cdot\Phi_n+(\cos x_0)\cdot\Psi_n\}$$
$$= c_n\alpha_1(t) + G\{(\sin x_0)\cdot\Phi_n+(\cos x_0)\cdot\Psi_n\}, \qquad (3.3)$$

where c_n is independent of t, and Φ_n and Ψ_n as defined in (2.5) are given by:

$$\Phi_2(x_1) = (x_1)^2/2, \quad \Psi_2(x_1) = 0,$$

$$\Phi_3(x_1, x_2) = x_1 \bullet x_2, \qquad \Psi_3(x_1, x_2) = (x_1)^3/6 \;,....$$

In general, Φ_n and Ψ_n are determined from the following relation:

$$\Sigma_2^{\infty} \delta^n \Phi_n = \Sigma_1^{\infty} (-1)^k [(\Sigma \delta^n x_n)^{2k}]/(2k)!,$$

$$\Sigma_3^{\infty} \delta^n \Psi_n = \Sigma_2^{\infty} (-1)^{k-1} [(\Sigma \delta^n x_n)^{2k-1}]/(2k-1)!, \tag{3.4}$$

First we use Lemma 1 to prove that the linear integral operators $G^{\pm}$ are bounded.

Lemma 1: The linear operators,

$$G^{\pm}\{y(t)\} = -\alpha_1(t)\int_0^t ds\alpha_2(s)y(s) + \alpha_2(t)\int_{\pm\infty}^t ds\alpha_1(s)y(s), \tag{3.5}$$

are bounded operators. Specifically, if y(t) is bounded on $0 \le t < \infty$, so

$$||y|| = \sup_{0 \le t < \infty} |y(t)| < \infty, \tag{3.6}$$

then

$$||G^{+}\{y\}|| \le 3||y||. \tag{3.7a}$$

Similarly, if y(t) is bounded for $t \le 0$, then

$$||G^{-}\{y\}|| \le 3||y||. \tag{3.7b}$$

Proof: Suppose y(t) is uniformly bounded for $t \ge 0$. Then

$$||G^{+}\{y(t)\}|| \le ||y|| \bullet \sup_{0 \le t < \infty} \{\alpha_1(t)\int_0^t ds\alpha_2(s) - \alpha_2(t)\int_{\infty}^t ds\alpha_1(s)\}.$$

But

$$\alpha_1(s) \le 1,$$

$$s \bullet (\text{sech } s) \le 2s \bullet [\exp(-s)] \le 2e^{-1} \le 1,$$

so

$$\alpha_1(t)\int_0^t \alpha_2(s)ds \le (1 + \text{sech } t + t \bullet \text{sech } t)/2 \le 3/2,$$

$$|\alpha_2(t)\int_{\infty}^t \alpha_1(s)ds| \le -[(\sinh t + 1)/2]\int_{\infty}^t 2e^{-s}ds \le 3/2,$$

and

$$||G^{+}\{y(t)\}|| \le 3||y||.$$

The proof of (3.7b) is similar. Q.E.D.

Recall that " $f(\delta) = O(\delta)$ " means that there is a fixed $W < \infty$ and a $\delta_0 > 0$, such that

$$|f(\delta)| \leq W\,\delta, \quad \text{for } 0 < \delta \leq \delta_0.$$

(i) We want to prove: $(y-x_1) = O(\delta)$. From (3.2) and (3.3),

$$(y-x_1) = (c-c_1)\alpha_1 + G\{f(y, t, \delta)\}.$$

Because by (2.6),

$$c - c_1 = O(\delta),$$

and the fact

$$(1): |\cos(\delta y)-1| \leq (\delta y)^2[\exp(\delta^2 y^2)],$$
(3.8)

$$(2): |\sin(\delta y)-\delta y| \leq |\delta y|^3[\exp(\delta^2 y^2)], \qquad (3.9)$$

then by Lemma 1,

$$||G\{f(y, t, \delta)\}|| \leq 3\delta\cdot(||y||^2+\delta||y||^3)\cdot\exp(\delta^2||y||^2).$$

So for $t \geq 0$ and $\varepsilon > 0$,

$$(y-x_1) = O(\delta). \qquad (3.10)$$

(ii) We want to prove: $\{y-x_1-\delta x_2\}/\delta = O(\delta)$. For any analytic function, $F(z)$, let $[F(\Sigma_1^\infty \delta^k a_k)]_n$ denote the nth truncation of its δ-power series:

$$[F(\Sigma_1^\infty \delta^k a_k)]_n = [\,\Sigma_1^\infty \delta^k b_k\,]_n = \Sigma_1^n \delta^k b_k. \qquad (3.11)$$

$$\{y-x_1-\delta x_2\}/\delta = \{c-c_1-\delta c_2\}\alpha_1/\delta + G\{f(y,t,\delta) - (\sin x_0)[\cos(\delta x_1)-1]_2/\delta\}/\delta.$$

By (2.6),

$$\{c-c_1-\delta c_2\}\alpha_1/\delta = O(\delta).$$

From (2.3b),

$$f(y,t,\delta) = (\sin x_0)(\delta y^2/2)+O(\delta^2).$$

Also

$$[\cos(\delta x_1)-1]_2/\delta = \delta x_1^2/2,$$

so that by (3.10)

$$\{f(y,t,\delta) - (\sin x_0)[\cos(\delta x_1)-1]_2/\delta\}/\delta = (\sin x_0)(y-x_1)(y+x_1)/2 + O(\delta) = O(\delta).$$

Then using Lemma 1 shows that

$$\{y-x_1+\delta x_2\}/\delta = O(\delta). \tag{3.12}$$

(iii) We want to prove: $\{y-x_1-\delta x_2-\delta^2 x_3)\}/\delta^2=O(\delta)$.

$$\{y-x_1-\delta x_2-\delta^2 x_3\}/\delta^2$$

$$= (1/\delta^2)\Big([c-c_1-\delta c_2-\delta^2 c_3]\alpha_1-G\Big\{(\sin x_0)\{(\cos \delta y)-1-[\cos(\textstyle\sum_{n=1}^{2}\delta^n x_n)-1]_3\}$$

$$+(\cos x_0)\{(\sin \delta y)-\delta y-[\sin(\textstyle\sum_{n=1}^{2}\delta^n x_n)-(\sum_{n=1}^{2}\delta^n x_n)]_3\}\Big\}/\delta\Big).$$

But

$$(\cos \delta y)-1-[\cos(\textstyle\sum_{n=1}^{2}\delta^n x_n)-1]_3 = (\cos \delta y) - \cos(\sum_{n=1}^{2}\delta^n x_n) + O(\delta^4)$$

$$= 2\sin\{(\delta y - \textstyle\sum_{n=1}^{2}\delta^n x_n)/2\}\bullet\sin\{(\delta y + \sum_{n=1}^{2}\delta^n x_n)/2\} + O(\delta^4)$$

$$= O(\delta^3)\bullet O(\delta) + O(\delta^4),$$

and

$$(\sin \delta y) - \delta y - [\sin(\delta x_1) - (\delta x_1)]_3 = (\sin \delta y-\sin \delta x_1) - (\delta y-\delta x_1) + O(\delta^4)$$

$$= 2\sin\{(\delta y-\delta x_1)/2\}\cos\{(\delta y+\delta x_1)/2\} - (\delta y-\delta x_1)+O(\delta^4)$$

$$= 2\sin\{(\delta y-\delta x_1)/2\} - (\delta y-\delta x_1) + O(\delta^4) = O(\delta^4).$$

Then the result follows by using (2.6) and Lemma 1.

(iv) In the general case, we want to prove:

$$\{y-(\textstyle\sum_{n=1}^{m}\delta^{n-1}x_n)\} = O(\delta^m). \tag{3.13}$$

The proof is by induction. Assume that (3.21) holds for m = 1,2,...,k. Then

$$[y-(\textstyle\sum_{n=1}^{k+1}\delta^{n-1}x_n)]/\delta^k$$

$$=(1/\delta^k)\Big([c - \textstyle\sum_{n=1}^{k+1}\delta^{n-1}c_n]\alpha_1$$

$$- G\Big\{(\sin x_0)\{(\cos \delta y)-1-[\cos(\textstyle\sum_{n=1}^{k}\delta^n x_n)-1]_{k+1}\}/\delta$$

$$+ (\cos x_0)\{(\sin \delta y) - \delta y - [\sin(\textstyle\sum_{n=1}^{k-1}\delta^n x_n)-(\sum_{n=1}^{k-1}\delta^n x_n)]_{k+1}\}/\delta\Big\}\Big).$$

But we don't know the general representation of $[\cos(\sum_{n=1}^{k}\delta^n x_n)-1]_{k+1}$ and $[\sin(\sum_{n=1}^{k-1}\delta^n x_n)-(\sum_{n=1}^{k-1}\delta^n x_n)]_{k+1}$. We can add some high order terms, then subtract these terms in order to avoid the general representations, i.e.,

$$[\cos(\Sigma_{n=1}^{k} \delta^n x_n)-1]_{k+1} = \cos(\Sigma_{n=1}^{k} \delta^n x_n) - O(\delta^{k+2}),$$

and

$$[\sin(\Sigma_{n=1}^{k-1} \delta^n x_n)-(\Sigma_{n=1}^{k-1} \delta^n x_n)]_{k+1} = \{\sin(\Sigma_{n=1}^{k-1} \delta^n x_n)-(\Sigma_{n=1}^{k-1} \delta^n x_n)\}-O(\delta^{k+2})$$

So that

$$(\cos \delta y) - 1 - [\cos(\Sigma_{n=1}^{k} \delta^n x_n)-1]_{k+1} = (\cos \delta y) - \cos(\Sigma_{n=1}^{k} \delta^n x_n)+O(\delta^{k+2})$$

$$= 2\sin\{(\delta y-\Sigma_{n=1}^{k} \delta^n x_n)/2\} \sin\{(\delta y+\Sigma_{n=1}^{k} \delta^n x_n)/2\} + O(\delta^{k+2})$$

$$= O(\delta^{k+1})\bullet O(\delta) + O(\delta^{k+2}),$$

and

$$(\sin \delta y) - \delta y - [\sin(\Sigma_{n=1}^{k-1} \delta^n x_n) - (\Sigma_{n=1}^{k-1} \delta^n x_n)]_{k+1}$$

$$= (\sin \delta y) - \sin(\Sigma_{n=1}^{k-1} \delta^n x_n) - \{\delta y - (\Sigma_{n=1}^{k-1} \delta^n x_n)\} + O(\delta^{k+2})$$

$$= 2\sin\{(\delta y-\Sigma_{n=1}^{k-1} \delta^n x_n)/2\}\cos\{(\delta y+\Sigma_{n=1}^{k-1} \delta^n x_n)/2\}-\{\delta y-\Sigma_{n=1}^{k-1} \delta^n x_n\}+O(\delta^{k+2})$$

$$= O(\delta^{3k}) + O(\delta^{k+2}) = O(\delta^{k+2}).$$

Then the result follows by using (2.6) and Lemma 1. Q.E.D.

REFERENCES

[1] Y.Chang and H.Segur, An Asymptotic Symmetry of the Rapidly Forced qPendulum, forthcoming (accepted by Physica D).

[2] All references in [1].

[3] M.Kruskal and H.Segur, Asymptotics beyond all orders in a model of crystal growth, Stud. App. Math., to appear.

[4] Yi-Hua Chang, Ph.D thesis, SUNY/Buffalo, expected 1991.

[5] M.Kruskal, Asymptotic Theory of Hamiltonian and other Systems with all Solutions Nearly Periodic, J. Math. Phys. 3:806 (1962).

$$[\cos(\sum_{n=1}^{k}\delta^n x_n) - 1]_k = \cos(\sum_{n=1}^{k}\delta^n x_n) - O(\delta^{k+1})$$

$$[\sin(\sum_{n=1}^{k}\delta^n x_n)(\sum_{n=1}^{k}\delta^n x_n)]_{k} = [\sin(\sum_{n=1}^{k-1}\delta^n x_n)(\sum_{n=1}^{k-1}\delta^n x_n)] + O(\delta^{k+1})$$

So that

$$[\cos\delta y - 1 - [\cos\textstyle\sum \ldots - 1]]_k = \cos\delta y - \cos(\sum_{n=1}^{k}\delta^n x_n) + O(\delta^{k+1})$$

$$= 2\sin(\tfrac{1}{2}(\delta y + \sum_{n=1}^{k}\delta^n x_n))\sin(\tfrac{1}{2}(\delta y - \sum_{n=1}^{k}\delta^n x_n)) + O(\delta^{k+1})$$

$$= O(\delta^{k+1})O(\delta) + O(\delta^{k+1})$$

and

$$[\sin\delta y\,\delta y - [\sin(\sum_{n=1}^{k-1}\delta^n x_n)(\sum_{n=1}^{k-1}\delta^n x_n)]]_k$$

$$= (\sin\delta y - \sin(\sum_{n=1}^{k-1}\delta^n x_n))\delta y + [\ldots](\delta y - \sum_{n=1}^{k-1}\delta^n x_n) + O(\delta^{k+1})$$

EXPONENTIALLY SMALL OSCILLATIONS IN THE SOLUTION OF AN ORDINARY DIFFERENTIAL EQUATION

J.G.Byatt-Smith and A.M.Davie

Department of Mathematics, University of Edinburgh
The Kings Buildings
Edinburgh, EH9 3JZ, SCOTLAND

1. INTRODUCTION

The solutions of the equation

$$\varepsilon^2 y'' = y^2 - t^2 - 1 \quad , \quad -\infty < t < \infty \tag{1.1}$$

are discussed in the limit as $\varepsilon \to 0$. This equation arises as a connection problem in the theory of resonant oscillations in a tank[1,2]. The solutions of (1.1) which oscillate about $-|t|$ as $t \to \pm\infty$, have asymptotic expansions whose leading terms are

$$y \sim -|t| + A_{\pm}|t|^{-1/4}\cos\left[\frac{2\sqrt{2}}{3\varepsilon}|t|^{3/2} + \varphi_{\pm}\right] , \tag{1.2}$$

where A_+, φ_+, A_- and φ_- are constants. These asymptotic expansions are valid as $|t| \to \infty$ and do not require A_+, A_- or ε to be small. The choice of A_+ and φ_+ determine a unique solution to (1.1) which for fixed ε has no singularity on the real line if A_+ is small enough. In fact [1,2] there exists, $A_s(\varepsilon)$, with $A_s \to \infty$ as $\varepsilon \to 0$ such that all solutions with asymptotic expansion (1.2) as $t \to +\infty$ are well behaved on the real axis provided $A_+ < A_s$. For values of $A_+ > A_s$ the solution can develop a singularity on the real axis whose leading term is a double pole. These results have not been proved rigorously but have been established by numerical methods[1,2] which also suggests that the minimum value A_+, for the symmetric solution, becomes exponentially small as $\varepsilon \to 0$.

To investigate this phenomenon we define the nonoscillating solution $W(t)$ of (1.1) on $[0,\infty)$ whose asymptotic expansion as $t \to +\infty$ is

$$W(t) \sim -t - \tfrac{1}{2}t^{-1} + 0(t^{-3}) . \tag{1.3}$$

This is a solution of (1.1) which corresponds to $A_+ \equiv 0$ in (1.2). Again we regard this as an asymptotic expansion for large t which does not require ε to be small. The existence of this nonoscillating solution has

Asymptotics beyond All Orders, Edited by H. Segur *et al.*
Plenum Press, New York, 1991

also only been shown numerically. However existence clearly follows if the solutions which have asymptotic expansions given by (1.2) exist for all $A < A_s$.

When ε is small we can define different asymptotic expansions for W and the small amplitude oscillatory solutions about W, which are valid as ε and A tend to zero. The expansion for $W(t)$ is given by

$$W(t) = -(1+t^2)^{1/2}\left\{1 - \frac{{}^1/_2\varepsilon^2}{(1+t^2)^{5/2}} - \varepsilon^4\frac{(9/8 - 5t^2)}{(1+t^2)^5}\cdots\right\} \qquad (1.4)$$

The other oscillatory solutions can be characterised by

$$y(t) = W(t) + \frac{A}{(1+t^2)^{1/8}}\cos\left[\int_0^t \frac{\sqrt{2}}{\varepsilon}(1+s^2)^{1/4}ds + \varphi\right]+\ldots \qquad (1.5)$$

where $\varphi = \varphi_+ - \int_0^\infty \sqrt{2}\ \varepsilon^{-1}[(1+s^2)^{1/4} - s^{1/2}\}ds$. Although we now regard (1.4) and (1.5) as asymptotic expansions for fixed t as ε and $A \to 0$, they are, of course, also valid for fixed ε and A as $t \to \infty$.

An apparent difficulty is immediate. Although $W(t)$ and hence $y(t)$ are initially only defined on $[0,\infty)$, when extended to $(-\infty,\infty)$ $W(t)$ is an even function of t to all orders in ε. Similarly all of the oscillatory solutions $y(t)$, when extended to $(-\infty,\infty)$, have $A_+ = A_- = A$, again to all orders of ε, independent of the choice of φ. This must be compared with the numerical evidence that suggests that there is no solution which is nonoscillating on the whole interval $(-\infty,\infty)$ and that among the solutions $y(t)$, only those that are even have $A_+ = A_- = A$. If these two results are consistent $W(t)$ must be not quite symmetric with a first derivative which is zero to all orders and presumably exponentially small as $\varepsilon \to 0$ and that similarly $|A_+ - A_-|$ is also presumably exponentially small.

With this difficulty in mind we define $W(t)$ as the nonoscillating solution of (1.1) on the interval $[0,\infty)$ which satisfies (1.3) as $t \to \infty$ and (1.4) as $\varepsilon \to 0$ and has A identically zero, as opposed to being zero to all orders as $\varepsilon \to 0$. The property which distinguishes $W(t)$ from the oscillating solution is of course the difference in the decay rate to $-(1 + t^2)^{1/2}$ as $t \to \infty$. We also denote the boundary values of W at $t = 0$ by

$$W(0) = \mu \quad \text{and} \quad W'(0) = -\lambda, \qquad (1.6)$$

anticipating that $W'(0)$ is negative. The purpose of this paper is to show that:

i) The value of λ is exponentially small as $\varepsilon \to 0$.

ii) The common value A, of the even solution to (1.1) has a minimum value, A_{min}, which is also exponentially small as $\varepsilon \to 0$ with $A_{min} \sim \lambda$.

iii) The nonoscillatory solution $W(t)$ defined by (1.3) and (1.4) which has $A_+ \equiv 0$, when extended to $(-\infty,\infty)$ has $A_- \sim 2\lambda$ as $\varepsilon \to 0$.

The explanation is that $W(t)$ is not quite symmetric and as a result develops exponentially small oscillations for values of $t < 0$. We first establish by asymptotic methods the results ii) and iii) assuming that $W(t)$ exists and that λ is exponentially small. We then evaluate λ using the Kruskal-Segur method[3], which extends the solution $W(t)$ into the complex $t = p + iq$ plane. The purpose of extending the solution into the complex plane is to locate points where (1.4) fails and then to rescale the dependent and independent variables in the differential equation (1.1) to obtain a locally valid inner solution near these points.

We show in section 3 that for this problem the inner solution requires a solution of the equation

$$\frac{d^2Y}{d\sigma^2} = Y^2 - 2i\sigma , \tag{1.7}$$

the equation for the first Painlevé transcendent. We also show in this section, using matched asymptotic expansions that

$$\lambda \sim \text{constant} \times \varepsilon^{-1/2} e^{-\varepsilon^{-1}\psi}, \tag{1.8}$$

where

$$\psi = \sqrt{2} \int_0^1 (1-t^2)^{1/4} dt . \tag{1.9}$$

We then go on to rigorously prove this asymptotic estimate. The proof involves converting (1.1) and (1.7) to integral equations. The existence and uniqueness of the solution to these integral equations are established in the complex plane, by use of the contraction mapping theorem. This also establishes the existence and uniqueness of the nonoscillating solution $W(t)$ on the positive real axis. We also prove that the appropriate solution to (1.7) provides a uniformly valid approximation to $W(t)$ over a suitably defined region of the complex plane.

The success of the Kruskal-Segur method, depends in this case, on the crucial fact that, as $\varepsilon \to 0$, $W(t) \equiv W(p+iq) = W_r(p,q)+iW_i(p,q)$ has an exponentially small imaginary part, $W_i(0,q)$ on a segment of the imaginary axis $-1 < q < 1$. We formulate the problem for $W_i(0,q)$ and establish the existence and uniqueness of the solution. This solution, which is found in section 5 with strict error bounds, then allows the determination of $\lambda = \frac{\partial W_i}{\partial q}(0,q)$ at $q = 0$.

2. THE SMALL AMPLITUDE CONNECTION PROBLEM

In this section we focus our attention on the small amplitude connection problem for the equation

$$\varepsilon^2 y'' = y^2 - t^2 - 1, \quad -\infty < t < +\infty \tag{2.1}$$

We assume the existence of the nonoscillating solution $W(t)$ on the interval $[0,\infty)$ which has boundary values defined by (1.6) and asymptotic expansion (1.4) as $\varepsilon \to 0$. The existence and uniqueness of $W(t)$ is established in section 4. We now suppose that U is the even solution of (2.1) with $U(0) = \nu$. To treat the small amplitude case we linearise about W. Thus we let $V = U - W$, so that to first order

$$\varepsilon^2 V'' = 2VW , \tag{2.2}$$

with

$$V(0) = \nu - \mu, \quad V'(0) = \lambda . \tag{2.3}$$

When ε is small we have $W \simeq -(1+t^2)^{1/2}$ so if we write

$$\nu - \mu = r \cos\theta , \quad \lambda = r \sin\theta , \tag{2.4}$$

we have

$$V \simeq r(1+t^2)^{-1/8} \cos(\tau - \theta), \tag{2.5}$$

where

$$\tau = \frac{\sqrt{2}}{\varepsilon} \int_0^t (1+s^2)^{1/4} ds . \tag{2.6}$$

Comparing (1.2),(1.5) and (2.5) we see that the even solutions have

$$\varphi \equiv \varphi_+ \quad - \frac{\sqrt{2}}{\varepsilon} \int_0^\infty [(1+s^2)^{1/4} - s^{1/2}] ds = -\theta \tag{2.7}$$

and

$$A_+ \simeq r = \{(\nu-\mu)^2 + \lambda^2\}^{1/2} . \tag{2.8}$$

For fixed μ and λ the minimum value occurs at $\nu = \mu$ and takes the value

$$A_{min} = \lambda . \tag{2.9}$$

For the asymmetric problem we use the symmetry of the differential equation (2.1) to convert the range $(-\infty,0]$ to $[0,+\infty)$. Thus we define a solution $y(t)$ of (2.1) of the form

$$\begin{aligned} y(t) &= U_+(t) \quad , \quad t > 0 , \\ &= U_-(-t) \quad , \quad t < 0 . \end{aligned} \tag{2.10}$$

The functions $U_+(t)$ and $U_-(t)$ will then both satisfy (2.1) on $[0,+\infty)$ and the continuity of y and dy/dt at $t = 0$ requires that

$$U_\pm(0) = U_\pm(0) \text{ and } U_+'(0) = - U_-'(0). \tag{2.11}$$

Again linearising about W we define $V_\pm = U_\pm - W$ to be the solution of (2.2) subject to

$$V_\pm(0) = \nu_1 - \mu, \quad V'_\pm(0) = \pm\nu_2 + \lambda, \tag{2.12}$$

so that the corresponding values $(r_\pm, \theta_\pm)$ defined by

$$\nu_1 - \mu = r_+\cos\theta_+ = r_-\cos\theta_- \ , \tag{2.13}$$

$$\nu_2 + \lambda = r_+\sin\theta \quad \text{and} \quad -\nu_2 + \lambda = r_-\sin\theta_- \ , \tag{2.14}$$

satisfy

$$A_+\cos\theta_+ = A_-\cos\theta_- \ , \tag{2.15}$$

and

$$2\lambda = 2A_{min} = A_+\sin\theta_+ + A_-\sin\theta_- \ , \tag{2.16}$$

where, as before $A_\pm \simeq r_\pm$.

This represents the general transformation from (A_+,φ_+) to (A_-,φ_-) valid in the limit $\varepsilon \to 0$ and small A. From (2.16) we can immediately obtain $(A_-,\theta_-) = (2\lambda,\pi/2)$ as the image of $A_+ = 0$. Thus the nonoscillatory solution $W(t)$ when extended to $(-\infty,\infty)$ develops oscillations which are characterised by the value A_- given by

$$A_- = 2A_{min} = 2\lambda \ . \tag{2.17}$$

3. THE ASYMPTOTIC FORM OF λ AS $\varepsilon \to 0$

In this section we show that the solution $W(t)$ defined by the asymptotic expansion (1.3) and (1.4) has an exponentially small value of $W'(0)$ as $\varepsilon \to 0$. In order to do this, we treat (2.1) as a differential equation defined in the complex t-plane. The methods of this section follow the lines of Kruskal and Segur[3].

The basis of the method is that (1.4) is not a uniformly valid expansion in the entire complex plane as $\varepsilon \to 0$. The nth term in the series takes the form $\{\varepsilon(1+t^2)^{-5/4}\}^{2n} P_n(t^2)$, where P_n is a finite degree polynomial in t^2 with $P(-1)$ not equal to zero. In the complex plane this is an asymptotic series for all $t^2 \neq -1$, but it is not uniformly valid for values of t such that $t^2 + 1$ is of order $\varepsilon^{4/5}$. In fact because of this the series cannot be continued into $\operatorname{Re} t < 0$. This is because the continuation of $W(t)$ along the negative real axis develops exponentially small oscillations corresponding to $A_- = 2\lambda$. These oscillations are not picked up by the asymptotic expansion (1.4) since they are exponentially small along the (negative) real axis. However of course these oscillations grow exponentially as we move up into the complex plane. As we will see they have their origin in the "shadow" region formed by the singularities of the solution $W(t)$ near $t^2 + 1 = 0$.

To find an appropriate expansion of W near (say) $t = i$ we choose new dependent and independent variables

$$Y = \varepsilon^{-2/5} W \quad \text{and} \quad \sigma = (t - i)\varepsilon^{-4/5} .$$

In terms of these variables (2.1) becomes

$$Y_{\sigma\sigma} = Y^2 - 2i\sigma , \tag{3.1}$$

where terms of order $\varepsilon^{4/5}$ have been neglected. This is the equation for the first Painlevé transcendent[4]. The solutions to (3.1) which have $(1+|\sigma|^{1/2})^{-1} Y$ bounded as $\mathrm{Re}\, \sigma \to \infty$ have the asymptotic behaviour

$$Y = Y_0(\sigma) + A\sigma^{-1/8} \exp\left\{\frac{4}{5}\, 2^{3/4} e^{5\pi i/8}\, \sigma^{5/4}\right\}$$

$$+ B\sigma^{-1/8} \exp\left\{-\frac{4}{5}\, 2^{3/4} e^{5\pi i/8}\, \sigma^{5/4}\right\} + \dots , \tag{3.2}$$

where Y_0 has the asymptotic expansion

$$Y_0 = -(1+i)\sigma^{1/2} - \frac{1}{8}\sigma^{-2} + \dots \qquad \text{as } \sigma \to \infty . \tag{3.3}$$

The coefficients A and B in (3.2) must, of course, be suitably chosen so as to eliminate the exponential with a positive real exponent. The asymptotic form (3.2) allows us to match the "inner" solution Y to the "outer" oscillatory solutions of (2.1) which have the form given by (1.5). For this purpose we reinterpret the solution $y(t)$ in the complex plane and write

$$y(t) = W(t) + \frac{A}{(1+t^2)^{1/8}} \cos\left[\int_{t_0}^{t} \frac{\sqrt{2}}{\varepsilon}(1+s^2)^{1/4} ds + \bar{\varphi}.\right] \dots \tag{3.4}$$

Where now t_0 is a complex constant and the path of integration is taken along the curve from t_0 which ensures that the integral is real. If t_0 is in the first quadrant this dictates that the path from t_0 is concave upwards and is asymptotically parallel to the real t-axis as $t \to \infty$. When

t_0 is real the path of integration is along the real axis and the representations of the solution (3.4) and (1.5) differ only in the definition of the constants φ and $\bar{\varphi}$ with $\bar{\varphi} = \varphi + \int_0^{t_0}\sqrt{2}\ \varepsilon^{-1}(1+s^2)^{1/4}ds$. The path of integration from $t_0 = i$ is initially along $\arg(t-i) = -\pi/10$ which correspond to the ray along which the inner solution Y has purely oscillatory behaviour, that is the arguments of the exponentials are pure imaginary. Amongst these solutions the only solution which will match to $W(t)$ is the one with $A = B = 0$. This condition is known [5,6] to give the unique solution of (3.1) which has $(1+|\sigma|^{1/2})^{-1}\ Y$ bounded in the sector $-4\pi/5 < \arg\sigma + \pi/10 < 4\pi/5$, apart from a finite number of double poles.

To calculate the value of $W'(0)$, we need to obtain the asymptotic expansion for Y_0 along $\sigma = -i\eta$ with η real and $\eta \to +\infty$. This must then be matched to a solution of (2.1) along $\mathrm{Re}\ t = 0$ as $t \to i$ from below. The appropriate asymptotic expansion of Y_0 is

$$Y_0 = -\sqrt{2}\eta + \eta^{-2} + \ldots + C\eta^{-1/8}\exp\left\{-\frac{4}{5}2^{3/4}\eta^{5/4}\right\}+\ldots \qquad (3.5)$$

The value of C is of course uniquely determined by the criterion $A = B = 0$ in (3.2). As $\eta \to \infty$ the exponential term is of smaller order than any of the algebraic terms. However, all of the algebraic terms are real. Thus the imaginary part Y_{0i} of Y_0 has an asymptotic expansion whose leading term is

$$Y_{0i} = C_i\eta^{-1/8}\exp\left\{-\frac{4}{5}2^{3/4}\eta^{5/4}\right\}. \qquad (3.6)$$

It turns out that the imaginary part is all that is required to determine $W'(0)$. This is no accident since all of the terms of the asymptotic expansion (1.4) are real when $\mathrm{Re}(t) = 0$ so that the imaginary part of $W(t)$ on the imaginary axis must be exponentially small as $\varepsilon \to 0$. To obtain the solution for $W(t)$ along the imaginary axis we write $t = p + iq$ and express (2.1) as two real equations for W_r and W_i the real and imaginary parts of W. Then along $p = 0$ we obtain

$$-\varepsilon^2 W_r'' = W_r^2 - W_i^2 + q^2 - 1 \qquad (3.7)$$

and

$$-\varepsilon^2 W_i'' = 2W_iW_r, \qquad (3.8)$$

where we have used the notation $'' \equiv d^2/dq^2 = -d^2/dt^2$ along $p = 0$. We presume that W_i is exponentially small, so that to first order in ε we may take

$$W_r = -\sqrt{1-q^2} . \tag{3.9}$$

This represents the same first order term as that occuring in (2.4) expressed in terms of the variable $q \equiv -it$. The higher order terms then effectively generate the same series as (1.4). There are of course other solutions to (3.7) when W_i is neglected. These take a similar form to the solutions (1.5) of (2.1) but with the oscillatory trigonometric terms being replaced by terms which are exponentially growing or decaying as q decreases from $q = 1$. The growing term must be excluded and the decreasing term matches the form of the exponential term in (3.5), but is beyond all orders in ε. Then (3.8) is solved in the same fashion as (2.5) to obtain

$$W_i = (1-q^2)^{-1/8}[D \exp\left\{\frac{\sqrt{2}}{\varepsilon}\int_0^q (1-s^2)^{1/4}ds\right\}$$

$$+ E \exp\left\{-\frac{\sqrt{2}}{\varepsilon}\int_0^q (1-s^2)^{1/4}ds\right\}] , \tag{3.10}$$

to leading order.

We assume that in the required solutions for W_i, D and E are constants of the same order of magnitude, so that when $q = (1 - \varepsilon^{4/5}\eta)$ we may expand (3.10) to obtain

$$\varepsilon^{-2/5}W_i = 2^{-1/8}D\eta^{-1/8}\varepsilon^{-1/2}\exp\left\{\frac{\sqrt{2}}{\varepsilon}\int_0^1 (1-\varepsilon^2)^{1/4}d\varepsilon - \frac{4}{5}\,2^{3/4}\eta^{5/4}\right\}+\ldots \tag{3.11}$$

Matching (3.11) with (3.6) gives

$$D = \varepsilon^{1/2}C_i 2^{1/8} I(\varepsilon) , \tag{3.12}$$

where

$$I(\varepsilon) = \exp\left\{-\frac{\sqrt{2}}{\varepsilon}\int_0^1 (1-s^2)^{1/4}ds\right\} . \tag{3.13}$$

The second exponential appearing in (3.10) increases as q decreases from zero and the coefficient E can be obtained by using a similar analysis of the behaviour of the solution $W(t)$ near $t = -i$. However, this is unnecessary, since we know that W is real along $q = 0$. Thus

$W_i(0) = 0$ so that E must equal $-D$.

The final expression is obtained from the Cauchy-Riemann equation as

$$\lambda = -W'(0) \equiv -\frac{\partial W_r(0)}{\partial p} = -\frac{\partial W_i(0)}{\partial q} . \tag{3.14}$$

With $E = -D$ in (3.10) we obtain

$$W_i(0 + iq) = 2D(1 - q^2)^{-1/8}\sinh\left[\frac{\sqrt{2}}{\varepsilon}\int_0^q(1-s^2)^{1/4}ds\right] . \tag{3.15}$$

Therefore

$$\frac{\partial W_i(0)}{\partial q} = \frac{2\sqrt{2}}{\varepsilon} D . \tag{3.16}$$

Hence

$$\lambda = -2^{13/8}\varepsilon^{-1/2} C_i I(\varepsilon) , \tag{3.17}$$

Where $I(\varepsilon)$ is exponentially small as $\varepsilon \to 0$ and is given by (3.13). The constant C_i has not been determined by the present analysis but has been shown[7] to be non zero. Hence (3.17) shows that $W(t)$ is not an even function of t .

4. EXISTENCE OF SOLUTIONS IN THE COMPLEX PLANE

The method outlined in section 3 is typical of the method of Kruskal and Segur[3] and allows exponentially small quantities to be evaluated. This method has been applied to a variety of different equations many of which can be found in this book. Exponentially small estimates have been established for a number of other problems. They occur naturally in the theory of adiabatic invariants [8], and in references to the Soviet literature [9] . Other examples arise in the theory of the delayed onset to stability for the Hopf bifurcation with slowly varying parameter [10,11] and the theory of dendritic growth [12]. This last example concerns the original problem studied by Kruskal and Segur[3] and establishes the same estimate of the exponentially small quantity using analytical methods without the aid of asymptotic expansions.

In the next sections we prove rigorously the results of section 3 by constructing a contraction mapping for the appropriate solutions of (2.1) and (3.1) in the complex plane. This establishes the existence and uniqueness of the solution $W(t)$.

To study the solution of the differential equation

$$\varepsilon^2 \frac{d^2y}{dt^2} = y^2 - t^2 - 1, \quad Re(t) > 0, \tag{4.1}$$

we make the change of variables

$$\zeta = \varepsilon^{-1}\{-i\psi+\sqrt{2}\int_0^t (1+s^2)^{1/4}ds\}, \tag{4.2}$$

and

$$y = -(1+t^2)^{1/2} + \varepsilon^{1/2}(1+t^2)^{-1/8}v, \tag{4.3}$$

where

$$\psi = \sqrt{2}\int_0^1 (1-s^2)^{1/4}ds = \frac{2}{3}K(\tfrac{1}{2}). \tag{4.4}$$

Here K is the complete elliptic integral of the first kind with parameter m equal to $1/2$. The constant of integration ψ has been chosen so that $\zeta = 0$ when $t = i$ and is the same as the constants arising in (2.7) and (3.13). In terms of these variables (4.1) can be expressed as

$$\frac{d^2v}{d\zeta^2} + v = \frac{1}{2}\varepsilon^{3/2}(1+t^2)^{-15/8} + \frac{\varepsilon^2(4-5t^2)}{32(1+t^2)^{5/2}} v + \frac{1}{2}\varepsilon^{1/2}(1+t^2)^{-5/8} v^2. \tag{4.5}$$

We consider t and ζ as complex variables and the transformation $\zeta(t)$ maps $\{t\colon Re(t) \geqslant 0, \mathcal{I}m(t) \geqslant 0, \mathcal{I}m(\zeta(t)) \leqslant 0\}$ on to the region $\{\zeta\colon Re(\zeta) \geqslant 0, -\varepsilon^{-1}\psi \leqslant \mathcal{I}m(\zeta) \leqslant 0\}$ which we denote by D_ε. We note also that the lines $\mathcal{I}m(\zeta) =$ constant, are the paths in the t plane along which the integral, defined by (3.4), remains real.

It is also convenient to define a scaled variable ρ by

$$\rho = \frac{-2\sqrt{2}i\varepsilon^{-1}}{5}(1+t^2)^{5/4}, \tag{4.6}$$

the branch of the root being determined so that $(1+t^2)^{5/4}$ is real and positive when t is real.

Then

$$\frac{d\rho}{d\zeta} = -it, \tag{4.7}$$

so that near $t = i$

$$\zeta = \rho(1 + O|t-i|). \tag{4.8}$$

Using the definition of ρ we can write (4.5) as

$$\frac{d^2v}{d\zeta^2} + v = 2^{5/4}(5i\rho)^{-3/2} + \left[\frac{9}{4}(5i\rho)^{-2} - 5\varepsilon^{4/5}2^{-16/5}(5i\rho)^{-6/5}\right]v + 2^{1/4}(5i\rho)^{-1/2}v^2$$

$$= G(\rho, v, \varepsilon) \ . \tag{4.9}$$

We also define a limiting differential equation obtained from (4.9) by replacing ρ by ζ on the right hand side and putting $\varepsilon = 0$. This yields

$$\frac{d^2V}{d\zeta^2} + V = 2^{5/4}(5i\zeta)^{-3/2} + \frac{9}{4}(5i\zeta)^{-2}V + 2^{1/4}(5i\zeta)^{-1/2}\, V^2$$

$$= G(\zeta, V, 0) \ . \tag{4.10}$$

This clearly represents a limiting differential equation valid near $\zeta = \rho = 0$, corresponding to values of t near $t = i$. The additional change of variables

$$\zeta = \frac{2\sqrt{2}}{5i}(2i\sigma)^{5/4}, \qquad Y = -(2i\sigma)^{1/2} + (2i\sigma)^{-1/8}\, V, \tag{4.11}$$

shows that (4.10) is equivalent to (3.1).

In order to compare the solutions of (4.9) and (4.10) we look for solutions which decay as $\mathrm{Re}(\zeta) \to +\infty$, that is we exclude the oscillatory solutions which are merely bounded as $\mathrm{Re}(\zeta) \to +\infty$ with $\mathcal{I}m(\zeta)$ finite. Thus we fix p in the range $1/2 < p < 3/2$ and consider the Banach Space X_ε of continuous complex functions v on $D_\varepsilon^a = \{\zeta : -\varepsilon^{-1}\psi \leqslant \mathcal{I}m(\zeta) \leqslant -a, \mathrm{Re}(\zeta) \geqslant 0\}$, where a is a positive number to be determined, such that $\|v\| = \sup\{|v(\zeta)|\ |\zeta|^p\} < \infty$. Similarly we define the space X of all continuous functions on $D^a = \{\zeta : \mathrm{Re}(\zeta) \geqslant 0, \mathcal{I}m(\zeta) \leqslant -a\}$ analytic on the interior with $\|v\| = \sup\{|v(\zeta)|\ |\zeta|^p\} < \infty$. For $v \in X_\varepsilon$ and $u \in X$ we define integral operators $K_\varepsilon v$ and Ku by

$$K_\varepsilon v(\zeta) = -\int_\xi^\infty G(\rho(\tilde{\zeta}), v(\tilde{\zeta}), \varepsilon)\sin(\xi - \tilde{\xi})d\tilde{\xi} \tag{4.12}$$

and

$$Ku(\zeta) = -\int_\xi^\infty G(\tilde{\zeta}, u(\tilde{\zeta}), 0)\sin(\xi - \tilde{\xi})d\tilde{\xi}, \tag{4.13}$$

where $\zeta = \xi + i\eta$ and $\tilde{\zeta} = \tilde{\xi} + i\eta$. We note here that solutions of these integral equations $K_\varepsilon v = v$ and $Ku = u$ are solutions of (4.9) and (4.10) respectively and are such that u and v tend to zero as $\mathrm{Re}(\zeta) \to \infty$ and as such provide nonoscillating solutions of these equations.

We show that K and K_ε are contraction mappings on the sets $\{v:\|v\| \leqslant b\}$ in the respective spaces, provided a is large enough and b is chosen appropriately. To do this we need the following readily obtainable estimates for all $\zeta \in D_\varepsilon$.

$$|\rho| \geqslant M_1|\zeta| \quad \text{and} \quad |\zeta-\rho| \leqslant M_2\, \varepsilon^{4/5}|\zeta|^{9/5}\ , \tag{4.14}$$

and for $k > 1$

$$|1-(\zeta/\rho)^k| \quad \text{and} \quad |1 - \frac{d\rho}{d\zeta}\,(\zeta/\rho)^k| \leqslant C_k\, \varepsilon^{4/5}|\zeta|^{4/5}\ , \tag{4.15}$$

$$\left|\frac{d\rho}{d\zeta}(\zeta/\rho)^k\right| \leqslant C_k\ , \tag{4.16}$$

and

$$\int_\xi^\infty |\tilde{\zeta}|^{-k} d\tilde{\xi} \leqslant C_k|\zeta|^{-k+1}\ . \tag{4.17}$$

Here and subsequently M_i denote universal constants and C_k denote constants depending on k only.

We now suppose that $\zeta \in D^a$ and $u,v \in X$; then

$$|Ku(\zeta)-Kv(\zeta)|=\left|\int_\xi^\infty \sin(\xi-\tilde{\xi})(G(\tilde{\zeta},u,0)-G(\tilde{\zeta},v,0))d\tilde{\xi}\right|$$

$$\leqslant M_3\|u-v\|\{|\zeta|^{-1-p}+(\|u\|+\|v\|)|\zeta|^{1/2-2p}\}, \tag{4.18}$$

using the definition of the norm and the last estimate (4.17).

Now if $a \geqslant 1$ and $1/2 < p \leqslant 3/2$, we obtain

$$\|Ku(\zeta)-Kv(\zeta)\| \leqslant M_3\ a^{1/2-p}\|u-v\|(1+\|u\|+\|v\|), \tag{4.19}$$

since $|\zeta|^{-1} \leqslant a^{-1} \leqslant a^{1/2-p}$.

Similarly, using the estimate $|\rho|^{-k} < M_1^{-k}|\zeta|^{-k}$ from (4.14) we can obtain, for $\zeta \in D_\varepsilon^a$ and $u,v \in X_\varepsilon$,

$$|K_\varepsilon u-K_\varepsilon v| \leqslant M_4\|u-v\|\{|\zeta|^{-1-p} + \varepsilon^{4/5}\ |\zeta|^{-1/5-p}+(\|u\|+\|v\|)|\zeta|^{1/2-2p}\}, \tag{4.20}$$

so that

$$\|K_\varepsilon u-K_\varepsilon v\| \leqslant M_4\|u-v\|(a^{1/2-p}+ \varepsilon^{4/5}\ a^{-1/5})\|u-v\|(1+\|u\|+\|v\|). \tag{4.21}$$

To determine a bound on $\|Ku\|$ and $\|K_\varepsilon u\|$ we also need to consider the action of K and K_ε on the zero function. An integration by parts shows that

$$2^{-5/4}\ K0(\zeta) = -\int_{\xi}^{\infty}\frac{15i}{2}(5i\tilde{\zeta})^{-5/2}\cos(\xi-\tilde{\xi})d\tilde{\xi} + (5i\zeta)^{-3/2}, \qquad (4.22)$$

which immediately gives, provided $a \geqslant 1$

$$|K0(\zeta)| \leqslant M_5|\zeta|^{-3/2} \quad \text{and} \quad \|K0\| \leqslant M_5 a^{p-3/2} \leqslant M_5. \qquad (4.23)$$

A similar proceedure using the estimates of (4.14) and (4.16) gives

$$|K_\varepsilon 0(\zeta)| \leqslant M_6|\zeta|^{-3/2} \text{and } \|K_\varepsilon 0\| \leqslant M_6 \ . \qquad (4.24)$$

If we put $v = 0$ in (4.19) we now obtain

$$\|Ku\| \leqslant \|K0\| + \|Ku-K0\| \leqslant M_3 a^{1/2-p}\|u\|(1+\|u\|) + M_5 \ . \qquad (4.25)$$

We now take $b = 2M_5$, so that if $\|u\| \leqslant b$ we obtain the inequality

$$\|Ku\| \leqslant 2M_3M_5(1+2M_5)a^{1/2-p} + M_5 \leqslant b \ , \qquad (4.26)$$

provided

$$2M_3M_5(1+2M_5)a^{1/2-p} \leqslant M_5 \quad \text{or} \quad a^{p-1/2} \geqslant 2M_3(1+2M_5) \ . \qquad (4.27)$$

Lastly if $\|u\| \leqslant b$ and $\|v\| \leqslant b$ we have from (4.19)

$$\|Ku-Kv\| \leqslant M_3 a^{1/2-p}\|u-v\|(1+2b) \leqslant \tfrac{1}{2}\|u-v\|, \qquad (4.28)$$

provided

$$a^{1/2-p} \geqslant 2M_3(1+2b) > 2M_3(1+2M_5). \qquad (4.29)$$

Hence provided a is large enough, K is a contraction mapping on the set $\{u \in X, \|u\| \leqslant b\}$. In exactly the same manner we can show that K_ε is a contraction mapping, and we get fixed points $\tilde{u} \in X$ and $\tilde{u}_\varepsilon \in X_\varepsilon$ for K and K_ε respectively. It is clear that $\tilde{u}$ and $\tilde{u}_\varepsilon$ are independent of the choice of p and as we can take $p = 3/2$, we have $\tilde{u}$ and $\tilde{u}_\varepsilon$ are both of order $|\zeta|^{-3/2}$ for large ζ. The fixed point $\tilde{u}_\varepsilon$ of K_ε satisfies the integral equation $K_\varepsilon\ \tilde{u}_\varepsilon = \tilde{u}_\varepsilon$ and hence also satisfies (4.5) and via the transformation (4.3) provides the nonoscillating solution of (4.1). Similarly the fixed point $\tilde{u}$ of K satisfies $Ku = u$ and (4.10) and again via the transformation (4.11) provides the solution of (3.1) subject to the asymptotic behaviour of (3.2) with $A = B = 0$. We also note that the contraction mapping argument shows that both of these solutions are unique.

5 RIGOROUS ESTIMATES FOR λ

To establish rigorous estimates for λ , whose value was calculated in section 3, we first need to prove that $\tilde{u}$ is a uniformly valid approximation to the solution $\tilde{u}_\varepsilon$ as $\varepsilon \to 0$. For technical convenience we work on a line $\mathcal{I}_m(\zeta) = \eta$ where η is fixed, with $-\varepsilon^{-1}\psi \leqslant \eta \leqslant -a$. We also take the exponent p to be 7/10 . So we define χ to be the space of continuous functions v on $\{\zeta = \xi + i\eta,\ \xi \geqslant 0\}$ with $\|v\|_\chi = \sup(|v(\zeta)||\zeta|^{7/10}) < \infty$. We can define integral operators K and K_ε on χ as in section 4 and the fixed points $\tilde{u}$ and $\tilde{u}_\varepsilon$ are just the restrictions of the fixed points obtained in section 4.

We now wish to show that, for $v \in \chi$, $\|K_\varepsilon v - Kv\|_\chi = o(1)$ as $\varepsilon \to 0$. The reason for choosing $p = 7/10$ can be illustrated by taking $v \equiv 0$. As integration by parts yields

$$K_\varepsilon 0 - K0 = 2^{5/4}\{(5i\zeta)^{-3/2} - (5i\rho)^{-3/2}\}$$

$$+ 2^{1/4}\int_\xi^\infty 15i\{(5i\tilde{\rho})^{-5/2}\frac{d\tilde{\rho}}{d\tilde{\zeta}} - (5i\tilde{\zeta})^{-5/2}\}\cos(\xi-\tilde{\xi})d\tilde{\xi} \qquad (5.1)$$

The estimates of (4.14) yield the two different estimates

$$|\zeta^{-3/2}\ \ \rho^{-3/2}| \leqslant (1+M_1^{-3/2})|\zeta|^{-3/2} . \qquad (5.2)$$

and

$$|\zeta^{-3/2} - \rho^{-3/2}| \leqslant M_7\varepsilon^{4/5}|\zeta|^{-3/2+4/5} = M_7\varepsilon^{4/5}|\zeta|^{-7/10}. \qquad (5.3)$$

Using (5.2) with p = 3/2 yield only the estimate $\|K_\varepsilon 0 - K0\|x_\varepsilon = 0(1)$ as $\varepsilon \to 0$, whereas we wish to use (5.3) to obtain $\|K_\varepsilon 0 - K0\|_\chi = 0(\varepsilon^{4/5})$. On the other hand we need to use the fact that $\tilde{u}$ and $\tilde{u}_\varepsilon$ are of order $|\zeta|^{-3/2}$ as $Re(\zeta) \to \infty$ so that we assume that $v \in X_\varepsilon \subset \chi$ with $K_\varepsilon v,\ Kv \in X_\varepsilon \subset \chi$ and evaluate $\|K_\varepsilon v - Kv\|_\chi$ assuming that $|v| < b\,|\zeta|^{-3/2}$ on D_ε^a .

Using this philosophy it is straightforward to show that

$$|K_\varepsilon v(\zeta) - Kv(\zeta)| \leqslant \tfrac{1}{2}M_8\ \varepsilon^{4/5}|\zeta|^{-7/10} \qquad (5.4)$$

and hence that

$$\|K_\varepsilon v(\zeta) - Kv(\zeta)\|_X \leqslant \tfrac{1}{2}M_8\ \varepsilon^{4/5}, \tag{5.5}$$

provided $\|v\|X_\varepsilon \leqslant B$.

In particular taking $v = \tilde{u}$ so that $K\tilde{u} = \tilde{u}$ we find that

$$\|K_\varepsilon\tilde{u} - \tilde{u}\|_X \leqslant \tfrac{1}{2}M_8\ \varepsilon^{4/5}. \tag{5.6}$$

Since the contraction mapping argument works for an arbitrary p in $1/2 < p < 3/2$ we have from the estimate on K_ε corresponding to (4.26)

$$\|K_\varepsilon\tilde{u}-\tilde{u}_\varepsilon\|_X \leqslant \tfrac{1}{2}\|\tilde{u}-\tilde{u}_\varepsilon\|_X \tag{5.7}$$

so that

$$\|\tilde{u}-\tilde{u}_\varepsilon\|_X \leqslant \|K_\varepsilon\tilde{u}-\tilde{u}\|_X+\|K_\varepsilon\tilde{u}-\tilde{u}_\varepsilon\|_X \leqslant \tfrac{1}{2}M_8\varepsilon^{4/5}+\tfrac{1}{2}\|\tilde{u}-\tilde{u}_\varepsilon\|_X. \tag{5.8}$$

Thus

$$\|\tilde{u}-\tilde{u}_\varepsilon\|_X \leqslant M_8\ \varepsilon^{4/5}, \tag{5.9}$$

so that

$$|\tilde{u}-\tilde{u}_\varepsilon| \leqslant M_8\ \varepsilon^{4/5}|\zeta|^{-7/10}, \tag{5.10}$$

which holds for $\zeta \in D_\varepsilon^a$, since η is arbitrary subject to the restriction $-\varepsilon^{-1}\psi \leqslant \eta \leqslant -a$.

We now consider the behaviour of $\tilde{u}_\varepsilon$ and $\tilde{u}$ on the interval of the imaginary axis, $\zeta = i\eta$, with $-\varepsilon^{-1}\psi \leqslant \eta \leqslant -a$. We write $\tilde{u}(i\eta) = q(\eta)+ir(\eta)$ and $\tilde{u}_\varepsilon(i\eta) = q_\varepsilon(\eta)+ir_\varepsilon(\eta)$. Since the imaginary ζ axis $(-\varepsilon^{-1}\psi < \eta < 0)$ is transformed into the imaginary t axis with $0 < \mathcal{I}_m t < 1$, the definition of ρ shows that on this axis $\mathrm{Re}(\rho) = 0$ with $\mathcal{I}_m(\rho)<0$. So we denote $i\rho$ by ν so that ν is real and positive. Then taking imaginary parts of (4.9) with $v = \tilde{u}_\varepsilon$ we obtain

$$\begin{aligned} -\frac{d^2r_\varepsilon}{d\eta^2} + r_\varepsilon &= \left\{\frac{9}{4}(5\nu)^{-2}-5\varepsilon^{4/5}2^{-16/5}(5\nu)^{-6/5}+2^{1/4}(5\nu)^{-1/2}q_\varepsilon(\eta)\right\}r_\varepsilon(\eta) \\ &= F(\nu,q_\varepsilon,\varepsilon)r_\varepsilon(\eta). \end{aligned} \tag{5.11}$$

Since $\tilde{u}_\varepsilon$ is real on the real t axis corresponding to $\mathcal{I}_m(\zeta) = -\varepsilon^{-1}\psi$ we have $r_\varepsilon(-\varepsilon^{-1}\psi) = 0$. Let $\beta = r_\varepsilon'(-\varepsilon^{-1}\psi)$, then

$$r_\varepsilon(\eta) = \beta\sinh(\eta+\varepsilon^{-1}\psi)-\int_{-\varepsilon^{-1}\psi}^{\eta}\sinh(\eta-\tilde{\eta})F(\tilde{\nu},q_\varepsilon,\varepsilon)r_\varepsilon(\tilde{\eta})d\tilde{\eta} \tag{5.12}$$

Similarly $r(\eta)$ is bounded as $\eta \to -\infty$ so we can write the equation for r in a similar form with

$$r(\eta) = De^{\eta} - \int_{-\infty}^{\eta} \sinh(\eta-\tilde{\eta})F(-\tilde{\eta},q(\tilde{\eta}),0)r(\tilde{\eta})d\tilde{\eta} \tag{5.13}$$

for some constant D. It has been shown[7] that $D \neq 0$ and we will assume this.

We now define the space Z_ε of all continuous functions s on the interval $[-\varepsilon^{-1}\psi,-a]$ with norm $\|s\|z_\varepsilon = \sup(e^{-\eta}|s(\eta)|) < \infty$. On Z_ε we define the linear operator $\mathcal{L}_\varepsilon$ by

$$\mathcal{L}_\varepsilon s(\zeta) = -\int_{-\varepsilon^{-1}\psi}^{\eta} \sinh(\eta-\tilde{\eta})F(\tilde{\nu},q_\varepsilon,\varepsilon)s(\tilde{\eta})d\tilde{\eta} \ . \tag{5.14}$$

From section 4 we have the estimate

$$|q_\varepsilon(\eta)| \leq |\tilde{u}_\varepsilon(i\eta)| \leq b|\eta|^{-3/2} \ , \tag{5.15}$$

and since $\sinh(\eta-\tilde{\eta}) \leq 0.5\exp(\eta-\tilde{\eta})$, we can obtain, using the methods and estimates of section 4, that

$$\|\mathcal{L}_\varepsilon s\|z_\varepsilon \leq \frac{1}{2}\|s\|z_\varepsilon(M_9a^{-1}+M_{10}\varepsilon^{4/5}a^{-1/5}) \tag{5.16}$$

Thus by increasing a if necessary, we can ensure that

$$\|\mathcal{L}_\varepsilon\|z_\varepsilon < 1/5 \ . \tag{5.17}$$

We denote the function $\sinh(\eta+\varepsilon^{-1}\psi)$ by $S_\varepsilon(\eta)$, then $\|S_\varepsilon\|z_\varepsilon \leq 0.5\exp(\varepsilon^{-1}\psi)$ and we can write (5.10) as

$$r_\varepsilon + \mathcal{L}_\varepsilon r_\varepsilon = \beta S_\varepsilon \ . \tag{5.18}$$

Hence

$$\|r_\varepsilon\|z_\varepsilon \leq \frac{5}{4}|\beta|\,\|S_\varepsilon\|z_\varepsilon \quad \text{and} \quad \|\mathcal{L}_\varepsilon r_\varepsilon\|z_\varepsilon \leq \frac{1}{4}|\beta|\,\|S_\varepsilon\|z_\varepsilon \ , \tag{5.19}$$

so that

$$|\mathcal{L}_\varepsilon r_\varepsilon(-a)| \leq \frac{1}{4}|\beta|\,\|S_\varepsilon\|z_\varepsilon e^{-a} \leq \frac{1}{8}|\beta|e^{\varepsilon^{-1}\psi-a}. \tag{5.20}$$

Now by reducing ε if necessary we can make $\varepsilon \leq \psi/(1+a)$ so that we can ensure that

$$S_\varepsilon(-a) = \frac{1}{2}(e^{\varepsilon^{-1}\psi-a} - e^{-\varepsilon^{-1}\psi+a}) \geq \frac{1}{4}e^{\varepsilon^{-1}\psi-a} \tag{5.21}$$

Hence

$$|r_\varepsilon(-a)|=|\beta S_\varepsilon(-a)-\mathcal{L}_\varepsilon r_\varepsilon(-a)|\geq \frac{1}{8}|\beta|e^{\varepsilon^{-1}\psi-a}. \tag{5.22}$$

Since from section 4 we know that r_ε has a bound which is independent of ε, we can deduce that

$$|\beta| \leq M_{11}e^{-\varepsilon^{-1}\psi}, \tag{5.23}$$

so that

$$\|r_\varepsilon\|z_\varepsilon \leq M_{11} \quad \text{and} \quad |r_\varepsilon| \leq M_{11}\, e^{\eta} \tag{5.24}$$

Since $r_\varepsilon(\eta) \to r(\eta)$ for any fixed η as $\varepsilon \to 0$ we can conclude that

$$|r(\eta)| \leq M_{11}e^{\eta} \quad \text{for all} \quad \eta \leq -a . \tag{5.25}$$

Using a similar argument to that used to derive (5.9) and the inequality $\sinh(\eta-\tilde{\eta}) \leq 0.5\exp(\eta-\tilde{\eta})$ we can show that

$$\|\mathcal{L}_\varepsilon r + r - 2De^{-\varepsilon^{-1}\psi}S_0\|z_\varepsilon \leq M_{12}\,\varepsilon^{4/5}. \tag{5.26}$$

where $0.5\exp(\varepsilon^{-1}\psi+\eta)$ is denoted by $S_0(\eta)$.

But since $S_\varepsilon(\eta) - S_0(\eta) = -0.5\exp(-\eta-\varepsilon^{-1}\psi)$ we can easily obtain the result

$$e^{-\varepsilon^{-1}\psi}\|S_\varepsilon(\eta) - S_0(\eta)\|z_\varepsilon = o(\varepsilon), \tag{5.27}$$

so that

$$\|\mathcal{L}_\varepsilon r + r - 2De^{-\varepsilon^{-1}\psi}S_\varepsilon\|z_\varepsilon \leq M_{13}\,\varepsilon^{4/5} \tag{5.28}$$

Substituting for S_ε using (5.17) we obtain

$$\left\|(1 + \mathcal{L}_\varepsilon)\left\{r - \frac{2D}{\beta}\, e^{-\varepsilon^{-1}\psi}\, r_\varepsilon\right\}\right\|_{z_\varepsilon} = M_{13}\varepsilon^{4/5}, \tag{5.29}$$

so that using (5.16) we get

$$\|r - \frac{2D}{\beta}e^{-\varepsilon^{-1}\psi}r_\varepsilon\|z_\varepsilon \leq \frac{5}{4}\, M_{13}\varepsilon^{4/5}. \tag{5.30}$$

In particular

$$|r(-a)- \frac{2D}{\beta}\, e^{-\varepsilon^{-1}\psi}r_\varepsilon(-a)| \leq \frac{5}{4}\, M_{13}e^{-a}\varepsilon^{4/5}, \tag{5.31}$$

and since from (5.9)

$$|r(-a) - r_\varepsilon(-a)| = O(\varepsilon^{4/5}), \tag{5.32}$$

we obtain

$$\beta = 2De^{-\varepsilon^{-1}\psi}(1 + O(\varepsilon^{4/5})). \tag{5.33}$$

Returning to the original equation (4.1), we find that for the solution y corresponding to the solution ũ of (4.1)

$$\frac{dy(0)}{dt} = \varepsilon^{\frac{1}{2}} \frac{dv}{d\zeta}(-i\varepsilon^{-1}\psi)\frac{d\zeta(0)}{dt} = \varepsilon^{\frac{1}{2}}\frac{dr_\varepsilon}{d\eta}(-\varepsilon^{-1}\psi).\sqrt{2}\varepsilon^{-1} = \sqrt{2}\varepsilon^{-1/2}\beta \tag{5.34}$$

Thus

$$\frac{dy(0)}{dt} = 2\sqrt{2}\varepsilon^{-1/2}De^{-\varepsilon^{-1}\psi}(1+O(\varepsilon^{4/5})), \tag{5.35}$$

allowing for the additional factor of $2^{-1/8}$ introduced by the transformation (4.11) this agrees with (3.17) thus establishing rigorously the result obtained by section 3.

REFERENCES

1. J.G.B.Byatt-Smith, Stud.Appl.Math. 79 (1988), 143.
2. J.G.B.Byatt-Smith, Stud.Appl.Math. 80 (1989), 109.
3. M. Kruskal and H. Segur, To be published.
4. E.L. Ince, Ordinary Differential Equations (New York: Dover 1956)
5. P. Boutroux, Ann.Sci.Ecole Norm.Sup.(3) 30 (1913), 255.
6. E.Hille, Lectures on Ordinary Differential Equations (New York: Addison Wesley, 1969),ch.12.4,693.
7. J.G.B.Byatt-Smith and A.M.Davie, Proc.Roy.Soc.Edin.114A (1990), 243.
8. P.B. Chapman and J.J.Mahony, SIAM.J.App.Math.34 (1978), 303.
9. V.I.Arnold, Geometric Methods in the theory of Ordinary Differential Equations, 2nd Edn.(Berlin: Springer, 1988).
10. A.I. Neistadt, Differential'nye Uraveniya 23 (1987), 2060: English Translation, J.Differential Equations 23 (1987) 1385.
11. A.I. Neistadt, Differential'nye Uraveniya 24 (1988) 226: English Translation, J.Differential Equations 24 (1987) 171.
12. C.J.Amick and J.B.McLeod, Arch.Rat.Mech. 109 (1990) 139

SINGULAR PERTURBATION OF SOLITONS

YVES POMEAU
Laboratoire de Physique Statistique
24, Rue Lhomond
75231, Paris Cedex 05, France

ABSTRACT

Long wavelength perturbation theory is at the heart of many physical approximations. It amounts to neglect derivatives and nonlinearities of higher orders, in a rational expansion. However, it may happen, that, this is formally consistent; yet the ensuing series is diverging, and so one has to investigate what really happens beyond all algebraic orders. Below, I look at a particular example of this situation, i.e. to the robustness of soliton-like solutions of the Korteweg-de Vries equation when higher order derivatives with respect to the space variable are included. This is an abridged version of a joint work with Alfred Ramani and Basile Grammaticos [1].

The Korteweg-de Vries (KdV) equation, famous for having been solved by inverse scattering methods [2], appears in various physical contexts, and is generic in the sense that it is the nonlinear equation always obtained by balancing nonlinearities with dispersion effects for weakly dispersive waves and for conservative systems in one space dimension(note that, for waves with arbitrary dispersion, the generic envelope equation is nonlinear Schrödinger). One interesting feature of this equation is precisely that it is parameterless, because it balances different physical effects with different scalings. Indeed, given a physical problem such as long gravity waves in shallow water, KdV is not exact and it is possible to continue the expansion beyond the order where Korteweg and de Vries stopped (see the contributions by Beale, VandenBroek and by Sun and Sen at this conference for more details on this). This yields subdominant terms, and they are many of them, even at the next order. In principle a calculation aiming at retaining the non perturbative terms, as presented below, should keep any term in the expansion beyond

Asymptotics beyond All Orders, Edited by H. Segur *et al.*
Plenum Press, New York, 1991

KdV. This is not what we shall do here, but we shall merely consider a model keeping one term only beyond the usual one, so that we shall sudy the partial differential equation:

$$\delta^2 \frac{\partial^5 u}{\partial x^5} + \frac{\partial^3 u}{\partial x^3} + u\frac{\partial u}{\partial x} + \frac{\partial u}{\partial t} = 0 \tag{1}$$

We wrote (1) in a form that emphasizes that it contains one parameter, δ, that is small in the KdV limit. One cannot get rid of this parameter by rescaling without changing other terms in the equation, which is obvious, and-most importantly-the sign itself cannot be changed. This means that, to be consistent, we should keep the two possibilities for this sign. We shall later comment on the present choice of sign.

As such this equation has no nontrivial and explicit solution, at least for arbitrary δ; such explicit solutions exist however for some special choices of δ together with convenient changes in the form of the equation itself(see [1] for the connection with the present model). Thus it seems natural to try a perturbation theory, for small δ. This perturbation theory is, in general, very complicated, even though one knows explicitly, via inverse scattering the solution of (1) at δ=0. We shall deal with a much more modest problem, that is the effect of δ upon the elementary one-soliton solution of (1) with δ=0. Owing to the simplicity of the equation at hand, the calculations can be made rather completely. So we want to investigate solutions of (1) that can be written in the form of traveling wave, i.e. they depend on x ant t through the combination ξ=x-Ct. For δ=0, there is a well-known continuum of such solutions:

$$u_0(x,t) = \frac{C}{2\mathrm{ch}^2[\frac{C^{1/2}\xi}{2}]} \tag{2}$$

In general, a traveling wave solution of (1) will be given by the solution of:

$$\delta^2 \frac{d^5 u}{d\xi^5} + \frac{d^3 u}{d\xi^3} + u\frac{du}{d\xi} = C\frac{du}{d\xi} \tag{3}$$

A localized solution of (3), as the one we are loooking at, has to decay to zero at ξ→±∞. Proceeding as Kruskal and Segur [3] for the geometrical model, we count the number of free parameters in (3), owing to this constraint. Asymptotically, the solution of (3) has to satisfy the equation linearized near u=0, that is:

$$\delta^2 \frac{d^5 v}{d\xi^5} + \frac{d^3 v}{d\xi^3} = C\frac{dv}{d\xi}$$

This can be integrated once, and depends on ξ as:

$v = e^{\pm\kappa\xi}$, or $e^{\pm ik\xi}$,

where κ^2 and $-k^2$ are respectively the positive and negative root of the polynomial in "a":

$$\delta^2 a^2 + a = C.$$

For δ small, those roots are $\kappa^2 = C - \delta^2 C^2 + ...$ and $k^2 = \delta^{-2} + ...$ To have a solution decreasing at $-\infty$, one must cancel three prefactors of the nondecreasing exponentials. similarly, at $+\infty$ one has three other conditions. This set of 6 conditions overdetermines the solution of a fifth order differential equation. The same sort of counting argument shows that by changing δ^2 into $(-\delta^2)$ one would lose enough conditions to recover a priori soliton-like solutions.

Thus the best one can hope with the original δ^2-term is a soliton-like solution but with an oscillating tail. Admitting this tail on either side relaxes 4 conditions, amply enough to allow such a solution.

We shall consider now the question of the amplitude of the tail on both sides of the soliton-like solution of (1) near $\delta=0$. We rely for computing this on the general method of Kruskal and Segur.

At any order of the regular perturbation expansion in powers of δ^2 any term oscillating as $\exp(\pm i\frac{\xi}{C})$ will be absent. This implies that the amplitude of those fast oscillations is a transcendental function of the small parameter δ^2. To compute it, we look at the region of the complex ξ–plane where the regular expansion and the transcendentally small terms on the real axis mix together, and make a boundary layer analysis of the neighborhood of this singularity to calculate the amplitude of the transcendentally small term. In the present case, it is also feasible to make a Borel-summation of the perturbation series, which yields this amplitude more directly.

As usual in this type of problem, the regular expansion becomes more and more singular at increasing orders near the singularities of the zeroth-order solution as given in (2), that is near the zeroes of $ch^2[\frac{C^{1/2}\xi}{2}]$, at $\xi_n = \frac{i\pi(2n+1)}{C^{1/2}}$, n integer. Near anyone of those singularities, one rescales u and ξ as:

$$\Xi = \frac{\xi - \xi_n}{\delta}, \quad U = \delta^2 u^2 .$$

This rescaling has two virtues. First, it is consistent with the $\frac{1}{\xi^2}$ behavior (without any δ dependance) of u_0 near the singularity, then it allows to get rid of one of the term in (3), that is the one proportional to C, and the inner (parameterless) equation reads:

$$\frac{d^5U}{d\Xi^5} + \frac{d^3U}{d\Xi^3} + U\frac{dU}{d\Xi} = 0 \qquad (4)$$

The dominant singularities of u_0 are nearest to the real axis at n=±1, or $\xi=\pm\frac{i\pi}{C^{1/2}}$.

One may define precisely the terms "beyond all orders" in the expansion of u, because the regular expansion is made of even powers of ξ with real coefficients, and is real on the imaginary axis. Thus any nonzero imaginary contribution to u on this imaginary axis represents this transcendentally small part of u. This avoids the difficulty of defining transcendentally small corrections to a diverging series...

As is usual in this matter, the asymptotic behavior of the solutions of the inner equation defines the amplitude of the high frequency-or WKB-part of the solution. On linearizing the nonlinear equation (4) about tye algebraic asymptotic behavior $-\frac{2}{\Xi^2}$ and looking for acceptable WKB solutions, one finds:

$U(\Xi)=-\frac{2}{\Xi^2}+...+A_\pm\, e^{\pm i\Xi}$, where $A_\pm$ keep constant values in sectors bounded by the Stokes lines Im Ξ =0. Coming back to the original problem, one realizes that-up to elementary transformations-those coefficients $A_\pm$ yield precisely the asymptotic oscillating behavior of the tail of the soliton-like solution of (3). As the inner equation (4) is of fifth order, five independent conditions can be imposed. The decay as $-\frac{2}{\Xi^2}$ at ±∞ imposes two conditions, as it could be changed into $-\frac{2}{(\Xi-\Xi_0)^2}$, Ξ_0 arbitrary. Furthermore only one of the $e^{\pm i\Xi}$ term can be retained at ±∞, because otherwise the tunneling factor would make appear transcendentally large (instead of small) corrections on the real axis. This imposes one condition at +∞ and one at -∞. One last free parameter can be chosen to be the amplitude of the wave at +∞ (or -∞) in which case the amplitude is automatically determined at -∞ (or +∞). In physical terms, one could say that in the frame of the soliton, one may have a kind of arbitrary balance between waves coming from +∞ or from -∞. Whence the free parameter.

All this boils down to the calculation of a pair of numbers, A_+ and A_-. This can be done numerically or by using the method invented by V. Hakim [4] that relies upon the Borel summation of the asymptotic solution of (4) in inverse powers. Let us give some indications on this last method. More details may be found in reference [1] as well as in the communication by Hakim at this conference.

We write first the Laurent series for the asymptotic behavior of (4) as:

$$U=-\frac{2}{\Xi^2}+\sum_{n=2}^{\infty}\frac{a_n}{\Xi^{2n}} \tag{5}$$

which yields a recursion relation for the a_n's. Indeed one is interested in the large n-behavior of this recursion. This is dominated by the terms coming from the linear part of the equation (5) and thus can be solved to give:

$a_n \approx K(-1)^n (2n-1)!$, and the number K itself can be determinated by numerically determining a_n up to a value large enough so that the asymptotic relation holds. Then the coefficient of $a_n \frac{(-1)^n}{(2n-1)!}$ will be K. A slightly refined version of this method gives $K \approx 19.969...$

Now we notice that the series (5) is Borel summable, in other terms U(.) can be formally written as:

$$U(\Xi) = \int_0^\infty dt\, e^{-t}\, V\left(\frac{t}{\Xi}\right) \qquad , \qquad (6)$$

where V(z) is given by series convergent inside the unit circle and with a logarithmic singularity at $z=\pm i$. This singularity prevents an easy ofthe integral on the right hand side of (6), when computed for imaginary Ξ, since then the integrand is singular at $t=\pm i\Xi$. One could see the Borel integral (6) by defining the solution of the inner equation (4) on the real Ξ-axis; then try to analytically continue this solution to the real x-axis, that is in the large negative imaginary direction for Ξ. But this cannot be done so simply because one does not know the analytic structure of V(.) for arbitrary arguments. We have thus to assume that the asymptotic behavior at large Ξ of the integral is dominated by the singularity of V (.) closest to the origin, that is the logarithmic branch point we alluded to already.

By deforming the contour in t, in order to reach the real x-axis (again a range of values of Ξ with a large negative imaginary part) one has to specify how to turn around the logarithmic singularity of the integrand. This singularity is of the form:

$$V = K \ln\left(1 + i\frac{t}{\Xi}\right).$$

The choice of the integration path for t on the Riemann sheet of the logarithm, allows to specify the asymptotic behavior of $U(\Xi)$, and thus of U(x).

As explained in ref. [1], if one wants for instance a solution without oscillating tail at $x=-\infty$, one has to keep the $i\pi$ contribution to the logarithm for t large positive. This yields the following asymptotic contribution to u for ξ large positive:

$$u \approx \frac{4\pi}{\delta^2} K \sin\left(\frac{\xi}{\delta}\right) \exp\left(-\pi\frac{\delta}{C}\right) \text{ for } \xi >> 1.$$

One could similarly impose the solution to be even, which gives:

$u' \approx \frac{2\pi}{\delta^2} K \sin(\frac{\xi}{\delta}) \exp(-\pi\frac{\delta}{C})$ for $\xi >> 1$ and minus the same expression for ξ large negative.

As noticed in ref. [1] these solitons with an infinite tail have an infinite energy and so cannot result from the decay of a localised solution with finite energy, because energy is conserved in the course of time. If one chooses such a localised function as an initial condition of (1) with δ small, it will form first well defined solitons of unperturbed KdV and then decay very slowly by radiating in an attempt to build their infinite "tails".

If one looks at the next order term of the dispersion relation for shallow water waves, one gets an equation similar to (1), but with the opposite sign in front of the fifth derivative. Counting then the number of free parameters, one finds that, contrary to the case studied here, a continuum of localised solitons exists, with a finite energy. It is well possible however that in other physical situations the sign is the same as studied in the present work.

REFERENCES

[1] Y. Pomeau, A. Ramani and B. Grammaticos, Physica D**31**, (1988), 127.

[2] A.C. Newell "Solitons in mathematics and physics", SIAM pub. Philadelphia (1984) and references therein.

[3] M. Kruskal and H. Segur, ARAP Tech. Memo, 85-25 (1985).

[4] T. Dombre, V. Hakim and Y. Pomeau, Comptes-Rendus de l'Académie des Sciences, **302** (1986) 803.

REFLECTION COEFFICIENT BEYOND ALL ORDERS FOR SINGULAR PROBLEMS

Jishan Hu * and Martin D. Kruskal **

* Department of Mathematics
West Virginia University
Morgantown, WV 26506

**Department of Mathematics
Rutgers University
New Brunswick, NJ 08903

ABSTRACT

This paper generalizes the Pokrovskii-Khalatnikov method to calculate the actual behavior of the reflection coefficient for singular problems, which vanishes to all orders in the small parameter ϵ. Two different classes of reflection coefficient problems are considered.

1. INTRODUCTION

In many physical problems, one needs to study the behavior of solutions of the equation

$$\epsilon^2 \frac{d^2y}{dx^2} + Q(x)y = 0, \tag{1.1}$$

where ϵ is a small positive parameter. Assume that the function Q is positive and analytic on the real axis with

$$Q(x) = Q_\pm + O(1/x^{1+\alpha}), \qquad x \to \pm\infty, \alpha > 0, \tag{1.2}$$

for some $Q_\pm \neq 0$ and let y be the solution of (1.1) satisfying the BC:

$$\begin{cases} y(x,\epsilon) \sim T(\epsilon) \cdot \exp\{-iQ_-^{1/2} x/\epsilon\}, & x \to -\infty, \\ y(x,\epsilon) \sim 1 \cdot \exp\{-iQ_+^{1/2} x/\epsilon\} + R(\epsilon) \cdot \exp\{+iQ_+^{1/2} x/\epsilon\}, & x \to +\infty, \end{cases} \tag{1.3}$$

for some arbitrary constants R, T, called the reflection and transmission coefficients, respectively. Classical analysis indicates that R and T are determinated uniquely as function of ϵ (rf. 1). For small ϵ, y can be represented by an asymptotic series in powers of ϵ to all orders, and $R(\epsilon)$ and $T(\epsilon)$ likewise. It has been known for decades that R has the expansion:

$$R = 0 + 0 \cdot \epsilon + 0 \cdot \epsilon^2 + \cdots. \tag{1.4}$$

Asymptotics beyond All Orders, Edited by H. Segur *et al.*
Plenum Press, New York, 1991

This is the so-called reflection coefficient problem. If we think x as the time variable, this problem is equivalent to the adiabatic invariant problem for harmonic oscillator. One of the challenging problems is to find the actual behavior of R. Many authors have expended a great deal of effort in this direction (rf. 2, 8, 10, 11).

A zero or singularity of the function Q is called a critical point of equation (1.1). In this paper, we will consider the case that the function Q has a finite number of critical points on the real axis. At singularities, equation (1.1) breaks down and we then have to define how to solve it. We avoid the singularities by solving equation (1.1) along a path in the lower-half complex plane. This definition is justified by some physical applications. For example, when we study large-scale waves in the atmospheric westerly winds, an investigation of barotropic instability problem leads to the form of equation (1.1) where the critical layer corresponds to the singularities of the function Q (rf. 3). The reason that we solve the differential equation slightly below the real axis has a clear physical meaning. Because we are looking for a solution which is exponentially growing in time, the phase speed therefore has a small but positive imaginary part. So, the singularities are located slightly above the real axis. As the imaginary part of the phase tends to zero, we have to indent the integration contour down to the lower half-plane.

Since we are interested in studying how waves propagate, we naturally assume either Q_- or Q_+ positive. Without loss of generality, in this paper we assume that the wave is being sent in from positive infinity. Thus we assume $Q_+ > 0$. Notice that we do not impose any restriction on the sign of Q_-. If $Q_- > 0$, by the classical theory of ordinary differential equations there is a solution of equation (1.1) satisfying the BC (1.3) (rf. 1). If $Q_- < 0$, the theory implies that there is a solution satisfying

$$\begin{cases} y(x,\epsilon) \sim T \cdot \exp\{(-Q_-)^{1/2} x/\epsilon\}, & x \to -\infty, \\ y(x,\epsilon) \sim 1 \cdot \exp\{-iQ_+^{1/2} x/\epsilon\} + R \cdot \exp\{+iQ_+^{1/2} x/\epsilon\}, & x \to +\infty \end{cases} \tag{1.5}$$

for some constants T and R. Now we have two types of well-defined reflection coefficient problems.

Our primary interest is to find the value of the reflection coefficient R. Just as in the "classical" case that the function Q is positive and analytic on the real axis, when Q is even negatively valued on the real axis, it turns out that the reflection coefficient has zero expansion in ϵ to all orders under some fairly general conditions. This can be done by solving equation (1.1) along any line underneath the real axis. Thus, to find its nontrivial behavior, analysis beyond all orders is required.

In 1061, Pokrovskii and Khalatnikov [PK] supplied a very simple approach to compute the reflection coefficient in the classical case (rf. 11). In this paper, we will generalize the PK method to allow the function Q some singularities on the real axis and/or some changes of sign. We will show that the amplitude of the reflection coefficient, for a large class of functions Q, does not depend on the singularities on the real axis, but rather on the "nearest" (in a certain sense) critical points in the lower half-plane.

2. STRUCTURE OF LEVEL LINES

We denote $Q(z)$ the unique analytic continuation of the real holomorphic function (except for a finite number of singularities on the real axis). As the same as the PK method, we will solve the equation in the complex plane in order to find R beyond all orders. To accomplish this goal, we must move the solutions down to the "level lines" of $\left|\exp\left\{\frac{i}{\epsilon}\int^z [Q(t)]^{1/2}\,dt\right\}\right|$ where both the WKB exponentials have the same magnitude:

$$Im\left\{\int^z [Q(t)]^{1/2}\,dt\right\} = \text{const}, \tag{2.1}$$

otherwise we will lose one of the exponentials. Furthermore, this line must pass through zeroes or singularities of the extended function Q, otherwise we cannot have the leading behavior of the reflection coefficient merely a bound on it. Passing through zeroes or singularities may produce some new information since one set of WKB coefficients of a local problem is valid only in a region bounded by Stokes lines emanating from these critical points.

To see geometric structure of level lines near a power-type critical point of equation (2.1), we have the following lemma.

Lemma 1 Suppose that in a neighborhood of z_0 the function Q is analytic except at z_0. Assume that

$$Q(z) \sim b(z - z_0)^{2\gamma-2}, \qquad z \to z_0, \tag{2.2}$$

where $b \neq 0$ is a constant and γ is a real number. Then, in the leading order, the structure of the level lines of (2.1) near $z = z_0$ can be divided topologically into four different classes:
(1) If $\gamma < 0$, they consist of rose curves, and the angle of each leaf is $\pi/|\gamma|$;
(2) If $\gamma > 0$, they consist of hyperbola-like curves, and the angle of each leaf is $\pi/|\gamma|$;
(3) If $\gamma = 0$,
  (a) if $Re\ b^{1/2} \neq 0$, they consist of an infinite number of spirals intersecting at $z = z_0$;
  (b) if $Re\ b^{1/2} = 0$, they consist of concentric circles centered at $z = z_0$. ∎

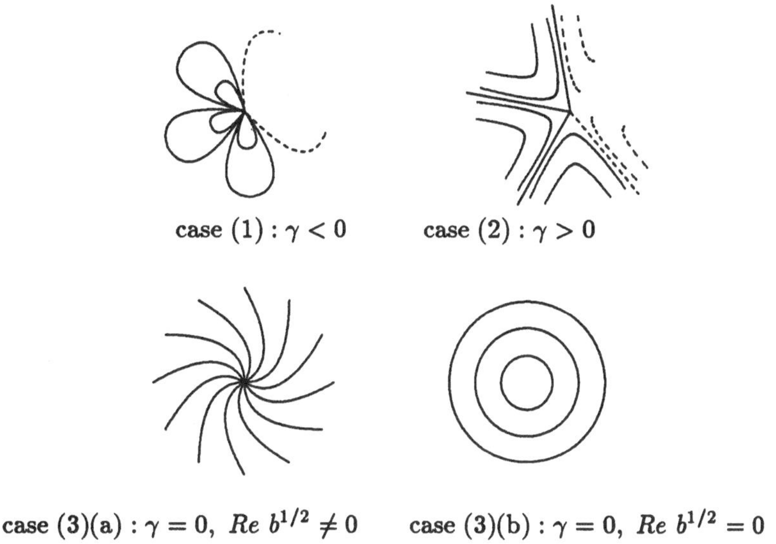

FIGURE 1 The structure of level lines near a critical point

In this paper, we only state the conclusions without proving. We will give proofs elsewhere (rf. 5).

Remark 1 In Lemma 1, if to higher order, the case (3)(b) may not be a family of circles anymore. Its graphics could consist of a family of infinite spirals centered at z_0 or a family of multi-revolution close curves.

It is convenient to call a critical point $z = z_0$ of Q with (2.2) regular if $\gamma > 0$. A critical point is irregular otherwise.

It is easy to see that on the real axis, if the function Q is positive at an analytic point, then the level lines (2.1) are parallel to the real axis in a neighborhood of this point; if negative, the level lines are perpendicular to the axis. In particular, near $+\infty$ the level lines are parallel to the axis since the function Q has a finite positive limit as $x \to +\infty$ and near $-\infty$, they are parallel or perpendicular to the axis accordingly as Q_- is positive or negative.

In general, we cannot extend a solution of equation (1.1), that is a combination of two exponentials across the level lines of (2.1). The reason is that as we cross the level lines, one exponential increases and the other decreases rapidly which makes one exponential negligible compared to the other and leads to a wrong asymptotic representation for trying to get information beyond all orders. However, there are a few places in the complex plane where we can do this extension. One of them is the place where the asymptotic series is convergent, for instance, at $-\infty$, and $+\infty$. Another way to move across level lines is in the case that a solution under consideration has only one exponential. In this case, we can always extend the solution across the level lines in the direction where the solution is increasing. This can be done because then the other exponential is decreasing and never appears in the sense of Poincaré's definition of asymptotic series.

To solve the reflection coefficient problems considered above beyond all orders, we shift the path along which we solve equation (1.1) from the real axis, down into the lower half-plane to a level line on which we first meet a critical point of equation (1.1). We call such a level line the nearest critical level line, and call a critical point major if it is on the nearest critical level line. Here we don't count any irregular critical point on the real axis as a major critical point because, the WKB approximation is valid in a neighborhood of such a point and, on the other hand, by prescription, we always have to leave the real axis and solve equation (1.1) in the lower half-plane. To help us understand this, we imagine that irregular singularities of this kind are located slightly above the real axis.

The structure of level lines in the complex plane can be very complicated. For a given function Q, there is not necessarily any major critical point. And if a major critical point does exist, it needs not unique.

In this paper, we make the following assumptions on the function Q.

Assumption:
(1) There is a unique nearest critical level line in the lower half-plane;
(2) There is no limit point of critical points of equation (1.1) on the nearest critical level line;
(3) All major critical points are regular.

3. CONNECTION FORMULA FOR SEPARATED CRITICAL POINTS

In this section, we will establish a connection formula of a solution of equation (1.1) near an isolated critical point. This result will show us how a solution changes from one branch of a level line of (2.1) to the other branch through a regular critical point.

Lemma 2 Let L_j, $j = 1, 2$, be two consecutive branches of a level line L. On L, there is a regular critical point, $z = z_0$, of equation (1.1) which connects L_1 and L_2. Near $z = z_0$

$$Q(z) \sim b(z - z_0)^{2\gamma-2}, \qquad \gamma > 0. \tag{3.1}$$

If on L_1 away from $z = z_0$ a solution of equation (1.1) has the form:

$$y(z,\epsilon) \sim A_1 \left[-Q(z)\right]^{-1/4} \exp\left\{+\frac{1}{\epsilon}\int_{z_0}^{z} \left[-Q(t)\right]^{1/2} dt\right\}$$
$$+A_2 \left[-Q(z)\right]^{-1/4} \exp\left\{-\frac{1}{\epsilon}\int_{z_0}^{z} \left[-Q(t)\right]^{1/2} dt\right\}, \tag{3.2}$$

and it is continuous in the open region $L_1 - L_2$ in the clockwise direction, then on L_2 away from $z = z_0$, we have

$$y(z,\epsilon) \sim B_1 \left[-Q(z)\right]^{-1/4} \exp\left\{+\frac{1}{\epsilon}\int_{z_0}^{z} \left[-Q(t)\right]^{1/2} dt\right\}$$

$$+B_2\,[-Q(z)]^{-1/4}\exp\left\{-\frac{1}{\epsilon}\int_{z_0}^{z}[-Q(t)]^{1/2}\,dt\right\}, \tag{3.3}$$

with

$$\begin{cases} B_1 = A_1 e^{-\pi i/(2\gamma)}, \\ \\ B_2 = \left[2A_1\cos\dfrac{\pi}{2\gamma} - A_2 e^{-\pi i/2}\right] e^{-\pi i/(2\gamma)-\pi i/2}. \end{cases} \tag{3.4}$$

∎

This lemma can be proved by matching WKB approximation with the local solution near critical points, which can be explicitly represented by Hankel functions. For more complicated situation, for instance, the function Q depending on ϵ and near singularities behaving like

$$Q \sim \frac{[(z-z_0)^\gamma + \eta]\,[(z-z_0)^\gamma + \zeta]}{(z-z_0)^2},$$

with $\eta, \zeta \ll 1$, we can use Wittaker functions to match WKB approximation (rf. 5, 6).

4. THE REFLECTION COEFFICIENT BEYOND ALL ORDERS

Now we are ready to compute the reflection coefficient beyond all orders for the two classes reflection coefficient problems stated in section 1.

Theorem 1 Suppose that Assumption in section 2 holds and $Q_- > 0$. Assume there are a finite number of major critical points at $z = z_j$, $j = 1, 2, \cdots, p$, on the nearest critical level line with

$$Q(z) \sim b_j (z - z_j)^{2\gamma_j - 2}, \qquad \gamma_j > 0, z \to z_j. \tag{4.1}$$

Then, for the solution of the boundary value problem (1.1) & (1.3), the nontrival leading behavior of the reflection coefficient R is

$$\left\{\sum_{j=1}^{p} 2I_j \cos\frac{\pi}{2\gamma_j}\right\} e^{2w^*/\epsilon}, \tag{4.2}$$

where I_j, $j = 1, 2, \cdots, p$, are constants of modulus unity independent of z and

$$w^* = Im\left\{\int_{x}^{z_j} [Q(t)]^{1/2} dt\right\}, \tag{4.3}$$

which is not larger than zero and independent of j. Here x is any real number larger than the largest singularity of Q on the real axis. ∎

Remark 2 The number of regular critical points on the nearest critical level line is different from that of geometric locations of those points. This can be easy to understand since level lines in general are on different Riemman sheets. At a regular critical point, by Lemma 1, we know that the nearest critical level line must be two branches of the asymptotes of its hyperbola-like curve. At a geometric location of one specific regular critical point, k critical points will be counted if there are $k+1$ nonjoint asymptotes (i.e. they do not form any close curves), including those two coinciding with the nearest critical level line in the region confined by the real axis and the nearest critical level line. Actually, when two asymptotes form a close curve, which is a part of the nearest critical level line, then the WKB approximation is valid on the curve away from the critical point.

Remark 3 It is easy to see that if w^* is strictly negative, then the reflection coefficient R is exponentially small. If the nearest level line coincides with the real axis at $+\infty$, then we can easily show that $w^* = 0$.

Figure 2 is a representative case of the structure of the level line in the complex plane. We solve equation (1.1) to the leading order along the nearest critical level line in Figure 2 to obtain the nontrivial leading behavior of the reflection coefficient R. To do this, we need to match the boundary condition (1.3) with the WKB approximation and use Lemma 2 in the neighborhood of regular critical points.

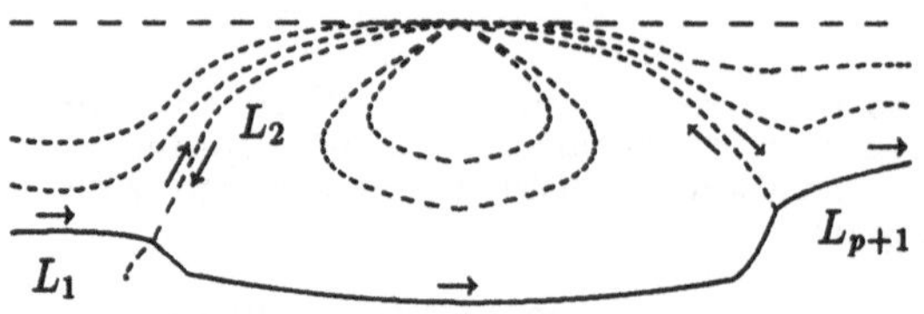

FIGURE 2 The path we solve equation (1.1) to obtain the reflection coefficient in the case that $Q_- > 0$.

Remark 4 In general, energy is not conservative for the harmonic motion (1.1) with Q singular, if we treat x as time variable. The value of the transmission coefficient changes drastically according to the regularness of the potential. In one case it can be of order 1. In another case, it can be exponentially large or small. It can be shown that the size of the transmission coefficient depends on the flux difference of the level lines between $-\infty$ to $+\infty$.

In a manner similar to Theorem 1, we can achieve the following result for the second class of the reflection coefficient problems.

Theorem 2 Suppose that Assumption in section 2 holds and $Q_- < 0$. Assume there are a finite number of major critical points at $z = z_j$, $j = 1, 2, \cdots, p$, on the nearest critical level line with (5.1). Then, for the solution of the boundary value problem (1.1) & (1.5), the nontrival leading behavior of the reflection coefficient R is (4.2). ∎

Figure 3 is a representative case of the structure of the level line in the complex plane.

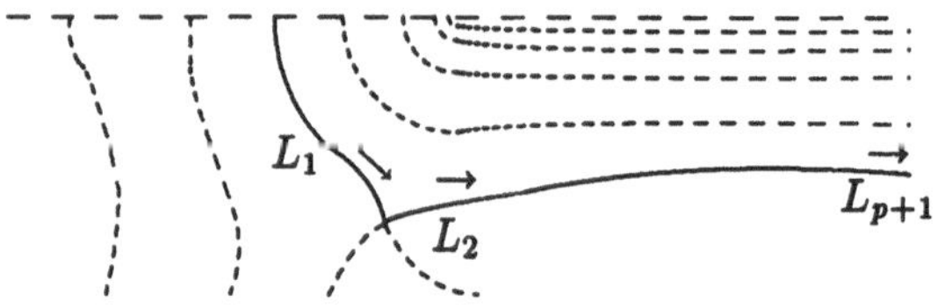

FIGURE 3 The path we solve equation (1.1) to obtain the reflection coefficient in the case that $Q_- < 0$.

ACKNOWLEDGEMENT

The authors would like to thank Dr. K. K. Tung for first informing us this problem. The second author would also like to thank for the support by DARPA under Contract No. N00014-86-K-0759 and by Department of Energy under Contract No. DE-FG0286ER25021.M004.

REFERENCES

1 C. M. Bender and S. A. Orszag, "Advanced mathematical methods for scientists and engineers", McGraw-Hill Book Co., New York (1978).

2 M. V. Berry, Semiclassically weak reflections above analytic and non-analytic potential barriers, J. Phys. A: Math. Gen., 15, 3693–3704 (1982).

3 R. E. Dickinson, Planetary Rossby waves propagating vertically though weak westerly wind wave guides, J. Atmos. Sci., 25, 984–1002 (1968).

4 I. S. Gradshteyn and I. M. Ryzhik, "Table of integrals, series, and products", Academic Press, New York (1980).

5 J. Hu and M. Kruskal, *Reflection coefficient beyond all orders for singular problems(I)*, to appear in September or October issue of J. Math. Phys., (1991).

6 J. Hu and M. Kruskal, *Reflection coefficient beyond all orders for singular problems(II)*, accepted for publication in J. Math. Phys., (1991).

7 M. Kruskal, Asymptotic theory of Hamiltonian and other systems with all solutions nearly periodic, J. Math. Phys., 3, 806–828 (1962).

8 J. J. Mahony, The reflection of short waves in a variable medium, Quart. Appl. Math., 25, 313–316 (1967).

9 R. E. Meyer, Adiabatic variation – Part I: exponential property for the simple oscillator, J. Appl. Math. Phys. (ZAMP), 24, 293–303 (1973).

10 R. E. Meyer, Quasiclassical scattering above barriers in one dimension, J. Math. Phys., 17, 1039–1041 (1976).

11 V. L. Pokrovskii and I. M. Khalatnikov, On the problem of above-barrier reflection of high-energy particles, Soviet Phys. JETP, 13, 1207–1210 (1961).

12 W. Wasow, "Linear turning point theory", Springer-Verlag, New York (1985).

LAMINAR FLOW IN A POROUS CHANNEL

J.B. McLeod
Mathematical Institute, University of Oxford, 24-29 St Giles, Oxford, OX1 3LB, U.K. and Department of Mathematics and Statistics, University of Pittsburgh, Pittsburgh, PA 15260, U.S.A.

INTRODUCTION

We consider the problem of steady incompressible viscous flow in a two-dimensional channel of infinite length, bounded by lines which we take to be the lines $y = \pm 1$. Thus the x-axis is along the centre of the channel. The walls of the channel are porous, and the problem can arise, for example, in situations where one wishes to cool a hot liquid flowing along the channel by allowing cooler liquid to enter through the walls (transpiration cooling) or where one seeks to separate two components in a mixture in the channel which may have different rates of diffusion through the walls.

If u,v are the components of velocity in the x and y directions respectively, then the boundary conditions are taken to be

$$u(x,1) = u(x,-1) = 0, \tag{1.1}$$

from viscosity, and

$$v(x,1) = V, \qquad v(x,-1) = -V, \tag{1.2}$$

due to porosity, it being assumed that the porosity of the two walls is the same. The constant V may be either positive (the diffusion example) or negative (transpiration cooling). In view of the symmetry in the conditions on the walls, we can (and will) look for solutions which are symmetric, in the sense that

Asymptotics beyond All Orders, Edited by H. Segur *et al.*
Plenum Press, New York, 1991

$$u(x,y) = u(x,-y), \qquad v(x,y) = -v(x,-y),$$

and we then have the corresponding boundary conditions on $y = 0$,

$$\frac{\partial u}{\partial y}(x,0) = 0, \qquad v(x,0) = 0. \tag{1.3}$$

If we use these, then we can restrict attention to the region $o \le y \le 1$.

The equations of motion are the steady Navier-Stokes equations together with the equation for continuity. The latter is

$$u_x + v_y = 0,$$

and implies the existence of a stream function ψ such that

$$u = \psi_y, \quad v = -\psi_x.$$

We shall be interested in similarity solutions of the form

$$\psi(x,y) = (U-Vx)f(y), \tag{1.4}$$

where U is a constant related to the velocity at $x = 0$, and the form (1.4) implies that $v = v(y)$. If we substitute (1.4) into the Navier-Stokes equations (with density ρ, pressure p)

$$uu_x + vu_y = -\frac{1}{\rho}p_x + \nu(u_{xx}+u_{yy}),$$

$$uv_x + vv_y = -\frac{1}{\rho}p_y + \nu(v_{xx}+v_{yy}),$$

and use the boundary conditions (1.1)-(1.3), we obtain the following boundary-value problem for f:

$$f^{iv} - R(ff''' - f'f'') = 0, \tag{1.5}$$

$$f(0) = f''(0) = 0, \tag{1.6}$$

$$f'(1) = 0, \ f(1) = 1. \tag{1.7}$$

Here $R = V/\nu$ is positive in the diffusion case, negative in the cooling case, and it is possible to integrate (1.5) once to obtain

$$f''' - R(ff''-f'^2) = \text{const.}$$

$$= f'''(0) + Rf'^2(0),$$

but the values of $f'''(0)$, $f'(0)$ are not of course a priori known.

The modelling of this problem was originally done by Berman[1], and subsequent investigations, amongst others, have been carried out by Terrill[2], Skalak and Wang[3] and Hastings, Lu and MacGillivray[4]. Our object is to review the position on the existence and uniqueness of solutions, and to begin an investigation of the behaviour of solutions as $R \to \infty$, which involves asymptotics beyond all orders.

2. A preliminary result

Before stating the existence and uniqueness theorems, we prove one very simple result which is crucial in all that follows.

Theorem 1. For any solution of (1.5)-(1.7), we have, in (0,1],

$$f^{iv} < 0 \text{ if } R > 0, \qquad f^{iv} > 0 \text{ if } R > 0.$$

Proof. Differentiation of (1.5) yields

$$f^{v} - Rff^{iv} = - Rf''^2,$$

so that

$$\{f^{iv}\exp(-\textstyle\int^t Rfds)\}' = - Rf''^2 \exp(-\textstyle\int^t Rfds)$$

$$\leq 0 \text{ if } R > 0, \quad \geq 0 \text{ if } R < 0. \qquad (2.1)$$

Since $f^{iv}(0) = 0$, integration of (2.1) gives the required result. (If f'' has zeros, then they must be isolated, in view of the differential equation satisfied by f^{iv}, unless $f'' \equiv 0$, which is impossible since it implies $f' \equiv 0$ and contradicts (1.6)-(1.7).)

3. Existence and uniqueness: $R < 0$

Theorem 2. If $R < 0$, any solution to (1.5)-(1.7) must have $f'(0) > 0$, $f'''(0) < 0$. Further there exists one and only one solution to (1.5)-(1.7), and it has the properties that, in (0,1),

$$f' > 0, \qquad f'' < 0, \qquad f''' < 0.$$

Proof. By the standard existence and uniqueness theorem for ordinary differential equations, the solution to (1.5)-(1.7) is certainly determined once $f'(0)$, $f'''(0)$ are known. There are *a priori* four possibilities, depending on the signs of $f'(0)$, $f'''(0)$.

Case 1. $f'(0) \geq 0$, $f'''(0) \geq 0$:

Then $f^{iv} > 0$ in $(0,1)$ implies $f''' > 0$, $f'' > 0$. f' increasing, which is a contradiction to $f'(1) = 0$. So Case 1 is impossible.

Case 2. $f'(0) < 0$, $f'''(0) \geq 0$:

Again $f^{iv} > 0$, $f''' > 0$, $f'' > 0$, and so $f'(1) = 0$ implies that $f' < 0$, which contradicts $f(0) = 0$, $f(1) = 1$. Thus Case 2 is impossible.

Case 3. $f'(0) \leq 0$, $f'''(0) < 0$:

Now $f^{iv} > 0$ implies that f''' has at most one zero in $(0,1)$, and it must have at least one since otherwise $f'' < 0$, $f' < 0$, and we cannot have $f(0) = 0$, $f(1) = 1$. So

$$f''' < 0 \text{ in } (0,x_3), \quad f''' > 0 \text{ in } (x_3,1],$$

$$f'' < 0 \text{ in } (0,x_2), \quad f'' > 0 \text{ in } (x_2,1], \quad x_2 > x_3,$$

and we see that $f'(1) = 0$ implies that $f' < 0$, contradicting $f(0) = 0$, $f(1) = 1$. So Case 3 is impossible.

This leaves only

Case 4. $f'(0) > 0$, $f'''(0) < 0$:

If there is a solution in this case, then clearly we must have that f''' has at most one zero in $(0,1)$,

$$f''' < 0 \text{ in } (0,t_3), \; f'''(t) > 0 \text{ for } t > t_3.$$

(It is possible at this stage that $t_3 > 1$.)

Then

$$f'' < 0 \text{ in } (0,t_2),\ f''(t) > 0 \text{ for } t > t_2,\ t_2 > t_3.$$

Also, f' has at most two zeros, t_1, t_1^*, with $t_1 < t_2$, $t_1^* > t_2$. But $f^{iv} > 0$ implies that

$$ff''' - f'f'' < 0,$$

so that at t_3 we must have $f' < 0$. So $t_1 < t_3$. Further, at t_1^*, $f < 0$. Thus a solution in Case 4 is possible only if f' has precisely one zero, at 1, and

$$f''' < 0 \quad \text{in} \quad [0,1],$$

$$f'' < 0 \quad \text{in} \quad (0,1].$$

To show that there exists a solution, we rescale the problem setting

$$f = \alpha F, \quad t = \beta T, \tag{3.1}$$

so that, with $f' = dF/dT$,

$$F^{iv} - \alpha\beta R(FF'''-F'F'') = 0,\ F(0) = 0,\ F''(0) = 0,$$

$$F'(\beta^{-1}) = 0,\ F(\beta^{-1}) = \alpha^{-1}. \tag{3.2}$$

We now choose α,β so that (recalling $R < 0$)

$$F'(0) = 1 \quad \text{and} \quad \alpha\beta R = -1.$$

Thus

$$\frac{\alpha}{\beta} = f'(0), \quad \alpha\beta = -R^{-1},$$

and

$$F^{iv} + (FF'''-F'F'') = 0, \tag{3.3}$$

with

$$F(0) = 0,\ F'(0) = 1,\ F''(0) = 0,\ F'''(0) = \gamma,\ \text{say},\ \gamma < 0.$$

Now we know from the discussion of Case 4 that this solution has a first zero for F', at T_γ, say, where T_γ is clearly a continuous function of γ. What we have to show is that we can choose γ so that (3.2) is satisfied, which is equivalent to saying that

$$T_\gamma F(T_\gamma) = -R, \tag{3.4}$$

for any given R.

If we take $\gamma = 0$, then the solution of (3.3) is just $F' = 1$. This means that if we take the limit as $\gamma \uparrow 0$, we have $T_\gamma \to \infty$, and so

$$T_\gamma F(T_\gamma) \to \infty. \tag{3.5}$$

If we take γ large (and negative), we rescale again,

with

$$F = aG, \quad T = bX$$

and

$$a = (-\gamma)^{1/4}, \quad b = (-\gamma)^{-1/4},$$

so that

$$XG(X) = TF(T).$$

With $G' = dG/dX$, (3.3) becomes

$$G^{iv} + (GG''' - G'G'') = 0,$$

with

$$G(0) = 0,\ G'(0) = (-\gamma)^{-1/2},\ G''(0) = 0,\ G'''(0) = -1.$$

Thus for large γ, and $X \le \delta$, say, for some fixed small δ, independent of γ,

$$G'(X) = (-\gamma)^{-1/3} - \tfrac{1}{2}X^2 + \ldots.,$$

so that, if X_γ is the first zero of G', then

$$X_\gamma \sim 2^{1/2}(-\gamma)^{-1/6},$$

and so

$$X_\gamma G(X_\gamma) = T_\gamma F(T_\gamma) \to 0. \tag{3.6}$$

Now from continuity in γ and (3.5) and (3.6) it follows that there must exist some value of $\gamma < 0$ such that (3.4) holds, and the existence theorem is complete.

To establish uniqueness, suppose that there are two solutions f_1, f_2, to (1.5)-(1.7). By means of the transformation (3.1), with, of course, different values α_i, β_i for the two solutions, we can reduce the problem to two solutions of (3.3), with

$$F_i(0) = 0, \quad F_i'(0) = 1, \quad F_i''(0) = 0, \quad F_i'''(0) = \gamma_i.$$

Note that

$$\alpha_1\beta_1 = \alpha_2\beta_2 = -R^{-1},$$

so that, at the point T_i where $F_i' = 0$ (which corresponds to $t = 1$, $f_i = 1$), we have

$$T_i F_i(T_i) = \frac{1}{\alpha_i\beta_i} t_i f_i = -R. \tag{3.7}$$

But, if we suppose without loss of generality that $\gamma_1 > \gamma_2$, then

$$(F_1-F_2)(0) = 0 = (F_1-F_2)'(0) = (F_1-F_2)''(0), \quad (F_1-F_2)'''(0) > 0,$$

so that initially

$$F_1-F_2 > 0, \quad (F_1-F_2)' > 0, \quad (F_1-F_2)'' > 0, \quad (F_1-F_2)''' > 0.$$

If we now integrate (3.3), we obtain

$$F_i''' + F_iF_i'' - F_i'^2 = \gamma_i - 1,$$

so that by subtraction

$$(F_1-F_2)''' + F_1(F_1-F_2)'' = \gamma_1-\gamma_2 + (F_1'+F_2')(F_1-F_2)' - F_2''(F_1-F_2),$$

and this can be written as

$$\{(F_1-F_2)''\exp(\textstyle\int^T F_1 ds)\}' = \{\gamma_1-\gamma_2 + (F_1'+F_2')(F_1-F_2)' - F_2''(F_1-F_2)\} \exp(\textstyle\int^T F_1 ds).$$

In view of the fact that $F_2'' < 0$, $F_1' > 0$, $F_2' > 0$, we see that the right-hand side remains positive so long as $F_1 - F_2$, $(F_1-F_2)'$ remain positive, i.e. $(F_1-F_2)'' > 0$ so long as F_1-F_2, $(F_1-F_2)'$ are positive. This implies that we always have

$$F_1 - F_2 > 0, \quad (F_1-F_2)' > 0, \quad (F_1-F_2)'' > 0.$$

But then $F_1' > F_2'$ implies $T_1 > T_2$, and so

$$T_1F_1(T_1) > T_2F_1(T_2) > T_2F_2(T_2),$$

which contradicts (3.7) and forms uniqueness.

4. Existence: $R > 0$

Using essentially the same methods as in Section 3, we can establish existence theorems, although the only case that can be ruled out is $f'(0) \le 0$, $f'''(0) \le 0$. (Also, $f'''(0) = 0$ is impossible.) In fact, we have

Theorem 3.

(i) If $R > 0$, there exists a solution to (1.5)-(1.7) with $f'(0) > 0$, $f'''(0) < 0$. It has the properties, in (0,1), that $f' > 0$, $f'' < 0$, $f''' < 0$.

(ii) If $R > 0$ and sufficiently large, there exists a solution to (1.5)-(1.7 with $f'(0) > 0$, $f'''(0) > 0$. It has the properties in (0,1) that $f' > 0$, f'' has precisely one zero (simple), f''' has precisely one zero (simple).

(iii) If $R > 0$ and sufficiently large, there exists a solution to (1.5)-

(1.7) with $f'(0) < 0$, $f'''(0) > 0$. It has the properties in (0,1) that f' has precisely one zero (simple), f'' has precisely one zero (simple), f''' has precisely one zero (simple).

Remarks. 1. We can prove solutions (ii) and (iii) exist if and only if R is sufficiently large because, when we make the scaling transformation (3.3) and consider what happens when $\gamma \to \infty$, we find that the result corresponding to (3.6) is of the form

$$X_\gamma G(X_\gamma) = T_\gamma F(T_\gamma) \to K,$$

say, where K is a strictly positive number. This implies a positive lower bound for values of R for which there is existence.
2. There has been no analytical discussion of the uniqueness of the solutions in Theorem 3, although there are of course numerical results. The analysis seems difficult.

5. Behaviour as $R \to \pm\infty$

Setting $R^{-1} = \varepsilon$, we have to study the behaviour of solutions of

$$\varepsilon f^{iv} - (ff'''-f'f'') = 0 \tag{5.1}$$

as $\varepsilon \to 0$, with

$$f(0) - f''(0) = 0,\ f'(1) = 0,\ f(1) = 1. \tag{5.2}$$

If $\varepsilon = 0$, then the equation becomes

$$ff''' - f'f'' = 0, \tag{5.3}$$

which can be explicitly solved to give (if $f(0) = f''(0) = 0$)

$$f = \text{const. } x \text{ or } f = \text{const. } \sin\beta x \text{ or } f = \text{const. } \sinh\beta x,$$

for any constant β. If this is to satisfy the remainder of the conditions in (5.2), and the fact from Theorem 1 that f^{iv} is of one sign, then

$$f(x) = \sin\tfrac{1}{2}\pi x.$$

It is therefore a reasonable conjecture that the solution of (5.1)-(5.2) behaves, as $\varepsilon \to 0$, like $\sin\tfrac{1}{2}\pi x$, and this seems to be the case (although we do not go into it here) if $\varepsilon < 0$. However, if $\varepsilon > 0$, which is the more interesting case, then this is impossible, as is easily seen by observing that, from Theorem 1, $f^{iv} < 0$, while the fourth derivative of $\sin\tfrac{1}{2}\pi x$ is positive. From now on, we consider only $\varepsilon > 0$.

We recall from Theorem 3 that there are now at least three solutions to (5.1)-(5.2) as $\varepsilon \downarrow 0$. Let us denote solutions of types (i), (ii) and (iii) in Theorem 3 by f_1, f_2 f_3. (We make no assertion that f_1, for example, is unique.) The behaviour of f_i as $\varepsilon \downarrow 0$ is determined by giving the behaviour of $f_i'(0)$ and $f_i'''(0)$, and we are interested particularly in $f_i'''(0)$, since this is related to the relevant pressure. We can prove the following theorem.

<u>Theorem 4</u>. <u>As</u> $\varepsilon \downarrow 0$,

$$f_1'(0) \to 1, \quad f'''(0) \sim -(\tfrac{1}{2}\pi)^{-\frac{1}{4}}\varepsilon^{-7/4}e^{-1/4\varepsilon^{-\frac{1}{4}}}.$$

<u>Remarks</u>. 1. This implies that the solution $f_1(t)$ tends to t in compact sub-intervals of $[0,1]$, and that there must then be a boundary layer near $t = 1$ where f_1' changes rapidly from 1 to 0.
2. For f_2, the result is (at least formally)

$$f_2'(0) \to 1, \quad f_2'''(0) \sim +(\tfrac{1}{2}\pi)^{-\frac{1}{4}}\varepsilon^{-7/4}e^{-1/4\varepsilon^{-\frac{1}{4}}}.$$

The proof should follow closely the lines of the proof of Theorem 4.
3. For f_3, the result is (at least formally)[5] $f_3(t) \sim \dfrac{\sin\pi t}{\pi A(\varepsilon)}$ in compact sub-intervals of $[0,1]$, where

$$\varepsilon^4\sqrt{\frac{A}{\varepsilon}}e^{A}/^{2\varepsilon} \to \frac{1}{\pi^4}\sqrt{\frac{1}{2\pi}}\,.$$

A rigorous proof of this remains to be supplied.

<u>Proof of Theorem 4.</u> We will sketch only the main steps in the proof.
<u>Step 1.</u> We know that $f^{iv} < 0$, and indeed

$$f^{v}Rff^{iv} = -\,Rf''^{2}, \tag{5.4}$$

so that $f^{v} < 0$ and f^{iv} is negative decreasing. Thus f^{iv} is bounded (as $\varepsilon \downarrow 0$) except near $t = 1$, since, if it were unbounded, we could certainly prove successively that f''', f'', f', f are unbounded, which is untrue for f.

Step 2. The equation (5.4) tells us that

$$f^{iv}\exp(-\int^{t}\frac{f}{\varepsilon}ds) \text{ is negative decreasing,}$$

and so

$$|f^{iv}(\tau)| \le |f^{iv}(t)|\exp(-\int_{\tau}^{t}\frac{f}{\varepsilon}ds) \text{ if } \tau < t.$$

This immediately implies that $f^{iv}(\tau)$ is exponentially small, i.e. of the form $O(e^{-a/\varepsilon})$, for some $a > 0$, except near $\tau = 1$, and it is an easy argument to conclude that $f'''(0)$ is exponentially small. It then follows that $f^{iv}(t), f'''(t), f''(t)$ are all exeponentially small except near $t = 1$, and that $f'(t)$ is close to 1, and $f(t)$ close to t, except near $t = 1$.

Step 3. By integrating the governing equation for f, and using the above information, we can obtain

$$f''(1) \sim -\frac{1}{\varepsilon}, \quad f'''(1) \sim -\frac{1}{\varepsilon^2}, \quad f^{iv}(1) \sim -\frac{1}{\varepsilon^3}.$$

Step 4. We now return to Step 2, and do the estimates more carefully, using the information we now have from Step 3. We can obtain, for example, that

$$f^{iv}(t) = f^{iv}(1) \exp\,(-\int_{t}^{1}\frac{f}{\varepsilon}ds)\{1 + O(\varepsilon)\} \tag{5.5}$$

+ terms uniformly exponentially small in [0,1].

Successive integrations of this lead to

$$f(t) = f'(0)t + \frac{\varepsilon^4 f^{iv}(1)}{\{f(t)\}^4}\ \exp\,(-\int_{t}^{1}\frac{f}{\varepsilon}ds)\{1 + O(\varepsilon)\} \tag{5.6}$$

+ terms uniformly exponentially small in [0,1],

this last being valid for $t \ge \delta$, say, for any fixed $\delta > 0$, because of the factor f^{-4} with $f(0) = 0$. Evaluating (5.6) at $t = 1$ gives

$$f'(0) = 1 + \varepsilon + O(\varepsilon),$$

and another integration of (5.6) gives

$$\int_{0}^{t} f ds = \tfrac{1}{2}f'(0)t^2 + O(\varepsilon^2).$$

Step 5. Finally, if we go back once again to (5.4) and integrate over (0,1), we have

$$f^{iv}(1) = -\frac{1}{\varepsilon}\int_0^1 f''^2 \exp\left(\int_t^1 \frac{f}{\varepsilon} ds\right) dt. \tag{5.7}$$

We can obtain an expression for f'' (involving $f'''(0)$) by integrating (5.5) twice. If we substitute this in (5.7) and keep only the leading terms on both sides, we achieve the required formula for $f'''(0)$.

It will be seen that the proof is simply a matter of moving the analysis back and forth between the boundary layer (which drives the problem) and the outer region, and making successively better estimates each time. Although other problems in asymptotics beyond all orders seem to necessitate an excursion into the complex plane, that does not appear to be relevant here.

References

1. A.S. Berman, Laminar flow in channels with porous walls, J. Appl. Phys. 24 (1953), 1232-1235.
2. R.M. Terrill, Laminar flow in a uniformly porous channel, Aero. Quart. 15 (1964), 299-310.
3. F.M. Skalak & C.-Y. Wang, On the nonunique solutions of laminar flow through a porous tube or channel, SIAM J. Appl. Math. 34 (1978), 535-544.
4. S.P Hastings, C. Lu & A.D. MacGillivray, A boundary value problem with multiple solutions from the theory of laminar flow, preprint.
5. A.D. MacGillivray, private communication.

EXISTENCE AND STABILITY OF PARTICLE CHANNELING IN CRYSTALS ON TIMESCALES BEYOND ALL ORDERS

H. Scott Dumas

Department of Mathematical Sciences
University of Cincinnati
Cincinnati, OH 45221

INTRODUCTION

Particle channeling in crystals is an important aspect of particle-solid interactions for which I have developed a mathematical theory using modern techniques in Hamiltonian perturbation theory. This article describes that theory in broadest terms; ample discussion and detailed proofs have been or will be shortly published elsewhere (cf. Dumas 1988, 1989, 1991, and Dumas and Ellison 1991). A number of people helped me develop these ideas, but I am especially grateful to my thesis advisor, Jim Ellison, whose extensive work on mathematical aspects of channeling was the starting point for what is described here. I am also indebted to Pierre Lochak and to other members of the CMA at the École Normale Supérieure (Paris) for their assistance during the year 1986–87. It was there that I learned many of the mathematical tools used to formulate the channeling theory: a collection of results comprising KAM theory and the related but less familiar theory due to Nekhoroshev (1971, 1977, 1979) and, more recently, to Bennetin, Galgani, and Giorgilli (1985) and to Bennetin and Gallavotti (1986). The latter group of mathematical physicists first recognized the advantages—in principle at least—of Nekhoroshev's approach in applications to physical systems where rigorous results on long-time stability of motion are desired; this approach is closely related to some of the techniques discussed in Professor Marsden's article (these proceedings) on adiabatic invariance.

PARTICLE CHANNELING IN CRYSTALS

When a beam of energetic charged particles is directed at a crystalline target in a random direction, the beam and crystal interact strongly: beam particles are backscattered, matter is ejected from the crystal; nuclear reactions may even take place. Radiative and collisional energy loss eventually bring many particles to rest, so that beam matter is implanted in the crystal. But if the crystal is now repositioned so that the beam is incident in a "non-random" direction—in the direction of a low-order crystal axis or plane—the results observed are very different. The average depth of penetration into the crystal is greatly increased, and the rate of particle backscattering may decrease by as much as two orders of magnitude. The interaction of the beam with the crystal in this way is called *channeling*, and for beams with positive charge, it is not entirely naïve to imagine particles streaming through the channels between planes or rows of crystal nuclei, with soft collisions with these planes or rows guiding errant particles away from nuclei.

Channeling has proved to be a useful tool for understanding the properties of solids, and has had numerous technological applications. It has been used as a material analysis tool to study crystal defects, surfaces and interfaces, and to determine the location of crystal impurities. It has been used to measure nuclear lifetimes, to study the strain in "strained-layer superlattices," and to deflect high energy particle beams. In fact, at the Superconducting Supercollider (SSC) currently under construction in Waxahachie, Texas, researchers forsee the possibility of fixed target experiments using bent crystals to extract a small fraction of particles from the halo of the main circulating beam. In preliminary experiments at both CERN and Fermilab, bent crystals have already been used to deflect beams with energies in the hundreds of *GeV* (see Carrigan et al. (1990) for a discussion of results at Fermilab). Surprisingly, the ability of a crystal to effectively channel charged particles does not appear to be impaired noticeably by the radiation damage the crystal suffers while immersed in the beam (see the article by Baker in Carrigan and Ellison (1987)). These and many other applications of channeling—including more speculative possibilities such as monoenergetic gamma ray sources and cosmic-ray telescopes—are discussed in a large body of literature, much of which is cited in the bibliographies of Carrigan and Ellison (1987), Kumakhov and Komarov (1989), and in the excellent if by now somewhat early review article by Gemmell (1974).

LINDHARD'S CONTINUUM MODELS; THE PERFECT CRYSTAL MODEL

Since the early 1960's, theoretical investigations of channeling have relied on the so-called *continuum model*, wherein channeling is described as the motion of particles moving in a *continuum potential* obtained by averaging the crystal potential over the axis or plane with which the particles' incident direction is most nearly parallel. This model was introduced independently by a number of theoreticians, but the most convincing arguments for it were given by J. Lindhard (1965).

Despite its usefulness and Lindhard's physically convincing arguments for it, the continuum model has remained without firm mathematical foundation. This situation is partly remedied by the work outlined here, in which a generalized continuum model is mathematically deduced from the so-called perfect crystal model.

The (cubic) perfect crystal model used here is the classical Hamiltonian system

$$\mathcal{H}(p,q) = \frac{1}{2m}p^2 + V(q), \tag{1}$$

where $p, q \in \mathbf{R}^3$, m is the mass of a single particle, and the crystal potential V is real analytic with period d in each component of q. The requirement that V be analytic is not as severe as it might seem, as perfect crystal potentials are typically constructed by summing screened, thermally averaged Coulombic atomic potentials centered on the lattice sites in $\mathbf{R}^3$. It should be pointed out that while this is the periodic potential which best models the thermal lattice vibrations of the crystal, since it has no "hard core" it cannot faithfully represent the process of close encounter between charged particles and nuclei.

But this objection is removed by recalling that channeling trajectories are precisely those which *avoid* close encounters with nuclei: If we assume that the potential

V has been adjusted so that its minimum value is zero and its maximum value over $\mathbf{R}^3$ is $\mathcal{E}_M$, then for energies $\mathcal{E}_\perp$ ($0 \leq \mathcal{E}_\perp \leq \mathcal{E}_M$), we may consider the subsets of configuration space

$$\mathcal{B}(\mathcal{E}_\perp) = \{q \in \mathbf{R}^3 \mid V(q) \geq \mathcal{E}_\perp\}. \tag{2}$$

If, as is assumed here, the potential governs the motion of positively charged particles, then clearly for sufficiently large $\mathcal{E}_\perp < \mathcal{E}_M$, the set $\mathcal{B}(\mathcal{E}_\perp)$ is the disjoint union of (slightly deformed) balls centered on the lattice sites. By choosing a physically suitable value for $\mathcal{E}_\perp$, we may distinguish particle trajectories which come too close to nuclei to be governed by the thermally averaged potential as those which enter $\mathcal{B}(\mathcal{E}_\perp)$. More precisely, fix $\mathcal{E}_\perp$, and consider a solution $(p(\tau), q(\tau))$ of the equations of motion corresponding to (1). Such a solution is a *channeling solution on the time interval* $\mathcal{I}$ provided

$$q(\tau) \notin \mathcal{B}(\mathcal{E}_\perp), \quad \text{or equivalently} \quad V(q(\tau)) < \mathcal{E}_\perp \quad \forall \tau \in \mathcal{I}. \tag{3}$$

This is the *channeling criterion*, and it is assumed that the perfect crystal model is a good approximation for particle trajectories that satisfy it. A trajectory which first fails to satisfy this criterion at time t_1 is assumed to suffer a "close encounter" with a nucleus, and is not viewed as a good approximation to an actual particle trajectory for subsequent times $t > t_1$.

If there are indeed trajectories of (1) satisfying the channeling criterion, they maintain a kinetic energy of approximately $\mathcal{E}$, which is very large compared to the maximum potential $\mathcal{E}_\perp$ felt by channeled particles. We are thus led to introduce the small parameter $\epsilon = \mathcal{E}_\perp/\mathcal{E} \ll 1$ so that (1) takes the scaled form

$$H(I, \theta) = \frac{1}{2} I^2 + \epsilon W(\theta), \tag{4}$$

where we have introduced the scaled momentum (actions) $I \in \mathbf{R}^3$, the scaled position (angles) $\theta \in T^3$, the scaled potential W, and the scaled time t by means of the transformations

$$I = (m\mathcal{E})^{-1/2} p, \quad \theta = \frac{1}{d} q, \quad W(\theta) = \frac{1}{\mathcal{E}_\perp} V(\theta d), \quad t = \frac{1}{d}\sqrt{\mathcal{E}/m}\,\tau, \tag{5}$$

and where attention is now restricted to solutions of (1) with fixed energy $\mathcal{E}$.

The region of close encounter $\mathcal{B}(\mathcal{E}_\perp)$ described in equation (2) is immediately reformulated in terms of the scaled variables as

$$\mathcal{C}(1) = \{\theta \in T^3 \mid W(\theta) \geq 1\}, \tag{6}$$

and so the criterion (3) for a solution $(I(t), \theta(t))$ of Hamilton's equations corresponding to (4) to be a channeling trajectory on the time interval $\mathcal{I}$ becomes

$$\theta(t) \notin \mathcal{C}(1), \quad \text{or equivalently} \quad W(\theta(t)) < 1 \quad \forall\, t \in \mathcal{I}. \tag{7}$$

A NEKHOROSHEV-LIKE THEORY OF CHANNELING

Applications of perturbation theory to (4) must be interpreted with care. First, the KAM theorem applies to this model and says that for sufficiently small perturbations, the action variables of solutions with "highly nonresonant" initial directions

$I(0)$ will remain uniformly close, for all time, to the action variables of the unperturbed rectilinear trajectories $(I(t), \theta(t)) = (I(0), \theta(0) + I(0)t)$. On the other hand, it can be shown (cf. Dumas 1991, Dumas and Ellison 1991) that highly nonresonant initial directions are a subset of the *non*channeling directions, in the sense that particles with these initial directions quickly enter the region $C(1)$ for which (4) is assumed invalid. Nevertheless, an important result may be derived which says that, for a set of initial directions consisting of a "nice" set in I-space which contains and closely approximates the highly nonresonant directions, trajectories are nearly rectilinear until they encounter nuclei. These encounters occur quickly, within a time which may be estimated in terms of ϵ and the Diophantine conditions defining the highly nonresonant directions. Together, I call these results the ***spatial continuum model*** by analogy with the other continuum models; the interested reader may consult Dumas (1991) and Dumas and Ellison (1991) for details.

Looking beyond KAM theory, it turns out that the perturbation theory that is able to resolve channeling motions in the perfect crystal model is an adaptation of Nekhoroshev's theory. Without entering more than superficially into this theory or its motivations, it is possible to say that it is the culmination of classical perturbation theory for nearly integrable Hamiltonian systems. For more than a century, mathematicians and mathematical physicists have sought the stability times of such systems, in other words the longest possible time intervals on which the action variables remain close to their initial values. The KAM theorem establishes infinite stability times (i.e. invariant tori) for trajectories emanating from highly nonresonant initial conditions. (In systems with only two degrees of freedom, the infinite stability times hold for *all* initial conditions because trajectories are trapped between concentric tori.) Although the highly nonresonant initial conditions comprise a set of large relative measure in phase space, the complement of these initial conditions is a dense open set—a serious obstacle to applications of KAM theory to physical systems. In contrast, Nekhoroshev trades the infinite stability times for finite stability times, and gains in return a result which is true for *all* initial conditions. Though finite, the stability times are nonetheless beyond all orders in inverse powers of the small parameter measuring the strength of the perturbation; they are in fact exponentially long in an inverse power of the small parameter (unfortunately this power tends to zero with increasingly many degrees of freedom). Nekhoroshev's proof of his theorem for a generic subset of analytic Hamiltonians—the class of so-called ***steep*** Hamiltonians—greatly increased both the generality of his results and the complexity of their proofs, as the reader may verify in the original manuscripts or their translations (Nekhoroshev 1971, 1977, 1979). For specific systems arising in applications, such as the one here, geometric considerations stemming from convexity of the unperturbed Hamiltonian often allow for much simpler proofs on the order of those of the KAM theorem. This fact has been exploited by a group of Italian mathematical physicists in a continuing series of very interesting articles beginning with Benettin et al. (1985, 1986). In fact, the proof of the theorem announced below makes use of the "resonant normal forms" appearing in Benettin et al. (1985) and also of the idea of "trapping into resonance" used in Benettin and Gallovotti (1986) and appearing earlier in Nekhoroshev (1977). Therefore, from a mathematical point of view, channeling may be viewed as motion at resonance in the Hamiltonian (4), the long-time stability of this motion being guaranteed by the convexity of the level sets (spheres) of the unperturbed part $\frac{1}{2}I^2$ of (4).

To lend concreteness to all of this, I will state a shortened—but I hope understandable—version of the channeling theorem for the scaled perfect crystal Hamiltonian (4). This first requires a few definitions and notational conventions. In order to discuss resonances, we recall that a finite subset $\{k^{(i)}\}_{i=1}^m$ of $\mathbf{Z}^3$ generates a submodule of $\mathbf{Z}^3$ defined as $\{z \in \mathbf{Z}^3 \mid z = \sum n_i k^{(i)}, n_i \in \mathbf{Z}\}$. Each submodule of $\mathbf{Z}^3$ has dimension 0, 1, 2, or 3, in the sense that that the smallest vector subspace of $\mathbf{R}^3$ containing the submodule has that dimension. If $\{k^{(i)}\}_{i=1}^m$ generates a submodule of dimension n, then the maximal submodule generated by $\{k^{(i)}\}_{i=1}^m$ is the largest submodule of dimension n containing $\{k^{(i)}\}_{i=1}^m$. In the remainder of this article, the symbol $\mathcal{M}$ will refer to a 1- or 2-dimensional maximal submodule of $\mathbf{Z}^3$. Not surprisingly, every $\mathcal{M}$ is generated respectively by integer combinations of a 1- or 2-element basis in $\mathbf{Z}^3$; the *order* $|\mathcal{M}|$ of a maximal submodule is the smallest integer r such that $\mathcal{M}$ admits a basis of vectors with (taxicab) norm less than or equal to r.

To each $\mathcal{M}$ corresponds a resonance in action space, namely the orthogonal complement (in $\mathbf{R}^3$) of $\mathcal{M}$. Actions in the neighborhood of a resonance (and sufficiently far from other low-order resonances) are channeling directions; this is formalized by first defining the *resonant zone corresponding to* $\mathcal{M}$ by

$$\mathcal{Z}_{\mathcal{M}}(C,\alpha) \;=\; \{I \in \mathbf{R}^3 \mid |k \cdot I| \leq C\epsilon^{\alpha}|k|^{-p}\ \forall k \in \mathcal{M}, 0 < |k| \leq |\mathcal{M}|\} \tag{8}$$

for appropriate positive values of C, α and p. In the case where $\mathcal{M}$ is one dimensional, the zone $\mathcal{Z}_{\mathcal{M}}(C,\alpha)$ will also be denoted $\mathcal{Z}_k(C,\alpha)$, where k is the unique generator (up to inversion) of $\mathcal{M}$. (The case $\mathcal{M} = \{0\}$ corresponds to nonchanneling directions and is not discussed here.) Finally, given $\mathcal{M}$, its associated *resonant block of order* $N \geq |\mathcal{M}|$ is defined as the subset of action space

$$\widehat{\mathcal{Z}}_{\mathcal{M}}^{C,\alpha}(N, C_1, \alpha_1) = \mathcal{Z}_{\mathcal{M}}(C_1, \alpha_1) \;\setminus \bigcup_{\substack{k \notin \mathcal{M} \\ |k| \leq N}} Int\, \mathcal{Z}_k(C,\alpha), \tag{9}$$

where *Int* denotes interior. Action values in these blocks correspond to channeling directions.

We turn now to the relevant assumptions and definitions concerning the scaled potential W. Given $\mathcal{M}$ and given $\theta \in T^3 \approx \mathbf{R}^3/\mathbf{Z}^3$, write $\theta = \theta^* + \widehat{\theta}$, where θ^* and $\widehat{\theta}$ are the projections onto $span\mathcal{M}$ and $(span\mathcal{M})^{\perp}$ in $\mathbf{R}^3$. The resonant subseries $\Pi_{\mathcal{M}} W(\theta^*)$ corresponding to $\mathcal{M}$ is then just the average over the variables $\widehat{\theta}$ determined by $\mathcal{M}$; that is, $\Pi_{\mathcal{M}} W(\theta^*)$ is a continuum potential. We next consider regions in configuration space bounded by equipotential surfaces of the continuum potentials. Given $\mathcal{M}$ and any number $Q \geq 0$, define the closed subset $\mathcal{A}_{\mathcal{M}}(Q)$ as

$$\mathcal{A}_{\mathcal{M}}(Q) \;=\; \{\theta \in T^3 \mid \Pi_{\mathcal{M}} W(\theta^*) \leq Q\}. \tag{10}$$

Our assumption about W is then the following:

Assumption A. *There exists a critical order $M^* \geq 1$ such that for any $\mathcal{M}$ with order $|\mathcal{M}| \leq M^*$, there are numbers Q' and Q'', $0 \leq Q' < Q'' < 1$ with the property that given any $Q \in [Q', Q'']$, $\mathcal{A}_{\mathcal{M}}(Q) \neq \emptyset$ and $\mathcal{A}_{\mathcal{M}}(Q) \cap \mathcal{C}(1) = \emptyset$, where $\mathcal{C}(1) = \{\theta \in T^3 \mid W(\theta) \geq 1\}$.*

This physical assumption says that at sufficiently low order, there are equipotential surfaces of the continuum potential $\Pi_{\mathcal{M}} W$ which do not intersect the restricted region $\mathcal{C}(1)$. If $dim\mathcal{M} = 1$ ("planar" channeling), these surfaces are planes; if $dim\mathcal{M} = 2$ ("axial" channeling), the surfaces are cylindrical sheets or tubes. The assumption therefore says that at sufficiently low order, there are clear planar or axial pathways through the crystal which do not meet the close encounter region $\mathcal{C}(1)$.

Finally, we define the transverse energy of a trajectory with respect to a particular continuum potential by

$$E_\perp(I,\theta) \;=\; \frac{1}{2}(I^*)^2 \;+\; \epsilon\Pi_{\mathcal{M}}W(\theta^*), \tag{11}$$

where again I^* and θ^* denote the projections of I and θ onto $span\mathcal{M}$. We now state an abbreviated form of the

Channeling Theorem. *Let $\mathcal{M}$ be a 1- or 2-dimensional submodule of $\mathbf{Z}^3$ with order $|\mathcal{M}| \leq M^*$. Fix suitable positive values for E (related to $\mathcal{E}$, $\mathcal{E}_\perp$) and A (which determines the width of resonant zones), and assume certain consistency relations among the positive numbers α, τ, c, p which allow for a transformation of the Hamiltonian (4) to a certain resonant normal form. (An example of such numbers is $\alpha = 1/8, \tau = 1/72, c = 5/8, p = 5$.) Fix the maximum initial (scaled) transverse energy $Q \geq Q'$ and the maximum change in (scaled) transverse energy $\delta > 0$ such that $Q + \delta \leq Q''$, where $Q' < Q'' < 1$ are defined in Assumption A ; set $C = |\mathcal{M}|^{p+1}(2Q/3)^{1/2}$. Then there exists a sufficiently small $\epsilon > 0$ such that any initial condition (I_0, θ_0) for (4) with initial transverse energy*

$$E_\perp(I_0,\theta_0) \;\leq\; \epsilon Q \tag{12}$$

and with suitable initial direction

$$I_0 \;\in\; \widehat{\mathcal{Z}}_{\mathcal{M}}^{\frac{5}{2}A,\alpha}(N,C,1/2) \tag{13}$$

gives rise to a solution $\big(I(t),\theta(t)\big)$ of (4) which satisfies the channeling criterion (7) on the exponentially long time interval $[0,T_0]$, where

$$T_0 \;=\; \frac{\sigma}{48E}\epsilon^{\tau}e^{\epsilon^{-\tau/4}} \;-\; 1 \tag{14}$$

(here σ is an analyticity parameter associated with the potential W). This solution is approximated by a "generalized continuum model" solution in the sense that (4) is a near-identity transformation of a normal form that coincides, to leading order, with the continuum model with continuum potential $\Pi_{\mathcal{M}}W$. Finally, on the interval $[0,T_0]$, the longitudinal momentum $\widehat{I}(t)$ is nearly constant:

$$\|\widehat{I}(t) - \widehat{I}_0\|_\infty \;\leq\; \frac{3}{4}A\epsilon^c, \tag{15}$$

as is the transverse energy:

$$\frac{1}{\epsilon}|E_\perp(I(t),\theta(t)) - E_\perp(I_0,\theta_0)| \;\leq\; \delta. \tag{16}$$

A complete statement and a proof of this theorem—including the explicit tranformation to normal form—can be found in Dumas (1991). In summary, it is worthwhile to point out that the conclusions of the theorem embody the basic features of channeling that are observed experimentally. These include the avoidance of close encounters with nuclei, the approximation of actual trajectories by continuum model trajectories, and the approximate conservation of longitudinal momentum and transverse energy, as reflected in Eqs. (15) and (16) above.

ACKNOWLEDGEMENT

I would like to thank the organizers of the workshop, especially Harvey Segur, for inviting me to make this contribution.

REFERENCES

Benettin, G., Galgani, L., and Giorgilli, A., 1985, A proof of Nekhoroshev's theorem for the stability times in nearly integrable Hamiltonian systems, *Celestial Mech.* **37**: 1–25.

Benettin, G. and Gallavotti, G., 1986, Stability of motion near resonances in quasi-integrable Hamiltonian systems, *J. Stat. Phys.* **44**: 293–338.

Carrigan, R.A. and Ellison, J.A., Eds., 1987, *Relativistic Channeling*, Plenum, New York.

Carrigan, R.A., Toohig, T.A., and Tsyganov, E.N., 1990, Beam extraction from *TeV* accelerators using channeling in bent crystals, *Nucl. Instr. Meth.* **B 48**: 167–170.

Dumas, H.S., 1988, A mathematical theory of classical particle channeling in perfect crystals, Ph.D. thesis (University of New Mexico, Albuquerque), UMI, Ann Arbor, MI.

————, 1989, Nekhoroshev's theorem and particle channeling in crystals, in *Integrable Systems and Applications* (M. Balabane, P. Lochak, and C. Sulem, Eds.) Lecture Notes in Physics No. 342: 87–94, Springer-Verlag, New York.

————, 1991, to appear in *Dynamics Reported.*

Dumas, H.S. and Ellison, J.A., 1991, Nekhoroshev's theorem, ergodicity, and the motion of energetic charged particles in crystals, IMA Preprint Series # 775, to appear in *Foundations of Quantum and Classical Dynamics* (J.A. Ellison and H. Überall, Eds.).

Gemmell, D.S., 1974, Channeling and related effects in the motion of charged particles through crystals, *Rev. Mod. Phys.* **46**: 129-227.

Kumakhov, M.A., and Komarov, F.F., 1989, *Radiation from Charged Particles in Solids*, American Institute of Physics, Translation Series, New York.

Lindhard, J., 1965, Influence of crystal lattice on motion of energetic charged particles, *Mat. Fys. Medd. Dan. Vid. Selsk.* **34**, no. 14.

Nekhoroshev, N.N., 1971, *Fun. Anal. Pril.* **5** (4): 82–84,
English translation: Behavior of Hamiltonian systems close to integrable, *Funct. Anal.* **5**: 338–339 (1971).

————, 1977, *Usp. Mat. Nauk. SSSR* **32** (6): 5–66,
English translation: An exponential estimate of the time of stability of nearly integrable Hamiltonian systems, *Russian Math. Surveys* **32** (6): 1–65 (1977).

————, 1979, *Tr. Sem. Petrows.* **5**: 5–62,
English translation: An exponential estimate of the time of stability of nearly integrable Hamiltonian systems II, in *Topics in Modern Mathematics, Petrovskii Seminar No. 5* (O.A. Oleinik, Ed.), Consultants Bureau, London, (1980).

GRAVITY-CAPILLARY FREE SURFACE FLOWS

Jean-Marc Vanden-Broeck

Mathematics Department and
Center for the Mathematical Sciences
University of Wisconsin-Madison
Madison, WI 53705

ABSTRACT

This paper describes the effect of surface tension on various nonlinear free surface flow problems. Accurate numerical solutions are presented for the flow past a bubble in a tube, the cavitating flow past a curved obstacle and gravity capillary elevation solitary waves. Each flow is characterized by a continuum of solutions when surface tension is neglected. It is shown that there is a discrete set of solutions when surface tension is taken into account.

1. INTRODUCTION

Numerical computations have shown that some free surface problems, which are characterized by a continuum of solutions when surface tension is neglected, possess a discrete set of solutions when surface tension is taken into account. Furthermore it was found that this discrete set of solutions reduces to a unique solution as the surface tension tends to zero. Therefore an arbitrary small amount of surface tension can be used to remove the degeneracy of some free surface flow problems.

A well know example of such a flow is the Saffman-Taylor model for fingering in a Hele-Shaw cell. Saffman & Taylor (1958) derived an exact solution for the problem without surface tension. This solution leaves the ratio λ of the width of the finger to the width of the channel undetermined. Vanden-Broeck (1983a) considered the problem with surface tension and provided numerical evidence that solutions exist only for a countably infinite number of values of the parameter λ. This discrete set of solutions include as special cases those of McLean & Saffman (1981) and Romero (1982). Similar results were obtained by Tanveer (1986) for a bubble in a Hele-Shaw cell.

In order to compute the discrete set of solutions, Vanden-Broeck (1983a) used the following simple numerical technique. In the first stage a modified problem is defined by allowing the slope of the surface of the finger to be discontinuous at the apex. Solutions of this modified problems are computed for all values of λ. The discrete set

Asymptotics beyond All Orders, Edited by H. Segur *et al.*
Plenum Press, New York, 1991

of solutions is then obtained by selecting among the solutions of the modified problem those for which the slope is continuous at the apex.

In this paper we consider three other free surface flows for which solutions are also selected by introducing an arbitrary small amount of surface tension. In each problem there is a continuum of solutions when surface tension is neglected and a discrete set of solutions when surface tension is taken into account.

2. BUBBLES RISING IN A TUBE

Let us consider the steady two-dimensional potential flow of an inviscid incompressible fluid past a bubble in a tube of width h (see Figure 1). The pressure in the bubble is assumed to be constant. We introduce Cartesian coordinates with the origin at the top of the bubble and we assume that the bubble is symmetric about the x-axis. Gravity acts in the negative x-direction. As $x \longrightarrow +\infty$, the velocity approaches the constant U. We define dimensionless variables by taking U as the unit velocity and h as the unit length. The problem is characterized by the Froude number

$$F = U/(gh)^{1/2} \tag{2.1}$$

and the Weber number

$$a = \rho U^2 h/T. \tag{2.2}$$

Here g is the acceleration of gravity, T the surface tension and ρ the density of the fluid. Vanden-Broeck (1984a, 1984b, 1986) solved the problem numerically by series truncation. The solutions were selected by using the technique derived by Vanden-Broeck (1983a) in his investigation of the effect of surface tension on the shape of

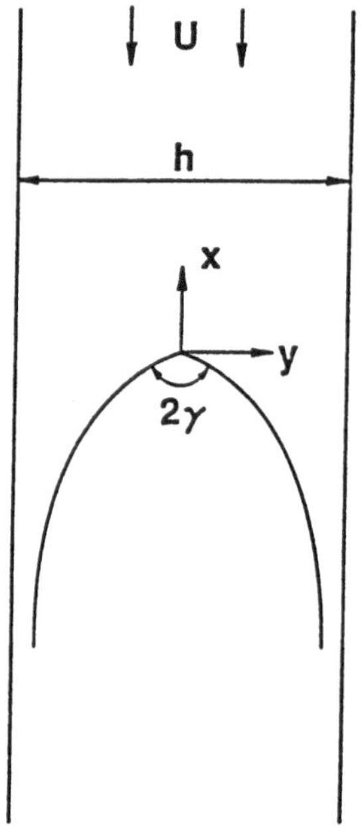

Fig. 1. Sketch of a bubble rising in a tube.

fingers in a Hele-Shaw cell. A modified problem is defined by allowing the slope to be discontinuous at the apex of the bubble. The angle at the apex is denoted by 2γ (see Figure 1). Here, γ is to be found as part of the solution. The modified problem is solved for all values of F. The solutions of the original problem are obtained by selecting among the solutions of the modified problem those for which $\gamma = \pi/2$.

When surface tension is neglected ($a = \infty$), the numerical computations show that

$$\gamma = 90° \qquad F < F_c \sim 0.36 \tag{2.3}$$

$$\gamma = 60° \qquad F = F_c \sim 0.36 \tag{2.4}$$

$$\gamma = 0° \qquad F > F_c \sim 0.36\,. \tag{2.5}$$

Relations (2.3) – (2.5) show that all solutions corresponding to $F < F_c$ are solutions of the original problem. This finding agrees with the analytical work of Garabedian (1957) who proved that a solution exists for all values of F smaller than a critical value. The solutions for $F \geq F_c$ are only solutions of the modified problem. They are characterized by a discontinuity in slope at the apex of the bubble.

Vanden-Broeck (1984b) solved the problem with surface tension ($a \neq \infty$). In Figure 2 we present values of γ versus F for $a = 10$. As F tends to infinity, γ tends to zero. As F approaches zero, γ oscillates often around $\pi/2$. Figure 2 suggests that there exists a countably infinite number of values of F for which $\gamma = \pi/2$. The solutions corresponding to these values of F are the solutions of the original problem. Similar results were found for other values of a. As a increases, the amplitudes and wavelengths of the oscillations in Figure 2 decrease. As $a \longrightarrow \infty$, the discrete set of solutions of the original problem reduces to a unique solution characterized by $F = F^* \sim 0.23$. Therefore a solution in the interval $0 < F < F_c$ is selected by introducing surface tension and then taking the limit as $T \longrightarrow 0$. The profile corresponding to $F = F^*$ and $T = 0$ is shown in Figure 3.

Collins (1965) performed some experiments and obtained the experimental value $F = 0.25$. In addition, he measured the ratio of the radius of curvature at the top of

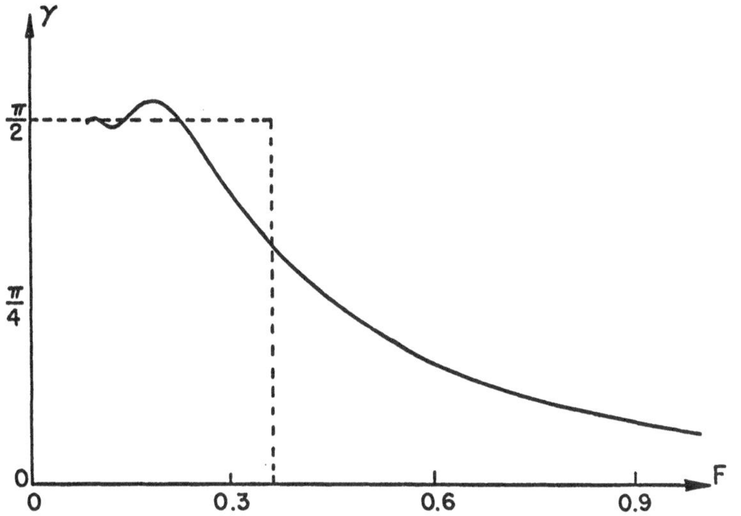

Fig. 2. Values of γ versus F for $\alpha = 10$. The broken line corresponds to the solution defined by (2.3) - (2.5).

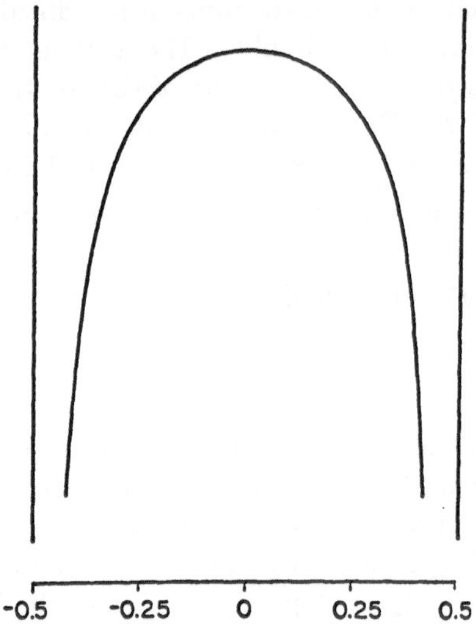

Fig. 3. Bubble profile for $F = F^* \sim 0.23$ and $T = 0$.

the bubble to the width h of the tube and obtained the value 0.305. The corresponding ratio for the theoretical profile of Figure 3 is 0.32. Similar results were obtained by Couet and Strumolo (1987) for a bubble rising in an inclined tube.

Levine and Yang (1990) and Vanden-Broeck (1991a) considered the corresponding axisymmetric problem. The results show that the axisymmetric problem is qualitatively similar to the two dimensional problem. In particular

$$\gamma = 90° \qquad F < F_d \sim 0.49 \tag{2.6}$$

$$\gamma = 65° \qquad F = F_d \sim 0.49 \tag{2.7}$$

$$\gamma = \ 0° \qquad F > F_d \sim 0.49\,. \tag{2.8}$$

Here F is the Froude number defined by (2.1) where h denotes now the diameter of the tube. Typical bubble profiles are shown in Figures 4 - 6. Relations (2.6) – (2.8) show that all solutions corresponding to $F < F_d$ are solutions of the original problem.

Levine and Yang (1990) investigated the effect of surface tension on the axisymmetric problem. Their results show that a unique bubble characterized by $F \sim 0.35$ can be selected by introducing surface tension and then taking the limit as the surface tension approaches zero.

3. CAVITATING FLOW OF A FLUID WITH SURFACE TENSION PAST A CIRCULAR CYLINDER

In this section we consider the cavitating flow past a circular cylinder of radius R (see Figure 7). We neglect the effects of viscosity and compressibility. At infinity the flow approaches a uniform stream with constant velocity U. We assume that the cavity extends to infinity and that the pressure in it is constant.

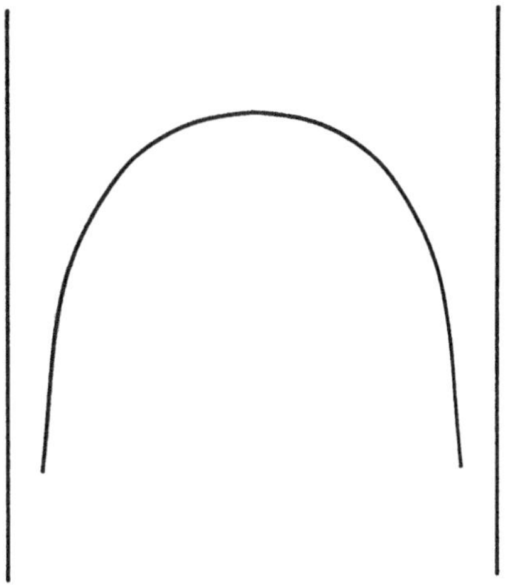

Fig. 4. Computed axisymmetric bubble profile for F = 0.35. The slope is continuous at the apex.

The problem is characterized by the angular position γ of the separation points A and B, the angle β between the surface of the cavity and the cylinder at the separation points and the Weber number

$$\alpha = \frac{\rho U^2 R}{T}\,. \tag{3.1}$$

Here ρ is the density of the fluid and T is the surface tension.

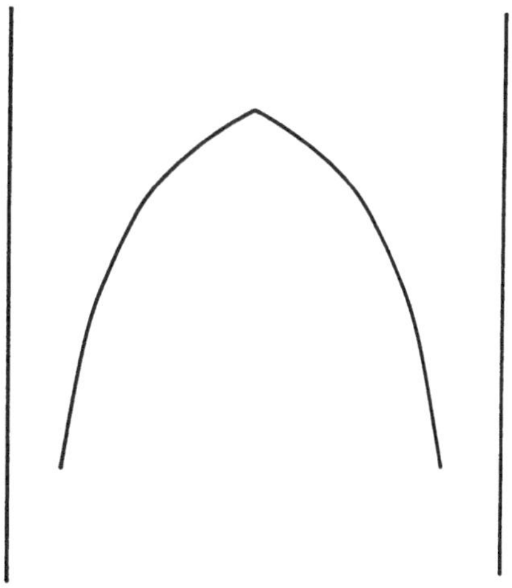

Fig. 5. Computed axisymmetric bubble profile for F = 0.49. The bubble is pointed with a 130° angle at the apex.

Fig. 6. Computed axisymmetric bubble profile for F = 1.1. There is a cusp at the apex of the bubble.

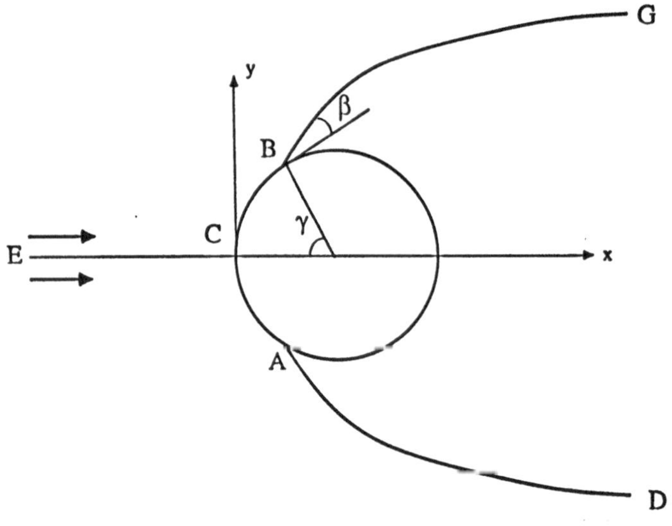

Fig. 7. Sketch of the cavitating flow past a circular cylinder.

When surface tension is neglected (i.e. $\alpha = \infty$), the angle β is equal to zero and the free surfaces BG and AD are tangent to the cylinder at the points B and A. It can be shown that there is a flow for each value of γ (Brodetsky (1923), Birkhoff and Zarantonello (1957), Vanden-Broeck (1983b)). A particular value of γ is usually selected by imposing the Brillouin-Villat condition which requires the curvature of the free surface to be finite at the separation points A and B. This determines a unique flow characterized by $\gamma = \gamma^* \sim 55°$.

The effect of surface tension on cavitating flows and on related flow problems has been considered by several previous investigators (Ackerberg (1975), Cumberbatch and Norbury (1979), Vanden-Broeck (1981, 1983b), Ackerberg and Liu (1987), and others). Vanden-Broeck (1981, 1983b) derived an asymptotic solution for α large. For the cavitating flow past a circular cylinder he found that

$$\beta \sim \frac{C}{2}\left(\frac{\pi}{\alpha}\right)^{1/2} \quad \text{as} \quad \alpha \longrightarrow \infty . \tag{3.2}$$

Here C is a constant which depends on γ. The angle β is counted positive when the free surface lies above the tangent to the cylinder at the separation point B. For $\gamma > \gamma^*$, $C > 0$ and the flow near each of the separation points is locally a flow inside an angle with a zero velocity at the separation point. For $\gamma < \gamma^*$, $C < 0$ and the velocity at the separation points is infinite.

As $\alpha \longrightarrow 0$ (i.e. $T \longrightarrow \infty$), the free surfaces approach two horizontal straight lines. Therefore

$$\beta \longrightarrow \gamma - \frac{\pi}{2} \quad \text{as} \quad \alpha \longrightarrow 0 . \tag{3.3}$$

Relation (3.3) shows that β is negative in the limit as $\alpha \longrightarrow 0$ when $\gamma < \frac{\pi}{2}$. On the other hand (3.2) shows that β is positive as $\alpha \longrightarrow \infty$ when $\gamma > \gamma^*$. Assuming that β is a continuous function of α, Vanden-Broeck (1983b) conjectured that there exists for each value of α, a particular value of $\gamma^* < \gamma < \frac{\pi}{2}$ for which the flow leaves the cylinder tangentially (i.e. for which $\beta = 0$). This conjecture was confirmed by the numerical calculations of Vanden-Broeck (1984c). He has computed solutions with $\alpha \neq \infty$ and $\beta = 0$.

Vanden-Broeck (1991b) computed solutions with $\beta \neq 0$ by series truncation. Accurate solutions were obtained for various values of α and γ.

In Figure 8 we present numerical values of β versus γ for various values of α. Each curve corresponds to a different value of α. As $\alpha \longrightarrow 0$, the free surfaces approach two horizontal straight lines. Therefore the curve corresponding to $\alpha = 0$ in Figure 8 is the straight line of equation

$$\beta = \gamma - \frac{\pi}{2} .$$

For $\alpha = \infty$, the angle β is equal to zero for all values of γ and the curve corresponding to $\alpha = \infty$ in Figure 8 is the γ-axis.

Figure 8 shows that for each value of $\alpha \neq \infty$, there is a particular value $\tilde{\gamma}$ of γ for which $\beta = 0$ (i.e. for which the free surfaces leave the obstacle tangentially). In Figure 9 we present numerical values of $\tilde{\gamma}$ versus of α^{-1}. As $\alpha \longrightarrow 0$, $\tilde{\gamma}$ approaches $\frac{\pi}{2}$ As $\alpha \longrightarrow \infty$, $\tilde{\gamma}$ tends to γ^*. Therefore the particular solution which satisfies the Brillouin Villat condition in the absence of surface tension can be viewed as the limit of the family of solutions of Figure 9 as the surface tension approaches zero.

4. ELEVATION SOLITARY WAVES WITH SURFACE TENSION

The influence of surface tension on waves propagating without change of form on the surface of a liquid above a horizontal bottom was considered by many previous investigators (Harrison (1909), Wilton (1915), Nayfeh (1970), Schwartz and Vanden-Broeck (1979), Hogan (1980), Chen and Saffman (1980) and Zufira (1987). These analytical and numerical calculations show that many different families of gravity-capillary waves exist when the Bond number

$$\tau = \frac{T}{\rho g H^2} \tag{4.1}$$

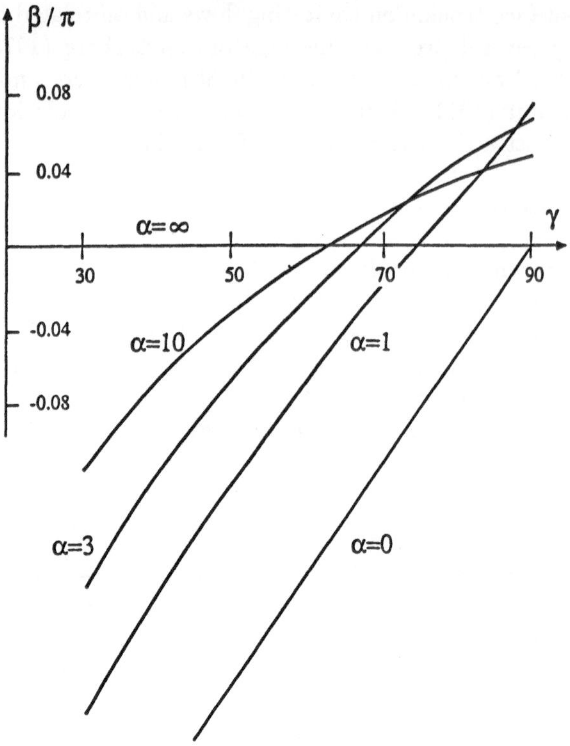

Fig. 8. Values of β versus γ for various values of α.

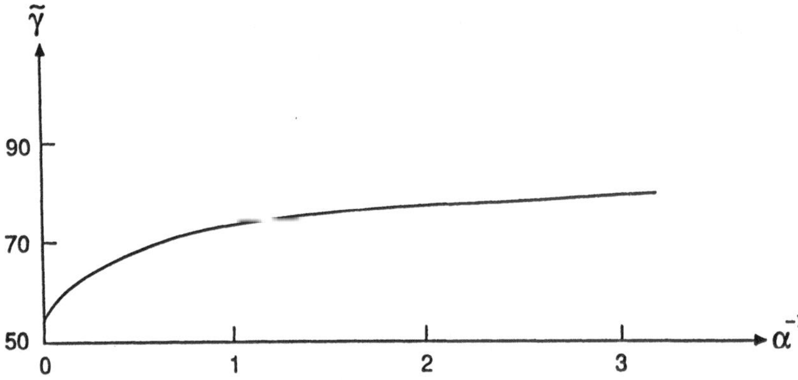

Fig. 9. Values of $\tilde{\gamma}$ versus α^{-1}.

is in the interval $0 < \tau < 1/3$. Here g is the acceleration of gravity, T is the surface tension, ρ the density of the fluid, and H the depth. On the other hand when $\tau > 1/3$ or $\tau = 0$ (i.e. in the absence of surface tension), there is only one family of solutions. For a given value of τ, each of these families is a two-parameter family of solutions. It is convenient to choose these parameters as the dimensionless amplitude of the wave and ratio

$$\ell = \frac{\lambda}{H}. \tag{4.2}$$

Here λ is the wavelength of the wave. In this section the variables are made dimensionless by choosing the phase velocity as the unit velocity and the depth as the unit length.

As the ratio ℓ tends to infinity, the periodic gravity-capillary waves approach solitary wave configurations. Gravity-capillary solitary waves were first considered by Korteweg and de Vries in their classical 1895 paper (see also Benjamin (1982) and Vanden-Broeck and Shen (1983)). By using a formal perturbation expansion, they derived an approximate differential equation for the unknown shape of the free surface. This equation is the well known Korteweg-de Vries equation: it predicts elevation solitary waves for $0 \le \tau < \frac{1}{3}$ and depression solitary waves for $\tau > \frac{1}{3}$.

Hunter and Vanden-Broeck (1983) reformulated the exact nonlinear problem as an integro-differential equation for the unknown shape of the free surface. This equation was then discretized and the resulting algebraic equations were solved by Newton's method. Their numerical results for $\tau > 1/3$ extend the result of Korteweg and de Vries to depression solitary waves of large amplitude. For $0 < \tau < 1/3$, they calculated periodic waves. As indicated before there are many different families of waves, each family depending, for a fixed value of τ, on two parameters: ℓ and the amplitude of the wave. Hunter and Vanden-Broeck (1983) calculated these families of waves for values of ℓ between 30 and 40. These rather long waves are characterized by a train of ripples in their troughs.

In recent years some new interesting analytical results have been obtained. Amick and Kirchgässner (1989) proved the existence of depression solitary waves when $\tau > 1/3$. Hunter and Scherule (1988), Beale (1991) and Sun (1991) proved the existence of elevation solitary waves when $0 < \tau < \frac{1}{3}$. These elevation solitary waves do not in general approach a uniform stream at infinity but are characterized by a train of ripples in the far field.

Vanden-Broeck (1991c) extended the numerical results of Hunter and Vanden-Broeck (1983) when $0 < \tau < \frac{1}{3}$ to values of ℓ as large as 120. These very long waves are used as approximations for solitary waves. The numerical results confirm that for each value of $0 < \tau < \frac{1}{3}$, there are many different families of solutions. Each family is characterized by two parameters. We choose these parameters as the dimensionless velocity u_0 at the crest of the solitary wave and the dimensionless wavelength ℓ. The parameter u_0 is a measure of the amplitude of the wave. Most of the computations were performed with 120 mesh points. We checked that the results are independent of the number of mesh points within graphical accuracy.

Values of the Froude number F versus ℓ for $\tau = 0.24$ and $u_0 = 0.99$ are shown in Figure 10. The curves (a) and (b) correspond to two computed families of solutions. Corresponding profiles of the waves for these two families are shown in Figures 11 and 12 respectively. These results show that there are ripples on the free surfaces. By comparing the profiles in Figures 11 and 12 we see that the number of inflexion points (and therefore the number of ripples) increase as we move from one family to another family further to the right in Figure 10. Furthermore, it follows from the symmetry of the flow that the last point of the graph in Figures 11 and 12 is either a crest or a trough of the ripples. These facts indicate that for ℓ sufficiently large, there is an infinite number of families of solutions. The corresponding curves in Figure 10 can be obtained from any particular curve by translating it horizontally by a multiple of the wavelength of the ripples. Solitary waves are then obtained by jumping from a curve to the next as we take the limit as $\ell \to \infty$. After each jump, two more crests or troughs appear on the free surface, one on the right and one on the left. In the limit as $\ell \to \infty$, we obtain a solitary wave with an infinite train of ripples in the far field. For

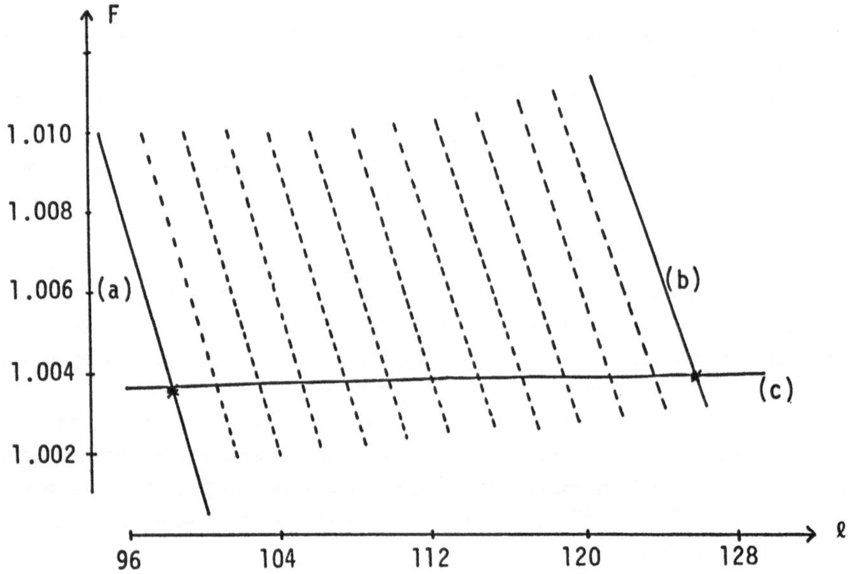

Fig. 10. Values of the Froude number F versus the dimensionless wavelength ℓ for periodic waves with $\tau = 0.24$ and $u_0 = 0.99$.

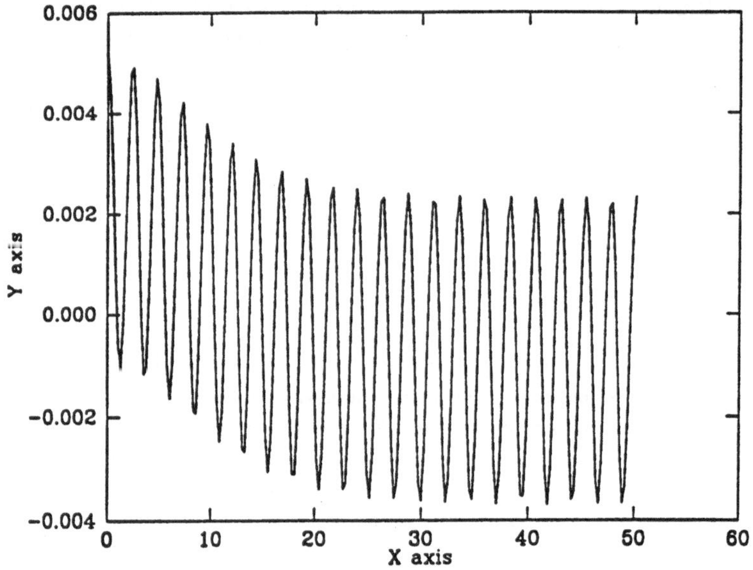

Fig. 11a. Computed free surface profile of a periodic wave with $\tau = 0.24$, $u_0 = 0.99$ and $F = 1.0005$. Only half of a wavelength of the wave is shown.

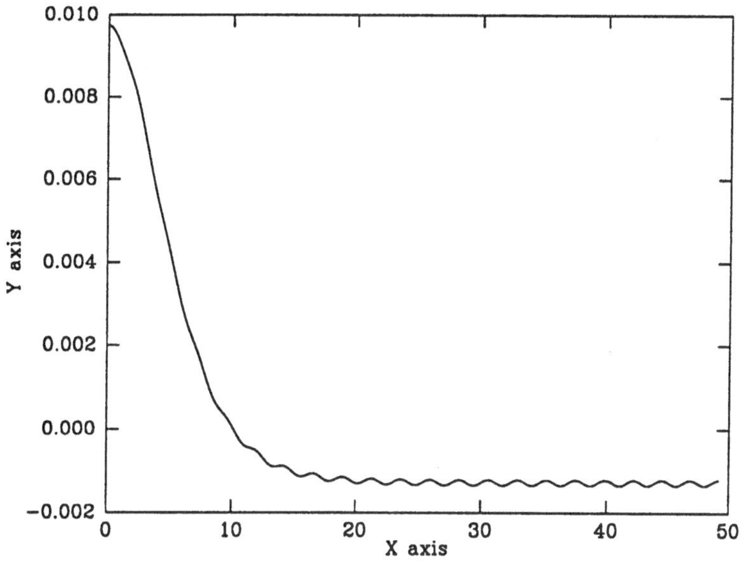

Fig. 11b. Same as in Fig. 11a with $F = 1.0035$.

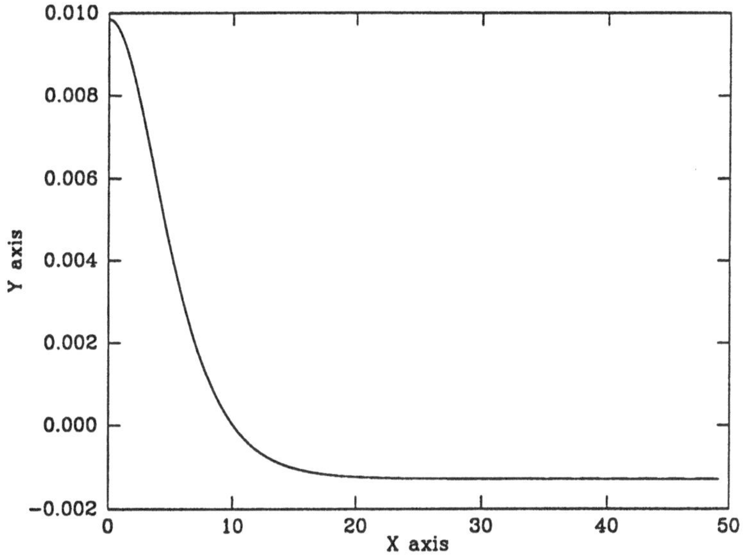

Fig. 11c. Same as in Fig. 11a with $F = 1.00358$. There are no ripples on the free surface. This solution corresponds to the cross on the left of Fig. 10.

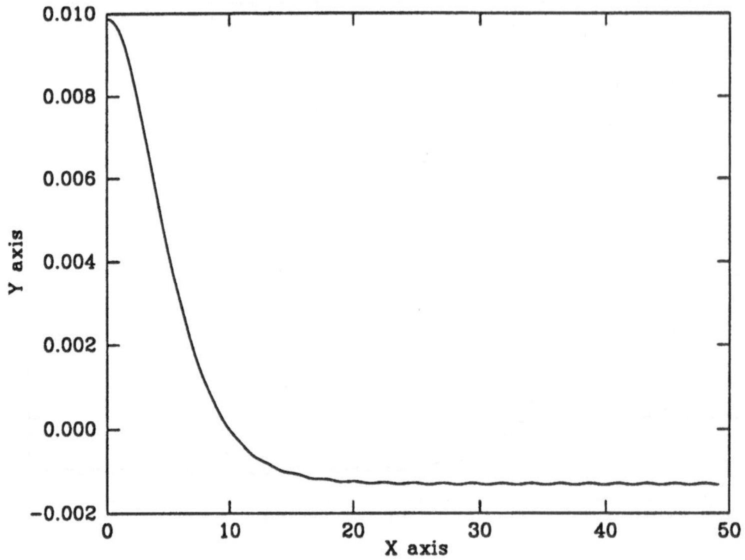

Fig. 11d. Same as Fig. 11a with F = 1.0036.

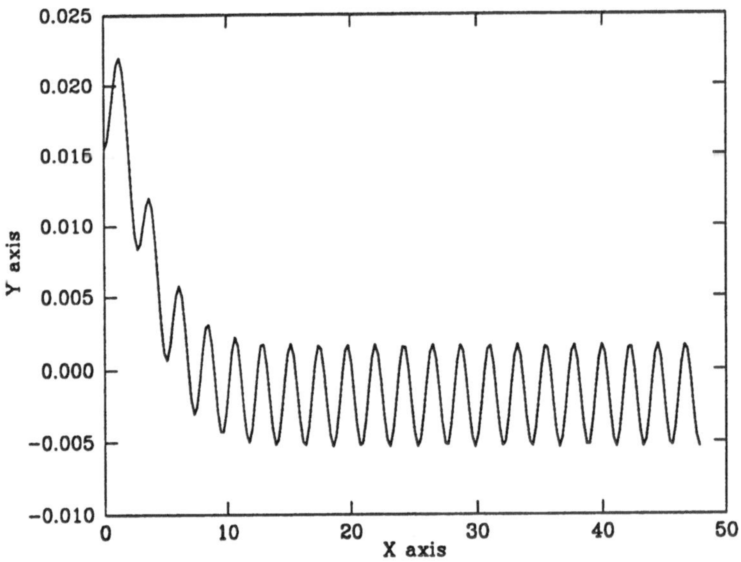

Fig. 11e. Same as Fig. 11a with F = 1.008.

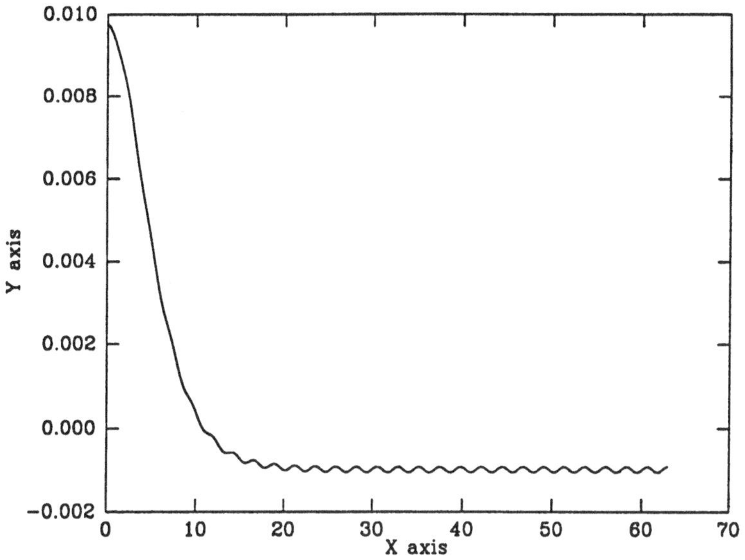

Fig. 12a. Same as Fig. 11a with F = 1.0038.

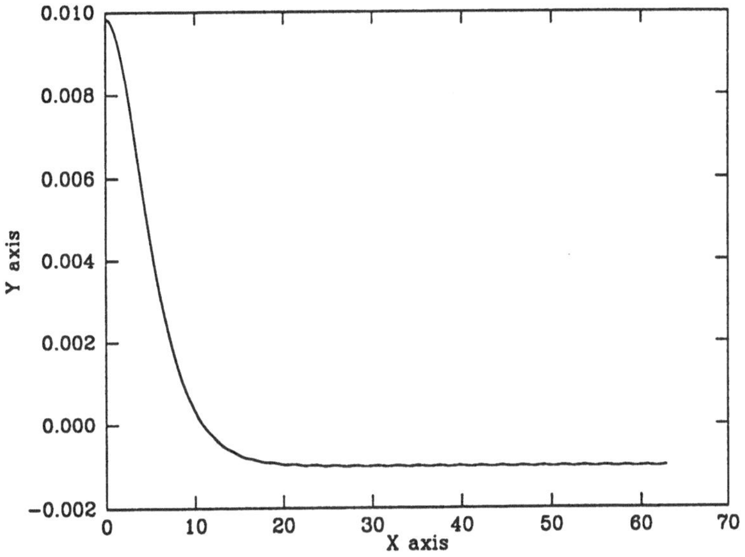

Fig. 12b. Same as Fig. 11a with F = 1.00387. This profile is close to the solution corresponding to the star on the right of Fig. 10.

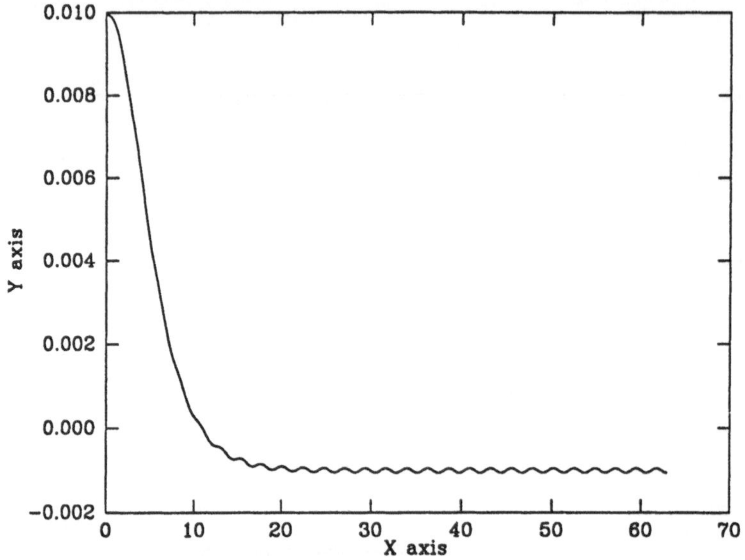

Fig. 12c. Same as Fig. 11a with F = 1.00395.

each value of $0 < \tau < \frac{1}{3}$ these solitary waves form a two-parameter family of solutions.

The profiles in Figures 11 and 12 show that the steepness of the ripples (i.e. the difference of height between a crest of the ripples and its trough divided by the wavelength of the ripple) is small for all the profiles presented. Therefore the dimensionless wavelength L of the ripples satisfies approximately the disperson relation of linear gravity-capillary periodic waves, namely:

$$F^2 = \frac{L}{2\pi}\left(1 + \tau\frac{4\pi^2}{L^2}\right)\tanh\frac{2\pi}{L}\,. \tag{4.3}$$

Starting with the curve (a) in Figure 10 we used (4.3) to calculate the value of L corresponding to each value of F. Then we performed the translation mentioned in the previous paragraph. The resulting curves are the broken lines in Figure 10. We note that the 12th broken line coincides within graphical accuracy with the curve (b). This constitutes a check on our calculations.

The solution corresponding to the profile in Figure 11a is very close to a train of periodic waves extending from $x = -\infty$ to $x = +\infty$. This shows that the solitary waves bifurcate from a train of periodic waves. As we move away from the bifurcation point, one crest of the train of periodic waves is progressively lifted.

Figure 11 shows that the amplitude of the ripples first decrease and then increase as we move along the branch of solutions. There is exactly one point on each curve in Figure 10 for which the amplitude of the ripples is minimum. These points are indicated by crosses in Figure 10 and the profile corresponding to the cross on the left is shown in Figure 11c. For values of u_0 sufficiently close to 1 (i.e. for sufficiently small solitary waves), the minimum amplitude of the ripples is zero within graphical accuracy (see Figure 11c). For larger solitary waves the minimum amplitude of the ripples is not zero. This is illustrated in Figure 13 where the wave with ripples of

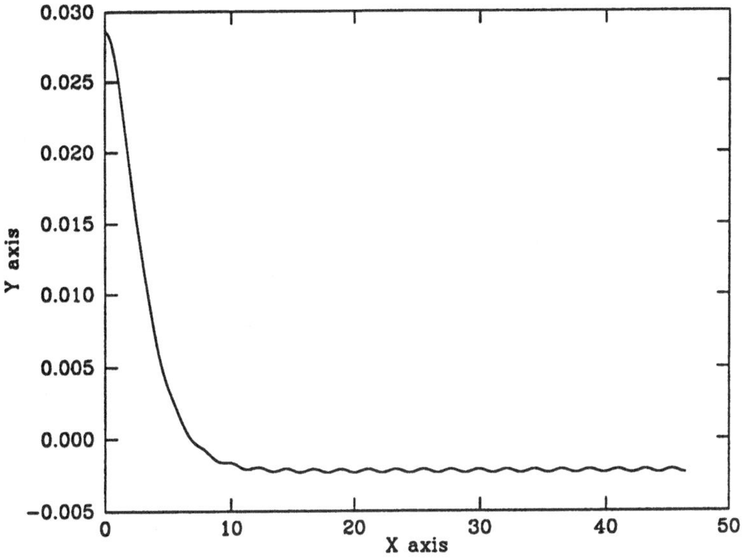

Fig. 13. Computed free surface profile with $\tau = 0.24$, u_0 - 0.97 and F = 1.00116.

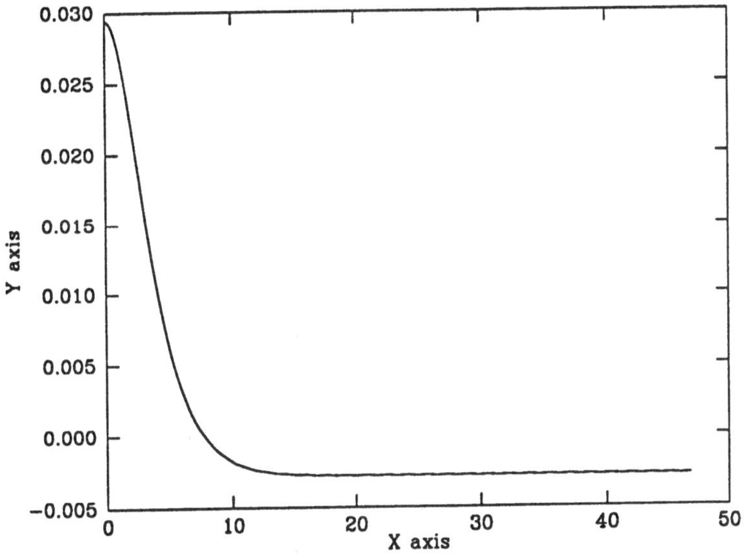

Fig. 14. Computed free surface profile with $\tau = 0.2$, $u_0 = 0.97$ and F = 1.0117. There are no ripples on the free surface.

minimum amplitude for $\tau = 0.24$ and $u_0 = 0.97$ is shown. However Figure 14 shows that a wave without ripples for $u_0 = 0.97$ can be obtained by decreasing τ to the value 0.2. As τ decreases, waves without ripples exist for larger values of the amplitude of the solitary waves.

Our results have shown that for u_0 sufficiently close to one, there are solitary waves for which the amplitude of the ripples is zero within graphical accuracy. These solutions form, for each value of $0 < \tau < \frac{1}{3}$, a one parameter family of solutions.

As mentioned earlier, Korteweg and de Vries (1895) derived an approximate equation to describe small gravity capillary waves in shallow water. This equation can be integrated as

$$2(F-1)\eta - \tfrac{3}{2}\eta^2 + (\tau - \tfrac{1}{3})\eta'' = \text{constant}. \tag{4.4}$$

Here $y = \eta(x)$ is the equation of the free surface.

Equation (4.4) admits periodic solutions which are known as conoidal waves. As the wavelength tends to infinity, the conoidal waves approach the solitary wave solution

$$\eta(x) = \alpha \operatorname{sech}^2 \frac{x}{b}$$
$$\alpha = 2(F-1) \quad b = \left[\frac{4(1-3\tau)}{3\alpha}\right]^{1/2}. \tag{4.5}$$

For $0 \leq \tau < \frac{1}{3}$, (4.5) is an elevation wave. For $\tau > \frac{1}{3}$, (4.5) is a depression wave. The solution (4.5) with $\tau > \frac{1}{3}$ provide an accurate description of the depression solitary waves of small amplitude (see Hunter and Vanden-Broeck (1983)). For $0 < \tau < \frac{1}{3}$, they do not provide an accurate description of the elevation solitary waves because they approach a uniform flow at infinity. Therefore the solitary waves with ripples are not accurately described by (4.5). However it is possible that (4.5) describes accurately our particular solutions without ripples (i.e. the solutions corresponding to the crosses in Figure 10). In order to check this idea, we computed the conoidal waves of (4.4) when $\tau = 0.24$ and $u_0 = 0.99$. For each value of τ, the conoidal waves form a family of solutions depending on the two parameters u_0 and ℓ. The values of F versus ℓ for $\tau = 0.24$ and $u_0 = 0.99$ are shown in Figure 10 (curve (c)). We see that the crosses lie on the curve (c). Furthermore the profile in Figure 11c was found to coincide with the conoidal wave within graphical accuracy.

ACKNOWLEDGMENT

This work was supported by the National Science Foundation under Grant No. DMS-8903083.

REFERENCES

Ackerberg, R. C., 1975, J. Fluid Mech. 70:333.
Ackerberg, R. C. and Liu, T. J., 1987, Phys. Fluids, 30:289.
Amick, C. J. and Kirchgässner, K., 1989, Arch. Rational Mech. Anal., 105:1.
Beale, J. T., Comm. Pure Appl. Math. (to appear).
Benjamin, T. B., 1982, Quart. Appl. Math., 37:183.
Birkhoff, G. and Zarantonello, E., 1957, "Jets, Wakes and Cavities", Academic Press, New York.
Brodetsky, S., 1923, Proc. R. Soc. London, Ser. A, 102:543.
Chen, B. and Saffman, P. G., 1980, Studies in Appl. Math., 62:95.
Collins, R., 1965, J. Fluid Mech., 22:763.
Couet, B. and Strumolo, 1987, J. Fluid Mech., 184:1.
Cumberbatch, E. and Norbury, J., 1979, Q. J. Mech. Appl. Math., 32:303.

Garabedian, R. R., 1957, Proc. R. Soc. London A 241:423.
Harrison, W. J., 1909, Proc. Lond. Math. Soc., 7:107.
Hogan, S. J., 1980, J. Fluid Mech., 96, 417.
Hunter, J. K. and Scherule, J., 1988, Physica D 32:253.
Hunter, J. K. and Vanden-Broeck, J.-M., 1983, J. Fluid Mech., 134:205.
Korteweg, D. J. and de Vries, G., 1895, Phil. Mag., 39:422.
Levine, H. and Yang, Y., 1990, Phys. Fluids A 2:542.
McLean, J. W. and Saffman, P. G., 1981, J. Fluid Mech., 102:455.
Nayfeh, A. H., 1970, J. Fluid Mech., 40:671.
Romero, L., 1982, Ph.D. thesis, California Institute of Technology.
Saffman, P. G. and Taylor, G. I., 1958, Proc. R. Soc. London A 245:312.
Schwartz, L. W. and Vanden-Broeck, J.-M., 1979, J. Fluid Mech., 95:119.
Sun, S. M., 1991, J. Math. Anal. Appl. (to appear).
Tanveer, S., 1986, Phys. Fluids, 29:3537.
Vanden-Broeck, J.-M., 1981, Q. J. Mech. Appl. Math., 34:465.
Vanden-Broeck, J.-M., 1983a, Phys. Fluids, 26:2033.
Vanden-Broeck, J.-M., 1983b, J. Fluid Mech., 133:255.
Vanden-Broeck, J.-M., 1984a, Phys. Fluids, 27:1090.
Vanden-Broeck, J.-M., 1984b, Phys. Fluids, 27:2604.
Vanden-Broeck, J.-M., 1984c, Phys. Fluids, 27:2601.
Vanden-Broeck, J.-M., 1986, Phys. Fluids, 29:1343.
Vanden-Broeck, J.-M., 1991a, Phys. Fluids, (in press).
Vanden-Broeck, J.-M., 1991b, Phys. Fluids, (in press).
Vanden-Broeck, J.-M., 1991c, Phys. Fluids, (submitted).
Vanden-Broeck, J.-M. and Shen, M. C., 1983, J. Appl. Math. and Physics, 34:112.
Wilton, J. R., 1915, Phil. Mag. 29:688.
Zufira, J. A., 1987, J. Fluid Mech. 184:183.

SOLITARY WATER WAVES WITH RIPPLES BEYOND ALL ORDERS

J. Thomas Beale

Department of Mathematics
Duke University
Durham, NC 27706

The famous solitary water wave, first observed and studied by Russell in the mid-nineteenth century, is best known in a mathematical context as a solution of the Korteweg de Vries equation, which describes long water waves of small amplitude. Solitary waves have received lasting attention because of their occurrence in several important physical settings, their striking persistence, and the remarkable structure of the KdV and related equations which has emerged over the last 25 years. Our interest here is in the extent to which actual solutions of the equations of water waves correspond to the soliton solution of the KdV. For the case in which surface tension is neglected, it was shown in the 1940's and 1950's by Lavrentiev and by Friedrichs and Hyers that there are exact progressing waves whose first approximation are the solitons of the KdV equation. The case with surface tension included is genuinely different, however, because of a resonance with linear waves of the same speed. The solitary waves have speed slightly greater than $C_0 = \sqrt{gh}$, where h is the depth of the water. Without surface tension, the linear waves of any wave number have speed $< C_0$, so that there is no overlap in the speeds of the two families. When surface tension is introduced, the dispersion relation changes, and the speed crosses C_0 as the wave number increases. Thus, in searching for a solitary wave of a certain speed, we may pick up the linear wave of the same speed. It appears that solitary waves can exist only in combination with the linear waves (called capillary waves or ripples), at least for most choices of the parameters. It is not known whether there are exact waves purely of solitary type, but there are ones which deviate only by a very small ripple at infinity. We will discuss here a mathematically rigorous construction of exact solitary water waves which at the center are well approximated by the KdV soliton, and approach at infinity an exact periodic wave whose first approximation is given by the linear theory. The amplitude of the periodic wave is of smaller order than any power of the amplitude of the main wave. Details and further references are given in [4]. A similar but different result was given by S. M. Sun [9]. He and M. C. Shen have also treated the cases of stratified fluids and two-fluid layers [10-14]. Recently Iooss and Kirchgässner [7] have obtained a variety of exact water waves with surface tension using a reduction process. They included waves connecting oscillations at infinity with waves of solitary type in between, but without a detailed description of the waves obtained.

Water waves are described by solutions of the Euler equations of inviscid, incompressible flow with appropriate boundary conditions. It is customary to assume theflow is irrotational, so that the velocity is the gradient of a harmonic potential. We take

Asymptotics beyond All Orders, Edited by H. Segur *et al.*
Plenum Press, New York, 1991

the fluid domain to be a strip $\{(X,Y): 0 < Y < S(X,T)\}$. The bottom $Y = 0$ is treated as a solid wall. On the free surface $Y = S$ we have the kinematic boundary condition, which merely says that the surface is preserved by the flow. Finally, the pressure on the free surface should differ from the atmospheric pressure above, taken to be zero, by a term proportional to the mean curvature of the surface. This condition is expressed through Bernoulli's equation on the surface.

The Korteweg-de Vries equation is derived from the water wave equations by perturbation methods in a limit for small amplitude waves on a length scale large compared to the depth, in a frame traveling at speed C_0. The equation is

$$y_t + \frac{3}{2}yy_x + \frac{1}{2}(\frac{1}{3} - \beta)y_{xxx} = 0 \tag{1}$$

where $y(x)$ is the surface profile and $\beta = \tau/\rho g h^2$ is the nondimensionalized coefficient of surface tension, τ being the dimensional one; β is often called the Bond number. The solitary wave of the KdV is the exact solution

$$y(x,t) = 4(\frac{1}{3} - \beta)\mathrm{sech}^2(x - ct), \qquad c = 2(\frac{1}{3} - \beta). \tag{2}$$

In the original variables, this solution corresponds to the surface height

$$S(X) = h + 4\varepsilon^2(\frac{1}{3} - \beta)\, h\, \mathrm{sech}^2(\varepsilon X/h) \tag{3}$$

traveling with speed $C = (1 + 2\varepsilon^2(\frac{1}{3} - \beta))\sqrt{gh}$. We have chosen ε so that the amplitude is $O(\varepsilon^2)$; note the stretched scale in the X-variable and the dependence of the speed on ε. It is apparent from (2) that for $0 \le \beta < \frac{1}{3}$ we have a wave of elevation, and for $\beta > \frac{1}{3}$ a wave of depression. For water waves under standard conditions β depends only on h, and $\beta < \frac{1}{3}$ corresponds to $h > .5$ cm. The resonance we have described occurs only for $\beta < \frac{1}{3}$, and we restrict our attention to this case.

In the linear theory of water waves it is found that waves of nondimensional wave number k (i.e., period $2\pi h/k$) have phase speed U_0 given by

$$\gamma k = (1 + \beta k^2)\tanh k \tag{4}$$

where $\gamma = U_0^2/gh$ is the Froude number. It turns out that for $0 < \beta < \frac{1}{3}$, for each $U_0 > C_0$ there is a unique root k of (4). It has long been known that the linear waves are genuine first approximations to exact periodic traveling wave solutions of the full system. Thus it is plausible that a solitary wave with $\gamma > 1$ might be carried by a periodic wave of the corresponding wave number.

We now describe the waves constructed in further detail. Since we are interested in traveling waves, we choose a frame of reference so that the waves are steady. We take the average depth h as fixed and construct waves with specified speed $U_0 > 1$ but close to 1. For $\varepsilon > 0$ we define the Froude number γ, and therefore U_0, by the relation

$$\gamma = 1 + 4\varepsilon^2(\frac{1}{3} - \beta),$$

which applies to the KdV soliton. Then for each $\varepsilon > 0$, sufficiently small, there is an exact, symmetric solution of the water wave equations with average depth h, average speed U_0 at $\pm\infty$, and surface profile

$$S(X) = h + 4\varepsilon^2(\frac{1}{3} - \beta)h\,\mathrm{sech}^2(\varepsilon X/h) + S^{(p)}(X) + O(\varepsilon^4 e^{-\varepsilon r X/h}) \tag{5}$$

for $X > 0$. The first term is just the KdV soliton. The part $h + S^{(p)}(X)$ is the profile of an exact periodic solution which to first approximation is

$$S^{(p)}(X) = \tilde{a} h \cos \tilde{k}(X/h + \delta) + O(\tilde{a}\varepsilon) \tag{6}$$

The amplitude $\tilde{a}$ and wave number $\tilde{k}$ satisfy

$$|\tilde{a}| \leq C\varepsilon^N, \qquad |\tilde{k} - k| \leq C\varepsilon^N.$$

where k is the wave number from the linear theory, satisfying (3). Here N is arbitrary; this is the "beyond all orders" effect. (So far as has been shown, the smallness condition for ε may depend on N.) The phase shift δ is arbitrary, except that a discrete set of values, including $\delta = 0$, must be excluded. It is evident from (5) that the KdV soliton is dominant for X within $O(1/\varepsilon)$, but as $X \to \infty$, the profile tends exponentially to the periodic wave.

While the construction shows that the periodic part is at most $O(\varepsilon^N)$ in amplitude for each N, the way in which the oscillation arises leads us to expect it is actually $O(\exp(-C/\varepsilon))$ for some C. The analogy with other beyond all orders problems will become more evident below. The argument does not show that the amplitude is nonzero. However, Hunter and Vanden-Broeck [6,15] have calculated cnoidal waves (the periodic waves of the same scaling as the solitary waves) which contain ripples; they seem to converge to solitary waves with ripples as the wavelength increases. There is evidence that the ripples disappear for special values of the parameters. A modification of the KdV equation has been found to have behavior like that described here for the full water wave equations, and more detailed information can be given; see references listed in [4]. This modified KdV can be thought of as a model for water waves with surface tension in the limit $\beta \to \frac{1}{3}$ from below.

We now sketch the construction of the exact waves, with emphasis on the appearance of the periodic waves at infinity. The first difficulty is that the domain of the fluid variables depends on the surface $Y = S(X)$, which is itself unknown. As has often been done, we use the velocity potential and stream function as independent variables; because the free surface is a streamline, the new domain is just a horizontal strip. As the unknown we take the angle of the velocity vector with the horizontal (cf. Stoker [8]). It should satisfy Laplace's equation with a nonlinear boundary condition on the top.

We wish to formulate the problem as a mapping in function spaces. However, it is somewhat awkward to work with the full unknown, which oscillates at infinity. For this reason we proceed in two steps. First we construct a family of exact periodic waves with fixed speed U_0, parametrized by amplitude a. Because the speed is fixed, the period varies slightly with the amplitude. we then write the unknown in the form $\theta = \theta^0 + \theta^1$, where θ^0 coincides with the periodic wave at $\pm\infty$ and θ^1 decays exponentially. Here each function has odd parity, and the phases of θ^0 at $\pm\infty$ are different; the two ends are connected by a specified cut-off function. Since θ^0 is known except for the amplitude, the full unknown θ is equivalent to the pair (θ^1, a). We now write the mapping in the form $F(\theta^1, a) = 0$ where $F : W \to Z$ for certain Banach spaces W and Z. Roughly speaking, the W-norm measures θ^1 in a Sobolev space with exponential weight, with, say, s derivatives, and a has weight ε^{-s}, i.e., $|(\theta^1, a)|_{W^s} = O(1)$ imples $a \leq O(\varepsilon^s)$. The elements of Z are also in Sobolev spaces with exponential decay. It is important that the limiting behavior at $\pm\infty$ of θ, given by θ^0, is an exact solution; as a consequence, $F(\theta^1, a)$ decays exponentially. A similar two-step procedure was used in [3]. The approach of Sun and Shen is similar in that the unknown is written as the sum of a decaying part and a periodic part, but somewhat opposite in that they allow general behavior in the periodic part of the unknown.

The mapping F has the form of an elliptic problem with nonlinear boundary condition. However, the horizontal variable is scaled appropriately for the KdV limit, so that the Laplace equation becomes

$$\varepsilon^2\theta_{xx} + \theta_{yy} = 0,$$

and the limit $\varepsilon \to 0$ is singular. The waves we are seeking form a branch bifurcating from the family of horizontal flows, intersecting at speed $C_0 = \sqrt{gh}$ or $\varepsilon = 0$. As $\varepsilon \to 0$, F becomes linear, with a certain null space N, and $QF = O(\varepsilon^2)$, where Q is a certain projection on Z. As in [1], we replace F by the modified mapping

$$\tilde{F} = \varepsilon^{-2}QF + (I - Q)F.$$

In the limit $\varepsilon = 0$, the equation $\tilde{F}(\theta, 0) = 0$ has a solution $\bar{\theta}$ which is just the KdV soliton. This modification is somewhat like the Lyapunov-Schmidt procedure. Next we check that the restriction of $D(Q\tilde{F})(\theta^1, a)$ to N is invertible as an operator from N to QZ for ε small and (θ^1, a) near $(\bar{\theta}, 0)$, where D means the Frechet derivative, or linearization, of the mapping. For $\varepsilon = 0$, $(\theta^1, a) = (\bar{\theta}, 0)$, this was essentially done in [5], and the inverse extends to the neighborhood by perturbation since there is no loss of derivatives.

In order to obtain an exact solution of the problem, we construct an inverse for the linearization $D\tilde{F}(\theta^1, a)$ for (θ^1, a) near $(\bar{\theta}, 0)$, estimate $D^2\tilde{F}$, and show that Newton's method converges in this case. Our remaining remarks will concern the inversion of $D\tilde{F}$, since this is the most crucial part. Keeping only the most important terms, we have to obtain estimates for problems like

$$(7) \qquad \begin{aligned} \varepsilon^2 w_{xx} + w_{yy} &= g, & 0 < y < 1, \\ \gamma w_y - w + \beta\varepsilon^2 w_{xx} &= h, & y = 1, \\ w &= 0, & y = 0. \end{aligned}$$

We can think of w as a correction to the last guess in the sequence approximating θ, the solution of the full problem, with the inhomogeneous terms arising from the error at the last stage. It is possible to introduce a set of eigenfunctions in the y-direction to separate variables (cf. [2]). If $w_n(x)$ is the coefficient of the nth mode, we find for all modes but one,

$$(8) \qquad \varepsilon^2 w_{n,xx} - \nu_n^2 w_n = g_n,$$

so that if g_n decays at infinity, w_n does also. The case $n = 0$ is special because $\nu_0 = O(\varepsilon)$, but this component turns out to correspond to the partial inverse on QZ discussed earlier, and consequently it does not create difficulties. However the remaining component satisfies

$$(9) \qquad \varepsilon^2 w_{-,xx} + k^2 w_- = g_-,$$

where k is again the wave number corresponding to the speed U_0 as determined by the linear dispersion relation (4). It is through this part that the oscillation at infinity enters. The important thing is to obtain estimates for the solution compatible with the way the unknowns have been written, so that the set of approximations for the full problem can be shown to converge. In keeping with the decomposition of the unknown described earlier, we write w_- as a decaying function plus an oscillation at infinity with specified phase, $w_-(x) = v(x) + \alpha v_0(x)$, with α a constant to be chosen, and $v_0(x) = \zeta(x)cos(kx/\varepsilon)$. The parity of the problem is such that g_-, w_-, v, v_0 should all be odd. We have chosen the phase shift $\delta = \pi/2k$ for simplicity; ζ is some smooth, odd function so that $\zeta = 1$ near ∞, $\zeta = 0$ near zero. We expect v to be exponentially decaying, provided the same is true for g_-, with α chosen to allow such decay in v.

In the Fourier transform (9) becomes

$$(k^2 - \varepsilon\xi^2)\hat{v} = \hat{g}_- - \alpha(k^2 - \varepsilon\xi^2)\hat{v}_0 \tag{10}$$

It is natural to try to choose α so that the right-hand side vanishes at $\xi = \pm k/\varepsilon$, the values at which the multiplier on the left is zero. It can be checked that

$$(k^2 - \varepsilon\xi^2)\hat{v}_0(\xi) \neq 0, \qquad \xi = \pm k/\varepsilon, \tag{11}$$

so that such a choice is possible. Note that if v_0 were a pure cosine, its transform would be a sum of δ-functions, and (11) would not hold. Because of the ζ, however, the transform contains additional terms with a singularity at $\xi = \pm k/\varepsilon$, which lead to (11). Having determined α, we can solve (10) for $\hat{v}$. Assuming $\hat{g}_-$ decays exponentially, the right-hand side of (10) is analytic in a strip. The choice of α implies that $\hat{v}$ is also analytic, the singularities being removable, and thus we obtain v in the appropriate class. We find that

$$\varepsilon^s|\alpha| + |v|_{X^s} \leq C\varepsilon^{-1}|g_-|_{X^s}, \tag{12}$$

where X^s is the Sobolev space of functions with s derivatives in $L^2(R)$ with exponential decay. When we finally obtain $D\tilde{F}^{-1}(\theta^1, a)$, the estimate for its norm is of order ε^{-1}, since the worst contribution comes from (12). Nonetheless, the error in satisfying the equation stays within $O(\varepsilon^2)$ for our iterates, and the quadratic convergence of Newton's method overcomes the poor estimate for the inverse.

It is to be expected that the approximate solutions are analytic in a strip in the x-plane, so that their transforms decay exponentially. Then in (10), $\hat{g}_-(\pm k/\varepsilon)$ should be $O(\exp(-C/\varepsilon))$. Therefore the oscillation at infinity, which is determined through α and v, should also be small of this order.

Acknowledgement: This research was supported by the National Science Foundation under Grant No. DMS-8800347.

References

1. J. T. Beale, *The existence of solitary water waves*, Comm. Pure Appl. Math. **30** (1977), 373–389.

2. J. T. Beale, *The existence of cnoidal water waves with surface tension*, J. Diff. Eqns. **31** (1979), 230–263.

3. J. T. Beale, *Water waves generated by a pressure disturbance on a steady stream*, Duke Math. Journal **47** (1980), 297–323.

4. J. T. Beale, *Exact solitary water waves with capillary ripples at infinity*, to appear in Comm. Pure Appl. Math.

5. K. O. Friedrichs and D. H. Hyers, *The existence of solitary waves*, Comm. Pure Appl. Math. **7** (1954), 517–550.

6. J. K. Hunter and J.-M. Vanden–Broeck, *Solitary and periodic gravity–capillary waves of finite amplitude*, J. Fluid Mech. **134** (1983),205–219.

7. G. Iooss and K. Kirchgässner, *Water waves for small surface tension - an approach via normal form*, preprint.

8. J. J. Stoker, *Water Waves*, Interscience, 1957.

9. S. M. Sun, *Existence of a generalized solitary wave solution for water with positive Bond number less than 1/3*, to appear in J. Math. Anal. Appl.

10. S. M. Sun and M. C. Shen, *A new solitary wave solution for water waves with surface tension*, to appear in Ann. Math. Pure Appl.

11. S. M. Sun and M. C. Shen, *Exact theory of solitary waves in a stratified fluid with surface tension, Part I. Nonoscillatory case*, preprint.

12. S. M. Sun and M. C. Shen, *Exact theory of solitary waves in a stratified fluid with surface tension, Part II. Oscillatory case*, preprint.

13. S. M. Sun and M. C. Shen, *A note on solitary wave and nonlinear oscillations in a stratified fluid with surface tension*, preprint.

14. S. M. Sun and M. C. Shen, *Solitary waves in a two-layer fluid with surface tension*, preprint.

15. J.-M. Vanden–Broeck, this conference.

GENERALIZED SOLITARY WAVES IN A STRATIFIED FLUID

M. C. Shen and S. M. Sun

Department of Mathematics
University of Wisconsin
Madison, WI 53706

1. INTRODUCTION

The problems considered in this paper deal with generalized solitary waves in a stratified fluid over a flat bottom in the presence of surface tension. By a generalized solitary wave we mean a solitary wave with a small amplitude oscillation at infinity. First let us briefly discuss some recent progress in the exact theory of solitary waves on water of constant density with surface tension. An approximate equation governing solitary waves with surface tension was originally derived by Korteweg and de Vries[1] and has been named after them. Let τ be the Bond number, a nondimensional surface tension coefficient, and F be the Froude number, the square of a nondimensional wave speed. The critical values of τ and F are respectively $1/3$ and 1. For $\tau > 1/3$, $F < 1$ but near 1, a solitary wave solution to this equation represents a wave of depression, and for $0 < \tau < 1/3$, $F > 1$ but near 1, the solution denotes a wave of elevation. The existence of a solitary wave of depression, which decays to zero at infinity, was proved by Amick and Kirchgässner[2]. For $0 < \tau < 1/3$, the linearized governing equations possess one positive eigenvalue and the corresponding eigenfunction is oscillatory. Numerical studies and other physical models due to Hunter and Vanden-Broeck[3], Zuferia[4] and Boyd[5] indicate that a generalized solitary wave may take place. The main difficulties are to isolate the oscillatory part and to estimate its amplitude. The existence proof of a generalized solitary wave has been given by Beale[6] and Sun[7] independently using different methods.

A systematic study of solitary waves in stratified fluids without surface tension was initiated by Peters and Stoker[8], and the existence of a solitary wave on a continuously stratified fluid with velocity shear was proved by Ter-Krikorov[9]. Benjamin[10] used a different approach to the solution of the latter problem and discussed various physical implications of the solution method. Recently a detailed study of interface waves in a two-layer fluid has been carried out by Amick and Turner[11,12], and numerical investigation of the problem has been presented by Turner and Vanden-Broeck[13]. Our main objectives in this paper are to study the effect of the interplay between density stratification and surface tension on a generalized solitary wave and to prove its existence. The approach used here is an extension of several ideas due to Peters and Stoker[8], Friedrichs and Hyers[14], Ter-Krikorov[9], Beale[6] and some results in references 7 and 15.

The contents of this paper are described as follows. In Section 2 the problem

Asymptotics beyond All Orders, Edited by H. Segur *et al.*
Plenum Press, New York, 1991

is formulated in terms of a streamline function and the fluid domain is transformed to a fixed strip by using the horizontal coordinate and stream function as independent variables. In the solution of the problem, we have to deal with two eigenvalue problems. One eigenvalue problem with $\nu = F^{-1}$ as the eigenvalue parameter yields infinitely many positive eigenvalues ν_n, $n = 0, 1, 2, \ldots$, corresponding to each of which a critical value τ_{cn} of τ is determined. For each ν_n, there is another eigenvalue problem associated with the linearized equations of the exact equations with μ as the eigenvalue parameter. The number of positive eigenvalues of the latter problem depends upon $\tau > \tau_{cn}$ or $\tau < \tau_{cn}$. To solve the equation in the horizontal strip, we need to construct a Green's function for the governing partial differential equation and boundary conditions. A solvability condition is imposed to deal with the contribution due to the zero eigenvalue of μ. However the positive eigenvalues of μ cause oscillatory terms in the Green's functions and must be eliminated by additional auxiliary conditions. The main results are presented in Section 3. At present no general result regarding the precise number of positive eigenvalues of μ is available. However Theorem 1 shows that there is one positive eigenvalue of μ for $0 < \tau < \tau_{c0}$ and no positive eigenvalue of μ for $\tau > \tau_{c0}$. Theorem 2 states the result for the case of only one positive eigenvalue of μ. The oscillation at infinity may be characterized by its amplitude A and phase shift δ. On the one hand, if A is fixed with the order of a positive power of a small parameter $\epsilon > 0$, δ can be determined. On the other hand, if δ is fixed, A can be of the order of any positive power of ϵ. We also note that the solvability condition as in the classical case[14] yields the approximate solitary wave solution, while an auxiliary condition yield an equation of nonlinear oscillations. Theorem 2 presents the results for the case of two positive eigenvalues k_1^2, k_2^2, $k_1 > k_2 > 0$. There are two possible solutions. One consists of a solitary wave solution plus a small amplitude oscillation with a period near $2\pi/k_1$. The other has a small amplitude oscillation with a period near $2\pi/k_2$ if $|k_1 - nk_2| > 0$ for any positive integer n. We remark that in this case we can only fix the order of the amplitude of the oscillation at infinity. Theorem 4 considers the case of three positive eigenvalues of μ. The crucial point in the proof is to generate sufficiently many constants to meet the solvability conditions for the equations of nonlinear oscillations obtained from auxiliary conditions, and some restrictions on the equilibrium state of the fluid flow must be imposed. There are three possible solutions consisting of a solitary wave solution plus an oscillation with a period near $2\pi/k_i$ if $|k_j - nk_i| > 0$ for any positive integer n, $i, j = 1, 2, 3$ and $i \neq j$, where k_i^2, $i = 1, 2, 3$ are the positive eigenvalues and $k_1 > k_2 > k_3$. The results can be extended to the case of more than three positive eigenvalues without much change. In Section 4, we consider two examples to show that generalized solitary waves can also appear in a stratified fluid without surface tension.

2. FORMULATION

We consider a two-dimensional wave of permanent type with constant velocity $c > 0$ in a layer of inviscid, incompressible fluid of variable density over a flat, horizontal bottom. A coordinate system moving with the wave is chosen so that in reference to the coordinate system the fluid flow is steady. The governing equations are the following:

$$(1) \quad \rho^*(u^* u^*_{x^*} + v^* u^*_{y^*}) = -p^*_{x^*} ,$$

$$(2) \quad \rho^*(u^* v^*_{x^*} + v^* v^*_{y^*}) = -\rho^* g - p^*_{y^*} ,$$

$$(3) \quad u^*_{x^*} + v^*_{y^*} = 0 ,$$

$$(4) \quad u^* \rho^*_{x^*} + v^* \rho^*_{y^*} = 0 ;$$

at the free surface $y^* = \zeta^*(x^*)$,

$$(5) \quad u^* \zeta^*_{x^*} - v^* = 0 ,$$

(6) $p^* = -T\zeta^*_{x^*x^*}(1+\zeta^*_{x^*})^{-3/2}$;

at the rigid bottom

(7) $y^* = 0, \quad v^* = 0$.

Here ρ^* is the density, (u^*, v^*) is the velocity, p^* is the pressure, g is the constant gravitational acceleration and T is the constant surface tension coefficient. Following Yih[16], we introduce two new variables $\tilde{u}, \tilde{v}$ such that

$$\tilde{u} = (\rho^*)^{1/2}u^*, \quad \tilde{v} = (\rho^*)^{1/2}v^*.$$

Then from (3), (4),

$$\tilde{u}_{x^*} + \tilde{u}_{y^*} = 0,$$

and a stream function $\psi^*(x^*, y^*)$ may be defined such that

(8) $\tilde{u} = \psi^*_{y^*}, \quad \tilde{v} = -\psi^*_{x^*}$,

and $\rho^* = \rho^*(\psi^*)$ only. From (1), (2) it follows that

(9) $\nabla^2\psi^* + g(d\rho^*/d\psi^*)y^* = dh^*/d\psi^*$,

where

(10) $h^*(\psi^*) = (\tilde{u}^2 + \tilde{v}^2)/2 + p^* + g\rho^* y^*$.

Finally we use the so-called streamline function f^* as the dependent variable, where $\psi^*(x^*, f^*) =$ constant along a streamline, and x^*, ψ^* as independent variables, and obtain from (8),

(11) $f^*_{\psi^*} = (\tilde{u})^{-1} = (\psi^*_{y^*})^{-1}, \quad f^*_{x^*} = \tilde{v}(\tilde{u})^{-1} = -\psi^*_{x^*}/\psi^*_{y^*}$.

We introduce dimensionless variables

$$x = x^*/H^*,\ \rho = \rho^*/\rho_0,\ \psi = \psi^*/((1+m)cH^*),\ f = f^*/H^*,$$

$$u = u^*/c,\ \psi_s = \psi^*_s/(cH^*(1+m)) = 1,\ h = h^*/c_1^2,$$

$$D = h^*(\psi^*_s)/c_1^2,\ \nu = gH^*/c = F^{-1},\ \tau = T/(\rho_0 c^2 H^*),\ c_1 = \rho_0^{1/2}c,$$

where H^* and ρ_0 are characteristic depth and density scales and m is a scaling constant, which measures the deviation of streamlines from the horizontal lines at infinity, and ψ^*_s is the value of ψ^* at the free surface. In terms of them, (9) and (10) become

(12) $(1/2)((1+f_x^2)/f_\psi^2)_\psi - (f_x/f_\psi)_x + \nu(1+m)^{-2}(d\rho/d\psi)f = (1+m)^{-2}dh/d\psi$,

(13) $(1/2)(1+m)^2((1+f_x^2)/f_\psi^2) + \nu\rho f - \tau f_{xx}(1+f_x^2)^{-3/2} = D(\nu)$ at $\psi = 1$,

(14) $f = 0 \quad$ at $\psi = 0$.

Consider an equilibrium state $f = \eta(\psi)$ as a solution of (12) to (14) with the right sides of (12) and (13) replaced by $(1+m)^{-2}dh_0/d\psi$ and $D_0(\nu)$ respectively. Then by (11),

$$\eta(\psi) = \int_0^\psi (\rho(t))^{-1/2} U(t)dt, \quad d\eta/d\psi = q(\psi) = \rho^{1/2}(\psi)U(\psi),$$

where $\rho(\psi)$, $U(\psi)$ are the density and velocity of the equilibrium state, and we assume $\rho(\psi) > 0$, $U(\psi)$ are sufficiently smooth, $q(\psi) > 0$, $\eta(1) = 1$. Let $f = \eta + w(\eta, x)$ as a function of x and η. Then from (12) to (14), the equations for $w(\eta, x)$ are

$$\begin{aligned}(15) \quad L_\nu(w) &= (q^2 w_\eta)_\eta + q^2 w_{xx} - \nu\rho' w \\ &= (-h + h_0)' + (3/2)(q^2 w_\eta^2)_\eta - (2m + m^2)q^2 w_{xx} + F(m, w), \\ &= \Phi_1(x, \eta),\end{aligned}$$

$$\begin{aligned}(16) \quad B_\nu(w) &= w_\eta - \nu\rho(1)q^{-2}(1)w + \tau q^{-2}(1)w_{xx} = D_1 + 2mw_\eta \\ &\quad -2m\nu\rho(1)q^{-2}(1)w + (3/2)w_\eta^2 + G(m, D_1, w) = \Phi_2(x),\end{aligned}$$

$$(17) \quad w = 0 \qquad \text{at } \eta = 0,$$

where $D_1 = q^{-2}(1)(D_0 - D)$, a prime means $d/d\eta$, q and p are functions of η and the expressions of $F(m, w)$, $G(m, D_1, w)$ are somewhat complicated and omitted.

To motivate existence results given later, we first derive the stationary K-dV equation by a formal asymptotic method. Let $\xi = \epsilon^{1/2}x$, $0 < \epsilon << 1$, and assume

$$w = \epsilon(w_0(\xi, \eta) + \epsilon w_1(\xi, \eta) + \cdots),$$

$$\nu = \nu_n(1 - \delta_n \epsilon).$$

Then $w_0 = S(\xi)z_n(\eta)$. Here $z_n(\eta)$ is the n-th eigenfunction of the eigenvalue problem

$$(q^2 z_\eta)_\eta - \nu_n \rho' z = 0,$$

$$z_\eta - \nu_n \rho(1)q^{-2}(1)z = 0 \text{ at } \eta = 1,$$

$$z = 0 \quad \text{at } \eta = 0.$$

It is known that there exist simple eigenvalues $0 < \nu_0 < \nu_1 < \cdots < \nu_n \to \infty$, and $S(\xi)$ satisfies

$$\gamma_n S_{\xi\xi} - (3/2)\alpha_n S^2 + \delta_n \beta_n S = 0,$$

where

$$\gamma_n = z_n^2(1)(\tau - \tau_{cn}), \quad \tau_{cn} = \int_0^1 q^2(\eta) z_n^2(\eta) d\eta z_n^{-2}(1),$$

$$\alpha_n = \int_0^1 q^2(\eta) z_{n\eta}^3(\eta) d\eta,$$

$$\beta_n = \int_0^1 q^2(\eta) z_{n\eta}^2(\eta) d\eta,$$

and a solitary solution is given by

$$S(\xi) = (\delta_n \beta_n / \alpha_n)\text{sech}^2((-\delta_n \beta_n / \gamma_n)^{1/2}(\xi - \xi_0)/2),$$

if $\gamma_n \neq 0$, $\alpha_n \neq 0$ and $\delta_n/\gamma_n < 0$. Here we note $z_n(1) \neq 0$, and call τ_{cn} the n-th critical value of τ.

To solve (15) to (17) for $\nu = \nu_n$, we have to impose a solvability condition since $w = z_0(\eta)$ is a solution of the homogeneous equations:

$$(18) \qquad \int_0^1 z_n(\eta)\Phi_1(x,\eta)d\eta - \Phi_2(x)z_n(1)q^2(1) = 0.$$

Next we construct the Green's function for (15) to (17) by a method due to Titchmarsh[17] so that we can convert them to an integro-differential equation. Let $\Phi_1 = \Phi_2 = 0$ in (15) and (16) and assume $w = X(x)N(\eta)$ to obtain

$$(19) \qquad X_{xx} + \mu X = 0,$$

$$(20) \qquad (q^2 N_\eta)_\eta - (\nu_n \rho' + q^2\mu)N = 0,$$

$$(21) \qquad N_\eta - \nu_n q^{-2}(1)\rho(1)N - \tau\mu q^{-2}(1)N = 0 \text{ at } \eta = 1,$$

$$(22) \qquad N = 0 \quad \text{at } \eta = 0.$$

Let the positive eigenvalues of (20) to (22) be $\mu_i = k_i^2$, $k_1 > k_2 > \cdots > k_r > 0$ and the corresponding eigenfunctions be Z_j, $j = 1, \ldots, r$. We denote the nonpositive eigenvalues by $\mu_{-j} = -k_{-j}^2$, $0 = k_0 < k_{-1} < k_{-2} < \cdots < k_{-n} \to \infty$, and the corresponding eigenfunctions by N_j, $j = 0, 1, 2, \ldots$. Then

$$(23) \qquad w(x,\eta) = \int_{-\infty}^{\infty}\int_0^1 G(x,\eta;\xi,\zeta)\Phi_1(\zeta,\xi)d\zeta d\xi$$
$$- \int_{-\infty}^{\infty} G(x,\eta;\xi,1)\Phi_2(\xi)d\xi.$$

where $G(x,\eta;\xi,\zeta)$ is the Green's function. The method of construction of $G(x,\eta;\xi,\eta)$ may be found in Ref. 7 The term due to $k_0 = 0$ in G can be removed by (18). However, the positive eigenvalues introduce oscillatory terms in G, which also have to be eliminated. Therefore we prescribe r auxiliary conditions:

$$(24) \qquad \int_0^1 Z_i(\eta)\Phi_1(x,\eta)d\eta - q^2(1)Z_i(1)\Phi_2(x) = 0$$
$$i = 1, \ldots, r.$$

Now we need to show that there exists a solution of small norm in some Banach space for (15) to (17), (18) and (24).

3. GENERALIZED SOLITARY WAVES

We remark in passing that if there is no positive eigenvalue we can set $m = D_1 = 0$, $(-h + h_0)' = 0$ in (15) and (16) and prove that $w_0 = S(\xi)z_n(\eta)$ indeed is an approximate solution to w. However a general result regarding how many positive eigenvalues the linearized equations possess for $\nu = \nu_n$ is still lacking. At present we only have the following result.

Theorem 1. For $\nu = \nu_0$, (20) to (22) possess no positive eigenvalue if $\tau > \tau_{c0}$ and only one positive eigenvalue if $0 < \tau < \tau_{c0}$.

First we consider the case of one positive eigenvalue k_1^2. Let $\nu = \nu_0(1-\delta_0\epsilon)$, $\delta_0 > 0$, $D_1 = h_1\epsilon$, $m = h_2\epsilon$, $\tau < \tau_{c0}$ and let

$$(25) \quad w = \epsilon(a(x)Z_1(\eta) + (S(\epsilon^{1/2}x) + \omega(x))Z_0(\eta) + \Theta(x,\eta)) .$$

Substitution of (25) for w in (15) to (17), (18) and (24) yields

$$L_{\nu_0}(\Theta) = \tilde{\Psi}_1(h_2, a, b, \Theta) = \Psi_1(x,\eta),$$

$$(26) \quad B_{\nu_0}(\Theta) = \tilde{\Psi}_2(h_1, h_2, a, b, \Theta) = \Psi_2(x) \quad \text{at } \eta = 1,$$

$$\Theta = 0 \qquad \text{at } \eta = 0,$$

$$(27) \quad \gamma_0\omega_{xx} + \epsilon\delta_0\beta_0\omega - 3\alpha_0\epsilon S(\epsilon^{1/2}x)\omega = \tilde{g}_1(S, h_1, h_2, a, \omega, \Theta) = g_1(x),$$

$$(28) \quad \Gamma_1(\tau)(a_{xx} + \kappa_1^2 a) = \tilde{g}_2(S, h_1, h_2, a, \omega, \Theta) = g_2(x),$$

where

$$\Gamma_1(\tau) = \tau Z_1^2(1) - \int_0^1 q^2(\eta)Z_1^2(\eta)d\eta > 0,$$

$$\kappa_1^2 = k_1^2 - \Gamma^{-1}(\tau)\left(\int_0^1 \nu_0\delta_0\epsilon\rho'(\eta)Z_1^2(\eta)d\eta - \rho(1)\delta_0\nu_0\epsilon Z_1^2(1)\right).$$

(26) to (28) may be converted to integro-differential equations by means of Green's functions. For a bounded periodic solution of (28) two solvability conditions must be imposed. We split a, b and Θ into a decaying part and an oscillatory part and require that they are even functions in x. Then we prove the existence of a solution via contraction map theorem. The oscillation at infinity is characterized by its amplitude A and phase shift δ. In addition, we have two more parameters h_1, h_2 at our disposal. Without loss of generality we choose $h_1 = 0$ and may use either h_2, δ or h_2, A to meet the solvability conditions. Two slightly different cases now appear as seen in

Theorem 2. If A is fixed and of order $\epsilon^{(j/2)+1}$, $j \geq 0$, $\nu = \nu_0(1-\delta_0\epsilon)$, $\delta_0 > 0$, $m = h_2\epsilon$, and $\tau < \tau_{c0}$, the sum of $\epsilon S(\epsilon^{1/2}x)Z_0(\eta)$ and a small oscillation with amplitude of order $\epsilon^{(j/2)+1}$, $j \geq 0$ provides an approximate solution to w in some appropriately chosen Banach space. If δ is fixed with $\sin\delta k_1 \neq 0$ for sufficiently small ϵ, then the amplitude of the small oscillation is of order $\epsilon^{(k+3)/2}$ for any positive integer k.

Remark. Definitions of various Banach spaces used in the proof of Theorem 2 may be found in Ref. 18. The result can be extended to the case that there is only one positive eigenvalue of μ corresponding to $\nu = \nu_n$ for any n if $\Gamma_1(\tau) \neq 0$.

For the case of two positive eigenvalues k_1^2, k_2^2, $k_1 > k_2 > 0$. We express

$$w(x,\eta) = \epsilon\left(\sum_{i=1}^{2} a_i(x)Z_i(\eta) + b(x)Z_0(\eta) + \Theta(x,\eta)\right).$$

From (24) we obtain two equations similar to (28) for $a_i(x)$, $i = 1,2$, and have enough parameters to meet the solvability conditions. We state the results as

Theorem 3. If $\nu = \nu_n(1 - \delta_n \epsilon)$, $\delta_n/\gamma_n < 0$, $D_1 = h_1\epsilon^3$, $m = h_2\epsilon$, and

$$\Gamma_i(\tau) = \tau Z_i^2(1) - \int_0^1 q^2(\eta) Z_i^2(\eta) d\eta \neq 0, \quad i = 1, 2,$$

then w consists of a solitary wave solution plus a small amplitude oscillation with a period $2\pi/\kappa_1$. If $|k_1 - nk_2| > 0$ for any positive integer n. There is another solution with the same properties except that the period of the oscillation is $2\pi/\kappa_2$, where

$$\kappa_i^2 = k_i^2 - \epsilon\nu_n\delta_n \left(\int_0^1 \rho'(\eta) Z_i(\eta) d\eta - \rho(1) Z_i^2(1) \right) \Gamma_i^{-1}(\tau), \quad i = 1, 2. \tag{29}$$

Remark. Here we can only fix the order of the amplitude of the oscillation at infinity.

The method can be extended to the case of more than two eigenvalues if there are enough parameters to meet solvability conditions. Therefore, we expand $(-h + h_0)'$ in (15) in terms of the complete set of eigenfunctions of (20) and (22), the coefficients of which yield enough parameters at our disposal. In fact, only the coefficients of the eigenfunctions corresponding to the positive eigenvalues except the largest one play significant roles in the proof. For the case of three positive eigenvalues we have

Theorem 4. Let

$$w(x, \eta) = \epsilon \left(\sum_{i=1}^{3} a_i(x) Z_i(\eta) + b(x) Z_0(\eta) + \Theta(x, \eta) \right),$$

$$(-h + h_0)' = \sum_{i=2}^{3} \lambda_i Z_i(\eta) q^2(\eta) \epsilon^3,$$

$$D_0 - D = \tau\epsilon^2(\lambda_2 Z_2(1) + \lambda_3 Z_3(1)), \quad m = h_2\epsilon.$$

If $\nu = \nu_n(1 - \delta_n\epsilon)$, $\delta_n/\gamma_n < 0$ for $\epsilon > 0$, then w is the sum of a solitary wave solution and a small amplitude oscillation with period $2\pi/\kappa_1$. If $|k_i - nk_j| > 0$ for any positive integer n and a fixed j where $i = 1, 2,\ \ j = 2, 3$ and $i \neq j$, then there is another solution with an oscillatory part of period $2\pi/\kappa_j$.

Remarks. The results here can be readily extended to the case of more than three eigenvalues. The detailed proofs of Theorems 1-4 are given in Refs 18-19.

4. GENERALIZED SOLITARY WAVES WITHOUT SURFACE TENSION

It is rather easy to construct an example to show that (20) to (22) possess positive eigenvalues even if $\tau = 0$. Let $\rho(\eta) = C^2 e^{-2\eta}$, $U = 1$ where $C = e/(e-1)$. Then μ satisfies

$$(\mu + 1 - 2\nu_k)^{1/2}(1 - \nu_k)^{-1} = -\tanh(\mu + 1 - 2\nu_k)^{1/2}$$

if $\mu + 1 - 2\nu_k > 0$, and

$$(2\nu_k - \mu - 1)^{1/2}(1 - \nu_k)^{-1} = -\tan(2\nu_k - \mu - 1)^{1/2}$$

if $2\nu_k - \mu - 1 > 0$. There are no positive eigenvalues for $\nu_0 = 1.354$ and $\nu_1 = 7.246$. The first positive eigenvalue $\mu_1 = 32.494$ appears when $\nu_2 = 22.179$. There are two positive eigenvalues $\mu_1 = 82.456$, $\mu_2 = 51.524$ when $\nu_3 = 46.885$. We have more positive eigenvalues as ν_n increases. Therefore, the effect of density stratification alone can also cause generalized solitary waves. Theorem 2-4 are still applicable to the case of $\tau = 0$.

Next we consider the two-layer fluid system as discussed by Peters and Stoker[8]. The governing equations can be found in Ref. 8 and are omitted here. Let h and H be the depths, and δ and Δ be the densities, of the lower and upper layers respectively. Let $r = H/h$, and $\rho = \Delta/\delta < 1$. There are only two eigenvalues of ν

$$\nu = \nu_{1,2} = ((1+r) \pm ((1-r)^2 + 4r\rho)^{1/2})/(2r(1-\rho)),$$

and μ satisfies

$$\mu = \nu_i \alpha_\pm(\mu), \quad i = 1, 2$$

$$\alpha_\pm(\mu) = ((\tanh\mu + \tanh\mu r) \pm ((\tanh\mu - \tanh\mu r)^2 + 4\rho \tanh\mu \tanh\mu r$$

$$\times (1 - (1-\rho)\tanh\mu \tanh\mu r))^{1/2})/(2(1 + \rho \tanh\mu \tanh\mu r)).$$

where ν_1, ν_2 be the values of ν corresponding to $+$ and $-$ respectively. Note that $0 < \nu_2 < \nu_1$. A solitary wave moves near the speed $(\nu_2 gh)^{-1/2}$ has the maximum amplitude at the free surface. It can be shown that there is no positive eigenvalue of μ. If $\nu = \nu_1$, the wave amplitude at the interface is much larger than the one at the top free surface. A positive eigenvalue of μ exists and a generalized solitary wave can appear.

ACKNOWLEDGMENTS: The research reported here was partly supported by National Science Foundation under Grant CMS-8903083.

REFERENCES

1. P. J. Korteweg and G. de Vries, On the change of form of long waves advancing in a rectangular canal, and on a new type of long standing waves, Phil. Mag., 39:422 (1895).
2. C. J. Amick and K. Kirchgässner, A theory of solitary water waves in the presence of surface tension, Arch. Rational Mech. Anal., 105:1 (1989).
3. J. K. Hunter and J. M. Vanden-Broeck, Solitary and periodic gravity-capillary waves of finite amplitude, J. Fluid Mech., 134:205 (1983).
4. J. A. Zufiria, Symmetry breaking in periodic and solitary gravity-capillary waves of finite amplitude, J. Fluid Mech., 184:183 (1987).
5. J. P. Boyd, Weakly non-local solitary waves, in: "*Nonlinear Topics in Ocean Physics*", Proceedings of the Fermi school, A. R. Osborne and L. Bergamasco, eds., North-Holland, Amsterdam, (1989).
6. J. T. Beale, Exact solitary water waves with capillary ripples at infinity, Comm. Pure Appl. Math., to appear.
7. S. M. Sun, Existence of a generalized solitary wave solution for water with positive Bond number less than 1/3, J. Math. Anal. Appl., to appear.
8. A. S. Peters and J. J. Stoker, Solitary waves in liquids having non-constant density, Comm.Pure Appl. Math., 13:115 (1960).
9. A. M. Ter-Krikorov, Théorie exact des ondes longues stationnaires dans un liquide hétéogéne, J. de Mécanique, 3:351 (1963).
10. T. B. Benjamin, Internal waves of finite amplitude and permanent form, J. Fluid Mech., 25:241 (1966).
11. C. J. Amick and R. E. L. Turner, A global theory of solitary waves in two-fluid systems, Trans. AMS, 298:431 (1986).
12. C. J. Amick and R. E. L. Turner, Small internal waves in two-fluid systems, Arch. Rational Mech. Anal., 108:111 (1989).
13. R. E. L. Turner and J. M. Vanden-Broeck, Broading of interfacial solitary waves, Phys. Fluids, 31:2486 (1988).

14. K. Friedrichs and D. Hyers, The existence of solitary waves, Comm.Pure Appl. Math., 3:517 (1954).
15. S. M. Sun and M. C. Shen, A new solitary wave solution for water waves with surface tension, Ann. Mat. Pure Appl., to appear.
16. C.-S. Yih, "Stratified Flows", 2nd ed., Academic Press, New York (1980).
17. E. G. Titchmarsh, "Eigenfunction Expansions Associated with Second-order Differential Equations II", 2nd ed., Clarendon Press, Oxford (1962).
18. S. M. Sun and M. C. Shen, Exact Theory of Solitary waves in a stratified fluid with surface tension, Part I. Nonoscillatory Case, Part II. Oscillatory case, submitted.
19. S. M. Sun and M. C. Shen, A note on solitary wave and nonlinear oscillations in a stratified fluid with surface tension, manuscript.

BENDING LOSSES IN OPTICAL FIBERS[1]

Ann Kahlow Hobbs[2], William L. Kath, and Gregory A. Kriegsmann[3]

Engineering Sciences and Applied Mathematics
McCormick School of Engineering and Applied Science
Northwestern University
Evanston IL, 60208

1. Introduction

A typical single- or few-mode optical fiber is a thin glass cylinder with an outer diameter of roughly $10^3\mu$. Most of the fiber makes up what is called the cladding, but at its center is an an inner core with a diameter of roughly 10μ containing glass with optical parameters slightly different from that in the cladding [1]. This is shown schematically in Figure 1. The inner core traps light because its index of refraction is larger than that in the surrounding cladding.

Typically, optical fibers are designed to accomodate light energy with a wavelength of roughly 1μ. Since this is a small fraction of the inner core diameter, one would normally expect a number of propagating modes to be present. This can be reduced to only a few, however, by making the optical fiber *weakly guiding*, that is, by making the index of refraction difference between the core and the cladding small, so that light is only weakly trapped. As a consequence of this weak trapping, the propagation characteristics of the optical fiber can be affected by perturbations which change the core's effective index of refraction. One such example of this is the perturbation induced when an optical fiber is bent; in this case the main effect is that the light energy is no longer perfectly trapped [2].

The failure of an optical fiber to perfectly trap light when bent has an intuitive explanation. As one moves out from the core in a direction away from the center of the bend, the phase fronts of the evanescent electromagnetic field in the cladding have an increasingly longer distance to travel. Eventually a point is reached (later this will be shown to be a caustic) where the velocity required for the wavefront to keep up becomes larger than the allowed velocity of light in the medium; at this point the wave changes from evanescent to propagating and is radiated away. This process is illustrated in Figure 2. Note that a similar phenomonon can occur for electromagnetic and acoustic surface waves [3, 4]. In a real optical fiber, this radiated energy propagates through the cladding and eventually hits the protective plastic jacket surrounding the optical fiber, but since this jacket is a good absorber no energy is reflected. Therefore, the cladding can be regarded in what follows as essentially infinite.

[1]Supported in part by grants from the Air Force Office of Scientific Research (Mathematical Sciences, Grant No. 90-0139) and the National Science Foundation (Applied Mathematics, Grant No. 9002951)

[2]Current address: Department of Mathematics, École Polytechnique Fédérale de Lausanne, CH–1015 Lausanne, Switzerland

[3]Current address: Department of Mathematics, New Jersey Institute of Technology, University Heights, Newark, NJ 07102

Of key interest is the calculation for the rate of energy loss by the bent optical fiber. Solutions for specific geometries have been obtained previously, such as for a slab waveguide [2], an optical fiber bent in a planar, circular path [5], and for an optical fiber bent in a circular helix (i.e., with constant curvature and torsion) [6]. Each of these solutions was obtained by patching together solutions of Maxwell's equations in different regions, and relied in part on being able to find exact solutions in at least some of the regions.

Here an analysis is presented which gives the energy loss rate for arbitrary geometries, i.e., for optical fibers with centerlines bent along paths with arbitrary curvature and torsion. In addition, the analysis clearly shows that the problem of calculating the exponentially small loss rate is one of asymptotics beyond all orders: while the perturbation induced by the bending does introduce corrections to the mode's propagation constant which are algebraic in the small parameter, none of them gives any energy loss. It is only after going beyond all of the algebraic corrections that one finally obtains a value for the rate of energy loss.

2. Governing equations

The starting point is Maxwell's time-harmonic curl equation for the electric field,

$$\vec{\nabla} \times (\vec{\nabla} \times \vec{E}) - \tilde{k}^2 n^2 \vec{E} = 0,$$

which is valid in the above form for non-magnetic materials such as silica-based glasses. Since

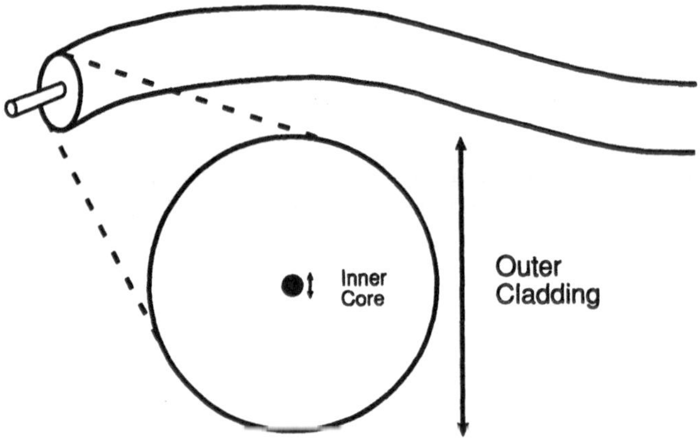

Figure 1. Optical fiber geometry.

the radius of curvature of the optical fiber centerline is very much larger than a typical wavelength, slowly varying coordinates following the fiber are used. It is assumed that bending moments only are applied to the fiber, and that no torsion is added. With this assumption it follows that no twisting results for any amount of bending [7]. A rotated Frenet-Serret coordinate system,

$$\begin{aligned} \vec{x} &= \vec{x}_0(s) + \tilde{\alpha}[\hat{n}\cos\theta - \hat{b}\sin\theta] + \tilde{\beta}[\hat{n}\sin\theta + \hat{b}\cos\theta], \\ &\text{where } \frac{d\theta}{ds} = \tau(s), \end{aligned}$$

is then the coordinate system which properly follows the fiber's orientation [8]; this also gives an orthogonal curvilinear coordinate system. Here $\vec{x}_0(s)$ is the curve describing the centerline of the fiber, while $\hat{n}$ and $\hat{b}$ are the principle unit normal and binormal vectors associated with it; κ (which enters below) and τ are the curvilinear curvature and torsion of the space curve $\vec{x}_0(s)$.

To specifically exploit the small curvature, all lengths will be scaled on the core radius, a. Then $k = \tilde{k}\, a\, n_{clad}$ will be a dimensionless wavenumber. The index of refraction will be assumed of the form $n^2 = n_{clad}^2[1 + f(\tilde{\alpha}, \tilde{\beta})/k^2]$ with f taken to be $O(1)$; this is the appropriate scaling for a weakly-guiding optical fiber [9]. It is then found that there are two dominant balances for the order of the dimensionless curvature: $1/k^4$ and $1/k^3$ [9]. Here only the results of using the former scaling will be discussed.

In addition, the paraxial or parabolic approximation is made [2] by assuming

$$\vec{E} = [A_1 \hat{t} + \vec{\psi}]e^{iks}.$$

Here $A_1 \hat{t}$ represents the longitudinal field, and $\vec{\psi}$ represents the transverse. When all of the above substitutions are made into Maxwell's equations, it is found that the longitudinal field is $O(1/k)$ smaller than the transverse field, and can be determined completely from it.

Therefore, the problem reduces to one for the transverse electromagnetic field,

$$\begin{aligned} &2i\frac{\partial \vec{\psi}}{\partial s} + \nabla_T^2 \vec{\psi} + f(\tilde{\alpha}, \tilde{\beta})\vec{\psi} \\ &\quad + \epsilon\left(2\kappa_1 \alpha\, \vec{\psi} + \vec{\nabla}_T(\vec{\nabla}_T f \cdot \vec{\psi}) + \frac{\partial^2 \vec{\psi}}{\partial s^2}\right) + \ldots \quad = 0, \qquad (1) \\ &\text{where } \alpha = \tilde{\alpha}\cos\theta + \tilde{\beta}\sin\theta \text{ and } \epsilon = 1/k^2. \end{aligned}$$

Here it is seen explicitly that the scaled curvature κ_1 acts as a perturbation to the refractive index profile. This perturbation is small in the core, but becomes large as one moves into

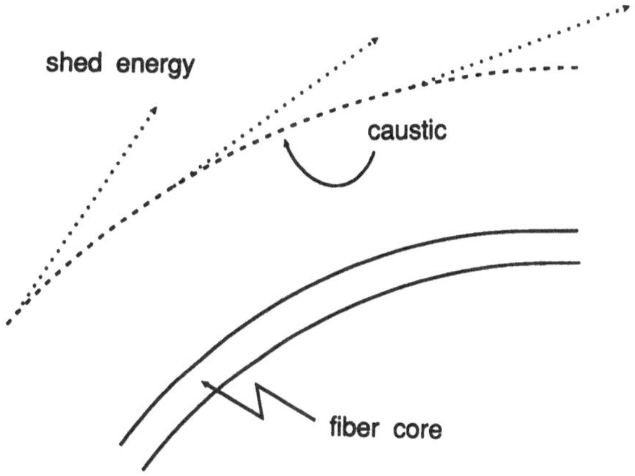

Figure 2: Heuristic explanation for energy loss due to bending; when the required velocity of the evanescent field outside the core becomes greater than the allowed velocity of light, the energy is radiated away.

the cladding, and therefore is a non-uniform perturbation. This is similar to the quantum mechanical Stark effect [10].

From the above coupled equations it will also be seen that the preferred radiation direction for the lost energy toward α positive [9]. Since this direction depends upon the angle θ, which changes very slowly with distance along the fiber, this means that in general it is possible for the loss rate to vary with θ.

The single-polarization (scalar) case will be used here to illustrate the method for calculating the rate of energy loss. Only minor modifications are needed for the coupled vector case (1), and they will be discussed in Section 4.

3. Loss rate calculation

A naive attempt at calculating the energy loss rate begins with the observation that when $\epsilon = 0$ one merely obtains the eigenvalue problem for the straight optical waveguide, and by assuming that these modes persist with only minor changes once the perturbation is applied. It then makes sense to look for solutions of the form $\psi = ye^{-i\lambda s/2}$, so that the decay rate will be $2\,Im\,\lambda < 0$. In the scalar case, one then obtains the eigenvalue problem

$$\nabla_T^2 y + f(\tilde{\alpha}, \tilde{\beta})y + \lambda y + \hat{\epsilon}\,\alpha\, y = 0 \tag{2}$$

with $\hat{\epsilon} = 2\epsilon\kappa_1$. If one then tries a regular perturbation expansion

$$\begin{aligned} y &= y_0 + \hat{\epsilon}\, y_1 + \hat{\epsilon}^2\, y_2 + \ldots, \\ \lambda &= \lambda_0 + \hat{\epsilon}\lambda_1 + \hat{\epsilon}^2\lambda_2 + \ldots, \end{aligned}$$

at leading order ($\hat{\epsilon} = 0$) one finds the straight eigenvalue problem, with $\lambda_0 < 0$. At the next order, one finds

$$\lambda_1 = -\frac{\iint \alpha |y_0|^2\, d\tilde{\alpha} d\tilde{\beta}}{\iint |y_0|^2\, d\tilde{\alpha} d\tilde{\beta}},$$

but unfortunately $Im\,\lambda_1 = 0$, so at this order in the expansion no energy loss is obtained. Continuing in this manner one finds that $Im\,\lambda_n = 0$ for *all* n, so that apparently no radiative loss is introduced by the perturbation, which is clearly not correct.

The failure of the regular expansion is clearly due to the non-uniform nature of the perturbation: it is small in the neighborhood of the core, but at large distances into the cladding (measured in terms of core radii) the perturbation becomes $O(1)$ and a regular expansion is no longer justified. Physically, this is the point at which the field changes from evanescent to propagating. Mathematically, the failure of the regular expansion is due to the singular nature of the perturbation, which is manifested in its effect upon the spectrum of the operator in (2) [10]. When $\hat{\epsilon} = 0$, the spectrum is mixed, with both discrete (i.e., the propagating modes) and continuous components, but for $\hat{\epsilon} > 0$ the discrete spectrum disappears. Its effect is still present, however, in that the spectral density has sharp peaks near the locations of each of the $\hat{\epsilon} = 0$ eigenvalues. The width of each of these spectral resonances gives the decay rate for solutions initially close to the unperturbed eigenmodes, and each width is determined by the distance between the real axis and a complex pole of the operator's resolvent [10]. These poles can also be thought of as eigenvalues of a related problem, where an outgoing radiation condition is used at infinity (which makes the problem non-selfadjoint) rather than the standard finite-norm condition.

A formal asymptotic resolution of the above failure has been given for the one-dimensional Schrödinger equation by Bender and Wu [11]. An outline of the method modified for this problem is as follows; a more careful discussion of the mathematical details of this method has been given by Paris and Wood [12]. One begins the Bender and Wu method by using the outgoing radiation condition to look for the quasi-eigenvalues and quasi-eigenfunctions described above. From (2) one calculates for these solutions that

$$Im\,\lambda = \frac{\frac{i}{2}\oint_C [y^* \frac{\partial y}{\partial n} - y\frac{\partial y^*}{\partial n}]\, ds}{\iint_R |y|^2\, dA},$$

where R is any region and C is its boundary. Note that interpreted physically this says that the rate of energy loss in R is equal to the flux out through its boundary. To use this relation one merely needs a good approximation for the quasi-eigenfunction y, which can be calculated with WKB methods.

To find this WKB solution, one takes C far from the core region, so it may be safely assumed that $f(\tilde{\alpha}, \tilde{\beta}) = 0$. The coordinates are rescaled so that $\hat{\epsilon}\,\alpha = \hat{\alpha}$ and $\hat{\epsilon}\,\beta = \hat{\beta}$, y will be assumed normalized so that $\iint |y|^2\, dA = 1$, and λ will be replaced by $\eta = -\lambda$. Then (2) becomes

$$\hat{\epsilon}^2 \hat{\nabla}_T^2 y + (\hat{\alpha} - \eta) y = 0,$$

and a WKB solution is sought in the form

$$y \sim a(\hat{\alpha}, \hat{\beta}) e^{-\varphi(\hat{\alpha},\hat{\beta})/\hat{\epsilon}},$$

which yields the eikonal and transport equations

$$(\hat{\nabla}_T \varphi)^2 = \eta - \hat{\alpha}$$

$$\hat{\nabla}_T \cdot (a^2 \hat{\nabla}_T \varphi) = 0.$$

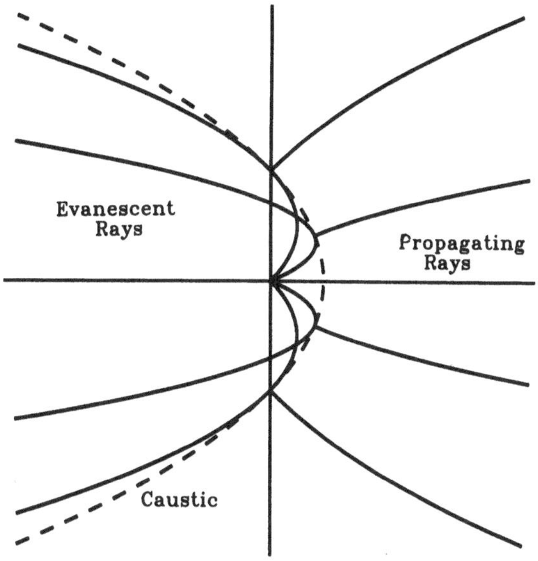

Figure 3: Diagram showing rays of the WKB solution.

Since in these scaled coordinates the size of the inner core is small, the characteristics of the eikonal start at the origin with $\varphi = 0$. The exact solution is then

$$\varphi = \frac{1}{3\sqrt{2}} \left\{ (\hat{\rho} - \hat{\alpha} + 2\eta)^{3/2} - (2\eta - \hat{\rho} - \hat{\alpha})^{3/2} \right\},$$

where $\hat{\rho} = (\hat{\alpha}^2 + \hat{\beta}^2)^{1/2}$ [8, 9]. This is the solution inside the caustic, which occurs on $\hat{\alpha} = \eta - \hat{\beta}^2/4\eta$. On the other side of the caustic, since the solution is known to possess only outgoing waves, one finds

$$\varphi = \frac{1}{3\sqrt{2}} \left\{ (\hat{\rho} - \hat{\alpha} + 2\eta)^{3/2} - i(\hat{\rho} + \hat{\alpha} - 2\eta)^{3/2} \right\}.$$

A schematic showing these rays is given in Figure 3. Since on the caustic $\varphi = \frac{2}{3}(\eta + \hat{\beta}^2/4\eta)^{3/2}$, $Re\,\varphi$ has a minimum on the $\hat{\alpha}$ axis, which means that the radiation comes out in a thin Gaussian sheet pointing in this direction.

Since the exponent is minimal on the $\hat{\alpha}$ axis, this means that the transport equation only needs to be solved on $\hat{\beta} = 0$. The result inside the caustic is

$$a(\hat{\alpha}, 0) = C \left[\frac{\sqrt{\eta} + \sqrt{\eta - \hat{\alpha}}}{\hat{\alpha}(\eta - \hat{\alpha})[\sqrt{\eta} - \sqrt{\eta - \hat{\alpha}}]} \right]^{1/4},$$

while the result outside the caustic is

$$a(\hat{\alpha}, 0) = C e^{\pi i/4} \left[\frac{\sqrt{\eta} - i\sqrt{\hat{\alpha} - \eta}}{\hat{\alpha}(\hat{\alpha} - \eta)[\sqrt{\eta} + i\sqrt{\hat{\alpha} - \eta}]} \right]^{1/4}.$$

These WKB solutions are substituted into the flux integral to calculate the energy decay rate $Im\,\lambda$. A circular contour beyond the caustic is used. One finds

$$Im\,\lambda \sim \frac{\hat{\rho}}{\hat{\epsilon}} \int_{-\pi}^{\pi} |a|^2 Im \frac{\partial \varphi}{\partial \hat{\rho}} e^{-2 Re\,\varphi/\hat{\epsilon}} \, d\phi,$$

and since $Re\,\varphi$ is minimum at $\phi = 0$, Laplace's method can be used to expand the integral asymptotically, giving

$$Im\,\lambda \sim -|C|^2 \left(\frac{2\pi}{\hat{\epsilon}} \right)^{1/2} e^{-4\eta^{3/2}/3\hat{\epsilon}} + \ldots.$$

The constant $C = C(\theta)$ is found by matching the WKB solution to the unit-normalized 'inner' eigenfunction (that found with the regular perturbation expansion). In addition, since $\eta = -\lambda_0 - \hat{\epsilon}\,\lambda_1 + \ldots = \eta_0 + \hat{\epsilon}\,\eta_1 + \ldots$, the exponential factor in the loss rate becomes

$$-2\eta_0^{1/2}\eta_1 - \frac{4\eta_0^{3/2}}{3\hat{\epsilon}};$$

this shows that the energy loss rate depends upon the eigenvalue perturbation. This is a key point, because although in the vector or two-polarization case (1) the outer WKB solution requires only minor changes, the details of the eigenvalue perturbation are quite different from the scalar case.

4. Modifications for two polarizations

As mentioned above, in the vector case the main changes from the scalar case are to the inner core perturbed eigenvalue problem; there are only slight modifications to the outer WKB solution (the equations decouple for large distances since $f = 0$ there). First of all, there are different cases to consider, based upon the symmetries present in the refractive index profile. If $f = f(\rho, \phi)$ then there is a two-fold degeneracy at leading order (since there are two polarizations), while $f = f(\rho)$ then there is a four-fold degeneracy (since each polarization has both $\cos\phi$ and $\sin\phi$ solutions) [9]. The perturbation splits the eigenvalue degeneracy at first order, and since the exponential factor in the loss rate depends upon the eigenvalue correction, this means that each split eigenfunction will have a different loss rate.

Rather than covering all possible cases here, it is more useful to give a few special cases. First of all, in the special case where $f = f(\rho)$, one finds that

$$Im\,\lambda \sim$$
$$-D_\nu^2 \frac{k\pi^{3/2}\kappa'^{1/2}}{2\eta_0^{3/4}} \exp\left(\frac{11\eta_0^{5/2}}{12\kappa' k^4} - \frac{\eta_0^{1/2}}{\kappa' k^4}\eta_1 \right) \exp\left(\frac{-2\eta_0^{3/2}}{3\kappa' k^2} \right),$$

where $\kappa' = \kappa\, a$, $k = \tilde{k}\, a\, n_{clad}$, and D_ν^2 is a normalization constant.

Table 1. Approximate ($\hat{\epsilon} = 0$) and corrected eigenmodes for a step index optical fiber.

ν	Approx.	Exact	η_1
0	$LP_{0\mu}$	$HE_{1\mu}$	$-\xi\,\eta_0\,J_0(\xi)/J_1(\xi)$
1	$LP_{1\mu}$	$TM_{0\mu}$	$-2\,\xi\,\eta_0\,J_1(\xi)/J_2(\xi)$
		$TE_{0\mu}$	0
		$HE_{2\mu}$	$\xi\,\eta_0\,J_1(\xi)/J_0(\xi)$
≥2	$LP_{\nu\mu}$	$EH_{\nu-1,\mu}$	$-\xi\,\eta_0\,J_\nu(\xi)/J_{\nu+1}(\xi)$
		$HE_{\nu+1,\mu}$	$\xi\,\eta_0\,J_\nu(\xi)/J_{\nu-1}(\xi)$

In the further special case of a step index profile, $f = f_0 H(1-\rho)$, one finds that the radial part of the solution is [9]

$$R_\nu(\rho) = \begin{cases} J_\nu(\xi\rho)/J_\nu(\xi) & (\rho < 1) \\ K_\nu(\zeta\rho)/K_\nu(\zeta) & (\rho > 1) \end{cases}$$

where $\nu = 0, 1, 2, \ldots$ is the angular frequency of the mode, $\xi = \sqrt{f_0 - \eta_0}$, $\zeta = \sqrt{\eta_0}$, and the eigenvalue condition is

$$-\xi\,\frac{J_{\nu-1}(\xi)}{J_\nu(\xi)} = \zeta\,\frac{K_{\nu-1}(\zeta)}{K_\nu(\zeta)}.$$

In addition, the normalization condition for this solution gives

$$D_\nu^2 = \left(\pi\,\frac{f_0}{\xi^2} K_{\nu+1}(\zeta) K_{\nu-1}(\zeta)\right)^{-1}.$$

Finally, the eigenvalue corrections for the various modes are given in Table 1. Note that the degenerate eigenmodes (from the $\hat{\epsilon} = 0$ problem) are traditionally called LP (linearly polarized) modes [2], and are linear combinations of the limiting cases of the exact HE, EH, TM and TE modes. The perturbation splits the degenerate LP modes into approximations of the more exact modes, and since the eigenvalue correction is different for each, they will have different loss rates.

5. Summary

Here a method has been presented for calculating the rate of energy loss caused by bending in an optical fiber. This loss can be interpreted as being due to a perturbation in the fiber's effective index of refraction, which adds an exponentially small imaginary part to the eigenvalues of a straight (unbent) fiber (the perturbation also adds algebraically small real perturbations, but these do not cause any loss). While this rate of energy loss is mathematically small, it can be important practically because of the potentially long propagation distances (measured in terms of core radii).

It has been shown that in cases where a leading-order degenerate eigenmode is split by the effective index perturbation, it is possible for the split modes to have different loss rates, with the ratio of the loss rates being some $O(1)$ quantity (not necessarily equal to one). In the case of of a step-index fiber, this means that a degenerate linearly polarized (LP) mode (which is the type of mode most easily produced with a laser) will not remain intact as it propagates down the fiber; the various exact HE, EH, TM or TE modes making up the LP mode will decay at different rates.

Finally, since the method presented here is a systematic one, it provides a procedure for including additional effects which may be present in optical fiber systems. This method is currently being used by other researchers in this context.

References

[1] A. W. Snyder and J. D. Love, *Optical Waveguide Theory*, Chapman and Hall, London, 1983.

[2] D. Marcuse, *Light Transmission Optics*, 2nd ed., Van Nostrand Reinhold, New York, 1982.

[3] M. V. Berry, Attenuation and focusing of electromagnetic surface waves around gentle bends, J. Phys. A: Math. Gen., **8** (1975), pp. 1952-1971.

[4] D. S. Jones, Acoustic tunnelling, Proc. Roy. Soc. Edinburgh, **81A** (1978), pp. 1-21.

[5] D. C. Chang and E. F. Kuester, Radiation and propagation of a surface-wave mode on a curved open waveguide of arbitrary cross section, Radio Sci., **11** (1976) pp. 449-457.

[6] D. Marcuse, Radiation loss of a helically deformed optical fiber, J. Opt. Soc. Am. **66** (1976), pp. 1025-1031.

[7] L. D. Landau and E. M. Lifshitz, *Theory of Elasticity*, Pergamon Press, Oxford, 1959.

[8] W. L. Kath and G. A. Kriegsmann, Optical tunnelling: radiation losses in bent fibre-optic waveguides, IMA J. Appl. Math., **41** (1988), pp. 85-103.

[9] A. K. Hobbs and W. L. Kath, Losses for full vector mode solutions of arbitrarily bent optical fibres, IMA J. Appl. Math., **44** (1990), pp. 197-219.

[10] B. Simon, Large orders and summability of eigenvalue perturbation theory: a mathematical overview, Int. J. Quant. Chem., **21** (1982), pp. 3-25.

[11] C. M. Bender and T. T. Wu, Anharmonic oscillator. II. A study of perturbation theory in larger order, Phys. Rev. D **7** (1973) pp. 1620-1636.

[12] R. B. Paris and A. D. Wood, A model equation for optical tunnelling, IMA J. Appl. Math. **43** (1989) pp. 273-284.

EXPONENTIAL ASYMPTOTICS AND SPECTRAL THEORY FOR CURVED OPTICAL WAVEGUIDES

Alastair D. Wood

School of Mathematical Sciences

Dublin City University, Dublin 9, Ireland

1 Introduction

The purpose of this paper is to present recent work on a class of eigenvalue problems for ordinary differential equations arising in optical tunnelling from a weakly-guiding fibre, and to discuss some generalisations which are of mathematical rather than physical interest. It is based on publications with colleagues at Dublin City University and with Dr. R. B. Paris of Dundee Institute of Technology. These seven papers fall into three distinct groups. The first group ([16], [5]) consists of one-dimensional models which seek to provide a deeper understanding of the mechanisms described by Kath and Kriegsmann [9] in determining radiation losses in bent fibre-optic waveguides. The second group considers related eigenvalue problems for singularly-perturbed ordinary differential equations ([4], [11]) and their connection to the theory of resonances in quantum mechanics [10]. The physical problem in the above papers is essentially one of radiation damping, a difficult area of transcendental asymptotics which has yet to be given a satisfactory general treatment, even for ordinary differential equations. For another non-quantum mechanics application, arising in the theory of surface waves trapped by round islands of small seabed slope, the reader is referred to Lozano and Meyer [12]. The third group ([17], [18]) is concerned with obtaining exponentially-improved asymptotic expansions for special functions which appear in the eigenvalue relations above.

2 Models for optical tunnelling

These models have their origins in the work of Kath and Kriegsmann [9] who discuss radiation losses in a weakly guiding optical fibre which is slightly bent. The equation for $y(\xi,\eta)$

$$\nabla^2 y + f(\xi,\eta)y - \lambda y + \varepsilon\alpha(\xi,\eta)y = 0 \tag{2.1}$$

is central to their analysis. Here (ξ,η) are coordinates orthogonal to the fibre in a torsion-free comoving coordinate system, $f(\xi,\eta)$ is the refractive index in the core relative the cladding, ε is a small positive quantity which measures curvature and α is a function linear in ξ and η. Together with appropriate boundary conditions, (2.1) gives an eigenvalue problem in the parameter λ whose imaginary part yields the energy loss from the fibre.

Asymptotics beyond All Orders, Edited by H. Segur *et al.*
Plenum Press, New York, 1991

Kath and Kriegsmann compute Im λ using WKB techniques and asymptotic matching. Because it is difficult to offer a completely rigorous analysis of the relatively complicated problem (2.1), a simpler model problem which can be solved explicitly and understood in detail is studied by Paris and Wood in [16]. The problem exhibits some of the same features as the boundary value problem based on (2.1), including an eigenvalue with exponentially small imaginary part, and consists of the ordinary differential equation

$$y''(x) + (\lambda + \varepsilon x)y(x) = 0 \quad \text{on} \quad (0, \infty) \tag{2.2}$$

with

$$y'(0) + hy(0) = 0,$$

where h is a positive constant, together with the condition that $y(x)$ has controlling behaviour $\exp[ip(x)]$ as $x \to \infty$, for some positive real-valued function $p(x)$. The unique function $p(x)$ is found via the Liouville-Green substitution $y(x) = e^{ip(x)}$ to be $p(x) = \frac{2}{3}\varepsilon^{1/2}x^{3/2}$. This corresponds to an outgoing wave condition in the original problem.

It is straightforward to show that the model problem has an eigenvalue at $\lambda = -h^2$ when $\varepsilon = 0$, but the perturbed problem is non-self-adjoint because of the form of the boundary condition at infinity. It is thus possible for the eigenvalue of the perturbed problem to be non-real, and we shall show that it has an imaginary part which is $O(e^{-1/\varepsilon})$ as $\varepsilon \to 0+$. We first observe that regular perturbation methods yield the asymptotic series

$$\lambda = -h^2 - \frac{\varepsilon}{2h} - \frac{\varepsilon^2}{8h^4} - \frac{5\varepsilon^3}{32h^7} - \frac{11\varepsilon^4}{32h^{10}} + O(\varepsilon^5). \tag{2.3}$$

This may be continued to as high an order as desired, but will never yield any information on Im λ. This is hardly surprising, since Im λ will turn out to be $o(\varepsilon^n)$ for any $n \in \mathbb{N}$.

The model problem (2.2) was chosen because an exact solution is possible in terms of Airy functions whose asymptotic properties are well-known. A fundamental set of solutions of (2.2) is provided by the functions $Ai(-z)$ and $Bi(-z)$, where $z = \varepsilon^{1/3}(x + \frac{\lambda}{\varepsilon})$ is a new scaled variable. The relation between the Airy and Hankel functions enables us to assert that $iAi(-z) + Bi(-z)$ is of the correct "outgoing wave" form as $x \to +\infty$. Substitution of this solution in the boundary condition at the origin, written in terms of the new z variable, gives the eigenvalue relation

$$-\varepsilon^{1/3}[Ai'(-\varepsilon^{-2/3}\lambda) - iBi'(-\varepsilon^{-2/3}\lambda)] + h[Ai(-\varepsilon^{-2/3}\lambda) - iBi(-\varepsilon^{-2/3}\lambda)] = 0. \tag{2.4}$$

To make progress, we replace the Airy functions and their derivatives by their asymptotic expansions for large H, where $H = -\varepsilon^{-2/3}\lambda$. We see that, as $\varepsilon \to 0+, |H| \to +\infty$ in a sector containing the positive real axis. If Im $\lambda < 0$ (as would be required for our boundary condition at infinity to correspond to an outgoing wave: for details see [16]), then Im $H > 0$, but it is also clear from (2.3) that arg $H \to 0$ as $\varepsilon \to 0^+$. Because of the irregular singular point at infinity in (2.2), the asymptotic behaviour as $|H| \to \infty$ will be sectorial in nature: in particular, the expansion of $Bi(H)$ will differ above and below the positive real axis.

The general solution of Airy's equation has expansion

$$c_1E_1(z) + c_2E_2(z) \qquad \text{as} \quad z \to \infty,$$

where the values c_1, c_2 depend on arg z and, writing $\zeta = \frac{2}{3}z^{3/2}$,

$$E_1(z) \sim z^{-1/4}e^{\zeta}\sum_{m=0}^{\infty}(-1)^m\frac{(1/3, m)}{(2\zeta)^m},$$

$$E_2(z) \sim z^{-1/4}e^{-\zeta}\sum_{m=0}^{\infty}\frac{(1/3,m)}{(2\zeta)^m}.$$

This sectorial dependence is characterised as follows. In $|\arg z| < \frac{\pi}{3}, E_2(z) << E_1(z)$ as $|z| \to \infty$, the maximum subdominancy occurring when $\arg z = 0$, which is known as the Stokes line. On the rays bounding this sector, the anti-Stokes lines, $E_1(z)$ and $E_2(z)$ are of the same order of magnitude. It follows that the values of the Stokes multipliers c_1, c_2 may change from sector to sector, $c_{1,2}$ jumping in value on the Stokes line corresponding the maximum subdominancy of $E_{1,2}(z)$ compared to $E_{2,1}(z)$.

Because the Airy function $Ai(z)$ is defined to be that solution which is exponentially small as $z \to +\infty$, the ray $\arg z = 0$ is not a Stokes line and we have $c_1 = 0$ so that

$$Ai(z) \sim \frac{1}{2\sqrt{\pi}}z^{-1/4}e^{-\zeta} \quad \text{as} \quad |z| \to \infty \text{ in } |\arg z| < \pi.$$

But the ray $\arg z = 0$ is a Stokes line for $Bi(z)$ and it is shown in [16] that the Stokes multipliers are

$$c_1 = \frac{1}{\sqrt{\pi}}, \quad c_2 = -\frac{i}{2\sqrt{\pi}} \quad \text{in} \quad -\frac{2\pi}{3} < \arg z < 0$$

$$c_1 = \frac{1}{\sqrt{\pi}}, \quad c_2 = \frac{i}{2\sqrt{\pi}} \quad \text{in} \quad 0 < \arg z < \frac{2\pi}{3},$$

being undefined on $\arg z = 0$. Thus

$$Bi(z) \sim \frac{1}{\sqrt{\pi}}z^{-1/4}e^{\zeta} \pm \frac{1}{2\sqrt{\pi}}z^{-1/4}e^{-\zeta}.$$

the upper sign being taken if $\arg z > 0$ and vice versa.

Applying these results to (2.4) and solving for λ gives the results, either

$$\text{Im } \lambda = 0 \tag{2.5}$$

or

$$\text{Im } \lambda = -\frac{4h^2}{e}\exp\left[-\frac{4}{3\varepsilon}h^3\right] \tag{2.6}$$

according as H is above or below the positive real axis. But we know that $\arg H \to 0$ as $\varepsilon \to 0+$. It was shown by Paris and Wood [16], by other methods, that the correct result is

$$\text{Im } \lambda = -\frac{2h^2}{e}\exp\left[-\frac{4}{3\varepsilon}h^3\right]. \tag{2.7}$$

The justification for this apparent averaging of (2.5) and (2.6) became clear with the publication of the paper [2] by M.V. Berry on smoothing of Stokes discontinuities. This is based on formal analysis of R.B. Dingle [6] using Borel summation of late terms of the asymptotic series, made rigorous by the results of Olver [14]. The net result is that the change in the multiplier across the Stokes line is not discontinuous, but occurs smoothly in a scaled variable representing distance from the line, the functional dependence being given by the familiar error function. It then follows that the value of the Stokes multiplier on the Stokes line is the average of these in the adjoining sectors. This leads to the correct result (2.7).

Regarded as a model for optical tunnelling, problem (2.2) shares some features with (2.1) but clearly lacks others. We may consider (2.2) as the limit as $\delta \to 0$ of the problem

$$-y''(x) + V(x)y(x) = \lambda y(x) \quad \text{on} \quad \mathbf{R} \tag{2.8}$$

where

$$V(x) = \begin{cases} -\varepsilon|x| & |x| > \frac{1}{2}\delta \\ -\frac{2h}{\delta} - \varepsilon|x| & |x| \le \frac{1}{2}\delta \end{cases}$$

and $y \in C^1(\mathbf{R})$. Reflection symmetry implies that the lowest eigenfunction is even (and its derivative odd), so that we may restrict attention to $x \in (0, \infty)$. In the limit as $\delta \to 0$ we obtain a delta function potential at $x = 0+$ and the jump condition $y'(0) + hy(0) = 0$ as before. This model has three major deficiencies. Firstly, the optical tunnelling problem is not symmetric (radiation goes out to one side only); secondly, the approximation for a weakly guiding fibre is obviously very poor for a delta function potential; and thirdly, the optical tunnelling problem (2.1) is two-dimensional. We consider below a more realistic one-dimensional model to which the first two criticisms do not apply. In response to the third point, the study of a one-dimensional model is justified by the physical fact that radiation is mainly in the plane of the bend.

In the more realistic model the potential is (2.8) is replaced by a slightly tilted square well potential

$$V(x) = \begin{cases} -\varepsilon x & |x| > \frac{1}{2}\delta, \\ -\frac{2h}{\delta} - \varepsilon x & |x| \le \frac{1}{2}\delta. \end{cases} \tag{2.9}$$

The boundary conditions at infinity are given by the appropriate WKB approximations, namely

$$y(x) \sim \frac{1}{\sqrt{\pi}}\varepsilon^{-1/12}x^{-1/4}\exp[(i(\frac{2}{3}\sqrt{\varepsilon}x^{3/2} + \frac{1}{3}\pi)] \quad \text{as } x \to +\infty \tag{2.10}$$

and

$$y(x) \sim \frac{1}{\sqrt{\pi}}\varepsilon^{-1/12}(-x)^{-1/4}\exp[-\frac{2}{3}\sqrt{\varepsilon}(-x)^{3/2}] \quad \text{as } x \to -\infty. \tag{2.11}$$

Our model problem now becomes one of finding values of λ for which (2.8) with potential (2.9) has $C^1(\mathbf{R})$ solutions which satisfy (2.10) and (2.11). Once again, the unperturbed problem ($\varepsilon = 0$) has a single negative real eigenvalue $\lambda_0 = -h^2 + \frac{2}{3}h^3\delta + O(\delta^2)$ for small δ. To treat the perturbed problem requires heavy manipulation of the asymptotic expansions of Airy functions, with correct use of the smoothing results of Berry [2]. The final result for the controlling behaviour of Im λ is

$$\exp\left[-\frac{4}{3\varepsilon}(-\lambda_0)^{3/2} + \sqrt{-\lambda_0}\delta\right].$$

Full details are given in [5]. In all of these models, the intimate connection between Stokes phenomenon and the appearance of eigenvalues with exponentially small imaginary part is made clear.

3 Generalisations of the optical tunnelling equation

It is a reasonable mathematical question to ask if the phenomenon of an eigenvalue with exponentially small imaginary part occurs for problems with a more general potential than (2.2). In this section we consider extensions to potentials of form εx^n for integers $n \ge 2$ and relate our results to the theory of resonances in quantum mechanics. The cases $n = 2$ and $n > 2$ receive totally different treatment. When $n = 2$ it is still possible to appeal to the asymptotics of special functions, although there is work to be done (see [18]) in making an exponential improvement to these. For $n > 2$ we have recourse to the method of matched asymptotic expansions.

We consider first the problem with $n = 2$, which may be stated as

$$y''(x) + (\lambda + \varepsilon x^2)y(x) = 0, \quad x \in (0, \infty) \tag{3.1}$$

with

$$y'(0) + hy(0) = 0, \quad h > 0 \text{ given}, \tag{3.2}$$

and the "outgoing wave" condition

$$\text{as } x \to +\infty, y(x) \text{ has controlling behaviour } e^{ip(x)}, \text{ for some } p(x) > 0. \tag{3.3}$$

As in (2.2), the function $p(x)$ is uniquely determined by the Liouville-Green approximation $p(x) = \frac{1}{2}\sqrt{\varepsilon}x^2$. In the same way as in §2, there is an unperturbed eigenvalue at $\lambda = -h^2$ and the perturbed problem fails to be self-adjoint because of the form of the boundary condition at infinity. In this case we make the transformation $z = e^{i\pi/4}\sqrt{2}\varepsilon^{1/4}x$ and $a = \frac{1}{2}\frac{i\lambda}{\sqrt{\varepsilon}}$ to obtain the parabolic cylinder equation

$$\frac{d^2y}{dz^2} = (a + \frac{1}{4}z^2)y \tag{3.4}$$

with boundary condition

$$e^{\frac{i\pi}{4}}\sqrt{2}\varepsilon^{1/4}\frac{dy(0)}{dz} + hy(0) = 0 \tag{3.5}$$

and the same "outgoing wave" condition at infinity. Noting that $z \to \infty e^{i\pi/4}$ as $x \to +\infty$, the required solution satisfying the condition (3.3) is

$$y(z) = \frac{1}{\pi}\Gamma(\frac{1}{2} + a)[U(a, -z) - ie^{i\pi a}U(a, z)] \tag{3.6}$$

where $U(a, z)$ is the Weber parabolic cylinder function, related to Whittaker's parabolic cylinder function by

$$U(a, z) = D_{-a-\frac{1}{2}}(z),$$

as given in Abramowitz and Stegun [1, p.687]. Substitution of (3.6) in the boundary condition (3.5) at the origin yields the eigenvalue relation

$$\frac{U(a,0)}{U'(a,0)} = \frac{e^{\frac{i\pi}{4}}\sqrt{2}\varepsilon^{1/4}(1 + ie^{ia\pi})}{h(1 - ie^{ia\pi})}, \tag{3.7}$$

which becomes

$$\frac{\Gamma(\frac{1}{4} + \frac{1}{2}a)}{\Gamma(\frac{3}{4} + \frac{1}{2}a)} = -\frac{2e^{i\pi/4}\varepsilon^{1/4}(1 + ie^{ia\pi})}{h(1 - ie^{ia\pi})} \tag{3.8}$$

on noting the values of the parabolic cylinder function and its derivatives at the origin. Recall that $a = \frac{1}{2}i\lambda/\sqrt{\varepsilon}$. Information on Re λ may be obtained by regular perturbation, similar to (2.3), as

$$\lambda = -h^2 - \frac{\varepsilon}{2h^2} - \frac{7\varepsilon^2}{8h^6} - \frac{121\varepsilon^3}{16h^{10}} + O(\varepsilon^4). \tag{3.9}$$

We conclude that Im $\lambda << \varepsilon^n$ for all $n \in \mathrm{N}$ and hence Im λ must be transcendentally small as $\varepsilon \to 0+$. Thus $a \to -i\infty$ as $\varepsilon \to 0+$. To make progress with the left-hand side of (3.8) we require an exponentially-improved version of the asymptotic expansion of a quotient of gamma functions: this gives rise to some interesting technical difficulties which we discuss in §4. Using the correct expansion and replacing λ by the first two terms of (3.9) gives the final result

$$\text{Im}\,\lambda \sim -2h^2 \exp\left[\frac{-\pi h^2}{2\sqrt{\varepsilon}}\right] \quad \text{as } \varepsilon \to 0+. \tag{3.10}$$

The details may be seen in Brazel, Lawless and Wood [4].

Comparison of the results (2.7) and (3.10) leads us to believe that in the case of general $n \geq 2$ there will be an n-dependence through a term $\varepsilon^{1/n}$ when equation (3.1) is replaced by

$$y''(x) + (\lambda + \varepsilon x^n)y(x) = 0, \quad x \in (0, \infty) \tag{3.11}$$

with same boundary condition (3.2) and the appropriate form of the "outgoing wave" condition (3.3). For $n > 2$ no special function solutions are available and we resort to the method of matched asymptotics. We outline the method detailed in [11]. We transform the variable x into a variable w, more convenient for matching, by $w = \delta(x - x_0)$ where $\delta = \varepsilon^{1/n}$ and $x_0 = (-\lambda/\varepsilon)^{1/n}$, with the branch of the root chosen and fixed to be the turning point closest to the positive real axis. This point will be exponentially close to the axis and tend to it as $\varepsilon \to 0+$. The boundary value problem is transformed into

$$\delta^2 \frac{d^2y}{dw^2} = -[(w + \delta x_0)^n + \lambda]y,$$

$$\delta \frac{dy}{dw}(-\delta x_0) + hy(-\delta x_0) = 0 \tag{3.12}$$

with the same "outgoing wave" condition at infinity. Now divide the w-axis into 3 regions:

Region I : $w >> \delta^{2/3}, \quad \delta \to 0+,$

Region II : $|w| << 1, \quad \delta \to 0+,$

Region III : Re $w < 0, -w >> \delta^{2/3}, \quad \delta \to 0+.$

In Region I we select the appropriate "outgoing wave" physical-optics WKB solution which we match to the Airy function approximate solution valid in Region II where we may approximate the potential by its linear part. We then match this combination of Airy functions to the WKB solution valid in Region III. It is this solution which we finally substitute into the boundary condition (3.12) to obtain the eigenvalue relation. It is interesting that the smoothing results of [2] are still required for the Airy functions in the neighbourhood of the turning point. The final result is

$$\text{Im } \lambda \sim -2h^2 \exp\left[-\frac{2h^{(n+2)/n} S(n)}{\varepsilon^{1/n}}\right] \tag{3.13}$$

where

$$S(n) = \frac{\Gamma(\frac{1}{n}+1)\Gamma(\frac{3}{2})}{\Gamma(\frac{1}{n}+\frac{3}{2})}.$$

The complex number λ, whose imaginary part is given by (3.13), is an eigenvalue in the sense that the differential equation (3.11) has a non-trivial solution which satisfies both boundary conditions when that particular value of λ is taken. It has no direct interpretation in linear operator theory. To provide an abstract setting for this result we must go to quantum mechanics and the theory of resonances or "pseudo-eigenvalues" as found in the opening chapter of Reed and Simon [19]. These resonances correspond to poles of the Titchmarsh-Weyl $m(\lambda)$ function situated just below the negative real axis: for a classical treatment see Titchmarsh [20, Ch XX], although the work "resonance" is not mentioned there. It is shown in a separate paper [10] that, for a more general class of potentials which includes (3.11), the non-real eigenvalues of the non-self-adjoint problems considered above correspond to the resonance poles of the self-adjoint problem where the "outgoing wave" condition is replaced by a L^2 boundary condition at $+\infty$. The proof is based on the work of Weyl and Titchmarsh and uses the fact that the eigenvalues of the self-adjoint problem are given by the poles of the Green's function, which is related in turn to the $m(\lambda)$ function. The perturbed self-adjoint problem does not admit eigenvalues in the sense of an operator in Hilbert space, but has a continuous spectrum on the real axis. The poles of the perturbed $m(\lambda)$ function are interpreted as resonances in the sense that they correspond to regions of high spectral density; for details see Ch. 12 of [19].

In the cases $n = 1, 2$ it is possible to construct the perturbed $m(\lambda)$ function explicitly from a knowledge of the asymptotics of the Airy and parabolic cylinder functions respectively. It is shown in [10] and [4] that the poles of these functions have imaginary parts corresponding to (2.7) and (3.10) respectively which were obtained by the direct approach of substituting into the boundary condition at zero the solution satisfying the boundary condition at infinity.

We observe that the mathematical theory of resonances in quantum mechanics developed by Harrell and Simon (see [7] and references therein) cannot be applied directly to the above problems. This is because the potentials arising in the model optical tunnelling problem and its generalisations do not fall within the class of potentials discussed in these papers.

4 Exponentially-improved asymptotic expansions

Although the asymptotics of special functions given in books such as Abramowitz and Stegun [1] are correct in the sense of Poincaré, they give no information about the subdominant exponential term and are not complete asymptotic expansions in the sense of Watson as described in the book by Olver [13]. In recent years there has arisen a whole new range of applications, including that to optical tunnelling above, where information concerning the subdominant terms, usually transcendentally small, is needed. Finding the correct multiplier for these recessive terms depends crucially on an understanding of the Stokes phenomenon described in §2. Since the discovery of the phenomenon by Stokes in 1857, this topic has always had an air of mystery, dispelled only by the results of Berry [2] in 1989. Using formal methods due to Dingle [6] he showed that, with sufficient resolution, the change in the multiplier of the subdominant term is in fact continuous near a Stokes line. At the 1989 Winnipeg meeting Olver [14] was able to put this theory on a rigorous mathematical footing by constructing uniform "exponentially-improved" asymptotics for a class of functions defined by Laplace integrals. These expansions possessed the greater accuracy than Poincaré-type expansions which was required in modern applications. Independently of Berry and Olver, Jones [8] announced similar results at the same conference. An alternative rigorous method has been given by Boyd [3] using Stieltjes, rather than Laplace, transforms.

More recent work of Paris [17] describes an alternative theory for the smoothing of Stokes discontinuities using Mellin-Barnes integrals. These are of form

$$I(z) = \frac{1}{2\pi i} \int_C g(s) z^{-s} ds \qquad (4.1)$$

where $g(s)$ is typically a product or quotient of gamma functions, perhaps with trigonometric functions, and C is an appropriate path in the complex s-plane. The domains of convergence of such integrals are discussed in Chapter 2 of the book by Paris and Wood [15]. The flexibility of such integrals and their wide domains of convergence allow uniform exponentially-improved expansions to be readily constructed. As in Olver, Jones and Boyd, this is done by using the exponential integral $T_\nu(x)$ (defined in terms of the incomplete gamma function as $e^{\pi\nu i}\Gamma(\nu)\Gamma(1-\nu,x)/2\pi i$) as an approximant to the remainder obtained by truncating the asymptotic expansion at its least term. We illustrate this method by applying it to the quotient of gamma functions in (3.8).

An expansion for a quotient of gamma functions is given in (5.02) of Olver's book [13], but this contains only algebraic terms and its use leads to an error by a factor of 2 in Im λ. Replacing a by z, observe that the left-hand side of (3.8) may be written

$$\frac{\Gamma(\frac{1}{4}+z)}{\Gamma(\frac{3}{4}+z)} = \frac{1}{\pi}\Gamma(\frac{1}{4}+z)\Gamma(\frac{1}{4}-z)\sin\pi(\frac{1}{4}-z). \tag{4.2}$$

Writing $z = iy$ for $y < 0$ (since we require the behaviour in (3.8) as $a \to -i\infty$), (4.2) becomes

$$\frac{1}{\pi}|\Gamma(\frac{1}{4}+iy)|^2 \sin\pi(\frac{1}{4}-iy) \sim -i|y|^{-1/2}e^{i\pi|y|}\{e^{i\pi(1/4-iy)} - e^{-\pi(1/4-iy)}\} \quad \text{as } |y| \to \infty$$

on using Stirling's approximation for the gamma function. In terms of z we obtain

$$\frac{\Gamma(\frac{1}{4}+z)}{\Gamma(\frac{3}{4}+z)} \sim z^{-1/2}(1 - ie^{-2\pi iz}) \tag{4.3}$$

as $z \to \infty$ along the negative imaginary axis.

This simple method depends on the reflection formula for the gamma function and works only when $z = iy, y \in \mathbf{R}$. For general z we require the representation for the logarithm of the gamma function in terms of the Riemann zeta function $\zeta(s)$

$$\log\Gamma(z+1) = z\frac{\Gamma'(1)}{\Gamma(1)} - \frac{1}{2\pi i}\int_{3/2-\infty i}^{3/2+\infty i} \frac{\pi z^s}{s\sin\pi s}\zeta(s)ds$$

valid in $|argz| \leq \pi - \varepsilon$. Here ε denotes an arbitrary positive real number. It is shown in [21, p.277] that displacement of the contour of integration to the left over the double poles at $s = 0$ and $s = 1$ and the simple poles at $s = -1, -3, \cdots, -2n-3$ then leads to the expansion, valid when $|\arg z| \leq \pi - \varepsilon$,

$$\log\Gamma(z) = (z - \frac{1}{2}\log z + \frac{1}{2}\log 2\pi + \sum_{r=1}^{n-1} \frac{B_{2r}}{2r(2r-1)z^{2r-1}} + R_n(z), \tag{4.4}$$

where B_{2r} are the Bernoulli numbers. The remainder function $R_n(z)$ resulting from truncation of the above expansion after $n-1$ terms is given by

$$R_n(z) = -\frac{1}{2\pi i}\int_{-2n+1+c-\infty i}^{-2n+1+c+\infty i} \frac{\pi z^s}{s\sin\pi s}\zeta(s)ds, \quad 0 < c < 2, \quad |\arg z| \leq \pi - \varepsilon. \tag{4.5}$$

In (4.4) n is an arbitrary positive integer which will subsequently be chosen to have its optimal value given by $n = O(|z|)$, when $|z|$ is large.

It is shown in [18] how this integral for $R_n(z)$ may be expressed as a series of functions of type $T_\nu(x)$, the approximants defined above, via the Mellin-Barnes type formula given in [17]

$$-2ie^{-\pi i\nu}T_\nu(x) = \frac{x^{-\nu}e^{-x}}{2\pi i}\int_{-c-\infty i}^{c+\infty i} \frac{\Gamma(s+\nu)}{\sin\pi s}x^{-s}ds, \quad \nu \neq 0, -1, -2, \cdots \tag{4.6}$$

valid in the wider sector $|\arg x| < \frac{3}{2}\pi$. The method depends on integrating the series expansion of the Riemann zeta function. The final result is that, writing $z = \rho e^{i\theta}$, as $\rho \to \infty$ in $|\theta| < \pi$,

$$\Gamma(z) \sim \sqrt{2\pi} z^{z-\frac{1}{2}} e^{-z} \left\{ 1 + \frac{1}{12z} + \frac{1}{288z^2} + \cdots + S(\theta) e^{\pm 2\pi i z} \right\}$$

where the upper or lower sign is taken according as z is in the upper or lower half-plane and the multiplier $S(\theta)$ is given by

$$S(\theta) \approx \frac{1}{2} \pm \frac{1}{2} \text{erf } [c(\theta \mp \frac{1}{2}\pi)\sqrt{\pi\rho}]$$

where

$$c(\phi) \approx \phi - \pi + \frac{1}{6} i(\phi - \pi)^2 - \frac{1}{36}(\phi - \pi)^3 + \cdots.$$

This is in accordance with the generic form of the Stokes multiplier established by Berry [2] and others. Its use in place of Stirling's formula enables the correct version of (4.3) to be established for z in the sector containing the Stokes line $\arg z = -\frac{1}{2}\pi$. It is this result which is used to establish (3.10).

The rigorous results obtained on exponential improvement to date have been for particular special functions or for functions having a certain integral representation. It is hoped that this survey of recent work, arising out of an optical tunnelling application, illustrates the need for a reappraisal by mathematicians of the formal methods of Dingle, and the smoothing due to Berry, with a view to establishing a rigorous theory of exponentially-improved asymptotics for a wide class of functions.

References

1 M. Abramowitz and I. Stegun (ed.) Handbook of Mathematical Functions (New York: Dover, 1970).

2 M.V. Berry. Uniform asymptotic smoothing of Stokes' discontinuities. Proc. Roy. Soc. Lond. A422, 7-21 (1989).

3 W.G.C. Boyd. Stieltjes transforms and the Stokes phenomenon. Proc. Roy. Soc. Lond. A439, 277-24 (1990).

4 N. Brazel, F. Lawless and A.D. Wood. Exponential asymptotics for an eigenvalue of a problem involving parabolic cylinder functions. To appear in Proc. Amer. Math. Soc.

5 J. Burzlaff and A.D. Wood. Optical tunnelling from one-dimensional square potentials. To appear in IMA Journal of Applied Math.

6 R.B. Dingle. Asymptotic expansions: their derivation and interpretation. (New York and London: Academic Press, 1973).

7 E. Harrell and B. Simon. The mathematical theory of resonances whose widths are exponentially small. Duke Math. J. 4 845-901 (1980).

8 D.S. Jones. Uniform asymptotic remainders, p.241-264 in Asymptotic and Computational Analysis, ed. R. Wong (New York: Marcel Dekker, 1990).

9 W.L. Kath and G.A. Kriegsmann. Optical tunnelling: radiation losses in bent fibre optic waveguides. IMA Journal of Applied Math., 41, 85-103 (1989).

10 F.R. Lawless and A.D. Wood. Resonances and optical tunnelling. Submitted for publication (1991).

11 J. Liu and A. D. Wood. Matched asymptotics for a generalisation of a model equation for optical tunnelling. To appear in European Journal of Applied Mathematics.

12 C. Lozano and R.E. Meyer. Leakage and response of waves trapped by round islands. Physics of Fluids, 19, 8, 1075-1088 (1976).

13 F.W.J. Olver. Asymptotics and Special Functions (New York: Academic Press, 1974).

14 F.W.J. Olver. On Stokes phenomenon and converging factors, p. 329-355 in Asymptotic and Computational Analysis ed. R. Wong (New York: Marcel Dekker, 1990).

15 R.B. Paris and A.D. Wood. Asymptotics of high order differential equations. Pitman Research Notes in Mathematics Series. Vol. 129 (London, Longman, 1986).

16 R.B. Paris and A.D. Wood. A model equation for optical tunnelling. IMA Journal of Applied Math. 43, 273-284 (1989).

17 R.B. Paris. Smoothing of the Stokes phenomenon using Mellin-Barnes integrals. Submitted for publication (1990).

18 R.B. Paris. and A.D. Wood. Exponentially improved asymptotics for the gamma function. Submitted for publication (1990).

19 M. Reed and B. Simon. Methods of modern mathematical physics. Vol. 4 Analysis of operators. (San Diego: Academic Press Inc., 1978).

20 E.C. Titchmarsh. Eigenfunction Expansions. Part 2. (Oxford: Clarendon Press, 1958).

21 E.T. Whittaker and G.N. Watson. Modern Analysis (Cambridge University Press, 1935).

SOLITARY-WAVES IN SELF-INDUCED TRANSPARENCY

Olivier Martin* and Spiros V. Branis#

*Department of Physics, City College, New York, NY, 10031
#Department of Physics, Emory University, Atlanta, GA 30322

I. INTRODUCTION

Numerous systems are known to give rise to velocity or wavelength "selection": crystal growth, fluid flow, shock waves, etc... Mathematically, the velocity or wavelength appear as non-linear eigenvalues[1,2]. It is often the case that the "unperturbed" problem (e.g., the zero-surface tension limit for crystal growth and fluid flow) has a continuous non-linear eigenvalue spectrum because of some underlying symmetry. The addition of a perturbation will usually break this symmetry and will lead to a discrete spectrum, that is to "selection". Exactly integrable PDEs provide a particularly interesting ground for selection studies. They typically have a continuum of soliton solutions with different velocities or amplitudes. Adding a generic perturbation destroys their exact integrability. Solitons should disappear or become solitary-waves under such perturbations. There are many examples where ordinary perturbations destroy the family of solitons, leaving a single solitary-wave[3]. The effects of singular perturbations are more subtle, but have been investigated in the last few years for the cases of the standard exactly integrable PDEs (KdV, NLS, SG; see several of the articles in these proceedings). The conclusion of these works is that higher derivatives destroy all solitary-waves; generally, steady-state solutions have capillary waves going out all the way to infinity from the main peak. The purpose of this article is to report on some work[4,5] on the coupled Maxwell-Bloch equations. In the slowly varying envelope approximation (SVEA), these PDEs reduce to an exactly integrable system which can be treated by the inverse scattering transform[6,7]. Also, to all orders in perturbation theory, there is a continuum of solitary-wave solutions. The underlying physical phenomenon of interest is called "self-induced transparency" (SIT): a large amplitude pulse can propagate in a frequency domain (the gap) which is forbidden in the linear theory (see section II). In our work, we go beyond the SVEA and beyond all orders in perturbation theory. We show that, for a carrier wave of given frequency, there is a discrete family of solitary-wave solutions of the <u>full</u> Maxwell-Bloch equations inside the gap. Outside the gap, there are no solitary-wave solutions at all.

This article is organized as follows: we begin by reviewing the Maxwell-Bloch equations which are the standard starting point for SIT studies. This is followed by an explanation of the SVEA and McCall and Hahn's work[8]. In Section III, following Bialynicka-Birula[9] and Akimoto and Ikeda[10], the complete Maxwell-Bloch equations for steady-state pulses are derived without assuming the SVEA. After mentioning previous methods for approximating the solitary-wave solutions, we present an electric field amplitude expansion to solve the coupled Maxwell-Bloch equations perturbatively. Setting the carrier frequency to be on resonance in our method gives the expansion introduced by Marth, Holmes, and Eberly[11]. However, perturbation expansions are misleading: they are not uniformly valid on the whole domain $(-\infty, +\infty)$, and they cannot determine whether there exist solitary-wave solutions. In Section IV, we show that the width τ of a steady-state pulse is not arbitrary, rather it can only take on certain selected discrete values. We find numerically that the solvability condition can be satisfied inside the gap, but not outside. Fig. 1 gives the location in parameter space of these selected solitary-wave solutions.

II. SIT AND THE MAXWELL-BLOCH EQUATIONS

Since the 1960's, intense and practically coherent monochromatic laser light has been available as a probe of optically resonant media. When the incident light lies inside a frequency gap occuring near resonance, it is strongly absorbed. In 1969, McCall and Hahn[8] discovered that above an intensity threshold, certain light pulses can propagate with anomalously low energy loss even near resonance. They coined the effect "self-induced transparency" (SIT) because the highly absorptive medium becomes essentially transparent when the light is sufficiently intense. The physical picture of SIT is that the front part of the pulse coherently excites the atoms in the medium, e.g., up to the state of complete inversion. The macroscopic polarization formed in this process then emits coherent radiation which joins the back part of the pulse. If the oscillators in the medium return to the ground state after this process, no light is absorbed and <u>steady-state</u> propagation is realized.

In general, the starting point of theoretical SIT analyses[7,12] is the semiclassical description given by the Maxwell-Bloch equations (Eqs. (II.1-4)). The absorbing medium (gas, semiconductor, etc...) is modeled as an ensemble of non-interacting two-level systems. The energy difference between the two levels of each such system is $\hbar\omega_t$. We consider only the case of no inhomogeneous broadening (i.e., all the two-level systems are degenerate). If one takes fields which are x and y independent, then the (classical) Maxwell wave equation becomes:

$$\left(\frac{\partial^2}{\partial z^2}-\frac{1}{c^2}\frac{\partial^2}{\partial t^2}\right)\mathbf{E}(t,z)=\frac{4\pi}{c^2}\frac{\partial^2}{\partial t^2}\mathbf{P}(t,z) \tag{II.1}$$

E is the electric field, and **P** is the polarization due to the dipoles. The medium has a density N of dipoles, each having a dipole moment d. Defining $\mathcal{E}$ to be the electric field magnitude, $\kappa=2d/\hbar$, there are functions $\theta(t, z)$, $u(t, z)$, and $v(t, z)$ such that

$$\mathbf{E}(t,z)=\mathcal{E}(t,z)\hat{\mathbf{a}}(t,z) \quad , \quad \hat{\mathbf{a}}=\hat{\mathbf{x}}\cos\theta(t,z)+\hat{\mathbf{y}}\sin\theta(t,z) \quad , \quad \theta=\omega t-Kz+\phi(t,z) \tag{II.2}$$

$$\mathbf{P}(t,z)=\frac{1}{2}N\hbar\kappa\left\{u(t,z)\hat{\mathbf{a}}+v(t,z)\hat{\mathbf{b}}\right\} \quad , \quad \hat{\mathbf{b}}=-\hat{\mathbf{x}}\sin\theta(t,z)+\hat{\mathbf{y}}\cos\theta(t,z) \tag{II.3}$$

ω is the frequency and K the wave-vector of the "carrier" wave. In the semi-classical limit, the time-dependence of **P** is given by the Bloch equations[12,13]

$$\frac{\partial u}{\partial t}=(\frac{\partial\theta}{\partial t}-\omega_t)v \qquad \frac{\partial v}{\partial t}=-(\frac{\partial\theta}{\partial t}-\omega_t)u+\kappa\mathcal{E}w \qquad \frac{\partial w}{\partial t}=-\kappa\mathcal{E}v \tag{II.4}$$

u and v are the in-phase (parallel to **E** or dispersive) and out of phase (orthogonal to **E** or absorptive) components of **P**, and w is the population inversion of the medium; they satisfy $u^2+v^2+w^2=1$. Eqs. (II.1-4) are the Maxwell-Bloch (M-B) equations.

The dispersion relation for small amplitude plane waves is given by linearizing these equations to give:

$$(\frac{cK}{\omega})^2=1+\frac{2\pi N\hbar\kappa^2}{\omega_t-\omega} \tag{II.5}$$

K varies steeply as one approaches the resonance frequency ω_t from below. Above ω_t there is a gap of size $\omega_{LT}=2\pi N\hbar\kappa^2$ in which propagation is forbidden (this is the polariton gap of semiconductors in the "local optics" picture). Low intensity light is strongly absorbed in this gap because K becomes imaginary there. However, for large electric field intensities, the non-linearities of the Bloch equations are important, and SIT shows that certain pulses can nevertheless propagate in the gap even though this is forbidden in the linear theory.

McCall and Hahn considered a circularly polarized pulse with a carrier wave of frequency ω and wave number K so that $\theta(t, z)=\omega t-Kz$. On the time and length scale of this carrier wave, the envelope is typically slowly varying. The slowly varying envelope approximation (SVEA) consists in taking an envelope function $\mathcal{E}(t, z)$ which satisfies

$$\left|\frac{\partial\mathcal{E}}{\partial z}\right|<<|K\mathcal{E}| \quad \text{and} \quad \left|\frac{\partial\mathcal{E}}{\partial t}\right|<<\omega|\mathcal{E}| \tag{II.6}$$

and neglecting all sub-leading terms in this limit. If one uses this approximation and forces the wave to be circularly polarized, most derivatives can be dropped and one obtains

$$\left((\frac{K}{k_0})^2-1\right)\mathcal{E}=2\pi N\hbar\kappa u \tag{II.7}$$

$$2\left(\left(\frac{cK}{k_0}\right)\frac{\partial}{\partial z}+\frac{\partial}{\partial t}\right)\mathcal{E}=2\pi N\hbar\kappa\omega v \tag{II.8}$$

$$\frac{\partial u}{\partial t}=(\omega-\omega_t)v \qquad \frac{\partial v}{\partial t}=-(\omega-\omega_t)u+\kappa\mathcal{E}w \qquad \frac{\partial w}{\partial t}=-\kappa\mathcal{E}v \tag{II.9}$$

Eq. (II.7) is a constraint equation. For the SVEA to make sense, the initial conditions must satisfy this relation to leading order. Then the SVEA is consistent, and the dynamics of ($\mathcal{E}$, u, v, w) are governed by Eqs. (II.8-9). For completeness, let us note that at resonance, $\omega=\omega_t$, this system becomes a sine-Gordon equation. Within the SVEA, McCall and Hahn found a family of *solitary-waves* of arbitrary pulse width τ with velocity $V=c/(1+2\pi\kappa\omega N d\tau^2)$:

$$\mathcal{E}(t,z)=\frac{2}{\kappa\tau}\operatorname{sech}\left[\frac{t-z/V}{\tau}\right] \quad \text{with} \quad \kappa\int_{-\infty}^{+\infty}\mathcal{E}(t,z)dt=2\pi \tag{II.10}$$

Such pulses have constant shapes and thus are transmitted with no loss (SIT). In 1971, Lamb[7] showed that the Maxwell-Bloch (M-B) equations in the SVEA (Eqs. (II.8-9)) form an exactly integrable system. This meant that the propagating hyperbolic secant pulses found by McCall and Hahn were in fact solitons. Lamb[14] showed using a series of variable transformations that the SIT equations reduce to the Zakharov-Shabat equations[15], and Ablowitz, Kaup, and Newell showed how to solve the general initial value problem[6]. A review of the inverse scattering transform as applied to SIT in the SVEA was given by Hauss[16].

What happens if one goes beyond the SVEA by keeping higher order terms in the Maxwell wave equation? A generic perturbation destroys exact integrability. Thus one expects that as soon as one goes beyond the SVEA, there are no more solitons. Since the perturbation of interest is singular, it would appear likely that there are no localized steady-states at all here just as for the KdV. Actually, we will see that there is an infinite discrete set of solitary-wave solutions inside the gap, but none outside. Thus the Maxwell-Bloch equations have steady-state solutions which are quite different from those of other perturbed integrable systems covered in these proceedings.

III. SOLITARY-WAVES AND PERTURBATION THEORY

Without assuming the slowly varying envelope approximation (SVEA) in dielectrics, we are looking for "travelling-wave" solutions for the electric field $\mathbf{E}(t, z)$ and the polarization density $\mathbf{P}(t, z)$. We search for pulses which are time-independent in a uniformly translating *and* rotating frame. Defining $\zeta=t-z/V$, this gives

$$\theta(t,z)=\omega t-Kz+\phi(\zeta) \tag{III.1}$$

and imposes that $\mathcal{E}$, u, v, and w be functions of ζ only. The wave number K and the pulse envelope velocity V can be derived by linearizing the Maxwell-Bloch equations and setting $\dot{\phi}=0$ and $w=-1$. In the tail of the pulse, the solutions are exponential, $\mathcal{E} \sim \exp[\pm\zeta/\tau]$. τ introduces a time scale which allows one to write everything in dimensionless form. Let us introduce the dimensionless frequency detuning $\Delta = (\omega-\omega_t)/\omega_{LT}$, ($\omega_{LT} = 2\pi N\hbar\kappa^2$), the dimensionless reciprocal pulse width $\Lambda = 1/\omega_{LT}\tau$, and $s = 1/\omega\tau = \Lambda/[\Delta+\omega_t/\omega_{LT}]$. The gap, where there is no propagation in the linear theory, corresponds to the range $0 < \Delta < 1$. Now we can write the Maxwell-Bloch equations for a solitary-wave in dimensionless form. Introduce the dimensionless electric field amplitude $E = \kappa\mathcal{E}/\omega_{LT}$, and the dimensionless time $\xi = \zeta/\tau = (t-z/V)/\tau$ where τ is the above defined time scale which will be essentially the pulse width. For a steady-state pulse, Eqs. (II.1-4) turn into coupled ODEs:

$$\gamma\ddot{E}-[\alpha+\beta\,\dot{\phi}+\gamma\dot{\phi}^2+\frac{s}{\Lambda}(\frac{s\Delta}{\Lambda}-2)w+\frac{s^2}{\Lambda}\dot{\phi}w]E=-(\frac{s\Delta}{\Lambda}-1)^2u \qquad \text{(III.2)}$$

$$\gamma\ddot{\phi}E+(\beta+2\gamma\dot{\phi}-\frac{s^2}{\Lambda}w)\dot{E}=-[(\frac{s\Delta}{\Lambda}-1)^2+(\frac{s}{\Lambda})^2E^2]v \qquad \text{(III.3)}$$

$$\Lambda\dot{u}=(\Delta+\Lambda\dot{\phi})v \qquad \text{(III.4)}$$

$$\Lambda\dot{v}=-(\Delta+\Lambda\dot{\phi})u+Ew \qquad \text{(III.5)}$$

$$\Lambda\dot{w}=-Ev \qquad \text{(III.6)}$$

where hereafter dots mean $d/d\xi$ (e.g. $\dot{w}=dw/d\xi$). The coefficients α, β and γ are given by

$$\alpha=\left(\frac{K}{k_0}\right)^2-1,\quad \beta=2s\left(\frac{cK}{Vk_0}-1\right),\quad \gamma=s^2\left[\left(\frac{c}{V}\right)^2-1\right] \qquad \text{(III.7)}$$

and are thus known functions of the three dimensionless parameters Δ, Λ and s (see reference 5). The problem is to solve Eqs. (III.2-6) under the boundary condition that as $\xi \to \pm\infty$, the electric field vanishes, $E=0$, and all the dipoles of the dielectric are in the ground state, so that $w=-1$.

There have been a number of studies for solving these equations perturbatively. Bialynicka-Birula[9] took as an expansion parameter $s=1/\omega\tau$. Her work was systematically expanded upon by Akimoto and Ikeda[10] to include various expansions whose parameters depended on the nature of the pulse. In all cases, the derived pulse shapes depend continuously on the pulse width τ which is an arbitrary parameter in the problem, just as McCall and Hahn found in the SVEA.

A slightly different approach for solving Eqs. (III.2-6) was developed by Marth, Holmes, and Eberly[11] who went beyond the SVEA for the on resonance case $\omega=\omega_t$. Their method of approximation is based on a series expansion in powers of the electric field rather than in a parameter. We have generalized[5] their amplitude expansion to the off-resonance case as follows. The phase equation (III.3) is a first-order linear differential equation for $\dot{\phi}$, from which one can derive

$$\gamma \dot{\phi} E^2 = -\frac{\beta}{2}E^2 + \Lambda(\frac{s\Delta}{\Lambda}-1)^2(w+1) + \frac{s^2}{\Lambda}E^2 w - \frac{s^2}{\Lambda}\int_0^E E'w(E')dE' \tag{III.8}$$

We expand $w(E)$ in a power series:

$$w(E) = \sum_{i=0}^{\infty} w_{2i} \cdot E^{2i} \tag{III.9}$$

With this ansatz, it is easy to see that $\dot{\phi}$ (resp. u) is an even (resp. odd) power series in E, and that v is equal to $\dot{E}$ times a power series in E. All the coefficients of these series can be determined recursively. For instance

$$w_2 = \frac{1}{2\Lambda}\frac{(\beta + \frac{s^2}{\Lambda})}{(\frac{s\Delta}{\Lambda}-1)^2} = \frac{1}{2(\Delta^2+\Lambda^2)}$$

$$u_1 = -2\Delta w_2 \quad and \quad u_3 = -\frac{1}{3}[4\Delta w_4 + 2\Lambda \phi_2 w_2] \tag{III.10}$$

Then from Eq. (III.2) one obtains an ODE for the electric field amplitude E:

$$\ddot{E} = E + \frac{(\beta+\frac{s^2}{\Lambda})\phi_2 + \frac{s}{\Lambda}(\frac{s\Delta}{\Lambda}-2)w_2 - (\frac{s\Delta}{\Lambda}-1)^2 u_3}{\gamma}E^3 + \cdots \tag{III.11}$$

Using this amplitude expansion, we have derived[5] the various limits of Akimoto and Ikeda in a unified way.

IV. SELECTED SOLITARY-WAVES

For all the above perturbative methods, one finds a continuous family of solitary pulses parametrized by their width τ or equivalently by their velocity V. However, these expansions do not guarantee the existence of solitary-wave solutions of the full M-B equations; in general, solitary-wave solutions of SIT will not exist. The higher derivatives in Eqs. (III.2-6) form singular perturbations which make the solitary-wave boundary value problem ill-posed. Eqs. (III.2-6) are equivalent to a system of six first order ODEs in E, $\dot{E}$, u, v, w, and ϕ. These equations have a six dimensional solution space. One dimension of this space corresponds to the solution (unique up to translations in ξ, and some discrete symmetries) which comes out of the E=0 and w=-1 point at ξ=-∞. The five other other dimensions correspond to modes which are "bad", i.e., which do not satisfy the boundary conditions as $\xi \to -\infty$. The same kinds of modes occur for ξ=+∞. In general, the continuation of the good solution from ξ=-∞ will have some of the bad modes as $\xi \to +\infty$. This is generically the case, so in general there are no solitary-wave solutions.

To find for which values of Λ, Δ, and s there are solitary-waves, we first make the problem well-posed. We follow the procedure developed for similar boundary value problems in other fields[17-19]. Eqs. (III.2-6) have translational symmetry in ξ and ϕ, and are invariant under $\xi \rightarrow -\xi$, $v \rightarrow -v$. Also, changing the sign of E, u, and v is a symmetry. Since the boundary conditions at $\xi = -\infty$ define the solution everywhere modulo the above translations and modulo the sign symmetry, it is not difficult to see that solitary-wave solutions (after shifting ξ so that the pulse peak is at $\xi = 0$) have w and $\dot{\phi}$ even in ξ, and E must be either even or odd. The standard hyperbolic secant 2π–pulses are even, so we will restrict ourselves to E even. (The analysis for odd pulses would proceed similarly). Then u is even and v is odd. Let us thus consider the ODEs on the interval $(-\infty, 0]$ with the same boundary conditions at $\xi = -\infty$ and the condition $v=0$ at $\xi = 0$. This new boundary value problem is well posed. If $\dot{E} = 0$ at $\xi = 0$ also, it is easy to see that one can construct a solitary-wave solution on the whole ξ-axis by reflection of the solution on the half-line with a change of sign for v. However, in general, $\dot{E} \neq 0$ at $\xi = 0$, corresponding to a solution which has some amount of growing (bad) modes as $\xi \rightarrow +\infty$. The amount of these bad modes vanishes to all orders in perturbation theory, but nonetheless can be non-zero. Using the above, we see that the condition for existence of a solitary-wave is that $\dot{E}$ and v vanish simultaneously; this condition can be interpreted as forbidding any cusp in E.

There are three parameters in Eqs. (III.2-6). s is not a singular perturbation, so the asymptotics as $s \rightarrow 0$ can simply be obtained by a perturbative expansion in powers of s. Δ and Λ play much more interesting roles as they give rise to non-analyticities. The Kruskal-Segur[18] analysis can be applied to the case where the small parameter multiplies the highest derivatives. It is readily seen in our problem that this requires Δ to be outside the gap and $\Lambda \rightarrow 0$ (long pulse). We found[5] that when $\Lambda \rightarrow 0$, the mismatch function $\dot{E}_{tip} = \dot{E}(0)$ behaves as $\dot{E}_{tip} \approx \exp[-\lambda(\Delta, s)/\Lambda]$. We were not able to carry the analysis to the point of obtaining a closed form formula for the function $\lambda(\Delta, s)$, but we found numerically that λ depends rather weakly on Δ far away from $\Delta = 0$ or 1, and that the asymptotic behavior sets in rather quickly. The conclusion of this is that there are no long solitary pulse solutions outside the gap, since $\dot{E}_{tip}$ does not vanish.

Inside the gap, the Kruskal-Segur analysis cannot be used, so we have studied the question numerically. We used the computer to integrate the system of Eqs. (III.2-6) beginning at large negative ξ. For such ξ, the solution can be obtained from the perturbative expressions for E, $\dot{E}$, $\dot{\phi}$, u, v, and w; these are used as initial conditions on the functions. We evolve forward and find the ξ or ξ's where $v=0$, and determine $\dot{E}$ there. Call this value $\dot{E}_{tip}(\Delta, \Lambda, s)$. In general $\dot{E}_{tip} \neq 0$, and solitary-waves do not exist for those values of Δ, Λ, s. However, inside the gap, we find that $\dot{E}_{tip}$ changes sign when the parameters are varied. Thus one can tune the parameters Λ, Δ, and s to obtain $\dot{E}_{tip} = 0$. Using a root solver, we determined the locus of the curves (Λ, Δ) for which there are solitary solutions at fixed ω_{LT}/ω_t. As can be seen in Figure 1, various branches of solutions rise from $\Delta=0$ and set at $\Delta=1$; note that these two points are also special in the linear theory. We have drawn some branches as stopping inside the gap because there appear multiple solutions to $v=0$. As one decreases Δ, the pulse shapes along these branches continue to develop more and more oscillations, eventually looking nothing like the original

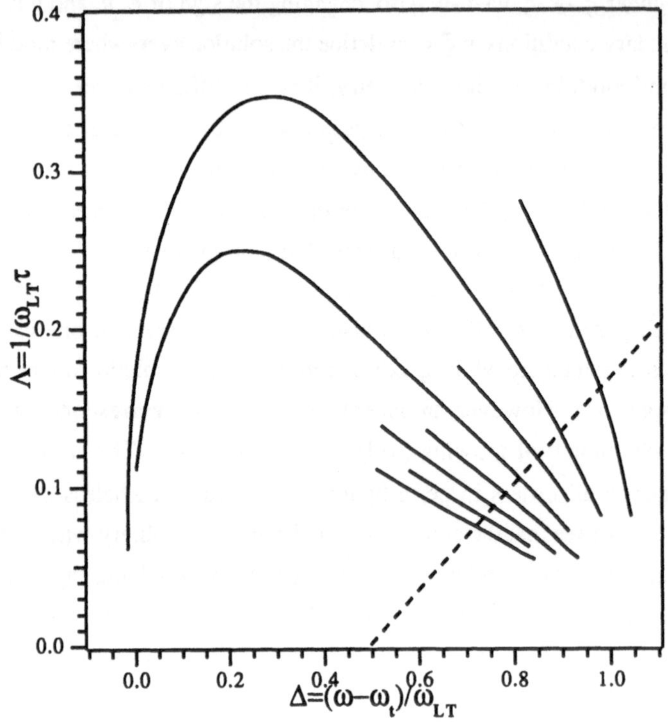

Fig. 1 Values of the parameters Δ, Λ for which steady-state solitary-wave solutions exist (solid curves). We have taken ω_t/ω_{LT}=1000.

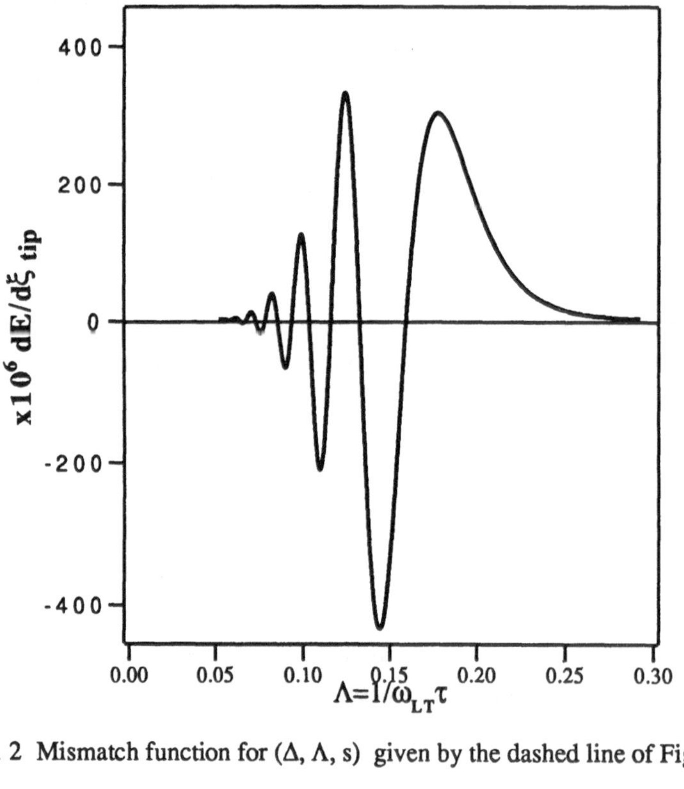

Fig. 2 Mismatch function for (Δ, Λ, s) given by the dashed line of Fig. 1.

hyperbolic secant shapes. In Fig. 2, we plot the mismatch function, i.e., the magnitude of the cusp of the solution, $\dot{E}_{tip}$, for the parameters given by the dashed line of Fig. 1. The oscillations are difficult to resolve all the way down to small Λ, but it seems that there are an infinite number of branches coming out of the point $\Delta=1$, $\Lambda=0$.

In summary, we first introduced a perturbation expansion to solve the coupled Maxwell-Bloch equations. General expressions for solitary-wave shapes were derived for an arbitrary incident carrier frequency ω and an arbitrary pulse width τ. However such perturbation series generally do not converge to the exact solutions, and thus they cannot be used to determine whether solitary-wave solutions exist. Then we showed that steady-state pulse solutions of the full M-B equations do not exist outside of the gap. Inside the gap however, solutions do exist for selected values of τ which in turn determine the propagation velocities. This selection principle dramatically changes the previous picture of SIT.

ACKNOWLEDGEMENTS

We thank Joseph L. Birman, Richard Friedberg, George Lamb, Sam McCall, and Harvey Segur for their suggestions and comments. This work was supported by awards from PSC-CUNY, in part (S.V.B.) by NASC # 19-87-G-0251, and in part (O.M.) by NSF-ECS-8909127.

REFERENCES

1. G.I. Barenblatt, "Similarity, Self-Similarity, and Intermediate Asymptotics" (Consultants Bureau, New York, 1978).
2. "Dynamics of Curved Fronts", edited by P. Pelcé, (Academic, New York, 1988).
3. V. Hakim, P. Jakobsen, and Y. Pomeau, Europhys. Lett. 11, 19 (1990).
4. S.V. Branis, O. Martin, and J.L. Birman, Phys. Rev. Lett. 65, 2638 (1990).
5. S.V. Branis, O. Martin, and J.L. Birman, Phys. Rev. A 43, 1549 (1991).
6. M.J Ablowitz, D.J. Kaup, and A.C. Newell, J. Math. Phys. 15 1852 (1974).
7. G. L. Lamb Jr., Rev. Mod. Phys. 43, 99 (1971).
8. S.L. McCall and E. L. Hahn, Bull. Am. Phys. Soc. 10, 1189 (1965); Phys. Rev. Lett. 18, 908 (1967); Phys. Rev. 183, 457 (1969).
9. Z. Bialynicka-Birula, Phys. Rev. A 10, 999 (1974).
10. O. Akimoto, and K. Ikeda, J. Phys. A 10, 425 (1977).
11. R.A. Marth, D.A. Holmes, and J.H. Eberly, Phys. Rev. A 9, 2733 (1974).
12. L. Allen and J.H. Eberly, "Optical Resonance and Two-Level Atoms" (Dover, New York, 1987).
13. F. Bloch, Phys. Rev. 70, 460 (1946).
14. G.L. Lamb Jr., Phys. Rev. Lett. 31, 196 (1973); Phys. Rev. A 9, 422 (1974).

15. V.E. Zakharov and A.B. Shabat, Zh. Eksp. Teor. Fiz. 61, 118 (1971) [Sov. Phys. JETP 34, 62 (1972)].
16. H.A. Hauss, Rev. Mod. Phys. 51, 331 (1979).
17. J-M. Vanden-Broeck, Phys. Fluids 26(8), 2033 (1983).
18. M. Kruskal, H. Segur, Aeronautical Research Associates of Princeton, Technical Memo, 85-25, 1985 (to appear in Studies in Applied Math., 1991).
19. D.A. Kessler, J. Koplik, and H. Levine, Adv. Phys. 37, 255 (1988).

EXPONENTIAL ASYMPTOTICS FOR PARTIAL DIFFERENTIAL EQUATIONS

R. E. Meyer

Center for the Mathematical Sciences
University of Wisconsin
610 Walnut Street
Madison, WI 53705

The principles are discussed by which asymptotics beyond all orders can be carried to genuine partial differential equations for which the variables cannot be separated. Access to solutions then depends on the canonical equations of the Hamiltonian operators, and their phase-space trajectories involve complex geometrical optics and branching. To handle the resulting Stokes phenomena reliably in several dimensions requires a connection method resolving them strictly locally at the branch points of individual trajectories. For linear problems, at least, solution approximations can then be formulated which are concrete and exponentially reliable, even if grossly nonuniform, and can be used to predict transcendentally small functionals of the operator that can be observed directly. The general theory is illustrated by examples of spectral degeneracy.

1. INTRODUCTION

Successful asymptotics beyond all orders may be said to have started with Littlewood.[1] The first concrete and rigorous predictions for linear oscillators were obtained in 1973[2,3] and the first fully nonlinear problem was resolved reliably in 1976[4]. Since then exponential asymptotics for ordinary differential equations has developed rapidly. Reality, however, it not one-dimensional and problems for which the variables can be separated depend on stringent, special symmetries a perturbation of which tends to preclude transcendentally precise results. It must therefore be an essential objective of the subject to discover an access to the transcendental asymptotics of nonseparable partial differential equations.

In its general sense, asymptotics beyond all orders is a dangerous swamp. To avoid drowning in it, the present Author has focused exclusively on the prediction of quantities that can be *observed directly*, but cannot be captured by conventional analysis. Observed quantities must needs be functionals of solutions and hence, functionals of the operator defining the solutions. The detour via the solutions is therefore not mandatory and for transcendental functionals, in particular, the efficiency and simplicity of the analysis depends on avoiding the solutions as far as possible.

The experience of the Author is that observable functionals then come on three levels of difficulty. The simplest are those, like some scattering coefficients[5], which

Asymptotics beyond All Orders, Edited by H. Segur *et al.*
Plenum Press, New York, 1991

have the null asymptotic expansion. The hard problems arise when functionals do have an asymptotic expansion, but also properties hidden behind it that can be observed directly. Such problems come sometimes with extenuating circumstances in the form of exact symmetries which can be used to filter our the relevant transcendentals. The hardest level arises in the absence of such luck, as in the non-selfadjoint problem of radiation damping[6,7]. A test of the new theory is that it should cope at least with the middle level, of which spectral degeneracy of selfadjoint operators is a typical and prominent example. A major part of the present approach, namely the description of the curious invariant toroids in complex phase space (Section 4) and the local resolution of Stokes phenomena (Section 6), is general enough to serve for nonlinear Hamiltonian operators. The last part, however, concerns minimal approximations and a rigorous basis for those is available as yet only for linear partial differential equations.

The method to be described combines and revises geometrical optics, connection theory and EBK quantization, and its many conceptual difficulties make it imperative to illustrate it by a simple example exhibiting all the essential features in a rather explicit form. Such a canonical example is the eigenvalue split of the Helmholtz equation with a symmetrical double-well potential[8] on a two-dimensional strip domain (Section 3). Another application of the method to near-degenerate continuous spectra is outlined in Sections 10-12.

2. WAVE EQUATIONS

This account concerns the spectra of Hamiltonian operators $H(-i\epsilon\nabla, \mathbf{x})$ on $L^2(\Omega)$ with domain $\Omega \subset \mathbf{R}^N$; ϵ is a traditional physical parameter and (until Section 10) need not always be small, especially in problems of physical interest. This is of practical importance because it takes the following well beyond a semiclassical theory; conversely, it helps to explain why shortwave results have been found often to have quantitative success far beyond the expectations supported by their mathematical rationale.

Such an operator has an associated wave equation

(1) $\quad H\psi = E\psi$

with eigenparameters E for wave functions ψ and also an associated Hamilton-Jacobi equation

(2) $\quad H(\nabla S, \mathbf{x}) = E$

for Hamilton's characteristic function $S(\mathbf{x})$ and canònical equations

(3) $\quad d\mathbf{x}/dt = \nabla_p H, \;\; d\mathbf{p}/dt = -\nabla_x H, \;\; \mathbf{p} \equiv \nabla S$

for the phase-space trajectories $(\mathbf{p}, \mathbf{x})(t)$ of the Hamiltonian. On those trajectories,

(4) $\quad H = const = E, \;\; dS/dt = \mathbf{p} \cdot \nabla_p H,$

and their projections onto coordinate space are the rays of geometrical optics.

It will be assumed that the Hamiltonian is integrable, so that the trajectories lie on recognizable invariant tori. This is a severe restriction, in principle, but recent numerical evidence [9-11] indicates that lack of integrability may have little influence on wave spectra. It is therefore possible that the main effect of the integrability restriction may be to permit us to cut out some still underdeveloped layers of analysis by talking about exact, rather than approximate, tori with exact, rather than approximate, action invariants. In any case, the canonical equations (3), (4) are essential for access to concrete solution structure when the variables cannot be separated.

The solutions ψ of the wave equation can always be represented as

$$(5) \quad \psi(\mathbf{x}) = A_e(\mathbf{x}) \exp\left(iS(\mathbf{x})/\epsilon\right)$$

provided that $A_e(\mathbf{x})$ satisfies the "transport equation" obtained by substituting (5) in (1) and using (2), (3). The following differs from conventional shortwave theory by interpreting S as the exact characteristic function, rather than an approximate phase function. One of the advantages of dumping all approximation issues on A_e is that geometrical optics and Stokes phenomena can be handled exactly even for nonlinear wave equations and the analysis need not be lost in a quicksand of approximations. All the same, the surprises and difficulties can be grasped much more easily at the hand of a simple, explicit example illustrable by figures.

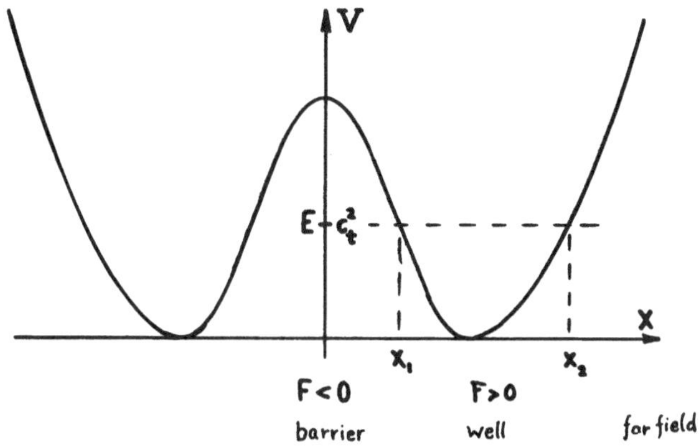

Fig. 1. Double-well potential with roots of $F(x; c_t)$ where $V(x) = E - c_t^2$.

3. DOUBLE WELL

Such an example is offered by the linear Helmholtz or Schroedinger equation

$$(6) \quad \epsilon^2 \nabla^2 \psi + [E - V(\mathbf{x})]\psi = 0$$

on a two-dimensional strip domain for a potential V of double-well type (Fig. 1). This is generated by the Hamiltonian $H(-i\epsilon\nabla, \mathbf{x}) = -\epsilon^2\nabla^2 + V(\mathbf{x})$ with Hamilton-Jacobi (or "eiconal") equation

$$(7) \quad H(\nabla S, \mathbf{x}) = |\nabla S|^2 + V(\mathbf{x}) = E$$

and transport equation

$$(8) \quad i\epsilon A_e \nabla^2 A_e = \operatorname{div}(A_e^2 \nabla S).$$

The potential of the example is taken to depend on only one coordinate x because the canonical equations (3), (4) then have an explicit integral

$$(9) \quad p_2 \equiv \frac{\partial S}{\partial y} = \text{const} = c_t, \qquad p_1 \equiv \frac{\partial S}{\partial x} = [F(x; c_t)]^{1/2}$$

$$(10) \quad F(x; c_t) \equiv E - V(x) - c_t^2$$

$$(11) \quad y(x) - y_0 = c_t \int_{x_0}^{x} [F(s;c_t)]^{-1/2} ds$$

$$(12) \quad S(x, y(x)) - S(x_0, y_0) = c_t[y(x) - y_0] + \int_{x_0}^{x} [F(s, c_t)]^{1/2} ds$$

in terms of the *parameter* x along the particular trajectory with "label" parameter c_t passing through $(x_0, y_0, [F(x_0; c_t)]^{1/2}, c_t)$. Admittedly, several distinct trajectory points can be associated with the same value of x (Section 4) and (9) - (12) must be interpreted on a Riemann surface over a complex x-plane, but that is an acceptable price for the simplicity of (9) - (12). (In a more general context, S itself is a more suitable parameter t in (3), (4), which adds an unfamiliar transformation without removing the multivaluedness.)

To simplify the example further, its potential $V(x)$ is taken to be even and real-analytic, to have just one maximum and two minima (Fig. 1) and to grow no more than a polynomial at large $|x|$. That turns out to imply the usual radiation condition.

$$(13) \quad \text{Im}\, S > 0 \quad \text{for sufficiently large} \quad |x|$$

for square-integrable solutions (5) of (6).

To make the example nontrivial and accommodate a situation prevalent in classical physics, the domain of the example will be taken to be a two-dimensional strip bounded by sidewalls $y = b_i(x)$, $i = 1$ or 2, also even and real-analytic in x, at which homogeneous boundary conditions, e.g.,

$$(14D) \quad \psi(x, b_i(x)) = 0$$

or

$$(14N) \quad \partial\psi/\partial\nu = 0 \quad \text{at} \quad y = b_i(x),$$

are specified, where ν denotes the direction normal to the sidewall. When $b_i(x) \neq$ const, the clash between the potential and the domain precludes separation of variables. The reason for the choice of a two-dimensional domain is merely that at least some degree of illustration by figures remains possible.

Any eigenfunction ψ of (6), (13), (14) can be resolved into its parts even and odd in x, and both are eigenfunctions. The symmetrical eigenfunctions satisfy

$$(15s) \quad \partial\psi/\partial x = 0 \quad \text{for} \quad x = 0, \; b_1(0) < y < b_2(0)$$

and the antisymmetrical ones satisfy

$$(15a) \quad \psi(0, y) = 0.$$

This makes it possible to confine most of the attention to the half-space $x > 0$.

4. INVARIANT TOROIDS

From (5) and (8) it is seen that construction of a solution ψ requires knowledge of S and ∇S to which the best access is offered by the canonical equations (3), (4) for the Hamiltonian trajectories when the variables cannot be separated. Surfaces in phase space formed by trajectory families are called invariant tori and those are relatively simple when the Hamiltonian is integrable. Nonetheless, there are surprises because the canonical equations (3) do not imply that the trajectories must be real. That purely real trajectories cannot be the rule is seen from Fig. 1 where the function F in (10) is negative on some x-intervals so that p_1 and y in (9), (11) are there complex. Such intervals are called "barrier" intervals of a trajectory because they inhibit real wave propagation. For a wave domain in $\mathbf{R}^N$, the phase space therefore lies in $\mathbf{C}^{2N}$ ...

Real trajectory segments are said to belong to a "well" of the potential and are familiar from geometrical optics[12-14]. They wind around phase-space tori of the familiar kind and it follows that the rays, which are the projection of trajectories on coordinate space, must form "caustic" envelopes[12,13], but those are singularities of only the projections, not of the trajectories or tori in phase space[15,16]. Over domains with sidewalls, as in the example, the trajectories have gaps where the real rays meet a sidewall because ∇S changes there discontinuously. Those gaps are bridged by the boundary conditions[13], which may be interpreted conveniently for (5) so as to leave A_ϵ continuous, but possibly change also S discontinuously. With the bridges, the real tori over wells can still be thought of as manifolds without boundary in $\mathbf{R}^{2N}$.

For spectral purposes, only tori are of interest on which H takes the same value E on all their trajectories. The physical requirement [12-14] that the solutions ψ must be single-valued in phase space adds the condition that the total change of S in (5) around a torus, from sidewall to sidewall and back, must be an integer multiple of $2\pi\epsilon$. In the example, that change may be evaluated at fixed x, e.g., as[13]

$$(16) \qquad \tfrac{2}{\epsilon}\int_{b_1(x)}^{b_2(x)} |\partial S/\partial y| dy + \mu_+ + \mu_- = 2m\pi, \quad m = 0, 1, 2, \ldots,$$

where μ_+, μ_- are the Keller indices[13] accounting for jumps in S at respective sidewalls. In problems without domain boundary, μ_+, μ_- are the more familiar Keller-Maslov indices [12-16]. The present account focuses only on tori on which $E - c_t^2$ lies between the minimum and maximum of V (Fig. 1), which restricts the tori to a finite set for $0 \leq m \leq M(\epsilon)$ in (16), but the set is large when ϵ is small. In any case, geometrical optics constructs all those admissible real tori over wells and attention may turn towards the issue which values of E on them can lie in the spectrum.

For problems involving more than one well, as do those of double wells and of periodic potentials, this spectral issue cannot be resolved without attention also to the complex parts of trajectories and tori. Such trajectory segments are said to lie over a "barrier". A connected trajectory segment cannot switch from a real subsegment to an adjacent complex segment without singularity at the border between well and barrier. The most typical case is that where a ray $y(x)$ has a branch point of square-root type. In the example, (11) shows that to occur at each simple root of $F(x; c_t)$ (Fig. 1). As a result, such a branch point is common to a pair of real trajectory branches and a pair of complex trajectory branches (Fig. 2). A complex barrier component of the invariant torus branches here off the real well torus, and this is a general insight not dependent on the particular root type of $y(x)$: trajectory branching is a typical property of any border between a well and a barrier.

When $b_i(x) \neq$ const in the example, the branch locus of the rays turns out[17] to differ from the envelope of the rays and this is a generic property of nonseparable problems. The branch locus, however, is a genuine singuarity of the Hamiltonian trajectories and tori and of the amplitude-phase representation (5) of the solutions. It appears more appropriate therefore to attach the term "caustic" to the branch locus, rather than to the envelope of the rays.

It is natural to ask which trajectory branch connects with which other branch at such a caustic point. On the basis of just the canonical equations, all of them connect, but the answer is quite different when the trajectories are interpreted in the light of the wave functions (5). It then turns out[17] that some wave functions carried by a real trajectory branch are carried into the central barrier (Fig. 1) on just one complex branch. Some other such wave functions, however, split onto both branches in the barrier. From the spectral point of view, the connectivity of the invariant tori is quite different from that associated normally with the term torus. Apart from its central barrier, the double well has also a pair of outer barriers, usually called far fields, in which $F < 0$. For wave representation, the radiation condition (13) suppresses one trajectory branch in a far field and that reduces the trajectory splitting at a border between well and far field.

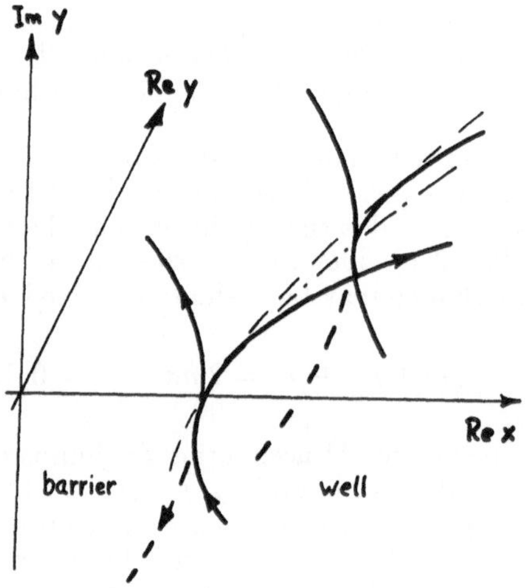

Fig. 2. Ray branches in the space of Re x and of y near the border between a well (to the right) and barrier (to the left). The thin lines indicate the envelope of the real rays over the well to the left of the locus of branch points.

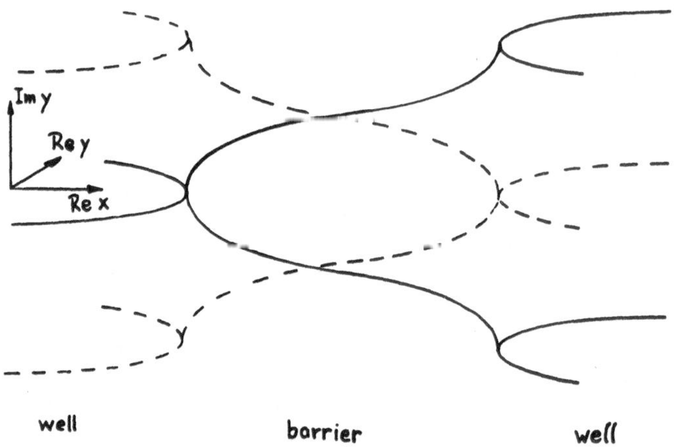

Fig. 3. The branching of real rays over a well and their continuation across a barrier into complex rays over the next well.

These unexpected aspects of the topology of phase-space tori are enhanced when one goes beyond questions of local branch structure to enquire about the more global nature of the complex components of invariant tori. It is not obvious that the complex rays over a barrier will meet the real sidewalls of a domain and in fact, they cannot be expected to do so. The example was chosen so as to yield specially explicit ray formulae illustrable by figures. In particular, since $F < 0$ in a barrier (11) shows that the complex rays $y(x)$ run there at fixed $Re\ y$ in a plane of $Re\ x$ and $Im\ y$ (Fig. 2). Even in the simple example, therefore, the barrier rays cannot intersect real sidewalls and boundary conditions like (14D,N) cannot be interpreted there in terms of wave reflection. By contrast to the real torus component, the gaps in the complex torus component cannot be bridged a priori.

It is also of some importance to ask how a trajectory will connect two wells across a barrier, and the answer must be that when a real trajectory branch over a well is followed through a barrier to the next well (Fig. 1), such a trajectory branch cannot be expected to emerge from the barrier as a real trajectory branch! This is seen easily already in the example: since F in (10) is negative and bounded in a barrier (Fig. 1), (11) shows $Im\ y(x)$ to be a strictly monotone function in the barrier. As a result (Fig. 3), the real torus component over a well branches across a barrier into a pair of torus components over the next well which are nonreal tori running in $\mathbf{C}^{2N}$ parallel to $\mathbf{R}^{2N}$. In a periodic potential (Section 10) with its infinite sequence of wells and barriers, this global branching of tori is repeated indefinitely; standard geometrical optics can construct one admissible torus over a well, but when this is followed further, it is seen to branch into an infinite tree of torus components in $\mathbf{C}^{2N}$.

This unfamiliar topology of phase-space trajectories impacts necessarily on spectral issues. To bear it in mind, it may help to use the term toroid, rather than torus, from now on.

5. SYMBOLS AND APPROXIMANDS

In the sense just described, geometrical optics serves to construct the exact characteristic function $S(x)$. To use it for spectral predictions requires three main further tools. First, a more semi-concrete symbol identifying solutions is needed than a mere letter ψ. The solutions involve usually transcendental growth or decay, and transcendentally small ("recessive") solutions are as important as transcendentally large ("dominant") ones for transcendentally precise spectral approximation. A unique identification is therefore achieved best on (anti-) Stokes manifolds L defined as those on which the solutions are "balanced", i.e., are free of transcendental growth or decay. Analytic continuation can then serve a unique identification elsewhere so as to avoid the nonuniqueness of dominant approximations that has bedeviled Mathematical Asymptotics. It will turn out that such analytic continuation can be achieved strictly locally in such a way that even the special example will not need the assumption of an analytic potential and domain shape.

Secondly, the characteristic function S is singular at its caustic branch manifolds and the "amplitude function" A_e in (5) has a corresponding singularity there because the solutions ψ are single valued. The exact singularity structure of A_e must be accounted for in order to resolve the Stokes phenomena anchored at those caustic manifolds reliably enough for transcendentally precise predictions of observed functionals.

Thirdly, solution approximations are needed which are fully reliable in those places where boundary, symmetry or radiation conditions must be satisfied. Uniformity of approximation, by contrast, is not relevant for the prediction of observed functionals and would complicate the analysis very heavily. While the first two objectives can already be achieved for nonlinear wave equations, the issue of approximation does not yet appear explored adequately in that wider context. In the narrower frame of linear equations, however, all three objectives can be achieved by approximations of WKB type, of which Maslov[18,19] has estabished the rigorous reliability by abstract Fourier analysis, provided they are used well away from singularities of S. Their main difficulty is their domain-dependence[20−22]. On the other hand, they have the practically important merit of the "double-asymptotic" property [21]: independently of the semiclassical parameters ϵ, they become arbitarily accurate on their proper domains at sufficient distances from any singularities of S. When they can be so employed (Section 7), they achieve much wider usefulness than the usual semiclassical rationale can suggest.

For the example of a Holmholtz or Schroedinger equation (6), these approximations satisfy the truncation

$$\mathrm{div}(A^2 \nabla S) = 0$$

of the exact transport equation (8), and this conservation law can be integrated[13] to

$$(17) \qquad A^2 J \partial S/\partial x = \quad \text{const} \qquad \text{on trajectories ,}$$

where J denotes a Jacobian of the transformation between Cartesian and ray coordinates. Accordingly, there are approximands

$$(18) \qquad \begin{aligned} \phi^{\text{out}} &= A^* F^{-1/4} J^{-1/2} \exp(iS/\epsilon) \\ \phi^{\text{in}} &= B^* F^{-1/4} J^{-1/2} \exp(-iS/\epsilon) \end{aligned}$$

which can also serve as solution symbols with the correct singularity structure, if the appropriate branches of the factors in (18) are used. The Stokes manifolds L are those on which S is real and the superscripts "out" or "in" then distinguish the approximands to outgoing waves in the sense of S increasing on trajectories from a branch point, when a factor $\exp(-i\omega t)$ is understood, from approximands to incident waves. The normalization constants A^{*2}, B^{*2} are the corresponding constants in (17). For some purposes, each function (18) must be split into a pair to distinguish the sign of the label c_t of the trajectory, but this distinction is omitted here because it would disappear again in the analysis outlined below.

A more important issue is the resolution of the Stokes phenomenon that the multivaluedness and nonuniformity of the approximands makes them valid only on restricted domains D_k associated with respective Stokes manifolds L_k. This generates the "connection problem"[20,21,23] of determining the branches in (18) and corresponding constants A^*, B^* which can make $\phi^{\text{out}} + \phi^{\text{in}}$ a consistent approximation to the same wave function ψ on different approximation domains D_k. This problem is more serious for partial than for ordinary differential equations and was solved heuristically in the pioneering work of Kravtsov[24] and Ludwig[25]. Their approach, however, is very difficult to bring to the reliability needed for transcendental precision and a fundamentally new method had to be developed.

6. CONNECTION

The new connection method[26] does not aim to approximate solutions but instead, focuses directly on the central issue of multivaluedness inherent in representations (5). It is intuitively plausible that multivaluedness is a global property of some degree of invariance. No quantitative measure of multivaluedness, however, appears to be known which possesses such global invariance in more than excessively restricted circumstances[21]. Nonetheless, a bridge can be builty very generally [20] which connects the singularity structure of singular points of ordinary differential equations with solution multivaluedness very far from such points. By this means, the resolution of the asymptotic Stokes phenomena far from a turning point, or even a highly irregular singular point of the differential equation,[27] can be accomplished by a strictly local analysis of the multivaluedness structure at the singular point itself.

For nonseparable partial differential equations the solutions (5) must be constructed from the canonical equations (3), (4) for the phase space trajectories. When the Hamiltonian is analytic, then on any connected trajectory segment free of singular points, the trajectory depends analytically on an appropriate parameter t along it in (3), (4). In the example, this is seen quite explicitly in (9)-(12). In any case, if the domain is two-dimensional for definiteness, then $(\mathbf{p}, \mathbf{x})(t)$ in (3) are a quartet of analytic functions of the one variable t. A caustic branch point of a trajectory is therefore a branch point of ∇S and x as analytic functions of t in the familiar sense of classical complex analysis. Their multivaluedness can accordingly be characterized in the strictest local sense at the branch point itself. Hence, if we choose any one generic trajectory and confine attention myopically to just that trajectory, then the Stokes phenomenon for partial differential equations, viewed in the light of the new connection method [26] does not differ from that for the ordinary differential system (3). On individual trajectories, the asymptotic connection relations are a direct sysmptom[17] of the local branch structure at branch points of analytic functions, independently of the number of domain dimensions of the partial differential equation.

It is of interest, moreover, that the analyticity of the Hamiltonian invoked by this argument if purely local. The new connection method[26] is based on a definition of the very differential equation on only an arbitrarily small neighborhood of its singular point. The bridge succeeds all the same,[26] as long as no further singular point interferes with the structure, because it is based on a Volterra integral equation. In practice, such local analyticity of the Hamiltonian is unlikely to fail at more than a few exceptional points of no significance for observed global functionals. In addition, the great generality[26] of the new connection method will fail rarely to cover the exceptional points as well. Furthermore, while the initial proof[26] applies to linear differential equations, its foundation on a local definition extends it also to nonlinear ones, such as the canonical equations (3), as long as a Frechet derivative is available.

For the determination of spectral properties, however, the connection problem concerns the approximation of eigenfunctions $\psi(\mathbf{x})$ at real $\mathbf{x}$, where the boundary, symmetry and radiation conditions apply. In the general context, the quantification of connection requires a considerable notational apparatus and it will be more helpful here to illustrate it at the hand of the double-well example formulated in Section 3. The connection is then needed from the central barrier to the far field (Fig. 1) and it is most convenient for the spectral predictions to connect the respective approximations at the same value of y. Since $V(x)$ and $b_i(x)$ in (6) and (14) are real-analytic, the solutions ψ of (6) are also analytic in the coordinate x at fixed y. The key to their approximation lies therefore in the connection of approximands (18) around individual branch points of S at fixed y. Figure 4 shows the complex plane of x at fixed real y with its caustic branch points x_1, x_2 (Fig. 1). Since $F(x; c_t)$ in (10) has simple roots at x_1 and x_2, (9) or (12) show the structure of S in this plane to be the same as at a "simple turning point"[21] in one dimension (which is the special case $c_t \equiv 0, J \equiv 1$ of the present analysis). By (12), S is real at fixed real y on three Stokes lines L_k issuing from each branch point (Fig. 4) (L_0 and L_3 coincide because the problem is selfadjoint, but no other Stokes lines intersect). Each Stokes line L_k is the backbone of an approximation domain D_k, which is bounded away from all other Stokes lines and from all singular points of S. The domains D_0 and D_1 are shown shaded in Fig. 4, from which approximation at real x in the barrier is seen to require approximation on D_1, while approximation at real $x > x_2$ in the far field is seen to require approximation on the analogous domain D_2 based on L_2. The immediate objective is therefore a connection from D_1 to D_2, which do not intersect, by the help of a connection of each with D_0, with which both overlap.

Since the branch points are simple turning points, this is not a novel problem and the details can be found elsewhere[17,21]. The appropriate denominations of F is (10) on D_1 and D_2 are related by $F_2 = e^{2\pi i} F_1$ and the corresponding "action function" branches of S, by $e^{-\pi i} S_2 = \Lambda - e^{\pi i} S_1$, where the phase integral

(19) $\Lambda = \int_{x_1}^{x_2} [E - \{c_t(x)\}^2 - V(x)]^{1/2} dx > 0$

is obtained during the real Geometrical Optics computation for the well[13]. The respective normalization constants are then related by

$$A_2^* = 2A_1^* cos(\Lambda/\epsilon) + iB_1^* \exp(i\Lambda/\epsilon),$$
(20)
$$B_2^* = iA_1^* \exp(-i\Lambda/\epsilon).$$

In the first place, their connection is at some chosen fixed y, but its results depends only on Λ and (9), (19) show 2Λ to be the total change in S along a path around the real well-component of the toroid, which is a homological invariant[12] of that toroid component and hence, cannot depend on the choice of y. Admittedly, since the domain boundaries $y = b_i(x)$ are curved, the domain may include values of y at which a straight path cannot be traced from barrier to barrier across a well, but this difficulty can be eliminated simply by extending the trajectories beyond the domain boundaries[28].

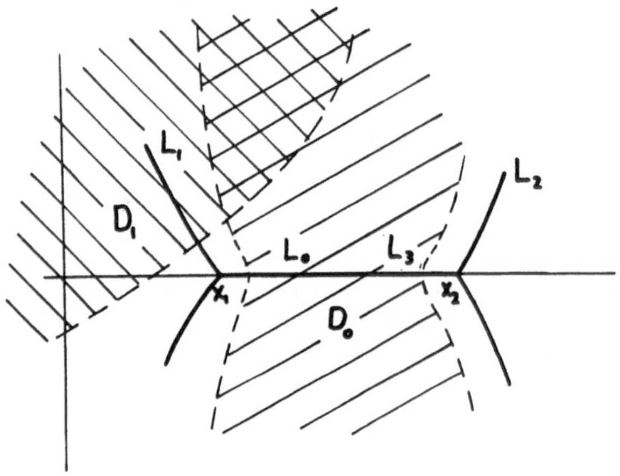

Fig. 4. Stokes lines L_k and approximation domains D_0, D_1 in the complex plane of x at fixed real y. The branch points x_i are numbered as in Fig. 1.

It is relevant that this analysis serves only to construct a rigorously founded approximation chain from the interior D_1 of one barrier across a well to the interior D_2 of the next barrier, but is quite independent of any issues relating to tunneling or symmetries.

7. EBK RULE FOR A DOUBLE WELL

The preceding sections have concerned key issues that need to be resolved in order to carry transcendental asymptotics to genuine partial differential equations. It is desirable to show that this is also essentially sufficient, for instance, for the example of Section 3. To characterize (by contrast to computing) the eigenfunctions of a double well (Fig. 1), it remains to apply the radiation condition (13) and symmetry conditions (15) to the approximation chain summarized by (19), (20).

In the far field $x > x_2$ (Figs.1,4), $Im\ S_2$ turns out [8] to be negative on D_2 to the right of L_2, so that ϕ_2^{out} in (18) is not compatible with (13). A nontrivial solution must therefore have

$$(21) \qquad A_2^* = 0, \quad B_2^* \neq 0,$$

still quite independently of the symmetry condition (15). For a single well, the same radiation condition (13) would apply also the left of the well (Fig. 1) and would demand similarly that $B_1^* = 0$, whence it would follow from (20) that

$$(22) \qquad \cos(\Lambda/\epsilon) = 0.$$

Together with (16), (22) is equivalent to the familiar Bohr-Sommerfeld quantization rule for a single well in two dimensions because (19) and (22) together read

$$(23) \qquad \frac{1}{\epsilon}\int_{x_1}^{x_2}[E - \{c_t(x)\}^2 - V(x)]^{1/2}dx + \mu_1 + \mu_2 = n\pi, \qquad n = 0, 1, 2, \ldots,$$

where $\mu_1 = \mu_2 = -\pi/4$ are the Keller-Maslov indices[12–16].

The application of the symmetry conditions (15) raises an obstacle in nonseparable wave problems because the only convenient access to eigenfunctions is by integration of the canonical equations along trajectories. However, when the real trajectories over a well are continued into the central barrier (Figs.2,3) they fail to intersect the line $x = 0, y$ real, at which the symmetry conditions (15) are specified. To obtain approximations (18) on that line would require complex geometrical optics computations of greater difficulty and much more labor than are involved in the real geometrical optics computation for a well. The obstacle can be overcome[8], however, by applying (15) at the complex point $(0, y(0; c_t))$ where the complex continuation of a real well-trajectory intersects the complex plane $x = 0$. The reason is that the complex toroid component over the right half of the central barrier (Fig. 2) of a symmetrical double well is necessarily accompanied by a symmetrical complex toroid component over the left half of that barrier (Figs. 3,5). Hence, if an approximation on a righthand trajectory component labeled c_t (Fig.5) satisfies (15a) at $(0, y(0; c_t))$, then this approximation possesses a continuous extension odd in x to the mirror image (labeled $-c_t$) of the trajectory component found over the left half of the central barrier (Fig. 5). The lefthand complex trajectory, in turn, carries this antisymmetry into the left well (Figs.1,5). In sum, application of (15a) at the complex point

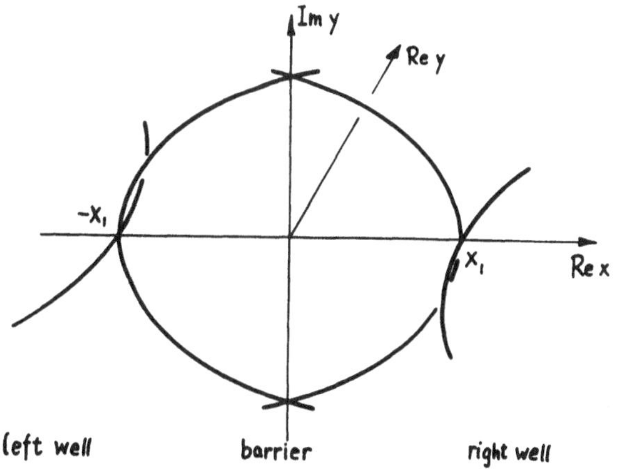

Fig. 5. Real and complex branches of a ray over Re $x > 0$ and their symmetrical counterparts over Re $x < 0$ in the space of Re x and of y.

$(0, y(0; c_t))$ of a trajectory guarantees a wave function approximation in x on that trajectory for $\mathrm{Re}x > 0$ and its mirror image for $\mathrm{Re}x < 0$. The condition (15a) turns out to be not the really relevant one for the antisymmetry.

When this fact is quantified carefully[8] for (12) and (18), the necessary and sufficient condition for antisymmetry of the approximation is seen to be

$$(24a) \qquad B_1^*/A_1^* = -\exp[-\Gamma(c_t)/\epsilon]$$

with

$$(25) \qquad \Gamma(c_t) = 2\int_0^{x_1(c_t)} |F(x; c_t)|^{1/2} dx > 0,$$

if $x = x_1(c_t)$ at the first righthand branch point of the trajectory (Figs. 1-5). The condition for even eigenfunctions is more delicate because the values of ∇S on a trajectory component and its mirror image do not match at $(0, y(0; c_t))$. The symmetry can be used, nonetheless, to show[8] that the necessary and sufficient condition for an even approximation is

$$(24s) \qquad B_1^*/A_1^* = +\exp[-\Gamma(c_t)/\epsilon].$$

In the first place, $\Gamma(c_t)$ in (25) is an integral with respect to the parameter x on a trajectory branch at fixed c_t. By (12), however,

$$2i\Gamma(c_t) = \int \nabla S \cdot ds$$

on a closed path $\gamma(c_t)$ tracing first the right upper trajectory branch (Fig. 5) and its mirror image and then the lower image branch and its righthand prototype once around the whole composite barrier toroid. Hence, $\Gamma(c_t)$ is an homological invariant of the torus and cannot depend on the trajectory and its label c_t. Furthermore, Γ is much easier to compute than the more familiar phase integral Λ in (19). Once a branch point x_1 on the lefthand border of the well is chosen and the value of c_t at that point is noted from the real geometrical optics of the well, then (10) makes (25) a trivial, real quadrature.

When (20), (21) and (24a,s) are combined,

$$(26a) \qquad 2e^{-i(\Lambda/\epsilon+\pi/2)} \cos\frac{\Lambda}{\epsilon} = e^{-\Gamma/\epsilon}$$

and

$$(26s) \qquad 2e^{-i(\Lambda/\epsilon+\pi/2)} \cos\frac{\Lambda}{\epsilon} = -e^{-\Gamma/\epsilon}$$

result, and together with (16), these are the basic EBK quantization[14,16] rules for the double well in two dimensions. When Γ/ϵ is at least moderately large, they are approximated also by

$$(27^1_s) \qquad \begin{aligned} &\Lambda/\epsilon + \mu_0 + \mu_1 + \mu_2 = n\pi, \quad n = 0, 1, 2, \ldots \\ &\mu_0 = \mp\exp(-\Gamma/\epsilon), \quad \mu_1 = \mu_2 = -\pi/4, \end{aligned}$$

respectively, with the upper sign in μ_0 for antisymmetrical eigenfunctions and the lower, for symmetrical ones; Λ is the familiar phase integral (19) and μ_1, μ_2 are the Keller-Maslov indices present already in (23), but μ_0 is a novel index representing the transcendental effect of the tunneling interaction between the two wells in the quantization rule. To the same order of approximation, the new rules can be written still more explicitly for the eigenvalues $E_{n,m,a}$ and $E_{n,m,s}$ of the odd and even eigenfunctions, respectively, as

$$(28a) \qquad \int_{x_1}^{x_2} |E_{n,m,a} - \{c_t(x)\}^2 - V(x)|^{1/2} dx = (n + \tfrac{1}{2})\pi\epsilon + \tfrac{1}{2}\epsilon e^{-\Gamma/\epsilon},$$

(28s) $\int_{x_1}^{x_2} |E_{n,m,s} - \{c_t(x)\}^2 - V(x)|^{1/2} dx = (n + \frac{1}{2})\pi\epsilon - \frac{1}{2}\epsilon e^{-\Gamma/\epsilon}$,

with

$$(\Gamma = 2\int_0^{x_1} |E_{n,m} - \{c_t(x_1)\}^2 - V(x)|^{1/2} dx,$$

where $E_{n,m}$ are the eigenvalues of the single well obtained from (16), (23) or any other preferred method for the computation of single-well eigenvalues.

8. PROOF AND APPROXIMATION

Asymptotics beyond all orders may involve delicate issues which make a proof of rigorous reliability desirable. In the present case, it can be based on Maslov's proof[18,19] that the WKB approximation on appropriate domains is a rigorous approximation even for linear partial differential euqations. It follows from the present analysis that (16), (26a,s) are also approximations in a rigorous sense. However, they lack error bounds, which leaves open whether the error on the lefthand sides of (26a,s) might not be larger than the whole righthand sides of those equations? It is not, in fact, possible to guarantee that the present analysis predicts all the eigenvalues $E_{n,ma}$ and $E_{n,m,s}$ with greater accuracy than the eigenvalues $E_{n,m}$ of a single well. The salient point, however, is that the lefthand sides of (26a,s) have been obtained here without any reference to symmetry or tunneling. The error on the lefthand sides is therefore the same in both (26a) and (26s), and when those two equations are subtracted from each other, that error cancels out exactly. What the analysis guarantees, therefore, are a rigorous approximation for the pair level $\frac{1}{2}(E_{n,m,a} + E_{n,m,s}) = E_{n,m}$ of the near-degenerate eigenvalue pairs and also a rigorous approximation for their split

(29) $E_{n,m,a} - E_{n,m,s} = \epsilon \exp(-\Gamma/\epsilon)$

But, those are just the two quantities which can be observed directly.

This caution is most relevant at the very lowest values of n, if $E_{n,m}$ is obtained from the semiclassical approximation, which is not highly accurate at those eigenlevels. By contrast, the split is predicted with very high accuracy indeed at just those levels. The reason is that the central barrier is there wide (Fig.1) and the symmetry conditions are applied at quite a large distance from the real branch points of S. In most practical cases, the potential generates further singularities of S, if any, only at large distances from real position space and then the double-asymptotic property[21] can be relied on to make the present approximation to (15a,s) quite accurate even if ϵ in (6) is not small. In such practical cases, therefore, the analysis of the eigenvalue split goes well beyond a semiclassical analysis!

As n increases, the semiclassical prediction of single-well eigenvalues gains accuracy rapidly and then the present analysis comes soon to give a good approximation even for $E_{n,m,a}$ and $E_{n,m,s}$ individually. This is useful for experiments involving a perturbation of the strict symmetry of a double well because the analysis then continues to apply, albeit at the price of a real geometrical optics computation for both wells.

However, the split increases with n and at levels of $E - c_t^2$ close to the maximum of the potential (Fig. 1), the split is so wide that the first approximation (29) to it may cease to be adequate and a modification of the analysis for barrier-top effects may become desirable. At levels above the maximum of the potential, the present analysis does not apply at all, but single-well results become soon accurate.

9. GENERALIZATIONS

While the special example has helped to keep the analysis concrete and explicit, the line of analysis is clearly not restricted to it. For instance, if the two-dimensional strip domain is replaced by a three-dimensional domain in a duct of rectangular cross-section, the analysis remains virtually unchanged. It applies equally to waves in a duct of more general cross-section, but the standard geometrical optics of the well is then more laborious. In double wells with more than one maximum of the potential, each maximum and minimum generates a transition in the spectral structure because the number of caustic branch manifolds changes across such levels of $E - c_t^2$. Except close to such levels, the analysis applies at the expense of a straightforward modification to account for connection across the additional wells and barriers[8]. This results in more elaborate quantization rules reflecting the changes in spectral structure.

The analyticity of the potential and domain shape has been specified to avoid diversion from the main topic, but has been used only in the immediate neighborhood of the caustic branch manifolds and even there, only to focus the discussion on the essence of the still unfamiliar approach to the resolution of the Stokes phenomena[26]. The connection method here used to achieve convincing "exponential precision" in nonseparable problems in several dimensions does not require analyticity at branch manifolds either, it copes also with highly irregular singular points[27] of the canonical equations.

The sidewalls of the wave domain have been introduced only to emphasize that the analysis applies also to classical problems despite the term "quantization rule", which has been carried to classical analysis by Keller[14] because there is no classical name for this ultimate objective of concrete, quantitative spectral analysis. Domain boundaries are not always germane and potentials may depend on several coordinates. The double-well analysis still applies, if the potential is even in one coordinate and grows sufficiently in the whole far field to assure adequate radiation conditions. The Keller indices for boundary conditions are then replaced by Keller-Maslov indices for the lateral caustics. The case of a potential even in more than one coordinate, so that it has quadruple or sextuple well structure, is also covered; a novel index μ_0 then arises in more than one of the equations spelling out the EBK rule to predict the more complicated eigenvalue split. The main difference to be anticipated with a potential dependent on several coordinates is that no single position coordinate x can serve as a particularly helpful parameter along trajectories. Instead, the characteristic function S will be found normally to be the most suitable such parameter. Indeed, it is the best parameter even in the simple example of Section 3 and has been avoided here only in order not to burden the discussion by an unfamiliar transformation.

A major motivation for the choice of the present approach has been that much of it is not restricted to linear wave equations. In particular, the use of real and complex geometrical optics is not so restricted, the canonical equations of the Hamiltonian operator are nonlinear is any case. Moreover, the strictly local connection method for the resolution of the Stokes phenomena remains applicable as long as a Frechet derivative is available almost everywhere on the branch manifolds. On the other hand, the details of the approximands here used, and their linear superposition, are peculiar to linear differential equations. Little appears known yet about good ways to approximate the transport equations in nonlinear problems.

10. PERIODIC POTENTIALS

It appears desirable to emphasize the flexibility of the method here sketched for carrying asymptotics beyond all orders to genuine partial differential equations further by showing that it can resolve degeneracies of continuous spectra as well as those of discrete spectra. The most prominent example arises for waves in periodic potentials relevant for crystal physics. The absence of complete far-field decay then rules out square integrable eigenfunctions, but does not preclude bounded solutions qualifying for a continuous spectrum.

For linear ordinary differential equations, waves in such potentials have been explored fully on the basis of Floquet's theorem[29] which implies in the generic case that the spectrum is continuous and has banded structure[30]. Numerical work on Mathieu's equation indicated that the bands can be very narrow and Harrell[31] gave the first proof that a degeneracy of exponential type occurs in the semiclassical case. A major step was taken by Weinstein and Keller[32], who used hard analysis with special functions based on Floquet's theorem to obtain a concrete, quantitative approximation for the bands and gaps of the spectrum of Hill's equation in the semiclassical limit. An analysis with special functions cannot be used for nonseparable partial differential equations, but those functions can be eliminated by means of the one-dimensional special case of the analysis sketched in the preceding sections. That simplifies the theory and makes it more flexible, but it still remains based totally on Floquet's theorem. Since no analog of that theorem for nonseparable partial differential equations appeared to be known, a different approach had to be found.

Shen[33] first showed that EBK quantization can be highly effective also for continuous spectra and the Author teamed up with him to apply the method sketched in the preceding sections to periodic potentials[28]. To fix the ideas, it is again helpful to begin with the simple example of Section 3, but now with a potential $V(x)$ and side-walls $y = b_i(x)$ periodic in x, with minimal period π, say, instead of the symmetry assumed in Section 3. Once this example has been treated, some of the generalizations in Section 9 will become obvious, but not all of them apply; in particular, this part of the theory has not yet reached partial independence of the shortwave parameter and the following is therefore based on the assumption that $\epsilon << 1$ in (6).

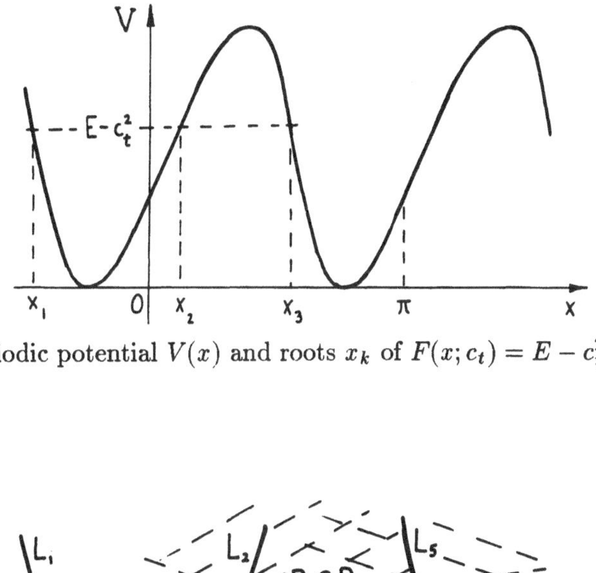

Fig. 6. Periodic potential $V(x)$ and roots x_k of $F(x; c_t) = E - c_t^2 - V(x)$.

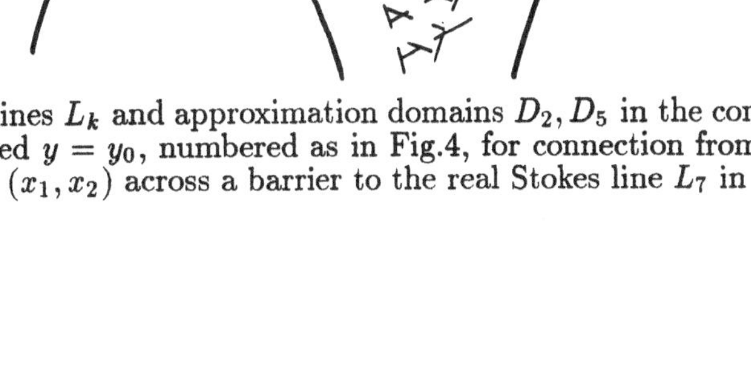

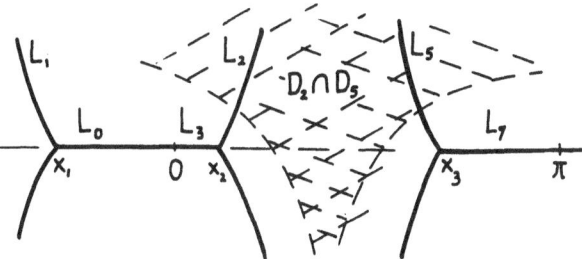

Fig. 7. Stokes lines L_k and approximation domains D_2, D_5 in the complex plane of x at fixed $y = y_0$, numbered as in Fig.4, for connection from the real well interval (x_1, x_2) across a barrier to the real Stokes line L_7 in the next well.

It is more convenient now to construct an approximation chain from the interior of a well across an adjacent barrier to the interior of the next well (Fig. 6) and to do so again at some generic real value y_0 of y. That involves different toroids (Fig. 3) already at this early stage, but otherwise, the connection analysis[28] is analogous to that in Section 6 and involves no reference at all to any periodicity. Figure 7 shows the complex plane of x at $y = y_0$, labeled as in Fig. 4; the approximation chain must now reach from the domain D_3 based on the real Stokes line L_3 via D_2 and D_5 to the domain D_7 based on the real Stokes line L_7. If If the origin of x is chosen somewhere in D_3 (Fig. 6) and x is scaled so that $x = \pi$ lies somewhere in D_7 (Fig. 6), then the result of the approximation calculation [28] is

$$\begin{aligned} &\begin{pmatrix} \phi_3^{\text{out}}(0) \\ \phi_3^{\text{in}}(0) \end{pmatrix} = i|F(0)|^{-1/4} J(0)^{-1/2} \begin{pmatrix} A_3^* \exp(iS_3(0)/\epsilon) \\ B_3^* \exp(-iS_3(0)/\epsilon) \end{pmatrix}. \\ (30) \qquad &\begin{pmatrix} \phi_7^{\text{out}}(\pi) \\ \phi_7^{\text{in}}(\pi) \end{pmatrix} = |F(\pi)|^{-1/4} J(\pi)^{-1/2} \begin{pmatrix} A_7^* \exp(iS_7(0)/\epsilon) \\ B_7^* \exp(-iS_7(0)/\epsilon) \end{pmatrix}. \end{aligned}$$

with

$$(31) \qquad S_3(0) = \int_0^{x_2} |F|^{1/2} dx > 0, \quad S_7(\pi) = \int_{x_3}^{\pi} |F|^{1/2} dx > 0,$$

$$\begin{aligned} &A_7^* = (e^{\Gamma/\epsilon} - e^{-\Gamma/\epsilon}) A_3^* + i e^{\Gamma/\epsilon} B_3^*, \\ (32) \qquad &B_7^* = i e^{\Gamma/\epsilon} A_3^* - e^{\Gamma/\epsilon} B_3^*, \end{aligned}$$

where

$$(33) \qquad \Gamma = iS_2(x_3) = \int_{x_2}^{x_3} |F|^{1/2} dx > 0$$

in analogy to (25).

As in Section 6, the possiblity of extending trajectories across the sidewalls assures that there is no restriction on the values of y_0 at which this approximation chain is available. For $\epsilon << 1$, moreover, if (30) approximate a solution ψ of (6), then from (9), $\partial\psi/\partial x$ is approximated[18,19] by

$$\begin{aligned} &\begin{pmatrix} \phi_3^{\text{out}\dagger}(0) \\ \phi_3^{\text{in}\dagger}(0) \end{pmatrix} = \tfrac{1}{\epsilon} |F(0)|^{1/4} J(0)^{-1/2} \begin{pmatrix} A_3^* \exp(iS_3(0)/\epsilon) \\ -B_3^* \exp(-iS_3(0)/\epsilon) \end{pmatrix}, \\ (34) \qquad &\begin{pmatrix} \psi_7^{\text{out}\dagger}(0) \\ \phi_7^{\text{in}\dagger}(0) \end{pmatrix} = \tfrac{1}{\epsilon} |F(\pi)|^{1/4} J(0)^{-1/2} \begin{pmatrix} A_7^\dagger \exp(iS_7(\pi)/\epsilon) \\ -B_7^* \exp(-iS_7(\pi)/\epsilon) \end{pmatrix}. \end{aligned}$$

11. STABLE TOROIDS

If the periodicity of the potential and domain are now introduced, one may ask Floquet's question whether there are approximations to solutions ψ of (6) with the property

$$\psi(x + \pi, y_0) = \rho(y_0)\psi(x, y_0)$$

for all x in the lefthand well (Fig. 5), at least at one y_0. Since the origin may be identified with any such x, the property implies

$$\psi(\pi) = \rho\psi(0), \quad \frac{\partial\psi}{\partial x}(\pi) = \rho\frac{\partial\psi}{\partial x}(0),$$

where reference to y_0 is suppressed for brevity. In terms of (30), (34), this reads

$$
(35) \qquad \begin{aligned} \alpha A_7^* e^{iS_7/\epsilon} + \alpha B_7^* e^{-S_7/\epsilon} &= i\rho A_3^* e^{iS_3/\epsilon} + i\rho B_3^* e^{-iS_3/\epsilon} \\ \beta A_7^* e^{iS_7/\epsilon} - \beta B_7^* e^{-iS_7/\epsilon} &= -i\rho A_3^* e^{iS_3/\epsilon} + i\rho B_3^* e^{-iS_3/\epsilon} \end{aligned}
$$

where S_3 and S_7 are understood at $(0, y_0)$ and (π, y_0), respectively, and

$$
(36) \qquad \binom{\alpha}{\beta} = [J(0)/J(\pi)]^{1/2} \binom{|F(0)/F(\pi)|^{1/4}}{|F(\pi)/F(0)|^{1/4}} .
$$

Since (32), (35) are a linear homogeneous system for $A_3^*, \ldots, B_7^*$, the condition for an approximation with the Floquet property is that their determinant vanishes. This can be checked numerically in any specific case from the geometrical optics computation, but general predictions about S_3, S_7, α and β are difficult because integrability tells us only that the trajectories wind around the toroids, but not where the rays through $(0, y_0)$ arrive at $x = \pi$. In any case, however, the determinant of (32), (35) is quadratic in ρ, in some analogy to Floquet's characteristic determinant[30].

A partial grip on the determinant can be obtained by studying the implications of the following notion[28].

Definition. A toroid will be called "stable at y_0" if the determinant has two distinct roots ρ_1, ρ_2 such that $|\rho_1| = |\rho_2| = 1$.

This means that the approximation can be phase-shifted over one period, but that its magnitude at $y = y_0$ is conserved over any number of periods. The solution ψ so approximated must then remain bounded over any finite number of periods, at least at $y = y_0$. Such stability implies that the term independent of ρ in the determinant of (32), (35) must have unit magnitude (when the determinant is scaled to make the coefficient of ρ^2 equal to unity). Furthermore, ρ_1 and ρ_2 must be complex conjugates and the coefficient of ρ in the determinant must be real. That is found[28] to imply that $\alpha = \beta = 1$ and that $J(\pi) = J(0)$ and hence from (36) and (10), $F(\pi) = F(0)$ and $c_t(\pi) = c_t(0)$, so that the toroid is geometrically periodic at $\mathrm{Re}\, y = y_0$ (apart from the complex shift in Fig.3). In addition, the stability condition $\alpha = \beta = 1$ reduces the determinant of (32), (35) to

$$
-2\rho^2 + 4\rho e^{\Gamma/\epsilon} \cos \tfrac{\Lambda}{\epsilon} - 2
$$

where

$$
\Lambda = \int_{x_3 - \pi}^{x_2} |F|^{1/2} dx > 0
$$

is the same as in (19). The condition for stability at y_0 is therefore

$$
(37) \qquad \rho^2 - 2\rho e^{\Gamma/\epsilon} \cos \tfrac{\Lambda}{\epsilon} + 1 = 0,
$$

which has the same stucture as Floquet's characteristic equation[30], but now depends on y_0.

Theorem. A toroid is stable at one y_0 if and only if it is stable at all y_0.

The proof follows from (37) by noting, as in Sections 6, 7, that Λ and Γ are homological invariants of the real and complex toroid components, respectively, which cannot depend on the value of $\mathrm{Re}\, y$ at which they are evaluated.

It follows that (37) is quite analogous to Floquet's characteristic equation[30] and indeed, in the one-dimensional special case of the present analysis, (37) is identical with the approximate characteristic equation of Weinstein and Keller[32]. A corollary of the theorem is therefore that *a theorem quite analogous to Floquet's basic theorem applies to the approximation on each individual toroid.* It is only a result on the approximation, rather than a theorem on the exact solutions, and it is of narrow scope, but it does give a strong clue to the direction in which an analog for partial differential equations of Floquet's theorem may be found.

12. SPECTRUM

It should be recalled from Section 4 that admissible toroids are characterized by a value E of the Hamiltonian common to all trajectories and by the invariant (16) with $m = 0, 1, 2, \ldots M(\epsilon)$. Boundedness of the approximation and, to a degree at least, of the corresponding exact solutions has been shown in the preceding section to require (37) with complex conjugate roots ρ_1, ρ_2 of unit magnitude. This suggests the name "approximate spectrum" of a toroid for the set of values of E for which (37) has such roots, because boundedness of eigenfunctions defines a continuous spectrum. The characteristic equation (37) admits also distinct real roots ρ, but one of those must have $|\rho| > 1$ so that the corresponding approximation is unbounded. The approximate spectrum of a toroid must therefore have banded spectrum, and for $\epsilon << 1$, the bands must be narrow because $\Gamma > 0$.

The transition from real to complex roots of (37) occurs where the roots coincide, i.e., where they are $\rho_+ = 1$ or $\rho_- = -1$, and the corresponding values $E = E^+$ and $E = E^-$ are spectral cut-offs of the toroid. By (37), they are given by

$$(38) \quad \begin{aligned} &\cos(\Lambda/\epsilon) = \exp(-\Gamma/\epsilon) \quad \text{for} \quad \rho_+ = -1, \\ &\cos(\Lambda/\epsilon) = -\exp(-\Gamma/\epsilon) \quad \text{for} \quad \rho_- = -1, \end{aligned}$$

which together with (16) are the EBK rule for the cut-offs. The resemblance of (38) to (26a,s) may be noted.

To cast this approximation for the cut-offs in an explicit from, recall from (22) that $\Lambda = (n + \frac{1}{2})\pi\epsilon$ is the EBK rule giving the semiclassical approximation $E_{n,m}$ to the eigenvalues of the single well when the interaction between different wells of the periodic potential is ignored. For $\epsilon << 1$, the cut-offs determined by (38) are

$$(39) \quad \begin{aligned} &E^+_{n,m} \sim E_{n,m} - (-1)^n \epsilon e^{-\Gamma/\epsilon}/\Lambda'(E_{n,m}), \\ &E^-_{n,m} \sim E_{n,m} + (-1)^n \epsilon e^{-\Gamma/\epsilon}/\Lambda'(E_{n,m}), \end{aligned}$$

where

$$\Lambda'(E) \equiv \partial\Lambda/\partial E = \tfrac{1}{2}\int_{x_1}^{x_2} |E - \{c_t(x)\}^2 - V(x)|^{-1/2} dx.$$

Since E cannot be in the spectrum of the m-th toroid when $E - c_t^2 < \min V(x)$ for all c_t on that toroid, the stability bands are

$$(40) \quad (E^+_{0,m}, E^-_{0,m}), (E^-_{1,m}, E^+_{1,m}), (E^+_{2,m}, E^-_{2,m}), \ldots$$

(In the standard notation,[30,32] $E^+_{n,m}$ is called $E_{n,m}$ and $E^-_{n,m}$ is called $E'_{n+1,m}$.) Each stable toroid has therefore an approximate spectrum of structure analogous to that of Hill's equation.[30] The bands here established are finite in number, however, because the present analysis focuses only on levels

$$E < \max V(x) + c_t^2 \quad \text{for all} \quad c_t \quad \text{on the toroid.}$$

The spectrum (6) for the periodic potential on a periodic strip is the union of the toroid spectra and is more difficult to discuss because it depends on too many parameters, namely ϵ, n, m, the potential and the domain shape. This helps to explain some of the difficulties encountered by Floquet theory for nonseparable partial differential equations. The only general inference emerging from a closer look[28] at the dependence on the parameters is twofold: First, the bands for the lowest values of n are so narrow that those for all $m \leq M(\epsilon)$ may fit without overlap into the gap between $(E_{0,0}^{+}, E_{0,0}^{-})$ and $(E_{1,0}^{-}, E_{1,0}^{+})$. Secondly, the bands widen as n increases and those for different m must soon begin to overlap. Most of the spectrum must therefore be expected to consist of wide bands with few gaps, if any, but with *many cut-offs*.

In any one specific problem, of course, all the details of the spectrum can be computed from the EBK rules (38) or (39). In particular, the last gap in the spectrum may be pin-pointed in that way.

ACKNOWLEDGEMENT

These investigations were supported in part by the National Science Foundation under Grants DMS-8521687 and DMS-8903083.

REFERENCES

1. J. E. Littlewood, Lorentz's pendulum problem, Ann. Physics 21:233 (1963).
2. R. E. Meyer, Adiabatic variation, Part I, Exponential property for the simple oscillator, J. Appl. Math. Phys. (ZAMP) 24:293 (1973); Part II, Action change for the simple oscillator, ibid: 517.
3. W. Wasow, Adiabatic invariance of a simple oscillator, SIAM, J. Math. Anal. 4:78 (1973).
4. R. E. Meyer, Adiabatic variation, Part V, Nonlinear near-periodic oscillator, J. Appl. Math. Phys. (ZAMP) 27:181 (1976).
5. R. E. Meyer, Gradual reflection of short waves, SIAM J. Appl. Math. 29:481 (1975).
6. C. Lozano and R. E. Meyer, Leakage and response of waves trapped by round islands, Phys. Fluids 19:1075 (1976).
7. R. E. Meyer, Quasiresonance of long life, J. Math. Phys. 27:238 (1986).
8. R. E. Meyer, On exponential asymptotics for nonseparable wave equations, Part II, EBK quantization, SIAM J. Appl. Math., to appear.
9. R. T. Skodje, F. Borondo and W. P. Reinhardt, The semiclassical quantization of nonseparable systems using the method of adiabatic switching. J. Chem. Phys. 82:4611 (1985).
10. R. T. Skodje and F. Borondo, On the use of adiabatic switching to locate quantized periodic orbits: Application to bound and reactive multidimensional problems, J. Chem. Phys. 84:1533 (1986).
11. C. Jaffé, Time-independent adiabatic switching in quantum mechanics, J. Chem. Phys. 86:4499 (1987).
12. J. B. Keller, Corrected Bohr-Sommerfeld quantum conditions for nonseparable systems, Annals Phys. 4:180 (1958).
13. J. B. Keller and S. I. Rubinow, Asymptotic solution of eigenvalue problems, Annals Phys. 9:24 (1960).
14. J. B. Keller, Semiclassical mechanics, SIAM Rev. 27:485 (1985).
15. V. P. Maslov, "Theorie des Perturbations," Dunod, Paris (1972).
16. I. C. Percival, Semiclassical theory of bound states, Adv. Chem. Phys. 36:1 (1977).
17. R. E. Meyer, On exponential asymptotics for nonseparable wave equations, Part I, Complex geometrical optics and connection, SIAM J. Appl. Math., to appear.
18. V. P. Maslov, The scattering problem in the quasiclassical approximation, Dokl. Acad. Nauk. SSSR 151:306 (1963) (Soviet Phys. Dokl. 8:666 (1964)).

19. V. P. Maslow, Nonstandard characteristics in asymptotic problems, Uspekhi Mat. Nauk. 38:6:3 (1983) (Russ. Math. Surv. 38:6:1 (1983)).
20. R. E. Langer, On the asymptotic solution of ordinary differential equations, Trans. Amer. Math. Soc. 33:23 (1931).
21. F. W. J. Olver, "Asymptotics and Special Functions," Academic Press, New York (1974).
22. R. E. Meyer, A simple explanation of the Stokes phenomenon, SIAM Rev. 31:435 (1989).
23. A. Zwaan, Intensitaeten im Ca-funkenspektrum. Arch. Neerland. Sci. Exactes *Natur.* 3A 12:1 (1929).
24. Yu. A. Kravtsov, A modification of the geometrical optics method, Radiofizika 7:664 (1964).
25. D. Ludwig, Uniform asymptotic expausions at a caustic, Commun. Pure Appl. Math. 19:215 (1966).
26. R. E. Meyer and J. F. Painter, Connection for wave modulation, SIAM J. Math. Anal. 14:450 (1983).
27. R. E. Meyer, Irregular points of modulation, Adv. Appl. Math. 4:145 (1983).
28. R. E. Meyer and M. C. Shen, "On exponential asymptotics for nonseparable wave equations, Part III, Approximate spectral bands of periodic potentials on strips," SIAM J. Appl. Math. to appear.
29. G. Floquet, Sur les equations differentielles lineaires a coefficients periodiques, Ann. Ecole Norm. ser. 2, 12:47 (1883).
30. W. Magnus and S. Winkler, "Hill's Equation," Wiley-Interscience, New York (1966).
31. E. M. Harrell, The band structure of a one-dimensional, periodic system in a scaling limit, Ann. Physics 119:351 (1979).
32. M. I.; Weinstein and J. B. Keller, Hill's equation with a large potential SIAM J. Appl. Math. 45:200 (1985).
33. M. C. Shen, R. E. Meyer and J. B. Keller, Spectra of water waves in channels and around islands, Phys. Fluids 11:2289 (1968).

PROBLEMS OF EXISTENCE OF NONTOPOLOGICAL SOLITONS (BREATHERS) FOR NONLINEAR KLEIN-GORDON EQUATIONS*

Vladimir Eleonsky

Lukin's Physical Research Institute
Moscow, USSR

1. The main goal of the present lecture is to present the results obtained by our group on investigations of the soliton-type states for nonlinear wave equations of the form

$$u_{tt} - u_{xx} - g(u) = 0,\ g(0) = 0,\ g_u(0) < 0\ , \qquad 1.1$$

and to discuss a number of unsolved problems of common interest. In spite of the fact that historically, investigations of these problems were led by analysis of corresponding asymptotic expansions, I think that a quantitative theory of dynamical systems must be the basis for a correct interpretation of asymptotic and numerical-analysis results. In the lecture only references to English-language publications of our group are presented.

As a first problem let us consider the problem of existence of nontopological solitons (breathers) for equation (1.1) and for nonlinearity of the kind

$$g(u) = -u + G(u),\ G(u) \sim 0(u^2) \quad \text{as} \quad u \to 0. \qquad 1.2$$

For a clearer statement of our results, I will address my attention to the common case of polynomial nonlinearity, assuming $G(u) = G_N(u,\ C)$, where $G_N(u,\ C)$ is a polynomial of degree $N \geq 2$ with arbitrary coefficients C, and to the simplest case $G(u) = u^3$.

So, for an equation (1.1) let us consider a problem of existence of soliton-type solutions that satisfy

$$\lim_{x\to\pm\infty} u(x,t) = 0,\ u(x, t + \frac{2\pi}{\omega}) = u(x,t) \qquad 1.3$$

for some continuous set of ω values.

Several approaches to solve this problem are possible. Let us exclude an "evolutional" approach, related to an analysis of the corresponding Cauchy problem. Such a decision was made on the basis of reasoning close to that stated by J. Boyd (1990).

The next step we made was to identify a dynamical system related to problem (1.1), (1.3). A choice of such a system is not unique. Let us consider further a dynamical system (more exactly, a one-parameter set of hamiltonian systems) generated by a class of solutions of equation (1.1) that are periodic with respect to t:

$$v_X = -\delta H/\delta u, \qquad u_x = \delta H/\delta v;$$

*** This paper was edited without the author's knowledge after repeated attempts to contact the author failed.**

Asymptotics beyond All Orders, Edited by H. Segur *et al.*
Plenum Press, New York, 1991

$$H = \int_{-\pi}^{+\pi} d\phi \left(\frac{1}{2} v^2 + \frac{1}{2} \omega^2 u_\phi^2 - \frac{1}{2} u^2 + U(u) \right) ; \qquad 1.4$$

$$\phi = \omega t, \quad \frac{dU}{du} = G(u)$$

The hamiltonian system (1.4) is defined on an infinite-dimensional phase space $\Gamma_\infty(\omega)$ of pairs of functions ($v = u_x$, u) periodic in t with period $2\pi/\omega$. For the case $G(-u) = -G(u)$, one can pick out a subclass of t-periodic solutions admitting a Fourier expansion

$$u(x,t) = \sum_{n \geq 0} u_{2n+1}(x) \cos(2n+1)\phi . \qquad 1.5$$

This subclass of solutions corresponds to an invariant manifold, on which the hamiltonian system (1.4) permits the following representation:

$$\frac{dv_{2n+1}}{dx} = -\frac{\partial H}{\partial u_{2n+1}}; \quad \frac{du_{2n+1}}{dx} = \frac{\partial H}{\partial v_{2n+1}}; \qquad 1.6$$

$$H = \frac{1}{2} \sum_{n \geq 0} \{ v_{2n+1}^2 - [(2n+1)^2 \omega^2 + 1] u_{2n+1}^2 \} + U(\ldots u_{2m+1} \ldots)$$

$$U(\ldots u_{2m+1} \ldots) = \int_{-\pi}^{+\pi} d\phi \, U(u);$$

In a particular case, for $G(u) = u^3$:

$$U(\ldots u_{2m+1} \ldots) = \frac{3}{8} (u_1^4 + u_3^4 + u_5^4 + \ldots) + \frac{1}{2} u_1^3 u_5 +$$
$$+ \frac{3}{2} (u_1^2 u_3^2 + u_1^2 u_5^2 + u_3^2 u_5^2 + \ldots) + \frac{3}{2} (u_1^2 u_3 u_5 + u_1 u_3^2 u_5 + \ldots) \qquad 1.7$$

The hamiltonian system (1.6) is defined on an infinite-dimensional phase space Γ $\{(v_1, u_1, \ldots, v_{2n+1}, u_{2n+1}, \ldots)\}$ with a countable basis. The next and principal step we made (V. Eleonsky et al, 1984, 1988) and suggested independently by A. Weinstein (1985), is a reduction of the problem of existence of nontopological solitons to the problem of existence in the corresponding phase space of trajectories which are doubly-asymptotic, as $x \to \pm\infty$, to a singular point $0(v = 0, u = 0)$ or $0(\ldots, v_{2n+1} = 0, u_{2n+1} = 0, \ldots)$ - a zero solution of (1.1). Using the terminology introduced by Poincaré, let us call such objects homoclinic loops of the singular point 0. For dynamical systems (1.4) and (1.6) one can formulate the following

Statement: The point 0 is a singular point of saddle-center type. The dimensions of its stable ($W_\omega^s(0)$) and unstable ($W_\omega^u(0)$) saddle manifolds are equal and finite for all $\omega^2 > 0$, while its center manifold is infinite-dimensional for all $\omega^2 > 0$.

For example, for the hamiltonian system (1.6), if

$$(2n+1)^{-2} < \omega^2 < (2n-1)^{-2} \quad , \quad n = 1, 2, \ldots \qquad 1.8$$

then its stable and unstable manifolds are each of dimension n.

Thus, homoclinic loops of 0, generated by intersections of finite -dimensional manifolds W_ω^s and W_ω^u, are images of nontopological solitons in Γ_∞. But this situation– intersection of finite-dimensional manifolds in an infinite-dimensional space– is not the case of general position. Since the number of free parameters, numbering the trajectories of the saddle manifolds, is finite, then conditions of existence of homoclinic loops of 0– conditions of a merging of semi-trajectories which are asymptotic to the singular point 0 as $x \to -\infty$ and as $x \to +\infty$– are essentially overdetermined and unsolvable in the case of arbitrary nonlinearity $g(u)$. In fact, further analysis of the problem on the basis of asymptotics beyond all orders (H. Segur, M. Kruskal, 1987), of a sequence of finite-dimensional models (V. Eleonsky et al, 1984, 1988) or of other approaches (for example, P.A. Vuillermot, 1986) can be reduced to answering the following question: Does the problem considered correspond to a case of general position of finite-dimensional objects (in this case, of saddle manifolds $W^S_\omega(0)$ and $W^u_\omega(0)$) in an infinite-dimensional phase space?

The investigations carried out by our group lead to the

Statement: A realization for nonlinearity $g(u)$ of a degenerate situation– merging of saddle manifolds of the singular point 0 and, correspondingly, solvability of problem (1.1), (1.3)– is a sign that the nonlinear wave equation (1.1) belongs to the list of completely integrable equations.

One must notice the following circumstance: for nonlinearity $g(u) = \sin u$, corresponding to the integrable case, it is known that the problem (1.1), (1.3) is solvable. Justification of this statement and its consequences was verified for all values $0 < \omega^2 < 1$ (V. Eleonsky et al, 1988). However, for nonlinearity $g(u) = \sinh\ u$, also corresponding to an integrable case, problem (1.1), (1.3) is not solvable. Is there a contradiction with the statement stated above? No, because a generalized problem

$$\lim_{x \pm\infty} u(x,t) = u^{\pm}_{\text{vac}}(t), \ u(x,\ t + 2\pi/\omega) = u(x,t) \qquad 1.9$$

is solvable. Here $u^{\pm}_{\text{vac}}(t)$, solutions of (1.1) that are homogeneous with respect to x, differ only by a shift in t. Corresponding to problem (1.1), (1.9), the solution is

$$u(x,t) = 4 \ \operatorname{arctanh} \left[k^{1/2} sn(\frac{t}{1-k}, k) \tanh(\Lambda x)\right],$$
$$v^{\pm}_{\text{vac}} = 4 \ \operatorname{arctanh} \left[k^{1/2} sn(\frac{t}{1-k}, k)\right], \ \Lambda^2 = \frac{4k}{(1-k)^2}, \qquad 1.10$$

where sn is the Jacobian elliptic sinus.

In phase space, a nonisolated set of singular points on a nonzero level of hamiltonian H corresponds to solutions $u_{\text{vac}}(t)$, and to the solution (1.10)– a homoclinic loop, belonging to merged saddle submanifolds of the set of nonisolated singular points (for details, see V. Eleonsky et al, 1987). Thus, one can exclude the nonlinearity $g(u) = \sinh\ u$ from a list of counterexamples (P.A. Vuillermot, 1986) to our statement. The main idea of it is related to the fact that a merging of finite-dimensional saddle manifolds of special objects (singular points, cycles and tori) in Γ_∞ of an arbitrary dynamical system is one of the signs of integrability of the nonlinear field equations. Concrete formulations of this statement are related to a class of special solutions (for example, (1.3), (1.9) and their generalizations). ¿From this point of view, the investigations of P.A. Vuillermot (1988, 1989) are very interesting. May one treat any formulation of our statement as proved (in terms of pure mathematics)? My answer is negative.

I have mentioned above that a choice of dynamical system corresponding with the problem (1.1), (1,3) is not unique. Indeed, let us consider first from conditions (1.3) the class of solutions which decrease rapidly as $x \to \pm\infty$. In the phase space $\tilde{\Gamma}_\infty$ of pairs of functions $(w = u_t, u)$ that are rapidly decreasing in x, equation (1.1) generates a hamiltonian system

$$w_t = -\delta H/\delta u, \quad u_t = \delta H/\delta w$$
$$H = \int_{-\infty}^{+\infty} dx \left(\frac{1}{2}w^2 + \frac{1}{2}u_x^2 + \frac{1}{2}u^2 - U(x)\right) \qquad 1.11$$

Images of nontopological solitons– solutions of the same problem (1.1), (1.3)– in $\tilde{\Gamma}_\infty$ are closed phase trajectories (cycles), belonging to center manifolds of singular points $0(w = 0,\ u = 0)$ in hamiltonian level H. The value of the constant H-level determines the period of the cycle $2\pi/\omega$. Consequently, images of the same essentially nonlinear object (nontopological soliton) of the field may be different objects of dynamical systems in $\tilde{\Gamma}_\infty$ and Γ_∞ (for example, homoclinic loops– in Γ_∞, and cycles– in $\tilde{\Gamma}_\infty$). Then descriptions of a physical object in different images ("hypostases") may have complementary character. We used this circumstance (G. Alfimov et al, 1990) in investigations of self-localized structures of a field based on a nonlinear equation of elliptic type

$$u_{xx} + u_{yy} + g(u) = 0 \quad (x,y) \in R^2 \qquad 1.12$$

Not having the possibility for a detailed statement of the results obtained by our group, I mention only that the problem of existence of solutions satisfying

$$\lim_{x\to\infty} u(x,y) = 0, \quad u(x, y + 2\pi/k) = u(x,y) \tag{1.13}$$

for equation (1.12) is solvable for a wide enough class of nonlinearities $g(u)$ (1.2). One of the main causes for the essential distinction between the solvability of problem (1.2), (1.3), and the solvability of problem (1.1), (1.3) is related to the following circumstance: a class of solutions of (1.12), periodic with respect to y, generates a dynamical system with an infinite-dimensional saddle manifold of the singular point $0(v \equiv u_x = 0, \quad u = 0)$ in an infinite-dimensional phase space. Here the infinite number of free parameters of these manifolds is sufficient to satisfy the existence of conditions for a one-parameter set of homoclinic loops of 0.

Comparison of these problems seems to be productive.

2. Let us return again to the problem (1.1), (1.3), and its analysis on the basis of the hamiltonian system (1.6). I shall state results of quantitative and numerical analysis of its finite-dímensional models, on which basis we have formulated a

Statement: Problem (1.1), (1.3) is not solvable for arbitrary polynomial nonlinearities $G_N(u, C)$. In other words, in these cases the nonlinear wave equation (1.1) does not permit solutions of the type of nontopological solitons ("breathers").

Let us consider the simplest case, $G(u) = u^3$, and $1/9 < \omega^2 < 1$. The singular point $0(\ldots v_{2n+1} = u_{2n+1} = 0, \ldots)$ of hamiltonian system (1.6) possesses one-dimensional saddle and infinite-dimensional center manifolds.

As a first step let us consider a model with one degree of freedom, arising by assuming in hamiltonian (1.6) that $v_{2n+1} = u_{2n+1} = 0$ for $n > 1$. In this case the system

$$-\frac{dv_1}{dx} + (1-\omega^2)u_1 = \frac{3}{4}u_1^3, \quad \frac{du_1}{dx} = v_1 \tag{2.1}$$

permits the existence in level $H = 0$ of a simple homoclinic loop of 0 for all $\omega^2 < 1$. The corresponding solution

$$u_1(x,t) \sim u_1(x)\cos\omega t = \left(\frac{8}{3}\right)^{1/2} \epsilon \frac{\cos\omega t}{ch\ \epsilon x}, \epsilon^2 = 1 - \omega^2 \tag{2.2}$$

for $\epsilon^2 \ll 1$ is an original by constructing both of formal solutions on a base of power asymptotic expansions, and of asymptotics beyond all orders.

Our next step is a model with two degrees of freedom, arising by assuming that $v_{2n+1} = u_{2n+1} = 0$ for $n \geq 2$:

$$\begin{aligned} -\frac{dv_1}{dx} + (1-\omega^2)u_1 &= \frac{3}{4}\, u_1^3 + \frac{3}{4}\, u_1^2 u_3 + \frac{3}{2}\, u_1 u_3^2, \quad \frac{du_1}{dx} = v_1 \\ -\frac{dv_3}{dx} + (1-9\omega^2)u_3 &= \frac{1}{4}u_1^3 + \frac{3}{2}u_1^2 u_3 + \frac{3}{4}u_3^3, \quad \frac{du_3}{dx} = v_3 \end{aligned} \tag{2.3}$$

In the phase space $\Gamma\{v_1, u_1, v_3, u_3)\}$ of system (2.3), $0(v_1 = u_1 = 0, \quad v_3 = u_3 = 0)$ is a singular point of "saddle-center" type. Numerical investigations of hamiltonian systems (2.3) showed that there are not in level $H = 0$ simple homoclinic loops of 0 with one zero of the function $v_1(x)$, *i.e.* with one point of extremum for $u_1(x)$. However, there exist not more than a countable set of loops of 0 with three zeroes of $v_1(x)$ and seven zeroes of $v_3(x)$, corresponding to three and seven points of extremum for the functions $u_1(x)$ and $u_3(x)$ respectively. Loops of 0 with such a structure are realized for a pointing sequence of eigenvalues ω^2 with probable condensing point by $\omega^2 = 1$. Thus, on the second step a degeneration of continuous spectrum of eigenvalues ω^2 in a pointing spectrum arises. Numerical analysis of the next step– to the model with three degrees of freedom with $v_{2n+1} = u_{2n+1} = 0, \quad n \geq 3$)– showed that there is no necessity to search for nonsymmetric loops of 0, because one can prove that if the problem (1.3), (1.6) is solvable, then it is solvable only in the class of symmetric functions $(u_{2n+1}(-x) = u_{2n+1}(x), \quad n \geq 0)$.

Let us discuss a situation taking place for $1/25 < \omega^2 < 1/9$. In this case saddle manifolds of 0 are two-dimensional, and central manifolds are infinite-dimensional. For the model with two degrees of freedom, (2.3), point 0 is a singular point of "saddle-saddle" type. Here there exists sets of both symmetric and nonsymmetric homoclinic loops of 0. The spectrum of eigenvalues ω^2 is continuous. By dealing with a model with three degrees of freedom, nonsymmetric homoclinic loops disappear, but not more than a countable set of symmetric loops of 0 survive. On that step a degenerating of continuous spectrum in a pointing spectrum of eigenvalues ω^2 takes place. At last, by transition to a model with four degrees of freedom symmetric loops of 0 disappear too. A cause is obvious: for fixed values of ω^2, increasing the dimension of the model leads to growth of the number of relations that determine conditions of stable and unstable manifolds of 0 merging, while the number of free parameters remains fixed. The last fact follows because the dimension of the saddle manifolds of 0 remains constant in these conditions.

Transition to arbitrary polynomial nonlinearities $G_N(u, C)$ cannot change this revealed tendency. Indeed, a finite number of complementary free parameters C leads only to growth of model dimension, evidently pointing on realization of case of general position.

Thus, for all $\omega^2 > 0$ the same scenario is realized: namely, scenario of spoiling all good and beautiful especially.

One cannot exclude the possibility of the variant, when severe reality will cut the wings of lonely (solitary) "nanopteron" (J. Boyd, 1990) or will transform it to an inhabitant of "Through the Looking Glass". The solvability of the problem of a solitary "nanopteron"

$$\lim_{x \to \pm\infty} u(x,t) = W^{\pm}(x,t),\ W(x + \frac{2\pi}{k}, t) = W(x, t + \frac{2\pi}{\omega} = W(x,t) \qquad 2.4$$

is essentially related to the structure and dimension of saddle manifolds of a cycle, which is an image of solutions $W(x,t)$ in a phase space $\Gamma_\infty(\omega)$. I hope that investigations of P. Vuillermot will lead to a solution of that problem.

Let us pass to a discussion of the problem (1.1), (1.3), being generalized to the case of a few spatial variables:

$$u_{tt} - \sum_{1 \le i \le s} u_{x_i x_i} - g(u) = 0$$

$$u(x_i, \ldots, x_s;\ t + \frac{2\pi}{\omega}) = u(x_1, \ldots, x_s;\ t);\ \lim_{\Sigma x_i^2 \to \infty} u(x_1, \ldots, x_s, t) = 0 \qquad 2.5$$

To my mind, this little-investigated problem demands great attention. I shall cite now considerations pointing to a possible conclusion about its unsolvability, at least, for polynomial nonlinearities $G_N(u, C)$ and solutions with spherical symmetry

$$u(x_1, \ldots, x_s,\ t) = u(r,t),\ r^2 = \sum_{1 \le i \le s} x_i^2\ . \qquad 2.6$$

For odd N it is possible to pick out a class of solutions (1.5), for which equation (2.5) generates a dynamical system of the kind

$$\frac{dv_{2n+1}}{dr} + \frac{(s-1)}{r} v_{2n+1} = -\frac{\partial H}{\partial u_{2n+1}};\ \frac{du_{2n+1}}{dr} = \frac{\partial H}{\partial v_{2n+1}}\ . \qquad 2.7$$

Here H is defined by the second of relations (1.6), but it is not a hamiltonian for (2.7). System (2.7) is defined in an enlarged infinite-dimensional phase space $\Gamma\{(\ldots, v_{2n+1}, u_{2n+1}, \ldots; r)\}$. For $N = 3$ together with a "singular" point $0(\ldots v_{2n+1} = u_{2n+1} = 0 \ldots)$ there exist "singular" points $P_\omega^\pm$ $(\ldots, v_{2n+1} = 0,\ u_{2n+1} = \pm\bar{u}_{2n+1}(\omega), \ldots)$. In the last case, for $N \ge 0$, $u_{2n+1}(\omega)$ are solutions of an algebraic system $\partial H / \partial u_{2n+1} = 0$, $n > 0$, or, what is the same– Fourier components of spatially uniform, t-periodic solutions of the wave equation (2.5). Point 0 lies in function level $H = 0$, and points $P_\omega^\pm$– in level $H(\omega) < 0$. It is easy to see that a relation

$$\frac{dH}{dr} = -\frac{(s-1)}{r} \sum_{1 \le i \le s} v_{2n+1}^2 < 0 \qquad 2.8$$

is valid. Due to (2.8) arbitrary data of a Cauchy problem $(\ldots, v_{2n+1} = 0,\ u_{2n+1} = u_{2n+1}(0), \ldots; r)$, with $H(0) > 0$ will lead as $r \to \infty$, as a rule, to approach of trajectory to one of the points $P_\omega^\pm$ in level $H(P_\omega^\pm) < 0$, and in exceptional cases– to attainment of point 0. In the enlarged phase-space, the manifold $W^{(s)}(0)$, built up from trajectories that approach the singular point 0, is infinite-dimensional. However, we can formulate the main

Statement: For all $\omega^2 > 0$ there exist only its finite-dimensional submanifolds $W_{\exp}^{(s)}(0)$, on which all $v_{2n+1}, u_{2n+1},\ n > 0$ vanish exponentially as $r \to \infty$. For example, for (1.8) dimension of $W_{\exp}^{(s)}(0)$is equal to $n_\omega + 1$. Here

$$u_{2n+1}(r) \sim C_{2n+1} \begin{cases} \exp[(1-(2n+1)\omega^2)^{1/2}\ r, & n < n_\omega \\ r^{(s-1)/2}\cos[((2n+1)^2\omega^2 - 1)^{1/2} r + \phi_{2n=1}], & n > n_\omega \end{cases} \quad 2.9$$

It is obvious that manifolds $W_{\exp}^s(0)$ and $W^s(0) \backslash W_{\exp}^s(0)$ are analogs of saddle and center manifolds of the singular point 0 for the case $s = 1$. A condition of finiteness of the norm

$$\int_{R^s} dx u^2(r,t) < \infty$$

leads to a selection of trajectories that approach the point 0 along the finite-dimensional manifold $W_{\exp}^s(0)$. Due to such a selection, the problem under consideration is essentially overdetermined and unsolvable for $G(u) = u^3$ and $G_N(u, C)$. Certainly the reasoning stated above requires specifications of the set of notions and great accuracy of definitions. Unfortunately, we have not completed numerical investigations of this problem, treated as a nonlinear problem (2.5) with eigenvalue, and numerical investigations of the Cauchy problem for nonlinear wave equations, which we know (see, for example, R.K. Dodd et al, 1982), are hardly able to elucidate a solution of problem (2.5).

Finally, I wish sincerely to thank H. Segur for his invitation and persistence in overcoming communication obstacles, and NSF for financial support.

REFERENCES

1. G. L. Alfimov, V. M. Eleonsky, N. E. Kulagin, L. M. Lerman, and V. P. Silin, Existence of nontrivial solutions of equation $\Delta u - u + u^3 = 0$, Physica D, 44:168 (1990).

2. J. Boyd, A numerical calculation of a weakly nonlocal solitary wave: the ϕ^4 breather, Nonlinearity, 3:177-195 (1990).

3. R. K. Dodd, J. C. Eilbeck, J. D. Gibbon, and H. C. Morris, "Solitons and Nonlinear Wave Equations", Academic Press, Inc.(London), (1982).

4. V. M. Eleonsky, N. E. Kulagin, N. S. Novozhilov, and V. P. Silin, "Methods of asymptotic expansions and of finite-dimensional models in the theory of nonlinear waves", in L. Debnath, ed., Advances in Nonlinear Waves,2:286, Pitman, Boston, (1985).

5. V. M. Eleonsky, N. E. Kulagin, and V. P. Silin, "Solitons: phase portrait, bifurcations in plasma theory and nonlinear and turbulent processes in physics", ed. V.G. Baryakhtar, 1:377, World Scientific, Singapore, (1988).

6. V. M. Eleonsky, N. E. Kulagin, L. M. Lerman, and J. L. Umansky, Dynamic systems and solitons state of integrable field equations, Radiophysica, Gorky, 31:2:149 (1988).

7. M. Kruskal, H. Segur, Nonexistence of small-amplitude breather solutions in ϕ^4 theory, Phys. Rev. Lett., 58:747 (1987).

8. P. A. Vuillermot,

   Nonexistence of spatially localized free vibrations for a class of nonlinear wave equations, Comment. Math. Helvetici **64**, pp. 573-586 (1987);

Varietes lisses a certains systems dynamiques et solitons quasiperiodiques pour les equations de Klein-Gordon nonlineaires sur R^2, C.R. Acad. Sci., Paris, 307:I:639 (1988);

Problems de Cauchy multiperiodiques et solutions quasiperiodiques pour les equations de Klein-Gordon nonlineaires sur R^2, C.R. Acad. Sci., Paris 308:I:215 (1989).

9. A. Weinstein, Periodic nonlinear waves on half-line, Commun. Math. Phys. 99:385, (1985).

EXPONENTIALLY SMALL RESIDUES NEAR ANALYTIC INVARIANT CIRCLES

R.S.MacKay

Nonlinear Systems Laboratory, Mathematics Institute
University of Warwick, Coventry CV4 7AL, UK

ABSTRACT

For analytic area-preserving twist maps, we sketch a proof that the "residues" of "good" sequences of periodic orbits with rotation number converging to that of an invariant circle with analytic conjugacy to rotation converge to zero exponentially, with decay rate at least the "analyticity width" of the conjugacy. This confirms numerical observations of Greene and Percival, and provides an important part of a mathematical foundation for Greene's residue criterion, relating existence of an invariant circle of given rotation number to the behaviour of the residues of periodic orbits with rotation number converging to the given one.

1. INTRODUCTION

A fundamental problem of Hamiltonian dynamics is to find which invariant tori exist. Each compact component of a level set of the integrals of an integrable system is a torus on which the motion is conjugate to a rotation [Arn]. Each torus has a winding ratio which measures the average rate of rotation around a chosen basis of cycles. *KAM theory*, named after Kolmogorov, Arnol'd and Moser, shows that all those invariant tori with sufficiently incommensurate winding ratio persist, just slightly deformed, for systems near to non-degenerate integrable systems (e.g. App 8 of [Arn]). For analytic systems, the tori constructed have analytic conjugacy to rotation. On the other hand, *Converse KAM theory* shows that there are no invariant tori of the basic "rotational" class for systems far enough away (e.g. [MP] for area-preserving twist maps, [MMS] for higher dimensional symplectic twist maps, [Mac3] for optical Hamiltonian systems), specifically, close to non-degenerate "anti-integrable limits" [Aub,AAb]. Instead there are "cantori" (e.g. [Ban] for the case of area-preserving twist maps, [MM2] for higher dimensional symplectic twist maps).

The above theorems are rigorous results on existence or non-existence of invariant tori, but their application becomes very hard work when one wishes to locate where in phase and/or parameter space tori break up. In the case of area-preserving maps, and the essentially equivalent continuous time systems of $1\frac{1}{2}$ and 2 degrees of freedom, Greene proposed a criterion for existence or not of invariant tori, which is easy to apply numerically and appears to discriminate very sensitively between the two cases [Gr1,Gr2,Gr3]. It was based on heuristic ideas that there should be a relation between existence or not of an invariant torus of given winding ratio and the "residue" of periodic orbits with nearby winding ratio, a quantity which measures their linear stability.

Over the past eight years, through discussions with various people, I have realised that although there are counterexamples, large parts of Greene's criterion can be proved. The story is not yet complete but in this paper I describe one of the most important parts, namely exponentially small estimates for the residues in the case of an invariant circle with analytic conjugacy to rotation, which is appropriate to this workshop.

The outline of the paper is as follows. In Section 2, the residue criterion is recalled and our result stated. Section 3 is a sketch of the proof. We comment on the result in Section 4. Concluding remarks are given in Section 5.

2. THE RESIDUE CRITERION

There are many variants of the residue criterion (e.g. [Mac2]). The version I seek to prove is basically that originally formulated by Greene in [Gr1].

Let $\mathbb{T} = \mathbb{R}/\mathbb{Z}$ be the circle.

Definition: A map $f : (x,y) \mapsto (x',y')$ of the cylinder $\mathbb{T}\times\mathbb{R}$ to itself is said to be an *exact area-preserving twist map* if it is C^1, preserves the ends of $\mathbb{T}\times\mathbb{R}$, $\det Df = 1$, $\partial x'/\partial y > 0$, $|x'-x| \to \infty$ as $|y| \to \infty$, and $\int_{f(\gamma)} y\, dx = \int_\gamma y\, dx$ for all *rotational* (i.e. homotopically non-trivial) circles γ.

Let $F : \mathbb{R}^2 \to \mathbb{R}^2$ be a lift of f, i.e. $\pi \circ F = f \circ \pi$ where $\pi : \mathbb{R}^2 \to \mathbb{T}\times\mathbb{R}$ is defined by $\pi(x,y) = (x \bmod \mathbb{Z}, y)$. We denote orbits by $(x_{i+1},y_{i+1}) = F(x_i,y_i)$, $i \in \mathbb{Z}$.

The *residue* of a periodic orbit is defined to be

$$R = (2-\mathrm{Tr}\, Df^q_z)/4 = (2-\lambda-\lambda^{-1})/4 \qquad (2.1)$$

where q is its least period, z is any of its points, and λ,λ^{-1} are its *multipliers* (the eigenvalues of Df^q_z). The residue is 0 for periodic orbits of an *integrable system* (that is, a map conjugate to one of the form $F(x,y) = (x+\Omega(y),y))$. Elliptic points correspond to residue $0 < R < 1$, regular hyperbolic points to $R < 0$, and inversion hyperbolic points to $R > 1$.

A *periodic orbit of type (p,q)* is an orbit for which

$$x_{i+q} = x_i + p,\ y_{i+q} = y_i\ . \tag{2.2}$$

An orbit is said to be *well-ordered* if $\forall$ i,j,k $\in \mathbb{Z}$, $x_i + k < x_j \Leftrightarrow x_{i+1} + k < x_{j+1}$.

The *rotation number* of an orbit is the number

$$\omega = \lim_{i\to\infty} \frac{x_i - x_0}{i} \tag{2.3}$$

if the limit exists. It is p/q for orbits of type (p,q). It exists for every orbit on a rotational invariant circle and is the same for all its orbits. We shall be concerned mainly with existence of circles with irrational rotation number, as existence of a rational circle is codimension-infinite.

In perturbation theory, the residue of the sum resonance produced by two resonances is to leading order proportional to the product of the residues of the parents (e.g. [GM]). This suggests an exponential dichotomy on following a sequence of resonances whose rotation numbers converge to an irrational: either the residues decay exponentially or they grow exponentially, apart from a marginal case in between. In the case of exponentially growing residues, it is unlikely that an invariant circle could coexist with such strong instability. On the other hand, if the residues decay exponentially, it seems likely that the resonances will converge to an invariant circle. This is the basis for *Greene's residue criterion* [Gr1], which follows.

Conjecture: Let ω be irrational, p_n/q_n a "good" sequence of rationals converging to ω (Greene took the convergents of ω, i.e. the truncations of its continued fraction expansion; we will allow many other sequences). Let R_n be the residue of some well-ordered periodic orbit of type (p_n,q_n). Then

$$\mu(\omega) = \lim_{n\to\infty} q_n^{-1} \log |R_n| \tag{2.4}$$

exists, and $\mu(\omega) \le 0$ implies there is a rotational invariant circle of rotation number ω, $\mu(\omega) > 0$ implies there is none.

The quantity $\mu(\omega)$ is the logarithm of Greene's "mean residue". Unfortunately, the limit does not always exist, but we can prove a lot about the behaviour of the residues in many cases. In another paper [Mac1], I will give results on how the residues behave when there is no invariant circle of rotation number ω or when there is a circle without an analytic conjugacy to rotation. My purpose here is to announce a theorem in the case that there is an invariant circle with analytic conjugacy to a Diophantine rotation.

Let us explain the terms. Suppose F is a lift of an analytic area-preserving twist map and Γ is a rotational invariant circle with rotation number ω. It is said to have *analytic conjugacy to rotation* if it is the image of the real line under a real analytic map (X,Y): $D_\delta \to \mathbb{C}\times\mathbb{C}$, for some $\delta>0$, where $D_\delta \subset \mathbb{C}$ is the strip $|\mathrm{Im}\,\theta| < \delta$, such that

$$F\circ(X,Y) = (X,Y)\circ R_\omega \tag{2.5}$$

and $$(X,Y)(\theta+2\pi) = (X,Y)(\theta) + (1,0), \tag{2.6}$$

where $$R_\omega(\theta) = \theta + 2\pi\omega. \tag{2.7}$$

The maximum δ for which the conjugacy can be defined is called the *analyticity width* of the circle. The image of D_δ under (X,Y) is an invariant annulus in $\mathbb{C}\times\mathbb{C}$ on which the map is conjugate to R_ω, called a *Herman ring.*

A number ω is said to be *Diophantine* if there exist $C > 0$, $\tau \geq 1$ such that

$$|q\omega-p| \geq C/q^\tau,\ \forall\ q \in \mathbb{N},\ p \in \mathbb{Z}. \tag{2.8}$$

We say a sequence of rationals p/q (in lowest terms) converging to ω is *good* if there exists $N \geq 2$ such that $T(p,q) \to 0$, where

$$T(p,q) = \max(q^4|\omega-p/q|^N, q^2|\omega-p/q|^{N-1}). \tag{2.9}$$

Theorem: If f is analytic, ω is Diophantine, and there is a rotational invariant circle of rotation number ω with analytic conjugacy to rotation of analyticity width δ, then for all sequences of periodic orbits of type (p,q) for which p/q is a good sequence converging to ω,

$$\bar{\mu}(\omega) - \limsup_{n\to\infty} q_n^{-1} \log |R_n| \leq -\delta. \tag{2.10}$$

3. IDEA OF THE PROOF

In fact, we prove a slightly more detailed result.

Proposition: Given an invariant circle of Diophantine rotation number ω with analyticity width δ, $0 < \rho < \delta$, and $N \geq 2$, there exists C such that the residues of all periodic orbits of type (p,q) with $T(p,q)$ small enough satisfy

$$|R| \leq C\, q^6|\omega-p/q|^N e^{-q\rho}. \tag{3.1}$$

Since ρ can be chosen as close to δ as one likes, this Proposition implies the Theorem.

The strategy of the proof of the Proposition is as follows.

1. By analytic coordinate changes in a neighbourhood of the piece $|\mathrm{Im}\ \theta| \le \rho$ of the Herman ring, the map can be brought into the following normal form (cf. [Dou,SZ]):

$$x' = x + \omega + \Omega_{N-1}(y) + O(y^N),\ \Omega_{N-1}(y) = b_1 y + b_2 y^2 + \dots b_{N-1} y^{N-1},\ b_1 > 0 \qquad (3.2)$$
$$y' = y\,(1 + O(y^N))$$

This depends on the assumption that ω is Diophantine.

2. The map (3.2) has generating function of the form

$$h(x,x') = H(\eta) + O(\eta^{N+1}) \qquad (3.3)$$

where $\eta = x' - x - \omega$. (3.4)

3. Periodic orbits of type (p,q) correspond to critical points of the function $W : P \to \mathbb{C}$, defined by

$$W(x) = \sum_{j=0}^{q-1} h(x_j, x_{j+1}). \qquad (3.5)$$

where $P = \{x \in \mathbb{C}^{\mathbb{Z}} : x_{j+q} = x_j + p\}$. (3.6)

4. Define approximate critical points $z(\theta) \in P$, for $\theta \in \mathbb{C}$, by

$$z(\theta)_j = \theta/2\pi + jp/q, \qquad (3.7)$$

and for $\theta \in \mathbb{C}$, define

$$E_\theta = \{x \in P : \sum_{j=0}^{q-1} (x_j - z(\theta)_j) = 0\}. \qquad (3.8)$$

Then P is foliated by $\{E_\theta : \theta \in \mathbb{C}\}$, and $z(\theta) \in E_\theta$.

5. Restricting to the submanifold E_θ, $D^2 W|E_\theta$ is non-degenerate near $z(\theta)$ and $DW(z(\theta))$ is small, so the implicit function theorem can be used to show that $W|E_\theta$ has a locally unique critical point $v(\theta)$ near $z(\theta)$. More precisely, there exists C′ such that for T(p,q) small enough, $W|E_\theta$ has a unique critical point $v(\theta)$ in

$$\|v - z(\theta)\| \le C' q^2 |\omega - p/q|^N, \qquad (3.9)$$

using the supremum norm on P.

6. Let $J(\theta) = W(v(\theta))$. (3.10)

If x is a periodic orbit of type (p,q) and $q^2|\omega-p/q|^N$ is small enough, then it is $v(\theta)$ for some θ such that $J'(\theta) = 0$.

7. The residue R of the periodic orbit $v(\theta)$ has the asymptotic formula

$$R \sim -\pi^2 q J''(\theta). \quad (3.11)$$

This comes from a formula of [MM1] relating the residue to the determinant of D^2W.

8. Let $J_0 = qH(p/q-\omega)$. Then for T(p,q) small enough, there exists C″ such that

$$|J(\theta) - J_0| \leq C'' q^3 |\omega-p/q|^N. \quad (3.12)$$

This comes from (3.3) and (3.9).

9. The function J is periodic with period $2\pi/q$:

$$J(\theta + 2\pi/q) = J(\theta). \quad (3.13)$$

This is a crucial step. It is clear that J is periodic with period 2π, but it is the fact that J has the much shorter period $2\pi/q$ which makes the next step give exponentially small estimates. To prove (3.13) note that $x \in E_\theta$ if and only if $\tilde{x} \in E_{\theta+2\pi/q}$, where $\tilde{x}_j = x_{j+q'} + p'$, and (p′,q′) is the Farey parent of (p,q), that is the penultimate truncation of the continued fraction expansion for p/q.

10. Fourier analyse $J(\theta)$ and estimate the coefficients by deforming the contour of integration to the curves Im $\theta = \pm\rho$, using analyticity of J in D_ρ and (3.12). From this we deduce that for $\theta \in \mathbb{R}$,

$$J''(\theta) = O(q^5 |\omega-p/q|^N e^{-q\rho}), \quad (3.14)$$

which together with (3.11) completes the proof of the Proposition.

The details are given in [Mac1]. Steps 4, 5, 8 and 10 were inspired by an analogous proof by Angenent on exponentially small splitting of separatrices [Ang].

4. COMMENTS

The possibility of obtaining exponential estimates from the analytic structure is an exciting new development (for another reference in the case of splitting of separatrices, see

[FS]). The ultimate goal, however, is to use the analytic structure to deduce the asymptotics, not just upper bounds. This has been achieved by Lazutkin, for example, for the splitting of separatrices of the standard map [LST].

What would be nice for our problem would be to find conditions under which

$$q_n^{-1} \log |R_n| \to -\delta \tag{4.1}$$

as observed numerically by Greene and Percival for $\omega = \gamma = (1+\sqrt{5})/2$ in the standard map [GP]. It would be worth investigating this further numerically, e.g. maybe there is some dependence not just on q_n but also on $|q_n\omega - p_n|^{-1}$, which are indistinguishable for convergents to γ, but may be different for non-convergents, or for sequences approaching other irrationals.

Equation (4.1) can not always hold. For example, if f is a map for which (4.1) holds, then let g be the map given by lifting f and then reducing by $2\mathbb{Z}$ (instead of $\mathbb{Z}$). Then an analytic invariant circle of rotation number ω and analyticity width δ for f becomes one of rotation number $\omega/2$ and analyticity width $\delta' = \delta/2$ for g. Each (p,q) orbit of residue R for f becomes a (p,2q) orbit of residue 4R(1-R) for g if p is odd, or two (p/2,q) orbits of residue R if p is even. The residues of the latter have decay rate $-2\delta'$ rather than $-\delta'$.

Similarly, one could consider the map f^2. An analytic invariant circle of rotation number ω and analyticity width δ for f becomes one of rotation number 2ω with the same analyticity width. Each (p,q) orbit of residue R becomes a (2p,q) orbit of residue 4R(1-R) if q is odd, or two $(p',q') = (p,q/2)$ orbits of residue R if q is even. The residues of the latter have decay rate $\lim_{q'\to\infty} q'^{-1}\log|R| = -2\delta$ rather than $-\delta$.

However, I believe that there may well be simple conditions under which (4.1) is true. One approach to proving this might be to understand the structure of the boundary of the Herman ring (cf [GP]), repeat the proof using $z(\theta)_j = X(\theta+2\pi jp/q)$ without using the normal form (3.2) which blows up near the boundary, and then to derive asymptotics for $J''(\theta)$ by pushing the contour of integration right onto the boundary.

5. CONCLUSION

We have sketched a proof of an important part of Greene's residue criterion, namely exponential decay of the residues of "good" sequences of periodic orbits when there is an invariant circle with analytic conjugacy to a Diophantine rotation.

There are several directions in which it would be nice to extend this work. Firstly, our result is an upper bound; it would be better to obtain asymptotics, and we indicated one way in which to attempt this. Secondly, it looks as if there might be some sort of analytic continuation between the analyticity width of an invariant circle and the Lyapunov exponent of a cantorus. Thirdly, it should be feasible to extend our result to higher dimensional symplecticmaps.

ACKNOWLEDGEMENTS

This idea for this paper was born on hearing a talk by Sigurd Angenent at the IMA workshop (Minnesota) on Twist Maps and their Applications, in March 90. It was written up during a visit to the University of Arizona, during a sabbatical leave from Warwick. I learnt the normal form (3.2) from John Mather in February 82. I would like to thank these people and institutions and organisers and sponsors for the benefit to me.

REFERENCES

[Ang] Angenent S, talk given at the IMA workshop on Twist Maps and their Applications, Minnesota, March 1990

[Arn] Arnol'd VI, Mathematical methods of classical mechanics (Springer, 1978)

[Aub] Aubry S, The concept of anti-integrability: definition, theorems and applications to the standard map, in Twist Maps and their Applications, ed Meyer KR, IMA conf proc, to appear (Springer)

[AAb] Aubry S, Abramovici G, Chaotic trajectories in the standard map: the concept of anti-integrability, Physica D 43 (1990) 199-219

[Ban] Bangert V, Mather sets for twist maps and geodesics on tori, Dynamics Reported vol. 1 (1988), eds Kirchgraber U, Walter HO (Wiley)

[Dou] Douady R, Applications du théorème des tores invariants, Thèse 3ème cycle, Univ Paris VII (1982)

[FS] Fontich E, Simo C, The splitting of separatrices for analytic diffeomorphisms, Ergod Th Dyn Sys 10 (1990) 295-318

[Gr1] Greene JM, A method for determining a stochastic transition, J Math Phys 20 (1979) 1183-1201

[Gr2] Greene JM, KAM surfaces computed from the Hénon-Heiles Hamiltonian, in Nonlinear Dynamics and the beam-beam interaction, eds Month M, Herrera JC, Am Inst Phys Conf Proc 57 (1980) 257-271

[Gr3] Greene JM, The calculation of KAM surfaces, Ann NY Acad Sci 357 (1980) 80-89

[GM] Greene JM, MacKay RS, An approximation to the critical commuting pair for breakup of noble tori, Phys Lett A 107 (1985) 1-4

[GP] Greene JM, Percival IC, Hamiltonian maps in the complex plane, Physica D 3 (1981) 530-548

[LST] Lazutkin VF, Schachmannski IG, Tabanov MB, Splitting of separatrices for the standard and semi-standard mappings, Physica D 40 (1989) 235-248

[Mac1] MacKay RS, On Greene's Residue Criterion, submitted to Nonlinearity

[Mac2] MacKay RS, Introduction to the dynamics of area-preserving maps, in Physics of particle accelerators, eds Month M, Dienes M, Am Inst Phys Conf Proc 153 (1987) vol.1, 534-602

[Mac3] MacKay RS, A criterion for non-existence of invariant tori for Hamiltonian systems, Physica D 36 (1989) 64-82

[MM1] MacKay RS, Meiss JD, Linear stability of periodic orbits in Lagrangian systems, Phys Lett A 98 (1983) 92-94

[MM2] MacKay RS, Meiss JD, Cantori for symplectic maps near an anti-integrable limit, submitted to Nonlinearity

[MMS] MacKay RS, Meiss JD, Stark J, Converse KAM theory for symplectic twist maps, Nonlinearity 2 (1989) 555-570

[MP] MacKay RS, Percival IC, Converse KAM: theory and practice, Commun Math Phys 98 (1985) 469-512

[SZ] Salamon D, Zehnder E, KAM theory in configuration space, Comment Math Helv 64 (1989) 84-132

ASYMPTOTICS OF PARTIAL DIFFERENTIAL EQUATIONS AND THE RENORMALISATION GROUP

Nigel Goldenfeld[1], Olivier Martin[2] and Y. Oono[1]

[1]Department of Physics
University of Illinois at Urbana-Champaign
1110 West Green Street
Urbana, IL 61801

[2]Department of Physics
City College
CUNY
New York, NY 10031

I. INTRODUCTION

It is well-known that the asymptotics of partial differential equations (PDEs) may often be found from consideration of similarity solutions. In the examples usually encountered, the combinations of variables making up the similarity variables may be deduced using dimensional analysis; typically, the similarity variables are products of variables raised to rational fraction powers. It is not so widely appreciated, however, that there is a large class of problems where the similarity variables cannot be deduced from dimensional analysis. As Barenblatt has emphasized, such problems are neither rare nor pathological, but occur in many situations of physical interest, for example, in continuum mechanics.[1]

The purpose of this article is to show how to obtain the asymptotics of PDEs, even in cases where dimensional analysis fails, using the renormalisation group (RG).[2] Renormalisation and the RG were originally developed to treat the divergences arising in the perturbation series of quantum electrodynamics, and, following the work of Kadanoff, Wilson and others, have found extensive application in later quantum field theories[3] and statistical mechanics.[4] Although it has been known for some time that field theories--be they quantum or statistical--are equivalent to stochastic partial differential equations[5], it is only recently that it was shown how to use RG techniques for partial differential equations without noise.[2] This article not only summarizes this development, but also emphasizes the connection with the problems of velocity selection in dendritic growth and asymptotics beyond all orders, discussed elsewhere in this volume. This connection arises because a travelling wave solution of a PDE in one space dimension, x, and time, t, of the form

$$u(x,t) = f(x - vt) \tag{1}$$

Asymptotics beyond All Orders, Edited by H. Segur *et al.*
Plenum Press, New York, 1991

may be mapped by the substitutions $x = \log X$, $t = \log T$ into a similarity solution of the form

$$u(X,T) = T^{\alpha} g(XT^{\beta}) \tag{2}$$

with $\alpha = 0$, $\beta = -v$. The goal of this work, then, is to calculate exponents such as α, β, v and the associated scaling functions f, g. Examples of physical interest occur in (e.g.) elasticity theory, shock wave dynamics, flame propagation, and flow in porous media.[1]

The application of the RG to partial differential equations has implications for the way in which systems approach thermodynamic equilibrium, and is almost certainly relevant to theoretical attempts to account for the prevalence of dynamical scaling. This connection has been discussed in a recent article which complements this one.[6]

2. ASYMPTOTICS OF THE DIFFUSION EQUATION

We begin with an elementary example, which shows explicitly how similarity solutions are relevant for asymptotics. Consider the initial value diffusion problem

$$\partial_t u = \frac{1}{2}\partial_x^2 u \,, \qquad -\infty < x < \infty \tag{3}$$

with $u \to 0$ as $|x| \to \infty$ and initial condition

$$u(x,0) = \frac{A_0}{(2\pi\ell^2)^{1/2}}\, e^{-x^2/2\ell^2} \,. \tag{4}$$

The solution after time t is

$$u(x,t) = \frac{A_0}{\left[2\pi(t+\ell^2)\right]^{1/2}}\, e^{-x^2/2(t+\ell^2)} \,. \tag{5}$$

Note that the dynamics conserves $\int u(x,t)\,dx$. For long times

$$u(x,t) \to \frac{A_0}{(2\pi t)^{1/2}}\, e^{-x^2/2t} \qquad \text{as } t \to \infty. \tag{6}$$

This long time behaviour can also be obtained by keeping t fixed, but letting the width of the initial distribution vanish:

$$u(x,t) \to \frac{A_0}{(2\pi t)^{1/2}}\, e^{-x^2/2t} \qquad \text{as } \ell \to 0. \tag{7}$$

We conclude that the long-time behaviour of the initial value problem is given by the degenerate limit of the initial value problem when $\ell \to 0$, i.e. the similarity solution corresponding to a delta-function initial condition.

In this example, the mathematical results follow "common sense" intuition: at very long times, the distribution should be insensitive to the initial condition. That is, when

$$\langle x^2 \rangle \equiv \int x^2 u(x,t)\, dx \gg \ell^2 \tag{8}$$

we expect that the solution u(x,t) should not depend on ℓ, and thus we can safely take the limit $\ell \to 0$ whilst maintaining the conservation of $\int u(x,t)dx$.

This sort of "common sense" argument is encountered frequently. Suppose a physical problem has been cast in dimensionless form, and the relationship between the dimensionless parameters $\Pi, \Pi_o, \Pi_1, \ldots \Pi_n$ is written as

$$\Pi = f(\Pi_o, \Pi_1, \ldots, \Pi_n). \tag{9}$$

Then "common sense" intuition states that as (e.g.) $\Pi_o \to 0$, $\Pi \to f(0, \Pi_1, \ldots \Pi_n)$. In the diffusion equation example,

$$\Pi = \frac{u}{Q}\sqrt{t}\,; \quad \Pi_o = \frac{\ell}{\sqrt{t}}\,; \quad \Pi_1 = \frac{x}{\sqrt{t}}\,. \tag{10}$$

3. ASYMPTOTICS OF THE SECOND KIND

Barenblatt[1] has pointed out that there are a wide class of problems where "common sense" intuition fails, because the limit $\Pi_o \to 0$ is singular. He uses the term "intermediate asymptotics of the first kind" to denote the case when the limit $\Pi_o \to 0$ is regular, and the term "intermediate asymptotics of the second kind" to denote the case

$$\Pi \sim \Pi_o^{-\alpha}\, g\left(\frac{\Pi_1}{\Pi_o^{\alpha_1}} \cdots \frac{\Pi_n}{\Pi_o^{\alpha_n}}\right) \tag{11}$$

as $\Pi_o \to 0$. All other possibilities are included under the category "third kind." The use of the term "intermediate asymptotics" has the connotation "prior to the final state of the system", which in the diffusion equation example is $u(x, \infty) = 0$.

When the asymptotics is of the second kind, a number of exponents, α, $\alpha_1, \ldots \alpha_n$ are introduced. These cannot be determined by dimensional analysis, since the Π's are already dimensionless. Usually, the exponents are found to satisfy a non-linear eigenvalue equation, obtained by seeking a solution to the governing PDE of the form of equation (11). In general, the exponents must be determined numerically, and the appropriate form of equation (11) obtained initially by guesswork.

In the following section, we consider a specific example with asymptotics of the second kind, and show that it is possible to determine systematically, using the RG, how many exponents are introduced, what their values are, and the form of the scaling function.

4. BARENBLATT'S EQUATION

When an elastic fluid flows through a porous medium which can expand and contract irreversibly, in response to the pressure u(x,t), the time evolution depends upon whether or not the pressure is increasing (medium expanding) or decreasing (medium contracting). The resulting equation for the pressure, using Darcy's Law, can be written as[1]

$$\frac{\partial u}{\partial t} = D \frac{\partial^2 u}{\partial x^2} \tag{12}$$

with $D = \frac{1}{2}$ for $\partial_x^2 u > 0$ and $D = \frac{1}{2}(1 + \varepsilon)$ for $\partial_x^2 u < 0$. The parameter ε depends upon the elastic constants of the fluid and the porous medium and the permeability. We consider only the initial value problem with $u \to 0$ as $|x| \to \infty$ and u(x,0) given by equation (4). The question we address is: what is the long-time behaviour of the Barenblatt equation?

The long-time behaviour cannot be of the form $u(x,t) \sim t^{-1/2} f(x\sqrt{t}, \varepsilon)$ for f having continuous second derivative. This can be seen by substituting this form into the Barenblatt equation; it is impossible to match across the point where $\partial_x^2 u = 0$. Nevertheless, uniqueness and existence of the initial value problem with continuous second derivatives in space have been proved.[7]

The renormalisation group approach to this initial value problem has six steps, each a direct counterpart of the procedure followed in quantum or statistical field theory. The first step is to construct a naive perturbation expansion in ε: it has the form, in the limit $\ell^2/t \to 0$,

$$u(x,t) \sim u_B(x,t) = \frac{A_0 e^{-x^2/2t}}{(2\pi t)^{1/2}} \left\{ 1 - \frac{\varepsilon}{(2\pi e)^{1/2}} \log \frac{t}{\ell^2} + O(\varepsilon^2) \right\} + \text{r.t.} \tag{13}$$

where "r.t." stands for terms which are regular in the limit, and the subscript "B" stands for "bare", in conformity with field theoretic usage. It is convenient to achieve the limit by keeping t fixed, and letting $\ell \to 0$, as we did for the diffusion equation.

The second step is to cure the logarithmic divergence of the perturbation series, by introducing the renormalised pressure

$$u_R(x,t) = Z(\ell/\mu)\, u_B(x,t), \tag{14}$$

where μ is an arbitrary length, about which we will say more shortly. The renormalised pressure, u_R, will eventually be found to be the correct asymptotic solution of the Barenblatt equation, as opposed to the naive perturbation expression u_B which is (incorrectly) divergent. The function Z is referred to as a renormalisation constant, and strictly speaking, it is associated with $A_0 = \int u(x,0)dx$. Since the Barenblatt equation does not conserve $\int u(x,t)dx$, A_0 cannot be deduced from knowledge of u(x,t) at long times (i.e. when the origin of time is indeterminate). In this sense, A_0 is unobservable at long times, in the same way that the bare electric charge is unobservable at long distances, according to quantum electrodynamics. The renormalisation constant depends on ℓ, so that as $\ell \to 0$, the divergence in u_B may be absorbed into Z to yield a finite u_R. In the procedure described below, the removal of the divergence in u_B occurs order by order in ε, so we will assume that Z has an expansion in powers of ε. However, Z is, by definition dimensionless, and therefore cannot depend solely on the dimensional parameter ℓ. For this reason, an arbitrary parameter μ with the dimensions of length must be introduced.

The third step of the renormalisation procedure is to renormalise the bare perturbation expansion. We expand

$$Z = 1 + a_1(\ell/\mu)\,\varepsilon + a_2(\ell/\mu)\varepsilon^2 + \ldots \tag{15}$$

and choose $a_1, a_2, \ldots$ to cancel order by order in ε the divergence in u_B as $\ell \to 0$. We find that

$$a_1 = \frac{1}{\sqrt{2\pi e}} \log\left[C_1\mu^2/\ell^2\right] \tag{16}$$

giving

$$u_R(x,t) = \frac{A_0 e^{-x^2/2t}}{\sqrt{2\pi t}}\left[1 - \frac{\varepsilon}{\sqrt{2\pi e}}\log\left(\frac{t}{C_1\mu^2}\right) + O(\varepsilon^2)\right]. \tag{17}$$

Here C_1 is an arbitrary number. Although this formula has two arbitrary parameters, it is more useful than it may seem; and it has the obvious virtue of being (trivially) finite as $\ell \to 0$, since ℓ is not present in the formula.

In fact, the formula (17) describes a family of solutions. Step four of the procedure is to chose a particular member of the family by requiring that (e.g.) at some time t^*, the value of u_R at the origin is some number Q:

$$u_R(0,t^*) = Q\,.$$

Then the corresponding solution, to $O(\varepsilon)$ is

$$u_R(x,t) = Q\left(\frac{t^*}{t}\right)^{1/2} e^{-x^2/2t}\left(1 - \frac{\varepsilon}{\sqrt{2\pi e}}\log\frac{t}{t^*} + O(\varepsilon^2)\right). \tag{18}$$

This expression will be referred to as the renormalised perturbation expansion. Note that to this order in ε, the constants C_1 and μ have dropped out. A proof that this occurs to all orders in ε for the arbitrary constants C_1, C_2, ... introduced by the renormalisation procedure would constitute a proof of perturbative renormalisability. Although we do not doubt that the Barenblatt equation is perturbatively renormalisable, we have not proven this. The work of ref. 7 is essentially a proof or renormalisability, beyond perturbation theory. The renormalised perturbation expansion is useful at best only for times such that $1 \gg \frac{\varepsilon}{(2\pi e)^{1/2}}\log(t/t^*)$. For $t \gg t^*$, the renormalised perturbation expansion breaks down. Nevertheless, we will now show that the arbitrariness of t^* enables the renormalised perturbation expansion to be improved.

In step 5, we use the renormalisation group argument, due to Gell-Mann and Low[8]: The renormalised perturbation expansion involves a parameter t^* not present in the original problem. How can the asymptotics depend upon such a parameter? The point is that Q must also depend upon t^*. After all, during the diffusion process, u(0,t) is expected to be a decreasing function of time. The dependence of Q on t^* can be found because its t^*-dependence must be that to cancel out the explicit t^*-dependence of the renormalised perturbation expansion. Thus

$$\frac{du_R}{dt^*} = \frac{\partial u_R}{\partial t^*} + \frac{\partial u_R}{\partial Q}\frac{dQ}{dt^*} = 0 . \tag{19}$$

The partial derivatives can be explicitly evaluated, at least to $0(\varepsilon)$, from equation (17). The result is

$$t^*\frac{dQ}{dt^*} = -Q\left[\frac{1}{2} + \frac{\varepsilon}{(2\pi e)^{1/2}} + 0(\varepsilon^2)\right] . \tag{20}$$

The final step is to solve this differential equation for Q. Substituting back into equation (17) and setting $t^* = t$, we finally obtain

$$u_R(x,t) = \frac{A}{t^{\alpha+1/2}}\, e^{-x^2/2t}\,(1 + 0(\varepsilon^2)) \tag{21}$$

with

$$\alpha = \frac{\varepsilon}{(2\pi e)^{1/2}} + O(\varepsilon^2). \tag{22}$$

Thus, the logarithmic terms in the perturbation expansion came from the expansion of $t^{-\alpha}$: the divergence of the perturbation series pointed the way to the correct asymptotics. We shall refer to α as the anomalous dimension. Is the expansion of $\alpha(\varepsilon)$ convergent within same non-zero radius of convergence? There is no obvious physical reason why the theory should be badly behaved for negative ε, and thus it is possible that $\varepsilon = 0$ is a regular point and the radius of convergence is non-zero. Finally, note that $\partial_t u_R$, $\partial_x u_R$ and $\partial_x^2 u_R$ are all continuous! This is surprising but correct: the discontinuity in D occurs where $\partial_x^2 u = 0$, and so $\partial_x u$ is continuous. On either side of this point, u_R is C^∞. Thus $\partial_x^2 u$ is continuous, as is $\partial_t u$.

5. GEOMETRICAL INTERPRETATION OF THE RG

So far, the exposition has been tied to perturbation theory. In fact, the RG approach is non-perturbative, and can be used as a procedure even when the expansion in ε is a poor approximation. We shall illustrate this here. Define the RG transformation on the space of functions $u(x,t_o)$ at a given value of t_o:

$$u'(x,t_o) \equiv R_{b,\phi}\left[u(x,t_o)\right] . \tag{23}$$

The RG transformation depends on 2 parameters and involves 3 steps: (1) Evolve the function $u(x,t_o)$ forward in time to $t_1 = bt_o$, $b > 1$, using the PDE. (2) Rescale x i.e. $x \to b^\phi x$, with ϕ as yet unspecified. (3) Rescale the function itself so that $u'(0,t_o) = u(0,t_o)$. The general idea is that similarity solutions, if they exist, are fixed points of the RG transformation. That is, we iterate the RG transformation and if the initial conditions are in the basin of attraction of a fixed point, then the fixed point will be reached after an infinite number of iterations. The real power of the RG derives from the fact that it is relatively easy to approximate step (1), because the evolution is over a finite time. The RG procedure then is capable of giving a good approximation to the long time behaviour.

It should also be noted that for generic values of ϕ, there may not exist fixed points of $R_{b,\phi}$. Thus, ϕ must be varied to search for fixed points. To illustrate the RG expressed in this form, we will use the renormalised perturbation expansion to approximate step (1), again starting from a Gaussian. We obtain

$$u'(x,t_o) = \frac{1}{Z(b)}\, u(b^{\phi}x, bt_o) \tag{24}$$

with

$$Z(b) = b^{-1/2}\left(1 - \frac{\varepsilon}{(2\pi e)^{1/2}} \log b\right) + O(\varepsilon^2) \tag{25}$$

The RG transformation forms a semi-group (semi, because $b > 1$):

$$R_{b_1,\phi}\; R_{b_2,\phi} = R_{(b_1 b_2),\phi}\,. \tag{26}$$

This implies $Z(b) = b^{y_Q}$ for any exponent y_Q, or

$$y_Q = \frac{\partial(\log Z)}{\partial(\log b)} = -\left(\frac{1}{2} + \frac{\varepsilon}{\sqrt{2\pi e}} + O(\varepsilon^2)\right) \tag{27}$$

Now let us determine ϕ. Performing one iteration on the initial condition, we obtain

$$u(b^{\phi}x, bt_o) = Q(t_o)\, b^{-1/2} \exp\left[-\, b^{2\phi}x^2/2bt_o\right] \times \left[1 - \frac{\varepsilon}{\sqrt{2\pi e}} \log b + O(\varepsilon^2)\right] \tag{28}$$

A fixed point is only possible if $\phi = 1/2$. At the fixed point

$$u^*(x,t) = b^{-y_Q}\, u^*(xb^{1/2}, bt)\ . \tag{29}$$

Choosing $b = 1/t$, we obtain the result.

$$u^*(x, t) = t^{-\alpha-1/2}\, u^*\!\left(\frac{x}{\sqrt{t}}, 1\right) \tag{30}$$

with $\alpha(\varepsilon)$ given as in the preceding section.

We conclude this section with several remarks. First, step (1) can in principle be carried out numerically[9]. There is no restriction to using perturbation theory, as we have done here for pedagogical purposes. Second, the origin of the anomalous dimension in PDE problems is precisely the same as in critical phenomena. Consider, for example, the two-point correlation function $G(k)$, of a scalar field at the critical point, as a function of wavenumber k: conventionally, $G(k) \sim k^{-2+\eta}$ as $k \to 0$, and η is an anomalous dimension. We assume, for concreteness, that the field is defined on the vertices of a regular lattice, with lattice spacing ℓ. It can be shown that G must have the dimensions of $(\text{length})^2$. How then, can it have the conventional form at the critical point? The answer is that even though the correlations in the system have infinite range, the lattice spacing ℓ is still important and cannot be neglected (i.e. set to zero). In fact, the correlation function is singular as $\ell \to 0$:

$$G(k) \sim \ell^{\eta} k^{-2+\eta} , \tag{31}$$

It is this singularity, combined with the necessity to respect dimensional analysis, which leads to the anomalous wavenumber dependence of G(k) at the critical point. Finally, the geometrical formulation of the RG given here is the counterpart of Wilson's method in statistical physics.[10]

6. SPECULATIONS

We conclude this article by offering a number of speculations[2] on future applications of this work. At the time of writing, these and other avenues of research are being actively followed.

Perhaps the most interesting application, from the point of view of this workshop, is to velocity selection in dendritic growth. We shall restrict ourselves to models such as the geometric model[11] (GM) or the boundary-layer model[12], where there is no doubt about the procedure to construct the steady states or needle crystals. In the GM, for example, the velocity of steady states is given by the non-linear eigenvalue problem

$$\left(\kappa + \frac{d^2\kappa}{ds^2}\right) = \frac{v \cos \theta}{1 + \varepsilon \cos (m\theta)} \tag{32}$$

with boundary conditions

$$\left.\frac{d\kappa}{ds}\right|_{s=0} = 0 \,; \quad \kappa \to 0 \text{ as } s \to \infty . \tag{33}$$

Here, $\kappa = d\theta/ds$ is the curvature of an interface whose normal is at an angle θ to the axis of symmetry of the needle crystal, at an arc length distance s from the tip. The degree of symmetry is m. Only for special values of v can the boundary conditions be satisfied. The solution corresponding to the largest value of v is the needle crystal which forms the tip of the dendrites in the GM. The steady state equation (32) is the analogue of the equation determining the scaling function f and the anomalous dimension α in Barenblatt's equation: substituting $u = t^{-(\alpha + 1/2)} f(\xi)$, $\xi = x/\sqrt{t}$ in Barenblatt's equation yields a non-linear eigenvalue equation for α and f. Indeed, this is typically how Barenblatt and others have solved problems with asymptotics of the second kind.

It is natural to conjecture that the scaling of v with ε can be obtained by studying the asymptotics of the initial value problem giving rise to (32), for small ε:

$$\left.\frac{\partial \kappa}{\partial t}\right)_{\theta} = - \left(\kappa^2 + \frac{\partial^2}{\partial \theta^2}\right) [1 + \varepsilon \cos (m\theta)] \left[\kappa + \frac{\partial^2 \kappa}{\partial s^2}\right] .$$

However, this is problematic because the time evolution for $\varepsilon = 0$ is quite different from that when $\varepsilon \neq 0$. In the Barenblatt equation, on the other hand, the expansion parameter ε had no qualitative effect on the time evolution--the perturbation is a marginal operator in the language of statistical mechanics. We do not yet know how to resolve this apparent difficulty.

Finally, it is of interest to apply the converse of our results to statistical mechanics: instead of determining critical exponents by performing successive renormalisation group transformations in space, which is the analogue of the initial value problem for PDEs, can one determine a non-linear eigenvalue problem for the critical exponents, which is the analogue of the steady state equation (or the equation for the scaling function f)?

ACKNOWLEDGEMENTS

NDG and YO are partially supported by the National Science Foundation through grant no. NSF-DMR-90-15791. One of us (NDG) thanks the organisers of the workshop for the opportunity to participate and present this work. He also gratefully acknowledges receipt of an Alfred P. Sloan Foundation Fellowship.

REFERENCES

1. G. I. Barenblatt, Similarity, Self-Similarity and Intermediate Asymptotics, (Consultants Bureau, New York, 1979).
2. N. Goldenfeld, O. Martin, Y. Oono and F. Liu, Phys. Rev. Lett. 64:1361 (1990); N. Goldenfeld, O. Martin, Y. Oono, J. Sci. Comp. 4:355 (1989).
3. D. J. Amit, Field Theory, The Renormalisation Group and Critical Phenomena, (McGraw-Hill, New York, 1978).
4. S.-K. Ma, Modern Theory of Critical Phenomena, (Benjamin/Cummings, Reading 1976).
5. See (e.g.) J. Zinn-Justin, Quantum Field Theory and Critical Phenomena, (Clarendon, Oxford, 1989).
6. N. Goldenfeld, in: Proceedings of the Institute of Mathematics and its Applications, Workshop on the Evolution of Phase Boundaries, M. Gurtin and G. McFadden (eds.), (Springer-Verlag, to appear).
7. S. L. Kamenomostskaya, Dokl. Akad. Nauk SSSR 116:18 (1957).
8. M. Gell-Mann and F. E. Low, Phys. Rev. 95:1300 (1954).
9. The work of M. Berger and R. Kohn, Comm. of Pure and Applied Math. 41:841 (1988) is closely related to the renormalisation group.
10. K. G. Wilson, Phys. Rev. B 4:3174 (1971); ibid. 4:3184 (1971).
11. D. Kessler, J. Koplik and H. Levine, Phys. Rev. A 31:1712 (1985).
12. E. Ben-Jacob, N. Goldenfeld, B. Kotliar and J. Langer, Phys. Rev. Lett. 53:2110 (1984).

INDEX